LAKE CHAMPLAIN: PARTNERSHIPS AND RESEARCH IN THE NEW MILLENNIUM

Edited by

Thomas O. Manley

Executive Director
Lake Champlain Research Consortium
Department of Geology
Middlebury College
Middlebury, Vermont

Patricia L. Manley

Department of Geology
Middlebury College
Middlebury, Vermont

and

Timothy B. Mihuc

Lake Champlain Research Institute
Plattsburgh State University of New York
Plattsburgh, New York

Springer Science+Business Media, LLC

Library of Congress Cataloging-in-Publication Data

Lake Champlain: partnerships and research in the new millennium/edited by Thomas O. Manley, Patricia L. Manley, and Timothy B. Mihuc.

p. cm.

Includes bibliographical references and index.

ISBN 978-1-4419-3449-9 ISBN 978-1-4757-4080-6 (eBook)

DOI 10.1007/978-1-4757-4080-6

1. Lake ecology—Research—Champlain, Lake—Congresses. 2. Champlain, Lake—Environmental conditions—Research—Congresses. I. Manley, Thomas Owen, 1952- II. Manley, Patricia Lee. III. Mihuc, Timothy B. IV. Lake Champlain Research Consortium.

QH104.5.C35L36 2004

363.739'4'0974754—dc22

2004044853

Proceedings of "Lake Champlain in the New Millennium," the Spring Symposium sponsored by the Lake Champlain Research Consortium, hosted by the Missisquoi Bay Watershed Corporation and held May 20–23, 2002, at the Fort Saint-Jean Campus, St. Jean, Quebec.

ISBN 978-1-4419-3449-9

Originally published by Kluwer Academic / Plenum Publishers, New York in 2004

Softcover reprint of the hardcover 1st edition 2004

http://www.kluweronline.com

10 9 8 7 6 5 4 3 2 1

A C.I.P. record for this book is available from the Library of Congress

LAKE CHAMPLAIN: PARTNERSHIPS AND RESEARCH IN THE NEW MILLENNIUM

Preface

Lake Champlain spans two states, Vermont and New York, as well as two nations, the United States of America and Canada. Such a large body of water facing a wide variety of environmental stresses cannot be put back on the path of restoration without the concerted efforts of motivated stakeholders. In order to be effective, however, these stakeholders must not only be dedicated to their own goals, but more importantly, be aware of, fully understand, and support the efforts of a myriad of other goals brought to the table by other dedicated groups. While easier said than done, Lake Champlain has and will continue to benefit from the cooperation and partnerships spanning international to local venues. This monograph represents the result of one of many partnerships between the United States and Canada; specifically the 2002 Spring Symposium sponsored by the Lake Champlain Research Consortium that was hosted by the Missisquoi Bay Watershed Corporation (Corporation Bassin Versant Baie Missisquoi) at the Fort Saint-Jean Campus, from May 20-23. The focus of this meeting was to bring together a wide variety of stakeholders to not only learn about new developments and research in the basin, but to forge new and stronger bonds for future cooperation and management of this truly great lake.

Tom Manley
Executive Director of the Lake Champlain Research Consortium

CONTENTS

MISSISQUOI BAY – AN INTERNATIONAL PARTNERSHIP TOWARDS RESTORATION

Martin Mimeault[1] and Tom Manley[2]

ABSTRACT

This paper presents an overview of Lake Champlain and Missisquoi Bay environmental issues as well as perspectives on cooperative agreements among the States of Vermont, New York, and the province of Québec in protecting these water bodies. While Lake Champlain is primarily confined within the United States, seven percent of its watershed lies within the Province of Québec and drains into the shallow Missisquoi Bay that is bordered by Vermont. In certain areas of Lake Champlain, such as Missisquoi Bay, excessive phosphorus loads have caused algal blooms and excessive growth of other aquatic plants, which were, and still are, detrimental to a myriad of activities associated with the use of these waters. In August 26, 2002, a formal agreement between Québec and Vermont was reached on reducing phosphorus loads into Missisquoi Bay based on a 1991 apportionment of 60% (Vermont) and 40% (Québec). This management plan focuses on agricultural non-point sources of phosphorus, by promoting best management practices (BMPs) of soil and water resources.

1. INTRODUCTION

Lake Champlain is widely recognized as one of the most beautiful bodies of water on the North American continent. Its basin extends from the summits of the Adirondacks (New York) to the Green Mountains (Vermont; Figure 1), and its northern section makes an incursion into Québec through Missisquoi Bay. The region has been inhabited by native communities for a very long time, and currently has over 600,000 residents (LCBP, 2002). Every year, the basin area accommodates millions of visitors who come to

[1] Ministère de l'Environnement, Direction régionale de la Montérégie, 201, place Charles-Le Moyne, 2nd Floor, Longueuil (Québec)

[2] Middlebury College, Dept. of Geology, Bicentennial Hall, Middlebury, VT 05753

take advantage of its lakes and other attractions of historical and natural interest. In 2000, tourist revenues for the entire area were estimated at $3.8 billion (LCBP, 2002).

The lake has been developed in various ways. It serves as a reservoir, providing drinking water for 35% of the area's population, and has recreational, agricultural and industrial uses. Lake Champlain is also endowed with numerous resources, and makes an important contribution to the region's economy. For this reason, significant steps had to be taken to address its environmental problems. In some parts of the lake, such as Missisquoi Bay, phosphorus loads are so high they generate excessive growth of algae, which cause the water to turn green; they also stimulate growth of aquatic plants, which make the water unusable. The introduction of harmful and exotic aquatic species, such as the sea lamprey, the water chestnut, the Eurasian watermilfoil, the zebra mussel and, more recently, the alewife, has had a negative effect on the fish and other biological resources of the Lake Champlain Basin. Each of these species has an impact on the aquatic ecosystem, and impedes recreational activities. Other factors that have a direct impact on the resources of Lake Champlain Basin include loss of wetlands and habitat, inadequate public access as well as the loss of heritage sites. As a result, both the United States (Vermont & New York) and Canada (Québec) considered it a necessity to begin a management strategy to protect and enhance these resources so as to meet current and future needs.

Missisquoi Bay is quite shallow (maximum depth of 4 m) even though it has a relatively large surface area of 77.2 km^2 (Meyer and Gruendling, 1978). Seven percent of Lake Champlain's 21,326 km^2 watershed is located within the province of Québec and a vast majority of this surface area (1,315 km^2) feeds directly into Missisquoi Bay via the Pike River. Of the total Missisquoi Bay watershed (3,130 km^2), 58% lies within the boarders of Vermont and is primarily drained by the Missisquoi River. (Figure 2; Missisquoi Bay Phosphorus Reduction Task Force, 2000).

Within its watershed, it has 20 municipalities and a permanent population of about 20,000 people. For several years, water quality in Missisquoi Bay has degraded as a result of proliferation of cyanobacteria (blue-green algae) in the summer. In 2001 and 2002, the Bay's public beaches were closed following a period of excessive growth of these algae that severely limited use of this body of water. The Montérégie Regional Health and Social Services Board issued health advisories recommending that the public avoid all direct contact with water in the bay. In addition to serving as an important vacation destination, Missisquoi Bay is the source of drinking water for the municipality of Bedford, and for the Philipsburg sector of Saint-Armand.

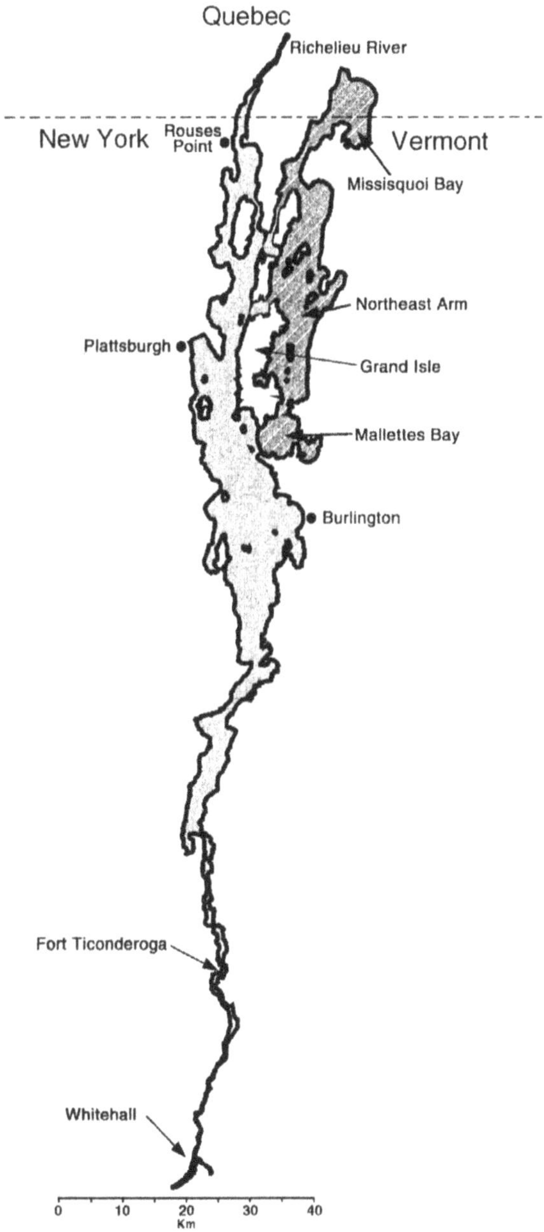

Figure 1. Missisquoi Bay is located at the northern end of Lake Champlain and falls within the boarders of Québec and Vermont. (adapted from Manley, this volume).

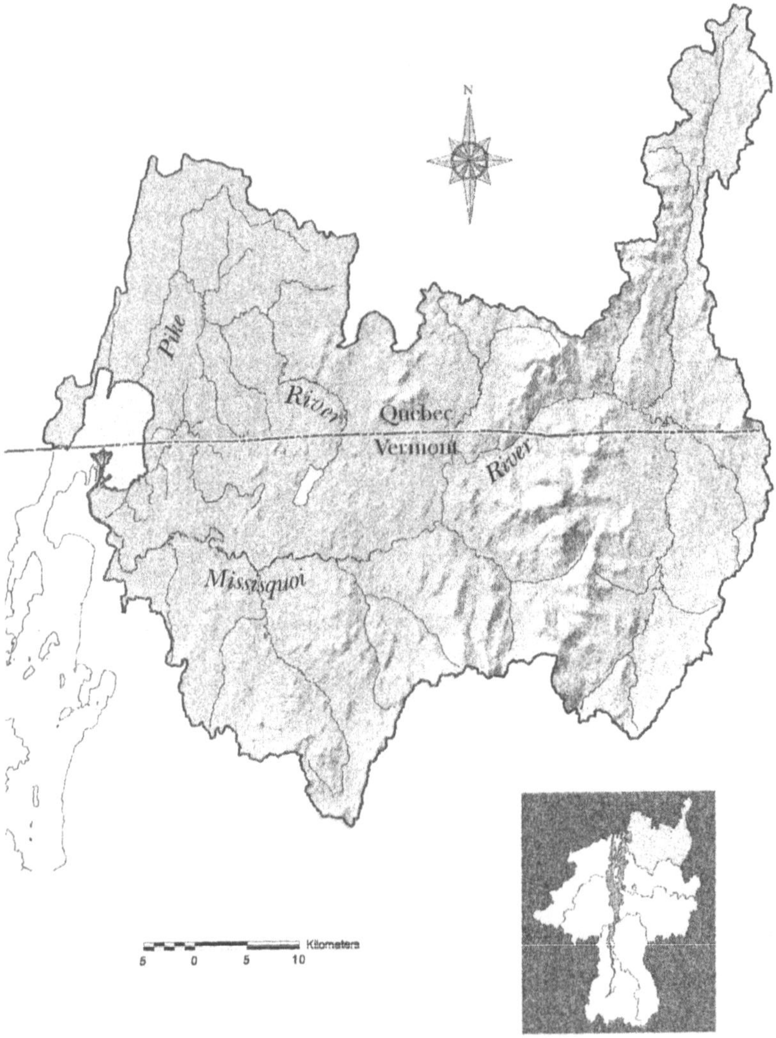

Figure 2. Missisquoi Bay and its associated watershed (adapted from the Lake Champlain Basin Program Office archives).

3. AN INTERNATIONAL FRAMEWORK FOR COOPERATION

Lake Champlain represents an eloquent example of integrated management of an international body of water, and the model of an extraordinary partnership that has passed the test of time. In 1988, the Premier of Québec and the Québec's Minister of the Environment signed a memo of understanding (MOU) relating to the management of

Lake Champlain with the States of New York and Vermont. The main objective of this agreement was to promote greater intergovernmental collaboration and harmonization of initiatives, goals, standards and programs affecting water quality in this drainage basin (Québec *et al.*, 2000). Québec participates officially in this program through the Lake Champlain Steering Committee with the Montérégie regional office of the Ministry of Environment acting as the province's representative on the panel as well as the Lake Champlain Basin Program (LCBP). The members of the panel meet four times a year to co-ordinate management of the lake. When the agreement was renewed in 1996, all three governments made a commitment to help implement the joint actions. Implementation was based on recommendations set out in the action plan, *Opportunities for Action* (LCBP, 2002), which covered a twenty-year period. The plan establishes targets for phosphorus loads in each area of the lake, with a view to enforcing the concentration criteria established in 1993 (Québec *et al.*, 1993).

In 1999, the Montérégie Regional Office and other agencies (Ministère de l'Agriculture, des Pêcheries et de l'Alimentation du Québec and the Ministère des Régions) met with local stakeholders to create the Corporation Bassin Versant Baie Missisquoi (CBVBM also know as the Missisquoi Bay Watershed Corporation). The CBVBM is a non-profit organization with a mandate to plan and co-ordinate local action that will improve the water quality of the Missisquoi Bay Watershed of Lake Champlain so as to preserve or recover its usage; its activities come within the scope of *Opportunities for Action*, which stems from various intergovernmental agreements. The CBVBM has rapidly become a forum for all individuals and groups concerned about the condition of the Bay and its watershed. The Corporation also represents the Comité Consultatif des Citoyens du Québec (Québec Citizen's Advisory Committee) on the Steering Committee.

4. INTERNATIONAL PHOSPHOROUS REDUCTION INITIATIVE

Faced with a severe environmental problem in Missisquoi Bay, and in spite of various urban and agricultural clean-up operations (primarily to control point source pollution), it was necessary to formulate actions that would not only complement conventional approaches but would control the environmental impacts of non-point source pollution from all activities in the Missisquoi Bay Watershed, both in Québec and in Vermont.

On August 26, 2002 the Government of Québec and the State of Vermont signed an agreement on phosphorus reduction for Missisquoi Bay. With this agreement, the two governments were committed to reducing phosphorus loads entering Missisquoi Bay, and reporting the progress made toward attaining the target concentrations that were mutually agreed upon. This agreement represented a turning point in efforts to reduce phosphorus in Missisquoi Bay. For the first time, it defined the responsibilities of each signatory in attaining the total target concentration of phosphorus originating in the Missisquoi Bay watershed: 40% for Québec and 60% for Vermont (Québec et Vermont, 2002). This agreement was the result of close collaboration between Québec and Vermont, who together created the Missisquoi Bay Phosphorus Reduction Task Force in 1997. The primary mandate of this task force was to assess the increase in phosphorus originating on both the Québec side and Vermont side of the Missisquoi Bay watershed, and to propose an equitable division of responsibilities between Vermont and Québec, and the measures that each party would have to take in meeting their objectives for reducing phosphorus in Missisquoi Bay.

In June 2000, the task force submitted a report entitled *A Division of Responsibility Between Québec and Vermont for the Reduction of Phosphorus Loads to Missisquoi Bay* (Missisquoi Bay Task Force, 2000). This report included an agreement, which has since been signed and is summarized as follows:

1) Total phosphorus reduction established for Missisquoi Bay was maintained at 0.025 milligrams per liter, the level to which the Task Force Québec–Vermont–New York on water quality) agreed in 1993;
2) In order to attain the desired concentration in 20 years (based on a 1996 starting date), the total target phosphorus load originating in the Missisquoi Bay watershed is set at 97.2 metric tons per year (mt/yr);
3) The allocation of responsibility is set at 40 % for Québec and 60 % for Vermont, based on their comparative 1991 contribution of phosphorus to Missisquoi Bay. Québec's and Vermont's target loads were set at 38.9 and 58.3 mt/yr, respectively. Measures to be taken to attain the phosphorus concentration and load objectives were identified by the LCBP office and in the MOU involving Québec, the State of New York and the State of Vermont;
4) An interim report on activities carried out in the Missisquoi Bay watershed and a joint Québec–Vermont program on monitoring water quality in Missisquoi Bay tributaries would be used to assess the degree to which the objectives have been attained

A monitoring program was set up in 1998 to identify, spatially and over time, the physio-chemical composition of the water in Missisquoi Bay's main tributaries. It was recently modified and improved to enable clear measurement of the phosphorus loads between Québec and Vermont, and to guide implementation of actions. The sampling frequency was increased, and flow stations were installed to determine when fixed targets were attained

The main purpose of the Québec-Vermont Agreement is to improve water quality and the aquatic ecosystem in the Missisquoi Bay Watershed of Lake Champlain so as to restore its uses and protect its environment. The improvement of water quality in the Bay is achieved primarily through a reduction at source of phosphorus originating throughout the Missisquoi Bay Watershed.

Seventy-nine per cent of phosphorus from non-point sources of pollution is attributable to agriculture. Thus, actions proposed within the framework of the Québec-Vermont Agreement aim to achieve environmental benefits by employing sustainable agricultural practices to reduce non-point source pollution of agricultural origin. (Hegman *et al.*, 1999). Non-point agricultural pollution is associated with erosion and surface runoff. Widespread adoption of agro-environmental practices and the introduction of stream bank restoration will facilitate a reduction of phosphorus and nitrogen in the aquatic environment.

5. A NEED FOR SUSTAINABLE DEVELOPMENT

The action plan must take into account the important economic role of agriculture in the Missisquoi Bay Watershed, so as to allow for the growth and continued vitality of this rural environment. They must also ensure that the approach taken is socially acceptable. This implies that all concrete initiatives will have to be compatible with the shift toward sustainable agriculture already undertaken by the basin's agricultural farmers through existing programs and new projects.

The vitality of Lake Champlain is essential in maintaining a vigorous regional economy. The action plan arising out of the Agreement between Québec and Vermont will have to recommend steps to protect and restore the watershed's natural resources so as to improve the condition of the Lake in the long term. Implementing the plan will include finding solutions to environmental problems while maintaining the basin's economic integrity. Determining viable measures that protect and improve the quality of the lake while maintaining the region's economic vitality will constitute a challenging task. The issue over the next few years is to take action that will protect, restore and conserve Lake Champlain and Missisquoi Bay and its watershed so that current and future generations can enjoy all that it has to offer.

6. REFERENCES

Hegman, W., D. Wang and C. Borer, 1999. *Estimation of Lake Champlain basin wide non-point source phosphorus export*, Technical Report No. 31, Lake Champlain Basin Program, United States, 81p.

Lake Champlain Basin Program (LCBP), 2002, *The Lake Champlain basin atlas CD*, Version 2.0, United States.

Meyer, G., and G. K. Gruendling, 1979, Limnology of Lake Champlain, Lake Champlain Basin study No. 30, New England river basins commission, Boston, Massachusetts 02109.

Missisquoi Bay Phosphorus Reduction Task Force Québec – Vermont, 2000, *A division of responsibility between Québec and Vermont for the reduction of phosphorus loads to Missisquoi Bay,*, Québec, Ministry of Environment, Montérégie Regional Office, 23 p.

Québec, Vermont and New York, 1993. Water quality agreement *Vermont-New York and Québec*, Ministry of Environment, Montérégie Regional Office, 20 p.

Québec, Vermont and New York, 2000. *Memorandum of understanding on environmental cooperation on the management of Lake Champlain among the State of New York, the State of Vermont and the Government of Québec*, Québec, Ministry of Environment, Montérégie Regional Office, 5 p.

Québec and Vermont, 2002. *Agreement between the government of Québec and the government of the state of Vermont concerning phosphorus reduction in Missisquoi Bay*, Ministry of Environment, Montérégie Regional Office, 4 p.

REFLECTIONS ON THE RELATIONSHIPS BETWEEN SCIENTIFIC RESEARCH AND A WATERSHED APPROACH TO WATER MANAGEMENT: THE CASE OF THE MISSISQUOI BAY WATERSHED CORPORATION

Missisquoi Bay Watershed Corporation

Chantal d'Auteuil*

1. INTRODUCTION

During the 1950s and 1960s, the Missisquoi Bay area of Lake Champlain was renowned as a recreational and tourist destination. It attracted large numbers of angling, boating, swimming, camping and birdwatching enthusiasts. Commercial fishing flourished and ice fishing was a popular pastime with both local residents and visitors. Biological productivity on the Québec side of Lake Champlain was so high that it was a source of supply for stocking walleye to other lakes in Québec between 1954 and 1957 and perch in 1974 and 1975 (Leclerc, 1975). Unfortunately, water quality in the area began to decline in the 1970s. A 1987 study of eutrophication of the Bay (Environnement Canada, 1987) revealed that it had already reached the last stages of eutrophication, and that there was no indication that this trend would change in the future. Development around the Bay and in the watershed had led to water nutrient enrichment and bacteriological contamination had become a major hindrance to swimming and water sports.

Bay area residents mobilized and set up a grassroots group called Conservation Baie Missisquoi in 1988. The group called upon the Québec government to conduct scientific studies of the water quality in the Bay, in which its members participated. When the environmental report card was issued for Missisquoi Bay in 1993, everyone's worst fears were confirmed: criteria of acceptability were exceeded for suspended solids, turbidity, phosphorus, and DBO_5 (Girard, 1993).

* Chantal d'Auteuil, Corporation Bassin Versant Baie Missisquoi, Saint-Armand, Québec, Canada
Translation: Michael Gilson, Tangerine Communications. Verification: Sandy Montgomery

An environmental diagnosis of the Pike River confirmed that problem, and four tributaries draining agricultural territory were identified as prime sources of non-point phosphorus loading: Ewing Creek, Beaver Creek, ruisseau aux Morpions and Wallbridge Creek (Caumartin and Vincent, 1994). The study sparked a sudden general awareness of the need for urgent action. Municipal wastewater treatment facilities were commissioned in Bedford (1994) and Venise-en-Québec (1996). The two facilities improved the water quality significantly by reducing concentrations of organic matter, nutrients and micro-organisms.

Around the same time, in 1996, the U.S.-based Lake Champlain Basin Program (LCBP) initiated "Opportunities for Action", a blueprint with a 20-year implementation span (Lake Champlain Management Conference, 1996). Québec signed this international agreement, as did New York and Vermont. From the very beginning, Québec citizens have played a role in this program through Citizen Advisory Committees. Stakeholders from the agricultural, environmental, municipal and tourism sectors then came together to create the Corporation Bassin Versant Baie Missisquoi (Missisquoi Bay Watershed Corporation) in February 1999. The Government of Québec pledged financial assistance over a three-year period and gave the Corporation the mandate to develop a detailed action plan for the entire watershed and to raise awareness among area stakeholders through a variety of activities.

The inception of the Corporation, spurred by the mobilization of local groups in response to scientific studies carried out in the watershed, enabled identification of the main causes affecting water quality deterioration and the loss of watershed usages. The mission of the Corporation thus came from the findings of those studies: to "promote the enhancement of water quality in the Missisquoi Bay watershed so as to render it usable again and enhance its usability from a sustainable-development perspective." The Corporation cannot carry out its mission without the contributions of scientific research and the support of experts at a variety of levels: the assessment of what is going on on the ground, the analysis of problems, the recommendation of solutions, and the development of action plans.

The relationship between scientific research and a watershed approach to water management poses some difficulties, which will be discussed here as they relate to five topics, with reference to Missisquoi Bay:

- Knowledge of the area and raising public awareness.
- Economic development and protection of waterbodies.
- Field surveys and decision making.
- Action plans and realistic timetables.
- Follow-up of outcomes of actions in the area.

2. KNOWLEDGE OF THE AREA AND RAISING PUBLIC AWARENESS

In-depth knowledge of the area, which is necessary for properly carrying out of a watershed approach to water management, is based essentially on scientific research that unfortunately is too often sector-based or fragmented; such knowledge must inevitably be expressed in plain language before it can be used to help raise public awareness.

For example, without scientific studies, the trophic state of Missisquoi Bay would not be understood. Eutrophication is a complex phenomenon that is not easily explained to the general public. Confronted by water quality survey results indicating that thresholds for, phosphorus, nitrogen and suspended solids, among others have been exceeded (Simoneau, 1993), citizens are perfectly aware that water quality has been adversely affected for the past 20 years or so, given the fact that the deterioration is quite evident (e.g. proliferation of aquatic plants, algal blooms, foul odors, greenish color of water, beach closings). Citizens, however, are not necessarily aware that this phenomenon will be exacerbated over time, and that it is urban and agricultural development that is at fault.

Although the eutrophication phenomenon is very well understood in the scientific realm, this is not enough to spur citizens to action. Locally relevant proof is needed to get them involved. Such proof makes it possible to properly explain the phenomenon and to seek solutions. Scientific research is thus essential, and must be conducted continually so as to maintain and improve public awareness. Messages must be repeated and convincing arguments made. Too often, citizens tend to believe that once a problem is no longer being talked about, it must be that it has been solved, and so they no longer need worry about it. They also have the impression that science will always find effective solutions to every problem—although they often believe that scientists fail to approach the search for solutions with enough urgency. The local citizens also feel that they have done enough—given that treatment of municipal wastewater has been inaugurated —and that it is now up to the farming community to adopt healthy practices. But the restoration of the waterbody will likely require another 20 years or more — a fact that citizens are not ready to accept. This issue is marked by a lack of information and education on the eutrophication phenomenon as it applies directly to Missisquoi Bay.

The first step in raising public awareness is properly informing citizens. It is never easy for the watershed committees to stay abreast of all of the scientific studies conducted on their territory. It would be beneficial for groups of researchers to inform the committees of their work as soon as it starts, and pass on the results of that work periodically. Similarly, research groups would do well to inform the municipalities implicated in their research. It is important that municipal officials be able to inform citizens. In fulfilling their educational mission, the watershed committees often have to deal with perceptions rather than facts. There is no point trying to convince citizens who have erroneous but well-entrenched perceptions of the issues, and who possess ready-made solutions that, in their opinion, are easy to apply, but that may well do more harm than good. Here again, citizens have to be shown, via scientific proof, that their perceptions are wrong before they can be oriented toward better-targeted actions. This is a thankless but vital task.

3. ECONOMIC DEVELOPMENT AND PROTECTION OF WATERBODIES

Recommendations to protect aquatic ecosystems and human health based on the outcomes of scientific studies are often at odds with the population's desire for economic development.

There are two major problems affecting Missisquoi Bay, and they are interdependent: excess phosphorus loading and the proliferation of cyanobacteria, which can release toxins and affect human health. Scientific research has recently shed some light on this public health issue, leading to the closing of beaches at Venise-en-Québec for the entire month of August 2001. Previously, restrictions on swimming were prompted by the presence of fecal coliform bacteria and the potential adverse health consequences of bacteriological contamination.

Although the problem of bacteriological contamination, identified thanks to scientific studies, has largely been resolved, another threat has emerged with the algal blooms: cyanobacterial toxins. Phosphorus is high on the list of suspects, with the principal source now identified as non-point phosphorus pollution from farming activity within the watershed. The search for solutions is now much more difficult. Actions to reduce the level of phosphorus confront the desire to exploit the highly productive lands within the watershed and to increase livestock production, both of which activities benefit from favorable market conditions. Agricultural productivity is key to the vitality of the entire region. Algal blooms, however, interfere with the economic activities and quality of life of Missisquoi Bay area residents. The risk that accompanies the presence of toxins is catastrophic from an economic perspective, and may even prompt a decline in the resident population, particularly those whose livelihood depends on recreation and tourism. This explains why citizens are constantly opposed to farming activity, and why residents are against beach closings: there have never been any cases of poisoning in the Bay, even though algal blooms have always existed.

Figure 1. Cyanobacteria and aquatic plants on the shore, Philipsburg, 2001.

Aquatic plants that wash up on shorelines are clearly a nuisance for residents, but one that pales in comparison to the economic impacts of cyanobacteria in algal blooms, which color beaches blue-green and release toxins.

Moreover, studies of the wetlands around the Bay (Gratton, 1995) have led to their protection by appropriate municipal zoning by-laws as well as creation of an ecological reserve at the mouth of Pike River. Such protection of marshes, swamps and floodplains, however, significantly constrains the potential for urban and tourist development, and this represents a grievance for municipalities. Farmers are also skeptical of the scientific studies findings on non-point phosphorus pollution and want garanties of their effectiveness before proceeding with any major changes to farming practices or restrictions on further increases in livestock populations.

Even when scientific research clearly identifies the sources of problems and/or health risks, citizens tend to challenge the validity of the studies and their outcome if they conflict with economic activity. This is all the more relevant if the solutions rely on the good will and cooperation of citizens, whether it imposes limitations on how they use their property or it entails increased municipal taxation. Economic development and the desire to maintain a population in a given area are perfectly legitimate aspirations, and it is pointless to try to prove otherwise. It is preferable to convince citizens of the need to manage development in a way that will prevent environmental degradation.

4. FIELD SURVEYS AND DECISION MAKING

The results of scientific studies do not necessarily aid the decision-making process, and the declarations that further studies are needed often leads to the impression, among citizens, that actions are being held up.

The decision to remove the causeways beside the Alburg-Swanton bridge as part of the project to replace it with a new bridge nearby seemed, on the face of it, to be a fairly obvious step. The two causeways, each 1,750 feet long, date back to the construction of the original bridge in 1939, and have resulted in a significant narrowing of Missisquoi Bay, impeding the free flow of water northward. Missisquoi Bay would, at last, be regenerated: phosphorus levels in the Bay would decline, algal blooms would cease, boating would be pleasant again, and tourists would again flock to the area.

But the scientific studies, required as part of the environmental assessment process related to the bridge project, would soon tarnish this idyllic vision. Some studies showed that a change in water flow dynamics could lead to problems downstream (Vermont Geological Survey, 1996), that phosphorus levels would not be reduced as much as had been hoped, and that an important hibernaculum for the endangered spiny softshell turtle population would disappear (Normandeau associates inc., 2001). The more the issue was studied, the more difficult it became to make a decision.

This explains why citizens began to cast doubt on the validity of the studies, which seemed to be conducted to prove that removing the two causeways would be harmful and would lead to environmental problems. The arguments built around protection of the spiny softshell turtle proved to be extremely prejudicial to the very protection of the species by citizens, because citizens now began to view the animals as a threat to the enhancement of water quality in the Bay and to concomitant recreational and tourist development. Moreover, the studies did not show beyond all reasonable doubt that there would be significant impacts in terms of water flow, phosphorus loading, and the turtle population. These three question marks generated even more skepticism on the part of the public.

Decision makers are almost always pleased to learn that a scientific study has been commissioned to look into an issue that concerns them directly. They place a great deal of faith in the decision-support value of the results of the research: "At last, someone who will clarify things for us!" This also helps appease citizens, who are generally impatient for decisions on the part of their elected officials. Unfortunately, they are usually disappointed with the studies results, especially when these reveal a high level of uncertainty or no recommendations are made. Both decision makers and citizens become extremely frustrated when told, "further studies will be necessary." It is at this point that people begin to feel that scientific studies are an unnecessary expense on the part of governments; that, at any rate, action should be taken and the results verified later, as this would be much more efficient and less costly. Some even go so far as to claim that the scientific studies are commissioned in order to delay decisions.

5. ACTION PLANS AND REALISTIC TIMETABLES

The recommendations issued as part of scientific studies usually involve timeframes that are too long to allow establishment of an action plan with local stakeholders, according to short-term timetables that would enable mobilization of citizens.

The Lake Champlain Basin Program action plan, dubbed "Opportunities for Action," was adopted in 1996 and is slated to continue through 2016. Scientific studies have shown that this period of time will be necessary to arrive at the results hoped for in the blueprint. Given a 20-year timetable, it is very difficult to set precise timetables for individual actions. For this reason, the action plan is an evolving plan, subject to review every five years. A review of the action plan led to the publication of a second version, which was then subjected to wide-ranging public consultation. Revised priorities for action were included in the document, which also included a progress report on actions already completed. This initial five-year period enabled local stakeholders to set realistic timetables and monitor the results of actions. This has facilitated the task of raising public awareness and that of explaining the outcomes of individual actions while taking into account the work to be done over the remaining 15 years.

To a scientist with a firm grasp of long-term ecological processes, a 20- or 25-year timeframe for effective change is a quite realistic one. Recommendations made pursuant to scientific studies are often founded on timetables of this nature. But for citizens and politicians who want to see actions taken locally, a five-year horizon is much more realistic. This does not mean deceiving the population by promising visible change after five years of effort; rather, it means that processes of change must be described according to five-year periods: "What can we expect to see after five years of work, and five years after that?" This is the principle applied by the LCBP and, after five years of application of the blueprint, it would seem that the results obtained in the area of phosphorus reduction have exceeded the targets set for Lake Champlain as a whole. Missisquoi Bay, however, is the only area of the Lake where the phosphorus-reduction threshold (25% of the target amount after the first five years) was not attained (Lake Champlain Basin Program, 2000). Because sources of phosphorus pollution are mostly non-point in the Missisquoi Bay watershed, efforts at reduction have failed to generate results as clear as those observed for reduction from point sources, which has already been achieved for Missisquoi Bay via municipal wastewater treatment.

Pike River

Missisquoi River

Figure 2. Two major sources of phosphorus in Missisquoi Bay: Pike River (right) and Missisquoi River (left) (photos: Faune et Parc Québec).

Sharing of responsibility for phosphorus reduction between Québec and Vermont is based on the amounts of phosphorus discharged by the Pike and Missisquoi rivers in Québec and Vermont respectively.

All this shows the importance of actions not being reduced to simply setting goals for phosphorus reduction. The actions to be taken must be properly described, and it must be explained that there will be a significant improvement at first, followed by gradually lesser improvement; that over time it will become increasingly difficult to reduce phosphorus levels. Citizens must be kept mobilized by first focusing on the most conclusive actions, and then persuading political decision makers to allocate more financial resources. Results will be more noticeable if efforts are focused on one type of action during the first five years, and all available resources are allocated to that one type of action. The five-year interval for actions does not pertain to local stakeholders only: during the initial five-year period directed toward the priority actions, scientists can prepare the next steps to be taken over the next five-year period, by conducting the studies necessary to properly chart those actions. Thus, even if local action is focused on short term priorities, research should be planned in a more global fashion so as to more effectively study individual issues and suggest concrete solutions.

6. FOLLOW-UP OF OUTCOMES OF ACTIONS IN THE AREA

The results of scientific studies usually do not impart knowledge of overall improvements in water quality in the watershed, or the effectiveness of efforts at restoring aquatic environments.

Readings have been taken along watercourses within the Missisquoi Bay watershed since 1979, enabling verification of improvements in water quality following the construction of the wastewater treatment plant for Bedford, including the Snyders agrifood plant in 1994 (Simoneau, 2002). In 1998, a network of eight sampling stations was implemented to monitor improvements in water quality, in particular the phosphorus reduction targets set by the LCBP blueprint. To improve monitoring, both the number of stations and the sampling frequency were increased in 2001. However, sampling at the Pike River and Missisquoi Bay stations is the responsibility of the U.S. program, and the data cannot be compared with those from sampling conducted by the Québec Ministry of the Environment. Nor is it possible to compare samples taken in the Pike River with data from the Missisquoi River, because samples are not collected on the same days, and thus climatic and hydrological conditions are not the same.

Québec's water quality sampling networks enable the collection of data over long periods of time, on a permanent basis. These networks therefore ensure stable monitoring of water quality, enabling comparisons to be made from one watershed to another and trends to be established for each parameter measured. Field studies round out the process of collecting invaluable information on water quality, on a local and specific basis. Despite the fact that an excellent sampling network is in place, there are significant delays in the production of water-quality reports, because of the time needed to analyse the data. Moreover, it is not possible to know which waterbodies are the sites of the most serious problems. What this means is that true reduction of phosphorus at the source cannot be achieved without attacking the smaller, most adversely affected waterbodies first. Also, other data collected as part of the scientific studies are not incorporated into the annual water quality report. Despite this profusion of water quality studies, it remains quite difficult for the watershed committee to properly interpret findings and be well positioned to raise public awareness.

All of this data, however, cannot provide an answer to the fundamental question asked by general public: "With all the work that's been done over the years, is it getting any better?" The problem is that it is difficult for scientists to pinpoint what "it" is. Are we talking about water quality, sediments, fish stocks, cyanobacteria? and so on. The scientists' answer will often be imprecise. For example, the following question is often asked: "Has the population of walleye increased over the years since the introduction of the fish sanctuary in the Pike River in the late 1960s?" (Le groupe-conseil enviram (1986) inc., 1988) What citizens really want to know is whether fishing restrictions may one day be curtailed, because as long as they are in place, area residents cannot legally practice this activity, which they enjoy a great deal. Although Québec's fish consumption guide, the *Guide de consommation du poisson*, is useful, citizens are not necessarily interested in knowing how many fish they can safely eat per week. They want to know what concentrations of toxic substances are present in the fish, and whether those concentrations are decreasing over time, which would show whether improvement has taken place.

Another key concern involves algae (a catch-all term used by citizens to describe aquatic plants, grass deposited onto beaches by wave action, foam on the water, decomposing sediments, cyanobacteria, and filamentous algae). Are there more or fewer algae than before? Why not get rid of them once and for all by conducting a huge cleanup of the Bay? When they ask whether the water is safe for swimming, they are told to consult the Québec *Programme-Plage* (a beach advisory service). But residents are not interested in knowing whether they can swim off the beaches at Venise-en-Québec (they already know the water is of good quality), but rather whether there is a risk for swimmers, or for youngsters wading, elsewhere in Missisquoi Bay, in the Pike River or in the Missisquoi River.

To properly address the key question of improving the aquatic ecosystem and water usage requires the ability to painstakingly analyse phenomena, using the right indicators. The general public's questions may be simple, but the answers often sidestep their concerns. Indeed, indicators are the only way of explaining to citizens the complexities of monitoring aquatic ecosystems and of informing them of water usages within the watershed. For an answer to be satisfactory, the indicators must speak for themselves.

7. CONCLUSION

The preceding overview of the interrelations between scientific research and grass-roots mobilization has made it possible to note that scientific research is truly necessary and that watershed-based management of water resources must be based on it. Nevertheless, for its results to be truly persuasive in the eyes of the general public and decision makers, scientific research must adjust to local needs. Planning of research in response to questions raised, hypotheses made, and solutions proposed would enable better definition of the environmental issues affecting populations.

In this way, scientists can also be sure of retaining their credibility within society. At the same time, it will be easier for them to secure the financial and human resources necessary to their research and to the monitoring of actions in local communities. Through the establishment of true partnerships with watershed committees and local groups, the scientific community would also have a way of becoming directly involved in improving water quality, protecting aquatic environments, and safeguarding the quality of life in their societies. The task of popularizing and promoting the findings of scientific studies, and of raising awareness among the population and decision makers, could be entrusted to the watershed committees.

The following recommendations are designed to help in the establishment of a lasting partnership between research groups and local stakeholders.

7.1. Knowledge of the area and raising public awareness

Proposal: Plan scientific studies in such a way that it makes it easier to understand the issues.

A watershed approach to water management would benefit from research that is planned in such a way that overall problems are well defined, to better gather evidence of causes and to suggest solutions. Studies of the trophic states of lakes and diagnoses of aquatic ecosystems are essential, and must be conducted periodically. Because such studies are large in scope, partnerships between research groups are advisable. Thus, instead of spreading research work over several years, it would be preferable to perform concentrated scientific studies over a short period so as to settle priority issues related to enhancement of water quality in the watershed. Moreover, such partnerships should encompass grassroot groups, who can also play a role in seeking funding by preparing grant applications. As fully fledged partners, the watershed committees could take responsibility for promoting the scientific studies and their popularization, as part of efforts to keep municipal officials and citizens better informed.

7.2. Economic development and protection of waterbodies

Proposal: Consult the population regarding the planning of scientific studies.

When formulating hypotheses as part of a scientific study, it is normal to analyse contrary hypotheses that could be raised by scientific colleagues as well, or consider the arguments that colleagues could use to invalidate the scientific results eventually obtained. When planning studies, scientists ought to consider hypotheses put forward by citizens or decision makers as well, even if these may appear somewhat improbable or even absurd by scientific standards. Scientists should at least be aware of them, and understand the motivations of all stakeholders to gather evidence that could invalidate their hypotheses, if need be. It is thus recommended that scientists meet with citizens' groups and with decision makers when developing scientific protocol, with a view to verifying their hypotheses and perceptions.

7.3. Field surveys and decision making

Proposal: Establish a scientific protocol that responds to all of the questions raised.

Proper planning of scientific studies is extremely important here. For a study to be a decision-support tool, one must be sure that it will answer the questions related to decision making by local stakeholders. Decision makers must have a firm grasp of the levels of uncertainty they will have to deal with, even if the study is conducted masterfully, with no problems as to instrumentation, sampling or climatic variations. It is thus important to allow enough time to conduct interviews with decision makers and even with citizens if need be. If the study does not provide answers to all the questions, decision makers and citizens must be informed of this right from the start, so as to avoid major disappointments in the end.

7.4. Action plans and realistic timetables

Proposal: Conduct field studies by sub-watershed, so as to test solutions.

Obtaining measurable local outcomes depends on focused actions. Actions spread out across a wide area are not effective because effort expended by certain stakeholders is negated by the non-participation of others. This is why everyone agrees that it is better to take small-scale actions by sub-watershed, rather than attempting to encompass the whole watershed. It is also easier to conduct follow-up scientific studies within a smaller sub-watershed. Mobilization is also more effective, because awareness-raising activities can be conducted with all residents in a sub-watershed. Working in accelerated fashion over a five-year period within one sub-watershed is sufficient to demonstrate the results of actions and apply corrective measures if need be.

7.5. Follow-up of outcomes of actions in the area

Proposal: Emphasize studies aimed at monitoring quality of water and aquatic environments.

Monitoring water quality is very costly in terms of both financial and human resources. If other types of indicators (i.e. biological and economic) are added to the mix, the costs will spiral even higher. Such monitoring of the status of aquatic ecosystems and groundwater has become vital in the majority of Québec's watersheds, especially in the Montérégie region south of Montréal. To keep costs under control, it should be possible to establish indicators with greater relevance to individual watersheds, based on the problems encountered there. Monitoring of certain indicators could also be entrusted to volunteers, under a program that would be supervised by the watershed committees and financed jointly by municipalities, university programs, subsidy programs for nonprofit organizations and—why not?—corporate sponsors. The participation of scientists is again crucial, so as to ensure the validity of monitoring by means of indicators, in partnership with local stakeholders.

It is hoped that these recommendations will enable continued discussion of the beneficial relationships that exist between scientific research and community interests. By taking part in the organization of the Lake Champlain Research Consortium's 2002 Spring Symposium, the Missisquoi Bay Watershed Corporation sought to establish a lasting relationship with scientific experts on both sides of the Canada–U.S. border. It is hoped that this will blossom into a true partnership, and the Corporation stands ready to participate in scientific research work to the best of its abilities, with a view to expanding local knowledge, analyzing problems, making choices, and developing more targeted actions pursuant to the results of scientific studies.

REFERENCES

Caumartin, Jean and René Vincent. 1994. *Diagnostic environnemental de la rivière aux Brochets.* Ministère de l'environnement et de la faune du Québec, Direction de l'aménagement des lacs et des cours d'eau, Direction des écosystèmes aquatiques. 101 p.

Environment Canada. 1987. Étude sur l'eutrophisation de la baie Missisquoi. 97 p.

Girard, Claude. 1993. *Diagnostic environnemental de la baie Missisquoi.* Ministère de l'Environnement et de la Faune du Québec, direction de l'aménagement des lacs et des cours d'eau. 33 p.

Gratton, Louise. 1995. *Caractérisation écologique de sites voués à la conservation de terres noires de la région du lac Champlain.* Conservation baie Missisquoi. 106 p.

Lake Champlain Basin Program. 2000. *Preliminary Evlauation of Progress Toward Lake Champlain Basin Program Phosphorus Reduction Goals*: A Lake Champlain Basin Program Internal Report. 42 p.

Lake Champlain Management Conference. 1996. *Opportunities for Action: An Evolving Plan for the Future of the Lake Champlain Basin* (First Edition).

Leclerc, Jean. 1975. *Rapport d'activités concernant le transfert de poissons de la baie Missisquoi.* Ministère du tourisme, de la Chasse et de la Pêche, Québec. 7 p., appendix.

Le groupe-conseil enviram (1986) inc. 1988. *Projet de dragage de l'embouchure de la rivière aux Brochets : Étude d'impact sur l'environnement.* 85 p.

Normandeau associates inc. 2001. *Spiny softshell turtle and map turtle study Missisquoi Bay Bridge Project.* Patrick galois. R-18552.000. 42 p.

Simoneau, Marc. 2002. *Portrait de la qualité des cours d'eau du bassin de la baie Missisquoi.* Plenary presentation at the Spring 2002 Symposium of the Lake Champlain Research Consortium, Saint-Jean-sur-Richelieu, Québec.

Simoneau, Marc. 1993. *Qualité des eaux du bassin de la rivière Richelieu, 1979-1992.* Ministère de l'Environnement du Québec, Direction de la qualité des cours d'eau. 126 p.

Vermont Geological Survey. 1996. *Hydrodynamic Modeling of Missisquoi Bay in Lake Champlain.* ASA Project 95-136. 101 p.

A PRELIMINARY INVESTIGATION INTO THE POSSIBLE EMISSION SOURCES FOR ATMOSPHERIC MERCURY FOUND IN THE LAKE CHAMPLAIN BASIN

Ning Gao[*1], Nathan G. Armatas[1], Benjamin Puchalski[1], Philip K. Hopke[2], and Richard L. Poirot[3]

ABSTRACT

Several types of source-receptor modeling methods have been applied to the atmospheric mercury data collected by the long term monitoring site at PMRC, Underhill, VT to infer the possible emission source types and source areas that may have significant contributions to mercury deposition into the Lake Champlain Basin. The methods employed were potential source contribution function (PSCF) analysis, residence time analysis (RTA), and multiple regression analysis against the factors of the positive matrix factorization (PMF) of the fine particulate data from the same site. The possible source areas are displayed by the PSCF maps and the RTA maps. Source contributions were calculated for vapor phase and particulate mercury.

1. INTRODUCTION

Mercury contamination is a prevalent problem for many inland lakes as well as coastal waters. Many sources emit mercury into the atmosphere and it cycles through the environment via different pathways and chemical forms. General types of Hg emission sources include coal-fired power plants, non-ferrous metal smelters (Porcella *et al.*, 1995a-b), gold and silver mining until the early 1900's (Nriagu 1993, 1994, 1996), chlor-alkali production (paper mills), incinerators and landfills (thermometers, fluorescent light

[*] Author to whom correspondences should be sent
[1] Department of Chemistry, St. Lawrence University, Canton, NY 13617
[2] Department of Chemical Engineering, Clarkson University, Potsdam, NY 13699
[3] Department of Environmental Conservation, State of Vermont

bulbs, alkaline batteries), oil refineries and combustion of their products (Wilhelm 2001), discharge of industrial effluent, discharge from municipal wastewater treatment facilities (US EPA 1997), soil (old Hg from atmospheric deposition and agricultural use of Hg-containing fungicide prior to the 1970's), and lake sediments (old Hg) (McIntosh 1994; Colman and Clark 1994; Landers *et al.*, 1998; Kamman and Engstrom, 2002). The estimated global mercury emissions could be found in a recent United Nations Environmental Program report (UNEP, 2002). Lake Champlain is suffering from the Hg contamination problem, which along with other pollutants such as PCB's, phosphorous, and nitrate have caused degradation of the water quality (McIntosh, 1994).

Previous studies have concluded that atmospheric deposition is the main pathway for mercury entering the Lake Champlain Basin (Scherbatskoy *et al.*, 1998; Shanley *et al.*, 1999). For the past two years, a multidisciplinary team has systematically evaluated the extent of the mercury pollution, the emission sources, the importance of both local and long-range transport, the mechanisms responsible for Hg transport to the lake, and transformations within the lake. This paper presents preliminary results of an investigation into the possible emission sources for atmospheric transport of mercury to the Lake Champlain basin. This study utilized three types of source-receptor modeling analysis, namely, potential source contribution function (PSCF) analysis on Hg data, residence time analysis (RTA) on Hg data, and multiple regression analysis of Hg data against the factors obtained from the positive matrix factorization (PMF) analysis of the multi-element data for fine particles collected at the same site. The modeling efforts are based on analysis of nearly a decade of air quality data on atmospheric deposition of mercury (both wet and dry) from the atmospheric station at Proctor Maple Research Center (PMRC) in Underhill, Vermont near the center of the basin.

2. MEASUREMENT DATA

Atmospheric Hg measurements for vapor-phase Hg, particulate Hg, and wet precipitation Hg were performed as part of a long-term Hg cycling study in the Lake Champlain Basin of Vermont (Keeler *et al.*, 1995; Burke *et al.*, 1995). The air samples were collected at the Proctor Maple Atmospheric Station in Underhill, VT (44.53 N, 72.86 W, 400 m elevation). Since December 1992, 24-hr vapor samples (switched at around 8 AM) were collected on Wednesdays and also on every sixth day weekly using Au-coated sand traps and then shipped overnight to the Air Quality laboratory at University of Michigan (UMAQL). The Hg° collected was analyzed using the dual-amalgamation cold vapor atomic fluorescence spectrometry (CVAFS) method (Keeler *et al.*, 1995; Fitzgerald and Gill, 1979).

Total particulate Hg was collected using an open-faced Teflon filter pack with 47 mm glass fiber filters for 24-hr at a flow rate of 30 L/min. The sampling schedule was the same as that of the vapor phase Hg. Filters were placed into acid-cleaned Petri dishes immediately after sampling. The dishes were Teflon-taped, shipped to UMAQL, and stored at -40°C until analysis. The filter extractions and analyses were performed in a Class 100 clean room. The filters were extracted in a 10% solution of a 70% nitric acid/30% sulfuric acid mixture in Teflon vials. After extraction, the solutions were oxidized with BrCl for 1 hr, during which all forms of Hg in the solutions were converted into inorganic Hg^{2+}. Then NH_2OH was added followed by $SnCl_2$ to reduce the

Hg^{2+} to Hg°. The resulting Hg° is volatile and can be released from solution by bubbling with Hg-free N_2. The Hg released was collected on Au-coated sand traps and subsequently thermally desorbed and analyzed using the CVAFS method (Burke *et al.*, 1995; Fitzgerald and Gill, 1979).

The event precipitation samples were collected with a MIC-B sampler equipped with a borosilicate glass funnel and a Teflon-coated carboy. The sample was picked up at 8 AM if there had been an event. However, if it was raining or snowing at 8 AM, then the sample would not be picked up until 8 AM the next day. The collected sample was poured into an acid-cleaned Teflon container and shipped overnight to UMAQL. A portion of the sample solution was saved for pH and anion analysis. The rest was oxidized with 1% BrCl solution overnight to convert all forms of Hg into inorganic Hg^{2+}. Treatment with NH_2OH and $SnCl_2$ converted the Hg^{2+} into volatile Hg°. The Hg was liberated by bubbling with Hg-free N_2, collected on Au-coated sand traps and subsequently desorbed thermally and analyzed using the CVAFS method (Burke *et al.*, 1995; Fitzgerald and Gill, 1979).

Fine particle measurements were conducted at the same site between September 1988 and June 1995 as a part of the NESCAUM Regional Particle Monitoring Network (Flocchini *et al.*, 1990; Poirot *et al.*, 1991). 24-hr samples (switched at midnight) were collected on Wednesdays and Saturdays and also on every sixth day on Teflon filters using one of the four IMPROVE modular samplers (Module A). Filters were subsequently analyzed at the Crocker Nuclear Laboratory, University of California, Davis for mass (by gravimetric method), light absorption (B_{abs}, by laser integrating plate, LIPM), elemental hydrogen (by proton elastic scattering analysis, PESA), and multiple elements with atomic number ranging from Na through Pb (initially by proton-induced X-ray emission, PIXE, and starting June 1992 by a combination of PIXE and X-ray fluorescence, XRF).

3. TRAJECTORY DATA

A set of backward air parcel trajectories (including horizontal and vertical location) had been previously calculated for the Underhill, VT site for the period of 1989 to 1996. These trajectories were calculated using the CAPITA Monte Carlo model (Schichtel and Husar, 1996) and an NGM meteorological driver (Rolph, 1996). The air mass histories were obtained by calculating the backward trajectory positions for each of 10 hypothetical particles released every 2 hrs from the receptor location and tracked backward in time for 5 days. This set of trajectory data was used in the recent source receptor modeling work of Underhill, VT fine particulate multi-element data using PSCF analysis and RTA modeling (Poirot *et al.*, 1999; Poirot *et al.*, 2001, Polissar *et al.*, 2001).

The 5-day back trajectories, longer duration trajectories than the 3-day back trajectories were used for the ensemble trajectory techniques RTA based on our belief that the longer duration trajectories do not have excessive uncertainties compared to the 3-day ones. An ensemble trajectory technique is based on hundreds, thousands (or in the current study tens of thousands of individual trajectory calculations). A fundamental characteristic of these ensemble trajectory techniques is that we do not in fact "trust" any individual trajectory calculation. Any error in each calculation (which could be described as a specific bias by comparison to the "true trajectory") is assumed to contribute a random form of "noise" when the spatial characteristics of thousands of individual trajectories are taken as a group. For these ensemble trajectories to exhibit a bias, there would need to be a systematic bias to all the individual trajectory calculations (e.g. they always turn too far to the left and always tend to reside inaccurately over a specific upwind location). Such systematic biases have not been identified in the CAPITA or HY-SPLIT trajectory models.

Schichtel and Wishinski (1996) conducted a detailed comparison of CAPITA and HY-SPLIT trajectory calculations, and noted that while there were often substantial differences between individual trajectory calculations from the two models, these differences essentially disappeared when multiple trajectories were employed in the ensemble RTA trajectory technique. The differences or relative biases between the 2 models were "washed out" when large numbers of trajectories were considered. In that case only a few hundred trajectories were employed (compared to tens of thousands in the current study), and the trajectory duration was 4.5 days.

This leads to the additional consideration that these ensemble trajectory techniques and their results are not in fact particularly sensitive to the trajectory duration. Other subjective trajectory options such as choice of trajectory model, meteorological driver, start height, vertical motion (or mixing) and trajectory frequency are substantially more important than trajectory duration for these ensemble trajectory results. A sensitivity analysis of the effect of trajectory duration on RTA results was conducted by Poirot & Wishinski (1996), by arbitrarily truncating trajectory durations in increments ranging from about 5-day to 2-day duration for a set of about 2500 trajectories and associated ozone concentrations. The results indicated minimal changes in the RTA results down to a duration of about 2.5-day. Trajectories of longer than 5-day duration also resulted in virtually no change in the resultant RTA patterns.

The ensemble trajectory techniques employ and require large numbers of trajectory calculations to draw inferences about contributing source locations. The multiple trajectory results are sorted and/or aggregated over a gridded domain. For these gridded results to be meaningful, it's important to have good "coverage" over the entire domain over which inference is drawn. Shorter duration trajectories reduce the spatial coverage. If the trajectory error is in the range of 20 to 30% of trajectory distance, the error for an individual trajectory at a given distance D from the receptor can be expressed as about D/4. However, when multiple trajectories are employed to calculate a gridded probability field, the probability that a specific trajectory will pass over a specific grid cell location at distance D is $1/D^2$. Thus, as D increases, the probability that trajectory error will cause a biased result in an RTA probability field decreases.

4. POTENTIAL SOURCE CONTRIBUTION FUNCTION ANALYSIS

To track down the emission source areas for Hg deposited into Lake Champlain Basin, the calculated backward trajectories were combined with the measured Hg concentrations using potential source contribution function (PSCF) analysis. For each twenty-four-hour sample, there are 14,400 1-hr trajectory segment endpoints associated with it. The locations of the endpoints of these trajectory segments were used for PSCF analysis. The possible source region was subdivided into a gridded array of cells, each the size of 1° latitude by 1° longitude. Let N represent the total number of trajectory segment endpoints during the study period. If n_{ij} endpoints fall into the ij-th grid cell, the probability of an event denoted as A, $P[A_{ij}]$, is given by

$$P[A_{ij}] = n_{ij} / N \quad (1)$$

In the same ij-th cell, there were m_{ij} endpoints from the trajectories that arrive at a receptor site when the pollutant concentrations were higher than some specified criterion value (In this study, the upper 60th percentile concentration values were used as the cut-off values for the four types of Hg data). The probability of this polluted event, B, is $P[B_{ij}]$, and is given by

$$P[B_{ij}] = m_{ij} / N \quad (2)$$

Then, PSCF, a conditional probability, for the ij-th cell is defined as

$$PSCF_{ij} = P[B_{ij}] / P[A_{ij}] = m_{ij} / n_{ij} \quad (3)$$

It will be a value between 0 and 1 and will reflect the probability that an area contributed pollutant to the receptor site (Gao *et al.*, 1996).

There are often problems in the PSCF analysis at the edges of the trajectory domain. In many cases, there can be entry of material into the field of view of the trajectories at a boundary as was seen in the analysis of data from Alert, Nanuvut, Canada (Cheng *et al.*, 1993; Hopke *et al.*, 1995). In other cases anomalous values can appear in these regions since there are generally small numbers of points in these regions and thus, there is a high degree of uncertainty in the PSCF values in such areas. In many recent studies and this study, such areas are more aggressively downweighted as was done by Polissar *et al.* (2001).

5. RESIDENCE TIME ANALYSIS

Another trajectory-based modeling method, residence time analysis (RTA) was also employed for the investigation of emission source areas for the three different forms of Hg data collected at Underhill, VT. There are many forms of RTA available. The one that was employed for this study was the "incremental probability RTA." The study region is subdivided into a network of gridded cells, with the cell size of 1° latitude by 1° longitude. Let N represent the total number of trajectory segment endpoints during the study period. If n_{ij} endpoints fall into the ij-th grid cell, the probability of an event denoted as "everyday probability," is given by

$$\text{Everyday Probability} = n_{ij} / N \quad (4)$$

Based on this equation, an everyday probability field could be calculated for the study region. In the same ij-th cell, there were m_{ij} endpoints from the trajectories that arrive at a receptor site when the pollutant concentrations were higher than some specified criterion value (In this study, the upper 60th percentile concentration values were used as the cut-off values for the four types of Hg data). Let M represent the total number of trajectory segment endpoints correspond to all the "polluted events," i.e., $M = \sum m_{ij}$, the probability of the polluted events associated with the ij-th cell is called "high-day probability," and is given by

$$\text{High-day Probability}_{ij} = m_{ij} / M \quad (5)$$

Based on this equation, high-day probability field can be calculated for various definitions of high concentration at the receptor.

Then, the "incremental probability" for the ij-th cell is defined as

$$\text{Incremental Probability}_{ij} = \text{High-day Probability} - \text{Everyday Probability}$$

$$= (m_{ij} / M) - (n_{ij} / N) \quad (6)$$

It may have a negative, zero, or positive value. A negative or zero value means that the geographical area represented by the ij-th grid cell is likely downwind from the receptor site when the detected pollutant concentration was high. A positive value means that the geographical area represented by the ij-th grid cell is likely to be upwind from the receptor site when the detected pollutant concentration was high. Since most of the incremental probability values are very small fractions, in practice, an arbitrarily chosen number, such as a value of 1000 used for this project, is frequently used as a multiplier to bring the incremental probability values to an easily recognizable range (Wishinski and Poirot, 1998; Poirot and Wishinski, 1998).

While the PSCF indicates the potential for a location to contribute if that area is upwind from the receptor, the RTA incremental probability reveals the most probable upwind locations if the source contribution is high. Therefore, an incremental probability field for an upper 60th percentile definition of high-day would differ from the PSCF results based on the same cut-off value. It is, thus, advantageous to utilize both modeling methods in a study to uncover as much information as possible on source-receptor relationships.

6. POSITIVE MATRIX FACTORIZATION

Positive matrix factorization (PMF) is a new type of factor analysis based on solving the factor analysis problem by least-squares using a data point weighing method, which produces a nonnegative solution. PMF utilizes error estimates of the data to provide an optimum data point scaling (Paatero and Tapper, 1994). A detailed discussion of the application of PMF method in air quality studies can be found in Polissar *et al.*, 2001. The multi-element fine particle data collected during 1988 to 1995 at Underhill, VT were analyzed by Polissar *et al.* and 11 emission source types were identified based on the source profiles (signature species) constructed from the PMF factors and the identification of the source areas were aided by PSCF analysis of each individual PMF factor. These 11 source types are: salt (signature species Na, Cl), aged marine aerosols (signature species Na, Br, S), Canadian non-ferrous metal smelting (signature species As, S), Cu smelting (signature species Cu, S, Zn), soil (signature species Al, Fe, Si, Ti), Canadian automobiles and/or steel production in the upper Midwest (signature species Mn, Pb), incinerator (signature species Zn, Pb), Midwest summer-coal (signature species S), east coast oil (signature species Ni, V), Midwest winter-coal (signature species Se, S), and woodsmoke (signature species K, BC). Average contributions from each of these 11 sources to $PM_{2.5}$ mass concentrations were calculated by a weighted multiple linear regression process, and accounted for 97% of the variance in the $PM_{2.5}$ mass at the Underhill site. The Midwestern coal sources, interpreted as representing the primary ("winter coal") and secondary ("summer coal") aerosol impacts from a common Midwestern source region, accounted for 60% of the $PM_{2.5}$ mass. Woodsmoke accounted for 15%, with much smaller fine mass contributions from the other 8 sources.

7. MULTIPLE REGRESSION ANALYSIS RESULTS

Multiple regression analysis was performed for each of the three types of the Underhill, VT mercury data, respectively, against the factors (source $PM_{2.5}$ contributions) obtained from the PMF analysis on the multi-element fine particle data mentioned above. Each of the three mercury data sets was sorted to use only the samples collected on the dates when fine particle data were available. Then based on the multiple regression results and the source contribution for each PMF factor derived by the PMF analysis of the fine particulate multi-elements, the source contribution for each of the likely vapor phase and particulate mercury emission source type identified above was estimated:

1) Source Contribution of Hg by A Source Type = Multiple Regression Coefficient for That Source Type x Source Contribution from PMF for That Source Type;

2) % Contribution to Average Hg Associated = Source Contribution of Hg by A Source Type / The Average of Hg data from The Same Sampling Dates.

The multiple regression analysis of the vapor phase mercury data against the PMF factors based on 119 common sampling days yielded the following solution:

Vapor Phase Hg = 1.49 + 0.000261 East Coast Oil
+ 0.000148 Canadian Gasoline / Midwest Steel
+ 0.000130 Soil
+ 0.000037 MW Summer Coal

R^2 = 33.2% ($P < 0.001$)

Source contributions of vapor phase Hg (% contributions to average Hg associated):

East Coast Oil 0.143 ng/m^3 (8.16%) Canadian Gasoline/Midwest Steel 0.025 ng/m^3 (1.44%)
Soil 0.0419 ng/m^3 (2.39%) Midwest Summer Coal 0.158 ng/m^3 (9.01%)

The multiple regression analysis has identified four out of the eleven PMF factors to be the likely emission sources for vapor phase mercury found in Lake Champlain Basin. The result shows that secondary aerosol formation from coal burning in the Midwest and the east coast oil-fired power plants are significant contributors of vapor phase mercury to Lake Champlain Basin. Natural emission from soils and gasoline consumption or Mn alloy production in and around Canadian metropolitan areas such as Montreal north of the border from Vermont and New York State and possible steel production in the upper Midwest have also been identified as likely sources. These four types of emission sources combined explained ca. 21% of the vapor phase mercury entered into Lake Champlain Basin. While these regression coefficients are "significant" statistically, it should also be noted that the overall R^2 indicates that only 1/3 of the Hg variance is accounted for by these sources of $PM_{2.5}$.

The multiple regression analysis of the particulate mercury data against the PMF factors based on 140 common sampling days yielded the following solution:

Particle Phase Hg = 2.88 + 0.0171 Aged Aerosols
+ 0.0143 Smelting
+ 0.00760 Canadian Gasoline / Midwest Steel
+ 0.00405 East Coast Oil
+ 0.00199 Incineration
+ 0.00147 Cu Smelting
+ 0.00104 MW Winter Coal

$R^2 = 42.9\%$ ($P < 0.001$)

Source contributions of particulate Hg:

Aged Marine Aerosol	0.887 pg/m^3 (9.56%)	Smelting	1.389 pg/m^3 (15.0%)
Canadian Gasoline/Midwest Steel	1.320 pg/m^3 (14.2%)	East Coast Oil	2.22 pg/m^3 (23.9%)
Incineration	1.164 pg/m^3 (12.5%)	Cu-Zn Smelting	0.179 pg/m^3 (1.92%)
Midwest Winter Coal	0.619 pg/m^3 (6.67%)		

The multiple regression analysis has identified seven out of the eleven PMF factors to be likely emission sources for particulate mercury found in Lake Champlain Basin. The most significant contributor is from east coast oil-burning concentrated in the area of Maryland to New York City. The second largest contributors are non-ferrous metal smelters (primarily in Quebec and Ontario), Canadian gasoline consumption/Midwest steel production, and municipal waste incineration in the major metropolitans across western New York State, Pennsylvania, Ohio, Indiana, and Illinois. The next group includes aged marine aerosols and winter coal burning in the Midwest. Copper-zinc metal ore smelting found in New York State, Pennsylvania, Maryland, West Virginia, and Indiana is the least significant type among all the identified particulate mercury contributing source types. These seven source types combined explained nearly 84% of the particulate mercury detected in Lake Champlain Basin. As with the vapor phase Hg, the PMF sources of fine mass accounted for less than half the variance in the concurrent particulate Hg samples.

The percentages of detected Hg explained by the PMF factors based on the multiple regression analysis performed for the vapor phase Hg and particulate Hg, respectively, are as expected. Vapor phase Hg in the form of elemental mercury (Hg^o) is the most volatile form of mercury in atmosphere, insoluble in water (i.e., little washout effect), and has a residence time of ca. 11/2 years. It is greatly affected by globally distributed sources and thus it is harder to extract the contributions from local and regional sources. Therefore, a 21% Hg accounted for (with R^2 of 33.2%) seems reasonable. On the other hand, Particulate Hg is mostly in the oxidized form (Hg^{2+}) deposited on airborne particle surface, less likely to re-volatilize and thus correlated well with atmospheric transport and deposition of fine particles. Therefore a high %Hg accounted for, 84% (with a higher R^2 value, 42.9%) is as expected.

The multiple regression analysis performed for the precip-Hg based on 93 common sampling days failed to produce a solution with all positive correlation coefficients for individual PMF factors. Also it has a low coefficient of variance ($R^2 = 22.8\%$). It is an indication that precipitation events tend to have a washout effect for particulate matter, and therefore, a positive correlation is not observed between precip-Hg and fine particulate elemental concentrations.

It should be noted here that sample size has an effect on the statistical significance of correlation coefficient, R^2. For example, when sample size equals to 200 and $R^2 > 0.0137$ (i.e., $R > 0.117$), then we can be 95% sure that R is really greater than zero and did not occur just by chance. But if the same value of correlation coefficient is observed for a much smaller sample size, we can conclude that the true correlation is null (Younger, 1979).

The above Hg contributions are presented here in "quantitative" terms, but we recommend they be taken as "qualitative" estimates that identify several likely contributing sources that should be evaluated by additional measurement and analysis techniques in the future. Three specific cautionary considerations are advised. First, the regression with the fine particle sources previously identified by Polissar *et al.* (2001) from PMF analysis of concurrent, collocated aerosol samples indicates statistically significant correlations between several of those sources and the various Hg measurements. However, none of these individual correlations is especially high, and the overall regression models accounts for less than half of the observed variance in the Hg components (vapor Hg: R^2=33.2%; particulate Hg: R^2 = 42.9%). Second, the regression-based estimates of Hg contributions are based on an implied assumption that the vapor and particle-phase Hg contributions from a source are linearly related to the fine particle mass contribution from the source. Substantial nonlinearities are probable for volatile Hg species, which could lead to understating or overstating the Hg contributions from these sources. Third, there were substantial uncertainties in the initial interpretations of several of the PMF sources. For example, the interpretation of the "aged marine aerosol" PMF source was presented tentatively by Polissar *et al.* (2001), and a comparable source was not identified in an application of the UNMIX receptor model to the same data (Poirot *et al.*, 2001). The "Canadian MV and/or upper Midwestern Steel" source was identified by both PMF and UNMIX models, but its actual origin was interpreted tentatively in both cases, and there were notable differences in the mass contribution from this source between the PMF and UNMIX results (Poirot *et al.*, 2001).

8. PSCF ANALYSIS RESULTS

The PSCF analysis results were mapped using ArcView 3.1, a GIS software equipped with spatial analyst options. A grid interpolation option was used to smooth out the transition of PSCF values in neighboring cells. Figures 1 through 4 show the PSCF maps based on analysis using the upper 60th percentile cut-off's for vapor phase Hg (ng Hg/m^3), particulate Hg (pg Hg/m^3), precip Hg (ng Hg/L of precipitation), and wet-dep Hg (ng Hg/m^2 deposition area), respectively. Two different forms of event precipitation Hg data were used because they are not identical with regard to the Hg information contained in each. For example, a high precip Hg value may be a result of the high concentration of Hg found in rain or snow. However, if the total amount of precipitation for that event is small, the wet-dep Hg value for that event would have a relatively low value. Therefore these two types of event precipitation Hg data may reveal different possible source areas. The vapor phase Hg data include 174 sampling days from December 16, 1992 to June 26, 1995. The particulate Hg data include 214 sampling days from December 16, 1992 to June 26, 1995. The precip Hg data and wet-dep Hg data are from the same 282 event days from December 17, 1992 to June 11, 1995.

Figure 1, the PSCF map of vapor phase mercury, shows the most likely sources (PSCF $\geq$ 0.6) are located in the east coast urban corridor, a pattern very similar to the east coast oil-burning source area identified by Polissar *et al.*, 2001 and Poirot *et al.*, 2001. The source areas with medium probabilities ($0.4 \leq$ PSCF < 0.6) are found across the Midwest and in eastern Quebec. The Midwest source area is very similar to that of the Midwest summer (and winter) coal source region (Polissar *et al.*, 2001, Poirot *et al.*, 2001) and the eastern Quebec source may be that of some smelting processes identified by Gao, *et al.* (1996). Overall the PSCF map of vapor phase mercury yielded results that are in very good agreement with those from the multiple regression analysis (above) and PSCF and RTA analysis of the PMF and UNMIX factors (Polissar *et al.*, 2001, Poirot *et al.*, 2001).

High probability source regions (PSCF $\geq$ 0.6) are absent except for one spot in eastern Quebec in Figure 2, the PSCF map of particulate Hg. The medium probability ($0.4 \leq$ PSCF $<$ 0.6) regions are located in eastern Quebec and also in isolated Midwest locations. These may represent the eastern Quebec smelter (Gao, *et al.*, 1996) and Midwest steel production. A combination of smelting, incineration, steel production, and Midwest winter coal-burning could explain the widespread low probability (0.2 ≤ PSCF < 0.4) areas in the eastern Quebec and the Midwest. Overall the PSCF analysis results for particulate Hg are consistent with those identified by the multiple regression analysis and the PSCF analysis of the PMF factors. However, there is a significant discrepancy in the PSCF result as it failed to identify east coast oil-burning as a significant source, which is in disagreement with the multiple regression analysis and the PSCF analysis of the PMF factors.

Figures 3 and 4, the PSCF maps of precip Hg and wet-dep Hg, respectively, have revealed very different source regions. It seems that east coast oil-burning, the non-ferrous metal smelter at Sudbury, Ontario, and possibly some Midwest steel productions have been identified as emission sources with medium probabilities (0.4 ≤ PSCF < 0.6) that had led to high concentrations of Hg in precip samples. On the other hand, incineration across the eastern and Midwest states, Midwest coal-burning and probably some steel productions produces high probability (PSCF ≥ 0.6) emission source areas that led to significant Hg wet deposition rate in the Lake Champlain Basin.

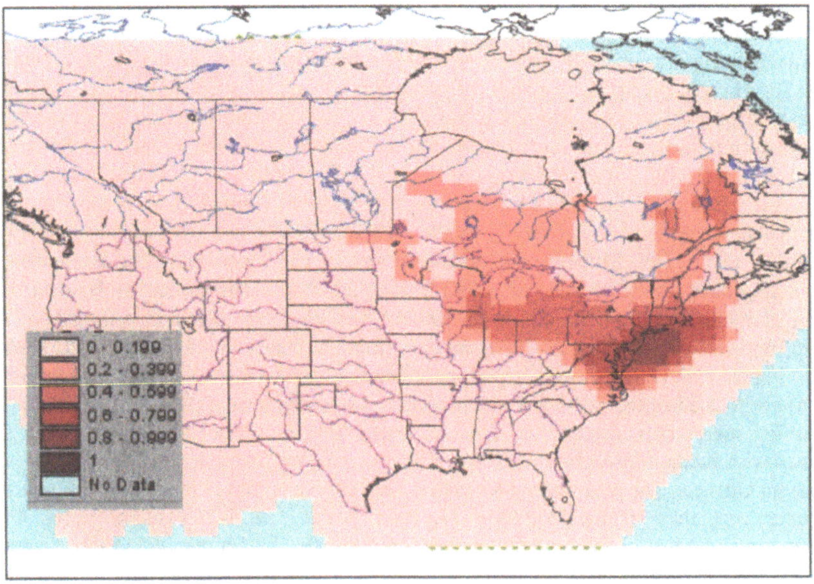

Figure 1. PSCF map of vapor phase Hg collected at Underhill, VT.

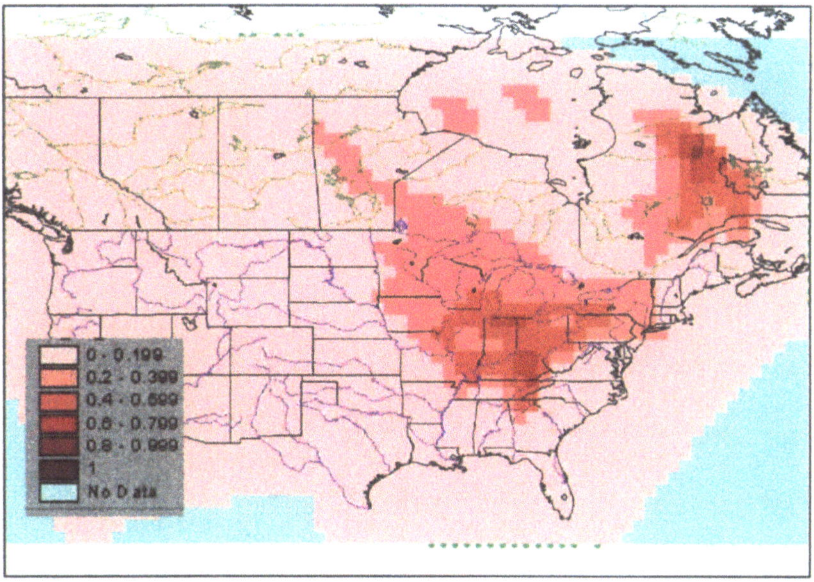

Figure 2. PSCF map of particulate Hg collected at Underhill, VT.

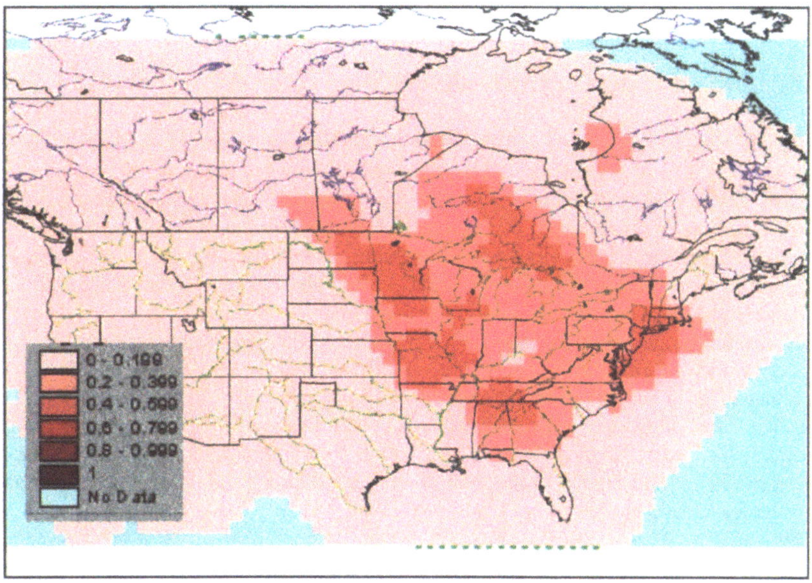

Figure 3. PSCF map of precip Hg collected at Underhill, VT.

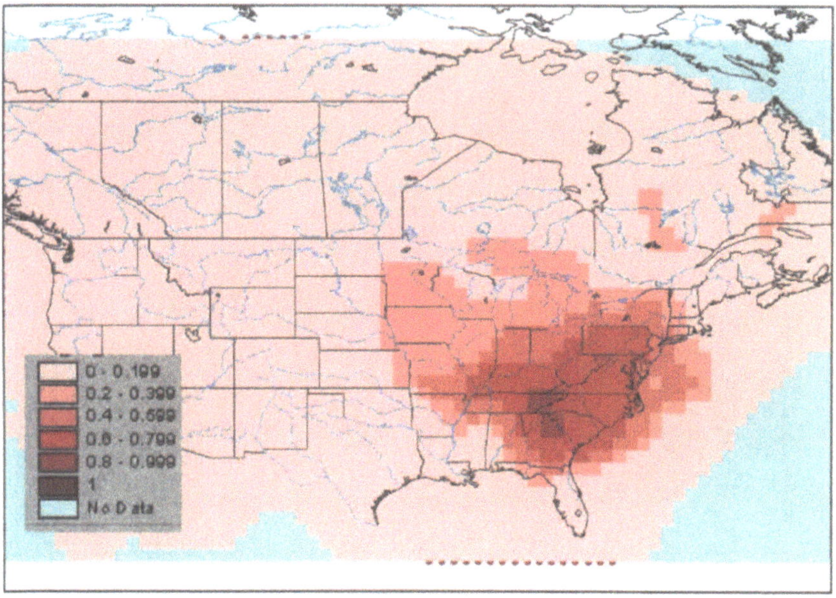

Figure 4. PSCF map of wet-dep Hg collected at Underhill, VT.

9. RESIDENCE TIME ANALYSIS RESULTS

The residence time analysis (RTA) incremental probability results were also mapped using ArcView 3.1. A grid interpolation option was used to smooth out the transition of RTA values in neighboring cells. Figures 5 through 8 show the RTA incremental probability maps based on analysis using the upper 60th percentile concentration cut-off's for vapor phase Hg, particulate Hg, precip Hg (ng Hg/L of precipitation), and wet-dep Hg (ng Hg/m^2 deposition area), respectively. Again, two different forms of event precipitation Hg data were used. The Hg data used for RTA analysis are identical to those used for PSCF analysis.

Figure 5 presents the RTA incremental probability map for vapor phase Hg that is in good agreement with the PSCF results (Fig. 1). The high probability areas identified by the PSCF map (Fig. 1) are located in the most likely upwind areas (RTA inc prob $\geq$ 1.00) in the RTA map. Most of medium source strength areas identified in the PSCF map also show up as moderately likely upwind areas (0.5 $\leq$ RTA inc prob < 1.0) in the RTA map. In addition, the RTA map shows one highly likely upwind region in eastern NY near Lake Champlain (Ticonderoga paper plant?).

Figure 6, the RTA incremental probability map for particulate Hg is in good agreement with the PSCF results (Fig. 2). The high probability areas shown in the PSCF map are identified as most likely upwind locations (RTA inc prob $\geq$ 0.5) in the RTA map. The RTA map also showed two extra most likely upwind regions that were not revealed by the PSCF map, one again in eastern NY near Lake Champlain and the other in the western NY state.

Most of the medium source strength regions in the PSCF map are also identified as moderately likely upwind locations (0.25 ≤RTA inc prob < 0.5).

Figures 7 and 8, the RTA incremental probability maps of precip Hg and wet-dep Hg, respectively, have more in common with each other compared with the PSCF maps (Figs 3-4). The most likely upwind locations (RTA inc prob ≥ 1.00) are rather similar, reflecting the east coast oil-burning effects. The moderately upwind locations (0.5 ≤ RTA inc prob < 1.0) probably include incineration and the Sudbury, Ontario smelter. The significant areas shown by these two RTA maps are more limited than those revealed by the PSCF maps.

While the PSCF indicates the potential for a location to contribute if that area is upwind from the receptor, the RTA incremental probability reveals the most probable upwind locations if the source contribution is high. Therefore, an incremental probability field (reflecting the frequency with which an area is upwind on high pollution days) can differ from the PSCF results (reflecting the potential for contribution if an area is upwind). In cases where sources with a high potential to contribute are also frequently upwind, the PSCF and RTA will produce similar results. When the PSCF and RTA patterns differ, additional information is provided about sources that may make large but infrequent contributions to the receptor as well as sources with contributions that are smaller but more frequent. The RTA incremental probability maps also contain one type of valuable information that is not found in PSCF maps, the likely downwind areas, i.e., areas with negative or zero RTA incremental values. A careful screening of these downwind areas will also be very beneficial to the source-receptor study.

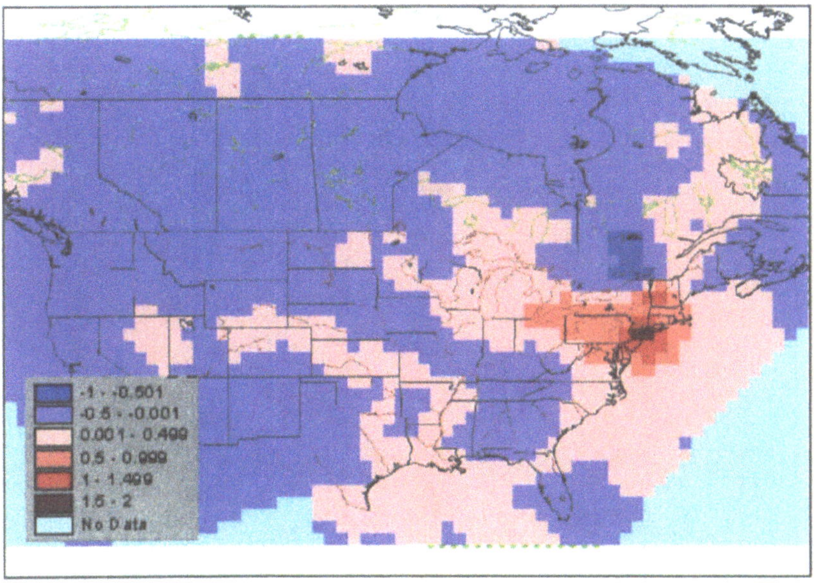

Figure 5. RTA map of vapor phase Hg collected at Underhill, VT.

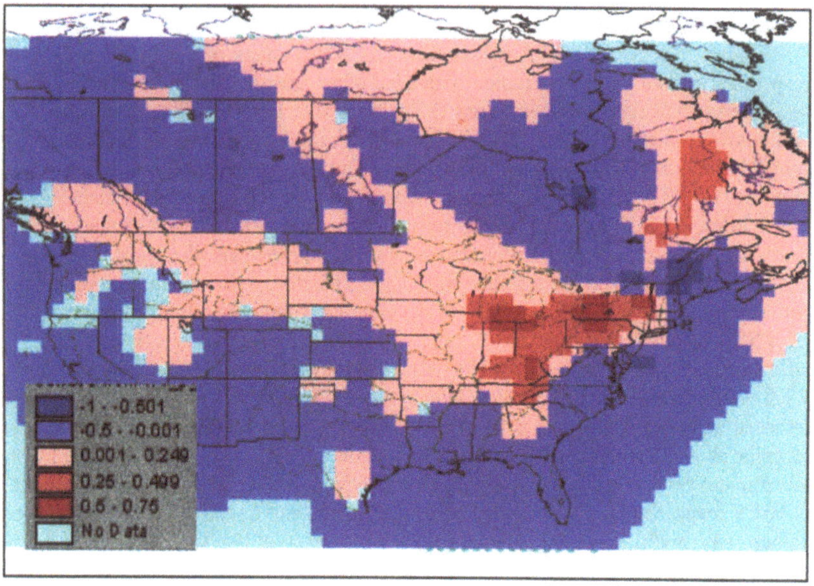

Figure 6. RTA map of particulate Hg collected at Underhill, VT.

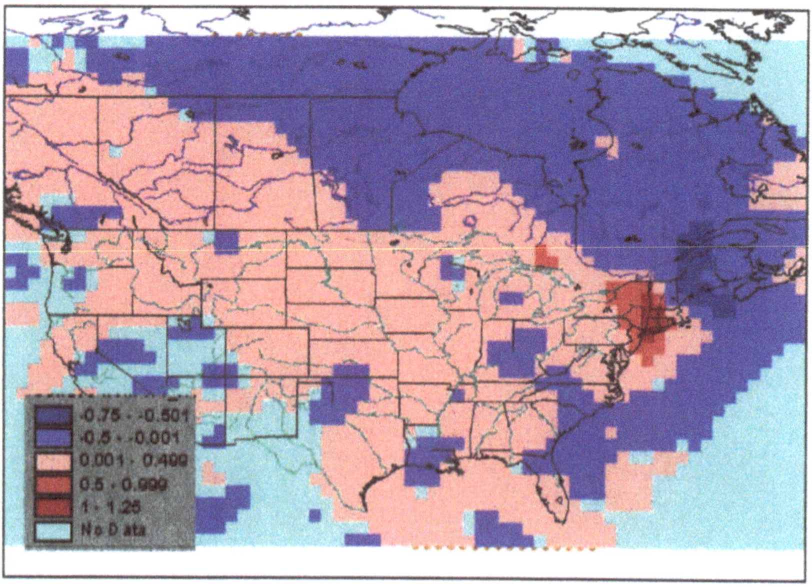

Figure 7. RTA map of precip Hg collected at Underhill, VT.

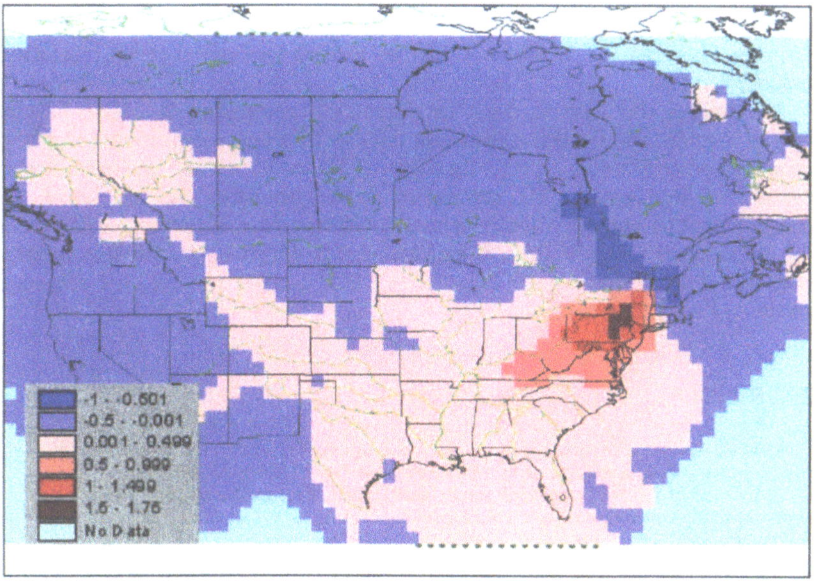

Figure 8. RTA map of wet-dep Hg collected at Underhill, VT.

10. CONCLUSIONS

This preliminary study on mercury found in the Lake Champlain Basin with multiple source-receptor modeling approaches has yielded interesting results with good correlations to each other. This study has identified several significant emission sources types/areas that appear likely to contribute mercury to the Lake Champlain Basin. Estimates derived for the source contributions suggest a broad range of contributing source categories, including: oil and coal combustion, primary and secondary smelting, incineration, steel production, mobile sources and natural (or anthropogenically enhanced) marine sea salt and crustal dust.

The authors and other collaborators are working on building a full set of missing-data free 5-day HY-SPLIT back trajectories for Underhill, VT up to 2001 to make it possible to include modeling analysis of more recent Hg data. PMF analysis directly based on a combined set of PM2.5 elemental data and Hg data is being conducted This source-receptor modeling study is also an integral part of a larger project, A Mass Balance Assessment for Mercury in Lake Champlain, an on-going project sponsored by CILER/NOAA/LCRC.

11. ACKNOWLEDGEMENTS

This publication is a result of work sponsored by the Cooperative Institute of Limnology and Ecosystems Research under cooperative agreement #MA67RJ0148 from the Environmental Research Laboratory, National Oceanic and Atmospheric Administration, U.S. Department of Commerce. The U.S. government is authorized to produce and distribute reprints for governmental purposes notwithstanding any copyright notation that may appear hereon. Supplemental fundings in the form of summer internships were also received from Department of Chemistry, St. Lawrence University and the Department of Environmental Conservation, Vermont.

The authors also wish to thank Dr. Timothy Scherbatskoy, Mr. Mike Tate at University of Vermont, and Dr. James Shanley at USGS-VT for their contributions in the long term monitoring program for atmospheric Hg at PMRC, Underhill, VT, and Dr. Gerald Keeler at University of Michigan for his contribution in the laboratory Hg analysis of the atmospheric deposition samples. The authors also wish to acknowledge the generous help from Dr. Alexandar Polissar at DEP-NJ and Mr. Eric Williams-Bergen of St. Lawrence University who shared with us their expertise in PSCF analysis and ArcView mapping, respectively.

12. REFERENCES

Burke, J., Hoyer, M., Keeler, G., and Scherbatskoy, T., 1995, Wet deposition of mercury and ambient mercury concentrations at a site in the Lake Champlain Basin, *Water Air Soil Pollut.* 80:353-362.

Colman, J. A. and Clark, S. F.,1994, Geochemical data on concentrations of inorganic constituents and polychlorinated biphenyl congeners in streambed sediments in tributaries to Lake Champlain in New York, Vermont and Quebec, 1992, US Geological Survey, Bow, New Hampshire, Open File Report 94-472.

Fitzgerald, W. F., and Gill, G. A., 1979, Subnanogram detection of mercury by two-stage gold amalgamation and gas phase detection applied to atmospheric analysis, *Anal. Chem.* 51:1714-1720.

Fluocchini, R. G., Cahill, T. A., Eldered, R. A., and Feeney, P.J., 1990, Particulate sampling in the Northeast: a description of the NESCAUM network. Transactions: A&WMA/EPA Specialty Conference on Visibility and Fine Particles; Mathai, C.V., ED.; Estes Park, CO.

Gao, N., Hopke, P. K., and Reid, N. W., 1996, Possible sources for some trace elements found in airborne particles and precipitation in Dorset, Ontario, *J. Air & Waste Management Assoc.* 46:1035-1047.

Kamman, N. C. and Engstrom, D. R., 2002, Historical and present fluxes of mercury to Vermont and New Hampshire lakes inferred from ^{210}Pb-dated sediment cores, *Atmos. Environ.* 36:1599-1609.

Keeler, G., Glinsorn, G., and Pirrone, N., 1995, Particulate mercury in the atmosphere: its significance, transport, transformation and sources, *Water Air Soil Pollut.* 80:159-168.

Landers, D. H., Gubala, C., Verta, M., Lucotte, M., Johansson, K., Vlasova, T., and Lockhart, W. L., 1998, Using lake sediment mercury flux ratios to evaluate the regional and continental dimensions of mercury deposition in arctic and boreal ecosystems, *Atmos. Environ.* 32: 919-928.

McIntosh, A. (ed.), 1994, Lake Champlain sediment toxics assessment program: an assessment of sediment associated contaminants in Lake Champlain, Phase I, Lake Champlain Basin Program Technical Report No. 5. Lake Champlain Basin Program, Grand Isle, Vermont.

Nriagu, J. O., Legacy of mercury pollution, *Nature* 363: 589.

Nriagu, J. O., 1994, Mercury pollution from the past mining of gold and silver in the Americas, *Science of the Total Environment* 149: 167-181.

Nriagu, J. O., 1996, A history of global metal pollution, *Science* 272: 223-224.

Paatero, P. and Tapper, U., 1994, Positive matrix factorization: a non-negative factor model with optimal utilization of error estimates of data values, *Environmetrics* 5: 111-126.

Poirot, R. L., Galvin, P. J., Gordon, N., Quan, S., Arsdale, A. V., and Flocchini, R. G., 1991, Annual and seasonal fine particle composition in the Northeast: Second year results from the NESCAUM monitoring network, 91-49.1, *Proceedings 84th Annual A&WMA Meeting*, Vancouver, Canada.

Poirot, R. and Wishinski, P., 1996, VT DEC air trajectory analysis of long-term ozone climatology, status report to OTAG Air Quality Analysis Workgroup, http://capita.wustl.edu/otag/Reports/vtdecair/vtdecair.html.

Poirot, R. L. and Wishinski, P.R., 1998, Long-term ozone trajectory climatology for the Eastern US, Part 2: Results", 98-A615, *Proceedings 91st Annual A&WMA Meeting*, San Diego, CA.

Poirot, R. L., Wishinski, P., Schichtel., B., and Girton, P., 1999, Air trajectory pollution climatology for the Lake Champlain Basin, in T Manley and P. Manley, eds., *Lake Champlain in Transition: From Research Toward Restoration*, Water Science and Application, American Geophysical Union, Washington, DC.

Poirot,R. L., Wishinski, P. R., Hopke, P. K., and Polissar, A. V., 2001, Comparative application of multiple receptor methods to identify Aaerosol sources in northern Vermont, *Environ. Sci. Technol.* 35: 4622-4636.

Polissar, A.V., Hopke P. K., and Poirot. R. L., 2001, Atmospheric aerosol over Vermont: Chemical composition and sources," *Environ. Sci. Technol.* 35: 4604-4621.
Porcella, D. B., Huckabee, J. W., and Weatley, B., 1995a, *Mercury as A Glaobal Pollutant.* Kluwer Academic Publishers, Boston.
Porcella, D. B., Chu, P., and Allan., M. A., 1995b, Inventory of North American Hg emissions to the atmosphere: Relationship to the global mercury cycle," In Proceedings of the NATO Advanced Workshop on Glabal Hg Cycle, Novosibirsk, Siberia, Russia. July 10-14.
Rolph, G. D. 1996, NGM Archive TD-6140, January 1991-June 1996; Prepared for National Climate Data Center (NCDC).
Scherbatskoy, T., Shanley, J. B., and Keeler, G. J., 1998, Factors controlling mercury transport in an upland forested catchment, *Water, Air, Soil Pollut.*, 105:427-438.
Schichtel, B. A. and Husar, R. B., 1996, Regional simulation of atmospheric pollutants with the CAPITA Monte Carlo Model, *J. Air Waste Manage. Assoc.* 47: 331-343.
Schichtel, B. and Wishinski, P., 1996, HY-SPLIT and CAPITA Monte Carlo Model Back Trajectory Comparison, http://capita.wustl.edu/otag/Reports/Trajcomp/trajcomp.html.
Shanley, J.B., Donlon, A. F., Scherbatskoy, T., and Keeler, G., 1999, Mercury cycling and transport in the Lake Champlain Basin, in T.O. and P.L. Manley, eds., *Lake Champlain in Transition: From Research Toward Restoration*, Water Science and Application, American Geophysical Union, Washington, DC, 1, 277-299.
United Nations Environmental Program, 2002, Global Mercury Assessment, UNEP, Geneva, Switzerland.
US-EPA, 1997, Mercury Study Report to Congress. Vol .1., EPA-600/R-97-003.
Wishinski, P. R. and Poirot, R. L., 1998, Long-term ozone trajectory climatology for the Eastern US, Part 1: Methods, 98-A615, *Proceedings 91st Annual A&WMA Meeting*, San Diego, CA.
Younger, M. S., 1979, *A Handbook for Linear Regression*, Wadsworth, Inc., California, pp. 246-247.

EFFECTIVENESS OF AGRICULTURAL BEST MANAGEMENT PRACTICES IN REDUCING PHOSPHOROUS LOADING TO LAKE CHAMPLAIN

William E. Jokela, John C. Clausen, Donald W. Meals, Andrew N. Sharpley *

1. INTRODUCTION

Phosphorus is an essential element for the growth of terrestrial and aquatic plants. But in P-limited freshwater lakes, increased P loading can accelerate eutrophication and an associated growth of undesirable algae and aquatic weeds (Carpenter et al., 1998). Eutrophication has been blamed for the decline in water quality in freshwater lakes and estuaries in the Northeast over the past decades. The issue has received increased public attention recently as a result of human health problems, such as the impaired drinking water supply for New York City and outbreaks of the dinoflagellate *Pfiesteria* in waters off the east coast (Sharpley et al., 2000). The role of excessive P levels in water quality degradation in Lake Champlain is well documented (LCBP, 1996; Smeltzer, 1992). Phosphorus concentrations in parts of Lake Champlain (Missisquoi and St. Albans Bays and the South Lake, in particular) are highly eutrophic (Medalie and Smeltzer, 2003), with P concentrations as high as those in the most polluted parts of the Great Lakes (e.g. western Lake Erie) in the 1970s (Smeltzer, 1992).

Increased P loading to surface waters has come from a range of point sources as well as urban and agricultural nonpoint sources. Estimates for Lake Champlain, for example, indicate that 30% of the P comes from point sources, primarily sewage treatment plants, and 70% from nonpoint sources in the Vermont-New York watershed (VT DEC and NY DEC, 1997). A more recent estimate suggests over 90% are from nonpoint sources (Medalie and Smeltzer, 2003). Agriculture is believed to contribute about two-thirds of the nonpoint P (Meals and Budd, 1998). Lake Champlain is especially sensitive to

* William E. Jokela, Plant and Soil Sci. Dept., Univ. of Vermont, Burlington, VT. John C. Clausen, Dept. of Natural Resources Mgmt. and Eng., Univ. of Connecticut, Storrs, CT. Donald W. Meals, New England Interstate Water Pollution Control Commiss., Lowell, MA. Andrew N. Sharpley, USDA-ARS Pasture Systems and Watershed Mgmt. Research Unit., University Park, PA.

nutrient inputs from the watershed because the land to water area ratio (19:1) is higher than that of many other lake watersheds (e.g. Great Lakes) (Watzin, 1992). The P loading estimates are comparable to those for the Chesapeake Bay watershed, in which over 80% of the nonpoint source and about half of the total P loading contributed by agriculture (Boesch et al., 2001).

Much of the attention in recent years has focused on the role of animal agriculture and land application of manure on P loading and water quality. The situation is especially problematic in areas with high animal densities associated with intensive poultry and hog operations. But even where there is adequate land for N-based manure application, such as on many Northeast dairy farms, the result is usually over application of P because the N:P ratio in manure is much lower than that of crop needs. The situation is further aggravated if fertilizer is applied without accounting for P applied in manure. This phenomenon has led to an accumulation of P over time and rising P soil test levels, with some states reporting greater than 60% of samples in the high or very high category (Sims, et al., 2000). While soil test P levels in the Champlain Valley are not as extreme as in some of the very high intensity livestock areas in other regions of the U.S. (Lander et al., 1998), summaries from Vermont show that about a third are high or excessive (i.e., above optimum level for crops), close to 40% in some counties with high dairy farm numbers. Concentrations of soluble P in runoff are well correlated with soil test P, as shown by experiments with either readily desorbed P or P in actual runoff (Pote et al,, 1996; Sharpley et al., 1996; Magdoff et al., 1999; McDowell and Sharpley, 2001), including recent research results from soils in the Champlain Valley of Vermont (Fig. 1; J. Tilley, W. Jokela, and F. Magdoff, 2002, unpublished data). Runoff from fields with recently applied manure or fertilizer P can cause even greater losses of P, overwhelming the effect of soil test P (Sharpley et al., 1998).

2. EFFECTIVENESS OF INDIVIDUAL BEST MANAGEMENT PRACTICES

Attempts to control nutrient runoff from agricultural fields have included implementation of a range of "best management practices", or BMPs, aimed at decreasing the amount or availability of P applied to the field or the transport of P from the field. A number of studies have been conducted to evaluate the effect of individual management practices on the loss of P in runoff. Most summarized here were conducted in Vermont or the Northeast region. In most cases, water quality effects were measured at the field edge, either from small field plots or from field-scale paired "watersheds" (Clausen and Spooner, 1993).

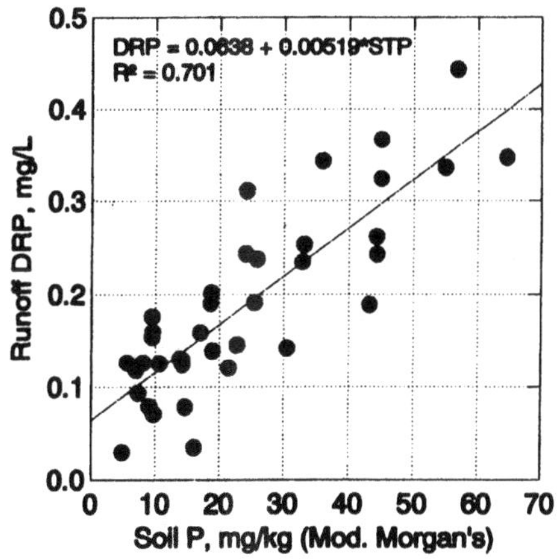

Figure 1. Concentration of dissolved reactive P (DRP) as a function of soil test P (Modified Morgan's) on soils in the Champlain Valley of Vermont.

2.1. Timing and Method of Manure Application

Paired watershed studies were carried out as part of the St. Albans Bay and LaPlatte River watershed monitoring projects in Vermont to evaluate the effect on P runoff of winter spreading of dairy manure on cornland compared to improved management (fall or spring, incorporated) (Clausen, 1991; Clausen, 1987). In both studies, winter spreading reduced both concentration and export of suspended solids, presumably as a result of the mulching effect. However, concentrations and mass export of total and soluble P were significantly higher from winter spreading compared to fall or spring application. Total P export was 30% higher from winter spreading in the LaPlatte and 11% higher in the St. Albans study. Overall, runoff losses of P from winter spreading were 5% and 15% of the manure P applied in the two studies. These results support the value of manure storage to allow farmers to avoid winter spreading of manure. Another Vermont study (Jokela et al., 1995) showed that applying dairy manure with a cultivator into the growing crop decreased total P runoff losses by over 80% compared to broadcast spring application, primarily because cultivation increased infiltration and decreased surface runoff volume.

2.2. Conservation Tillage

A paired watershed study was conducted in the Lake Champlain Basin of Vermont to compare runoff losses of P and N from cornland under conservation tillage (spring or fall disking; 40-60% residue cover) compared to a moldboard plow-disk system (Jokela and Clausen, 1994; Clausen et al., 1996). Implementation of conservation tillage lowered runoff volume and concentrations of total P, soluble P, and total N for individual runoff events and decreased cumulative mass export of P and N by 80 to 90%; most of the difference occurred in two events. Other studies in Wisconsin and Minnesota have shown less sediment and total P from conservation than conventional tillage, but effects on soluble P have varied (Andraski et al., 1985; Ginting et al., 1998; Sharpley and Smith, 1994).

2.3. Riparian Zone/Field-Edge Vegetative Buffers

A paired watershed study in Connecticut evaluated the effect of a 30-m riparian buffer of fescue with woody species near the stream edge adjacent to a field of silage corn (Clausen et al., 2000). Concentrations of total P and total suspended solids in overland flow were decreased by 73% and 92%, respectively. Another paired watershed study conducted in Vermont evaluated the effectiveness of field-edge buffers of managed mixed grass-legume in minimizing nutrient losses in runoff from a corn field (Jokela et al., 1999). Significant reductions in total P export in runoff occurred with implementation of both the 7.5-m (24%) and 15-m (72%) buffer strips. These results are consistent with results found elsewhere, for example in Virginia (Dillaha, 1989) and Europe (Uusi-Kämppä, 2000).

2.4. Stream Bank Protection/Fencing

The effect of stream bank and riparian zone protection/restoration and livestock exclusion with fencing on the water quality of streams was evaluated in a larger-scale paired watershed study in north central Vermont (Meals, 2001). Following three years of calibration, the riparian restoration/fencing treatment was implemented on approximately half of the pastured stream length in the treated watershed, which represented stream exclusion of 97% of the pastured livestock. Significantly lower concentrations of total P and N, total suspended solids, and bacteria were measured from the treated watershed, resulting in 30 to 50% decreases in mass export of N, P, and TSS to the stream.

2.5. Farmstead Sources of P

A monitored milkhouse waste vegetative filter strip was effective at decreasing the loss of P, N, and suspended solids to a nearby stream (Schwer and Clausen, 1989); however, a barnyard runoff filter strip was relatively ineffective because of hydraulic overloading (Schellinger and Clausen, 1992). To be effective, these strips must be carefully designed, installed, maintained, and monitored; otherwise, P or hydraulic overloading can limit their effectiveness.

2.6. Reduction of P levels in Feed

Dairy producers typically supplement rations with inorganic P to ensure top milk production and maintain reproductive health. Despite the fact that research has shown that lower levels of P may be fed to lactating cows without adversely affecting animal health and productivity, many dairy producers continue to overfeed P supplements (Wu et al., 2000; Valk et al., 2000). A survey of feed company representatives and consultants in Vermont found that cows were typically fed 20 to 50% higher than recommended P levels (Anderson and Magdoff, 2000). Excess P in the ration directly increases P excreted in manure and, when land-applied, enriches total P and, especially, soluble P in runoff (Eberling et al., 2002). Phosphorus in rations for hogs and poultry needs to be maintained at a relatively high level because of the low availability of phytic acid P; but use of the enzyme phytase and corn bred for low phytic acid show promise for increasing P availability and, therefore, allowing a decrease in P levels in feed (Poulsen, 2000; Ertl et al., 1998).

3. WATERSHED-SCALE MONITORING OF BMP IMPLEMENTATION EFFECTS

While a number of research studies have shown significant reductions in runoff P at the edge-of-field from implementation of individual BMPs, only a few studies have monitored long-term effects of watershed-scale BMP implementation on receiving waters. Monitoring studies were conducted in two watersheds in Vermont to document changes in water quality from BMP implementation over a ten-year period (Meals, 1990; VT RCWP Coord. Comm., 1991; Meals, 1996). St. Albans Bay and Shelburne Bay were identified as critical water quality problem areas in the 1970s because of excessive algae and macrophyte growth and high bacteria levels. The problems were attributed to runoff of P, bacteria, and sediment from cropland and barnyards, as well as P from a sewage treatment plant in each watershed.

Agriculture, primarily dairy, constituted 47 to 65% of the land area in the watersheds. Federally supported cost-share and technical assistance programs to support BMP implementation and intensive water quality monitoring to assess their effects were conducted in the LaPlatte River (the principal tributary to Shelburne Bay) and St. Albans Bay watersheds from 1979/1980 to 1989/1990 (Meals, 1990; VT RCWP Coord. Comm., 1991). BMP implementation, completed by the mid-point of the project periods, included manure storage, barnyard and milkhouse runoff control, and cropland erosion control systems. Long-term stream monitoring showed significant decreases in suspended sediment and bacteria concentrations (Fig. 2) and load with the implementation of BMPs. However, despite the potential BMP effectiveness based on edge-of-field trials (see above section), stream monitoring showed no significant decrease in P concentration (Fig. 2) or load. Small decreases in TP concentration were noted in the inner portion of the St. Albans Bay itself as a result of an upgrade of the sewage treatment plant, but P concentrations in the main part of the bay actually increased slightly.

The Chesapeake Bay situation presents another case where intensive efforts were made to implement BMPs intended to decrease nutrient inputs through a combination of education, cost-share, and technical assistance (Boesch et al,, 2001). BMPs have included stream bank protection, manure storage, and reduced tillage with an emphasis on decreasing sediment (and associated nutrients) in surface runoff. Nutrient management plans with field-by-field nutrient recommendations were developed for about one-fourth of the agricultural acreage by 1997. Despite these efforts, P loading from nonpoint sources was estimated to have been reduced by only 9% after ten years, compared to 58% of P from point sources (Boesch et al., 2001). In the German Branch watershed, which was targeted for aggressive BMP implementation, P concentrations have remained at elevated levels (Millard et al., 1997).

4. APPROACHES TO MORE EFFECTIVE CONTROL OF P RUNOFF

Why has long-term watershed-scale monitoring shown so little water quality benefit from BMP implementation programs when edge-of-field studies of individual BMPs have shown reductions in runoff P losses? What changes could increase their effectiveness? There are a number of possible explanations and opportunities for improvement.

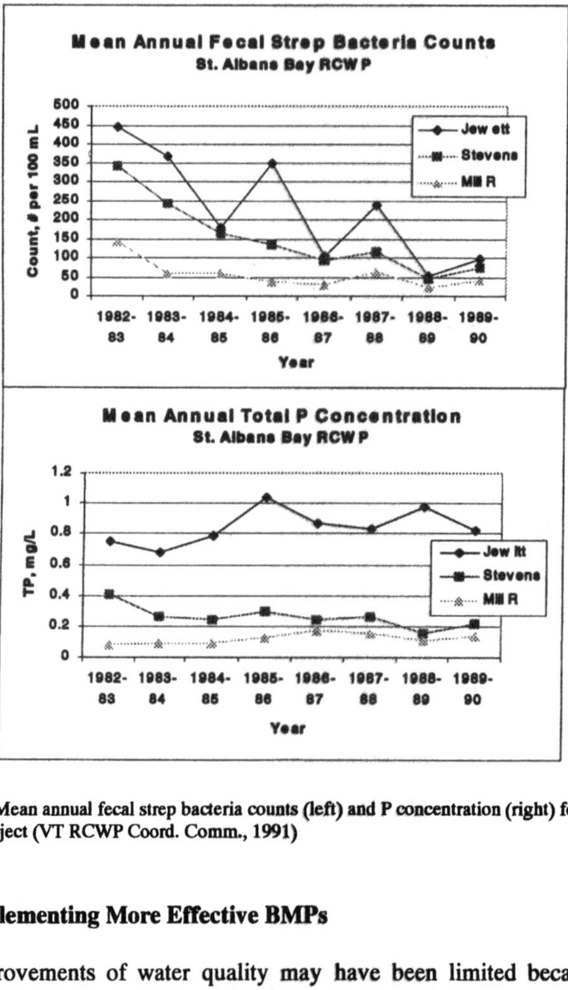

Figure 2. Mean annual fecal strep bacteria counts (left) and P concentration (right) for the St. Albans Bay RCWP project (VT RCWP Coord. Comm., 1991)

4.1. Implementing More Effective BMPs

Improvements of water quality may have been limited because some potentially effective BMPs were not included in the supported program or because implemented BMPs were not as effective under the circumstances as had been expected. Some examples are presented here.

4.1.1. Manure timing/Incorporation

The Vermont large watershed monitoring projects (Meals, 1996) provided cost-share for manure storage and required elimination of winter spreading for participants. Fall application of manure and incorporation within two weeks on annually tilled crops were acceptable practices. While these practices were likely an improvement over winter spreading, they may not have been as effective as spring application and quicker incorporation of manure.

4.1.2. Nutrient management

The waste utilization and fertilizer management BMPs included in the Vermont monitoring projects nominally included some aspects of nutrient management. However, recommendations were fairly general and not closely monitored. There was no field-by-field soil test-based approach to manure and fertilizer management and no use of nutrient management specialists or consultants. More intensive nutrient management could be effective in decreasing P loading and the resultant P runoff (Beegle et al., 2000). For example, evaluation of seven farms in the Lower Missisquoi Watershed Project in Vermont showed reductions in P inputs of over 30% as a result of consultant-assisted nutrient management planning, primarily due to decreased fertilizer P purchases (Fig. 3; Knight et al., 1998).

4.1.3. Conservation Tillage

Conservation tillage, while listed in the package of BMPs for the Vermont monitoring projects, had very low levels of implementation. This was, in part, because most corn in Vermont is harvested for silage, leaving very little residue and necessitating a cover crop to achieve the level of residue needed for erosion control. In the Chesapeake, where conservation tillage was heavily promoted and implemented, effectiveness may have been less than expected because soluble P is not necessarily decreased by conservation tillage and may even increase (Boesch et al., 2001).

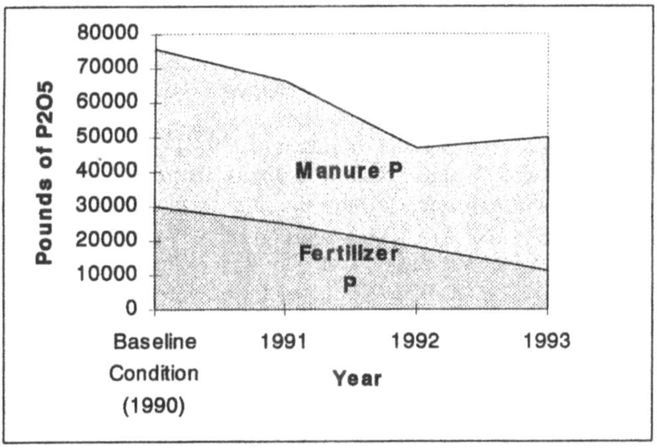

Figure 3. Manure and fertilizer P applied to fields on 7 farms in the Lower Missisquoi Watershed Project (Knight et al., 1998).

4.1.4. Field edge/Riparian Zone Buffers and Fencing

In the two Vermont monitoring studies, riparian zone/field buffer management was not included as a BMP option. Streambank protection was included but not widely implemented and fencing to exclude livestock was not a supported practice. While some farmers may consider such practices impractical, legislation requiring vegetative buffers on row crop fields adjacent to streams (where runoff occurs) was enacted in Vermont in 1996 (VDAFM, 1996). If vegetated field buffers are implemented, it is important that they conform to recognized recommendations for establishment and maintenance to assure effectiveness.

4.1.5. Reduction of Phosphorus Levels in Feed

Reduction of P levels in feed was not a BMP promoted in the Vermont watershed projects. Because of its potential to improve the whole-farm P imbalance, as well as reduce costs for the producer, this is a BMP that is now gaining acceptance and may have real water quality benefits in the future.

4.2. More Complete BMP Implementation

In the Vermont projects, farmer participation and BMP implementation were quite high, 60 to 80% depending on the BMP, meeting or surpassing target levels (Meals, 1996). But even with this high level of participation, there were a large number of farms without treatment, which may have limited overall program effectiveness. For example, winter spreading of manure by non-cooperators may have masked the effect of BMP manure management by the cooperating farmers. In the LaPlatte River watershed there was some indication of a threshold effect, i.e. need for some minimum BMP implementation level before a water quality effect occurs (Meals, 1996).

4.3. More Site-Specific Focus for BMP Implementation

Prioritization of land areas to receive assistance for BMP implementation has been based primarily on the scale of the watershed or subwatershed for the impaired water body, for example, watersheds of selected bays in Lake Champlain or subwatersheds of St. Albans Bay. This means that available resources are spread over a relatively large area, including portions of the watersheds that may have relatively little effect on the quality of the impaired water body. A more targeted approach might have a greater water quality effect and be more cost-effective.

4.3.1. Use of a Phosphorus Index

The Phosphorus Index is a tool developed to assess the potential for phosphorus runoff from individual fields based on soil and field characteristics and management practices. The details of a P Index vary by state but it typically includes several source factors (soil test P, rate and application method for fertilizer and manure P) and transport factors such as erosion, runoff class, distance to stream, vegetative buffer, and sometimes leaching or flooding frequency (Lemunyon and Gilbert, 1993; Sharpley, 2000; Jokela, 2000; Leytem et al., 2000). Use of a P Index provides a means of identifying fields that have high P runoff potential and, therefore, should be high priority for implementation of BMPs to control erosion and runoff or improve manure and fertilizer management. This allows farmers and government agencies with cost-share funds to focus on those sites with high risk of serious P runoff. Several states in the Northeast (PA, MD, DE, VT, and NY) have been particularly active in P Index development in the past two years.

While there has been very little field evaluation of the P Index, Eghball and Gilley (1999) and Sharpley et al. (2001) found relatively high correlations between P Index values and runoff P concentrations and losses using rain simulation on sites with a range of soil P and management conditions. Incorporation of a P Index into nutrient management planning on 11 Pennsylvania farms showed that it was the most flexible approach (compared to agronomic or environmental P threshold strategies) and that it had a relatively low impact on manure and P management on farms (McDowell et al., 2002b; Weld et al., 2002). These results have indicated a need for re-evaluation and modification of the current P Index.

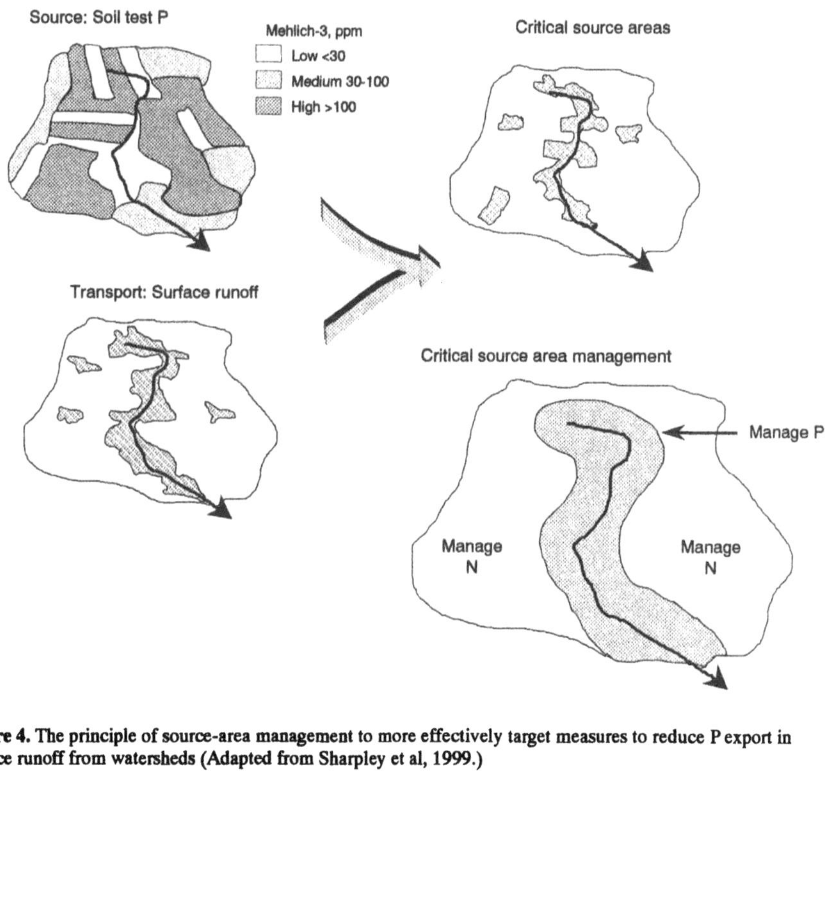

Figure 4. The principle of source-area management to more effectively target measures to reduce P export in surface runoff from watersheds (Adapted from Sharpley et al, 1999.)

4.3.2. Critical Source Area Approach

Addressing the problem of phosphorus in agricultural runoff is complicated by the fact that surface runoff is not generated uniformly across a field or watershed. Only a portion of the watershed contributes most of the runoff for a given storm, and the extent of this contributing area varies with season and size of storm, hence the term "variable-source-area" (Gburek and Sharpley, 1998). Portions of the watershed that are runoff-contributing areas *and* have a high level of available P, as indicated by excessive soil test P, are termed "critical source areas" (Fig. 4). They have the highest probability for P runoff and should be the focus of P reduction management practices. Incorporation of these hydrologic concepts into the transport function of the P Index could greatly improve its effectiveness in prioritizing sites for P runoff potential (Gburek et al., 2000). Most current P Indexes attempt to address the critical source area concept, at least on a field scale, by assigning the highest ratings to fields with high values for both transport (e.g. erosion and runoff) and P source (soil test P and applied P).

4.4. Lag time for Water Quality Effect

Even if implemented BMPs are effective in reducing nutrient additions or runoff and erosion from fields, there may a long lag time before water quality improvements are observed. Using a modeling approach Clausen et al. (1992) and Cassell and Clausen (1993) estimated the lag time for various BMP scenarios to lower field P export to a specified levels in the St. Alban Bay watershed based on a relationship between soil P and runoff P concentration. Estimated lag times for most scenarios were considerably longer than the 10-year monitoring period because of the large soil P reservoir and/or the net P additions to a field.

Equilibrium with the reservoir of P in sediments of impaired water bodies can likewise maintain an elevated P concentration even if inputs of P cease or are decreased (McDowell et al., 2002a). To a large extent, the lag time between a decrease in agricultural P inputs to surface waters and productivity response of the water is determined by recycling of P already in the aquatic system. McDowell et al. (2002a) examined the processes controlling sediment P release to the Winooski River, the largest tributary to Lake Champlain. The input and delivery of fine sediment enriched with P was influenced by adjacent land use. Specifically, iron-oxide strip P (algal-available P) of the river sediments adjacent to agricultural land (3.6 mg kg^{-1}) was greater than that of sediments adjacent to forested land (2.4 mg kg^{-1}). Notably, impoundment (731 mg kg^{-1}) and reservoir sediments (803 mg kg^{-1}) had greater total P concentrations than did river sediments (462 mg kg^{-1}). Internal recycling of P from reservoir sediments (e.g., release of P from iron-P compounds in an anaerobic hypolimnion) can be an important source of P with regard to total watershed management and TMDL implementation, especially during reservoir stratification (Baldwin et al., 2000; Golterman, 2001; Gray and Kirkland, 1986; Knuuttila et al., 1994). The results of this research clearly demonstrate that there is a strong influence of fluvial hydraulics on the properties of sediment within river systems (McDowell et al., 2002a). The fluvial sediment, particularly at the outflow of the river into Lake Champlain, represents a store of P, which has a long-term potential to release a large amount of P to overlying waters.

Cassell et al. (2002), working in the same (Winooski) river system, assessed the magnitude and spatial distribution of sediment and P storage across transects where stream bank erosion was active. Simulation modeling showed that both sediment and total P were accumulating within the river system. An estimated P residence time of over 2000 years indicates that P export response to watershed management programs can be expected to be very slow.

5. THE LONG-TERM SOLUTION: BALANCING THE P BUDGET AND IMPLEMENTING BMPS

Implementation of BMPs that reduce erosion and runoff or limit loss of P from a field can lead to improved water quality; but in the long term such improvements will be limited unless whole-farm P budgets are brought into closer balance. Estimates of whole-farm P balance on typical confinement dairy farms in the Northeast indicate that only about 20 to 40% of the P entering the farm (as feed, fertilizer, etc.) leaves the farm as milk, meat, or other products (Klausner, 1995; Allshouse et al., 1996), causing average net accumulations of over 35 kg P/ha/yr (Anderson and Magdoff, 2000). On more intensive poultry and hog operations P imbalances are even greater (Sims et al., 2000).

Watershed ecosystem modeling based on dynamic simulation of P mass balance has been successful in describing P flux and accumulation in large, complex watersheds in Vermont, Delaware, and elsewhere. Long-term simulations suggest that traditional BMPs for runoff and erosion control can have only a relatively short-term effect (e.g. 80 years in a Vermont example) on P loss to surface waters (Cassell et al., 1998; Cassell and Clausen, 1993; Aschmann et al., 2000). Long-term goals for reducing P export from watersheds can only be achieved by also addressing P import/export imbalance in the agricultural sector and, where appropriate, in the urban environment as well.

Some improvements in P balance can be achieved by changes in specific management practices by individual producers that reduce P inputs to the farm. Changes such as reducing P fertilizer purchases to account for manure P and eliminating unnecessary P feed supplements or feeding rations with increased P digestibility (Wu et al., 2000; Valk et al., 2000; Poulsen, 2000), can provide cost savings to the producer as well as environmental benefits. Other changes at the farm scale require more significant modifications in the farm production system. Increasing on-farm feed production by management-intensive grazing or a shift from purchased to home grown feed grains will decrease the need for purchased inputs of feed P. Lowering animal density, either by decreasing animal numbers or obtaining more land, will improve the P balance; but attention must be paid to economic impacts on individual farms or on rural infrastructures. More long-term changes on a society level could include re-integration of animal and crop production systems on the same farm or within geographical areas and a shift to diets less dependent on animal-based food production systems (Magdoff et al., 1997).

This is not meant to minimize the importance of traditional BMPs, but to emphasize that without an improved P balance BMPs alone will not be adequate to accomplish long-term goals for reduced P loading. What is needed is a comprehensive approach that combines effective, focused implementation of BMPs at the field, farm, and watershed scale and more long-term efforts to achieve a more balanced whole farm P budget.

6. REFERENCES

Allshouse, R.D., E.D. Thomas, C.J. Sniffen, K. Grimes, and C. Majewski. 1996. Characterization of on-farm phosphorus budgets and management in the Lake Champlain Basin. W. H. Miner Agric. Research Inst., Chazy, NY.

Anderson, B.H., and F.R. Magdoff. 2000. Dairy farm characteristics and managed flows of phosphorus. Amer. J. Altern. Agric.

Andraski, B.J., D.H. Mueller, and T.C. Daniel, T.C.1985. Phosphorus losses in runoff as affected by tillage. Soil Sci. Soc. Am. 49:1523-1527.

Aschmann, S.G., D.P. Anderson, R.J. Croft, and E.A. Cassell. 2000. Using a watershed nutrient dynamics model, WEND, to address watershed-scale nutrient management challenges. J. Soil and Water Cons. 54:630-635.

Baldwin, D.S., A.M. Mitchell, and G.N. Rees. 2000. The effects of *in situ* drying on sediment-phosphate interactions in sediments from an old wetland. Hydrobiologia 431:3-12.

Beegle, D.B., O.T. Carton, and J.S. Bailey. 2000. Nutrient management planning: justification, theory, practice. J. Environ. Qual. 29:72-79.

Boesch, D.F., R.B. Brinsfield, and R.E. Magnien. 2001. Chesapeake Bay eutrophication: scientific understanding, ecosystem restoration, and challenges for agriculture. J. Environ. Qual. 30: 303-320.

Carpenter, S., N.F. Caraco, D.L. Correll, R.W. Howarth, A.N. Sharpley, and V.H. Smith. 1998. Issues in Ecol., No. 3., Summer , 1998. Ecol Soc. Am. Web: http://esa.sdsc.edu/

Cassell, E.A., and J.C. Clausen. 1993. Dynamic simulation modeling for evaluating water quality response to agricultural BMP implementation. Water Sci. Tech. 28:635-648.

Cassell, E.A., J.M. Dorioz, R.L. Kort, J.P. Hoffmann, D.W. Meals, D. Kirschtel, and D.C. Braun. 1998. Modeling phosphorus dynamics in ecosystems: mass balance and dynamic simulation approaches. J. Environ. Qual. 27:293-298.

Cassell, A.E., R. Kort, T. Menees, and D. Newman. 2002. Sediment and phosphorus storage and transport in the Winooski River system. P. 35. *In* Proceed of symposium: Lake Champlain in the new millennium. Saint-Jean-sur-Richelieu, Quebec, May 20-23, 2002. Lake Champlain Research Consortium. Middlebury Coll., Middlebury, VT.

Clausen, J.C., K. Guillard, C.M. Sigmund, and K. Martin Dors. 2000. Water quality changes from riparian buffer restoration in Connecticut. J. Environ. Qual. 29:1751-1761.

Clausen, J.C. 1987. Bushey field study. *In* Meals, D.W. LaPlatte River Watershed water quality monitoring and analysis program – year 8. Program Report No. 9. Vermont Water Resources Research Center, University of Vermont, Burlington, VT.

Clausen, J.C. 1991. Best manure management effectiveness. P.193-197 *In* Vermont RCWP Coordinating Committee. St. Albans Bay Rural Clean Water Program, Final Report, 1991. Vermont Water Resources Research Center, University of Vermont, Burlington, VT.

Clausen, J.C., D.W. Meals, and E.A. Cassell. 1992. Estimation of lag time for water quality response to BMPs. Pages 173-180 *In* The National Rural Clean Water Program Symposium, September, 1992, Orlando, FL, US EPA ORD, EPA/625/R-92/006.

Clausen, J.C., W.E. Jokela, F.I. Potter III, and J.W. Williams. 1996. Paired watershed comparison of tillage effects on runoff, sediment, and pesticide. J. Environ. Qual. 25:1000-1007.

Clausen J. C. and J. Spooner. 1993. Paired watershed study design. US Environmental Protection Agency. EPA 841-F-93-009. Office of Water. Washington D. C.

Dillaha, T.A, R.B. Reneau, S. Mostaghimi, and D. Lee. 1989. Vegetative filter strips for agricultural nonpoint source pollution control. Trans.ASAE 32:513-519.

Ebeling, A.M., L.G. Bundy, J.M. Powell, and T.W. Andraski. 2002.Dairy diet phosphorus effects on phosphorus losses in runoff from land-applied manure. Soil Sci. Soc. Am. J. 66: 284-291.

Eghball, B. and J.E. Gilley. 1999. Phosphorus and nitrogen in runoff following beef cattle manure or compost application. J. Environ. Qual. 28(4):1201-1210.

Ertl, D.S., K.A. Young, and V. Raboy. 1998. Plant genetic approaches to phosphorus management in agricultural production. J. Environ. Qual. 27:299-304.

Gburek, W.J. and A.N. Sharpley. 1998. Hydrologic controls on phosphorus loss from upland agricultural watersheds. J. Environ. Qual. 27:267-277.

Gburek, W.J., A.N. Sharpley, L. Heathwaite, and G.J. Folmar. 2000. Phosphorus management at the watershed scale: a modification of the phosphorus index. J. Environ. Qual. 29:130-144.

Ginting, D., J.F. Moncrief, and S.C. Gupta. 1998. Interaction between manure and tillage system on phosphorus uptake and runoff losses. J. Environ. Qual. 27:1403-1410.

Golterman, H.L. 2001. Phosphate release from anoxic sediments of 'What did Mortimer really write?'. Hydrobiologia 450:99-106.

Gray, C.B.J., and R.A. Kirkland. 1986. Suspended sediment phosphorus composition in tributaries of the Okanagan Lakes, BC. Water Research 20:1193-1196.

Jokela, W.E. 2000. A phosphorus index for Vermont. p. 302-315. In Proc from Managing nutrients and pathogens from animal agriculture. Camp Hill, PA. 28-30 Mar., 2000. NRAES-130. Ithaca, NY.

Jokela, W.E., S. Bosworth, and D. Meals. 1995. Water quality and yield effects of sidedressed dairy manure on corn: first year results. *In* Animal Waste and the Land-Water Interface: Poster Proceed. Arkansas Water Resources Center, Fayetteville, AK.

Jokela, W.E., and J.C. Clausen. 1994. Tillage effects on P and N losses in surface runoff from cornland. p. 63. *In* Agronomy abstracts. ASA, Madison, WI.

Jokela, W.J., J.W. Hughes, D. Tobi, and D. W. Meals. 1999. Managed vegetative riparian buffers to control P runoff losses from corn fields. P. 286 *In* Agronomy abstracts. Amer. Soc. Agron. Madison, WI.

Klausner, S. 1995. Nutrient management planning. P. 383-392 In K. Steele (ed.), Animal waste and the land-water interface. CRC Press, Boca Raton, FL.

Knight, L.G., W.E. Jokela, and S. Cushing.1998. Dairy farmers save dollars and nutrients by participating in HUA crop management service. Plant and Soil Sci. Dept, Univ. of Vermont, Burlington, VT. http://pss.uvm.edu/vtcrops/LMWQ/Lmwq5.pdf

Knuuttila, S., O.P. Pietilainen, and L. Kauppi. 1994. Nutrient balances and phytoplankton dynamics in two agriculturally loaded shallow lakes. Hydrobiologia 276:359-369.

Lake Champlain Basin Program.1996. Opportunities for action: an evolving plan for the future of the Lake Champlain Basin. LCBP, Grand Isle, VT.

Lander, C.H., D. Moffitt, and K. Alt. 1998. Nutrients Available from Livestock Manure Relative to Crop Growth Requirements. Resource Assessment and Strategic Planning Working Paper 98-1 U.S. Department of Agriculture, Natural Resources Conservation Service, Washington, DC. (http://www.nhq.nrcs.usda.gov/land/pubs/nlweb.html).

Lemunyon, J.L., and R.G. Gilbert. 1993. The concept and need for a phosphorus assessment tool. J.Prod. Agric. 6:483-486.

Leytem, A.B., J.T. Sims, F.J. Coale. 2000. Adapting the phosphorus site index to the Delmarva peninsula: Deleware's experience. In Proc from Managing nutrients and pathogens from animal agriculture. Camp Hill, PA. 28-30 Mar., 2000. NRAES-130. Ithaca, NY.

Magdoff, F.R., C. Hryshko, W.E. Jokela, R.P. Durieux, and Y. Bu. 1999. Comparison of phosphorus soil test extractants for plant availability and environmental assessment. Soil Sci. Soc. Am. J. 63:999-1006.

Magdoff, F., L. Lanyon, and W. Liebhardt. 1997. Nutrient cycling, transformations, and flows: implications for a more sustainable agriculture. Adv. Agron. 60:1-73.

McDowell, R.W, and A.N. Sharpley. 2001. Approximating phosphorus release from soils to surface runoff and subsurface drainage. J. Environ. Qual. 30: 508-520.

McDowell, R.W., A.N. Sharpley, and A.T. Chalmers. 2002a. Land use and flow regime effects in phosphorus chemical dynamics in the fluvial sediment to the Winooski River, Vermont. Ecol. Eng. (in press).

McDowell, R.W., A.N. Sharpley, D. Beegle, and J. Weld. 2002b. Comparing phosphorus management strategies at the watershed scale. J. Soil Water Conserv. 56:306-315.

Meals, D.W. 1990. LaPlatte River Watershed water quality monitoring and analysis program: comprehensive final report. Program Report No.12. Vermont Water Resources Research Center, University of Vermont, Burlington, VT.

Meals, D.W. 2001. Lake Champlain Basin Agricultural Watersheds Section 319 National Monitoring Program Project, Final Project Report: May, 1994-- November, 2000. Vermont Dept. of Environmental Conservation, Waterbury, VT, 224 p.

Meals, D.W. 1996. Watershed-scale response to agricultural diffuse pollution control programs in Vermont, USA. Water Science and Technology 33(4-5):197-204.

Meals, D.W. and L.F. Budd. 1998. Lake Champlain nonpoint source phosphorus assessment. J. American Water Resources Assoc. 34(2):251-265.

Medalie, L., and E. Smeltzer. 2003. Phosphorus trends in Lake Champlain and its tributaries. (This Volume).pp. 215-247.

Millard, C.J., J.L. McCoy, N.L. Primrose, M.G. Dobson, S.E. Bowen, P.E. Sturm, and R.J. Windschitl. 1997. German Branch targeted watershed project: Biotic and water quality monitoring of restoration projects in German Branch. Maryland Dept. of Natural Resources, Annapolis, MD.

Pote, D.H., T.C. Daniel, A.N. Sharpley, P.A. Moore, D.R. Edwards, and D.J. Nichols. 1996. Relating extractable soil phosphorus to phosphorus losses in runoff. Soil Sci. Soc. Amer. J. 60:855-859.

Poulsen, H.D. 2000. Phosphorus utilization and excretion in pig production. J. Environ. Qual. 29:24-27.

Schellinger, G.R. and J.C. Clausen. 1992. Vegetative filter treatment of dairy barnyard runoff in cold regions. J. Environ. Qual. 21:40-45.

Schwer, C.B. and J.C. Clausen. 1989. Vegetative filter strip treatment of dairy milkhouse wastewater. J. Environ. Qual. 18:446-451.

Sharpley, A.N. 2000. The phosphorus index: assessing site vulnerability to phosphorus. p. 255-281. In Proc from Managing nutrients and pathogens from animal agriculture. J. Environ. Qual. 29-166.

Sharpley, A.N., R.W. McDowell, J.L. Weld, and P.J.A. Kleinman. 2001. Assessing site vulnerability to phosphorus loss in an agricultural watershed. J. Environ. Qual. 30:2026-2036.

Sharpley, A.N., and S.J. Smith. 1994. Wheat tillage and water quality in the Southern Plains. Soil Till. Res. 30:33-38.

Sharpley, A., T.C. Daniel, J.T. Sims and D.H. Pote. 1996. Determining environmentally sound soil phosphorus levels. Journal of Soil and Water Conservation 51: 160-166.

Sharpley, A.N., B. Foy, and P. Withers. 2000. Practical and innovative measures for the control of agricultural phosphorus losses to water: an overview. J. Environ. Qual. 29:1-9.

Sharpley A.N., T.C. Daniel, J.T. Sims, J. Lemunyon, R.A. Stevens, and R. Parry. 1999. Agricultural phosphorus and eutrophication. U.S. Department of Agriculture - Agricultural Research Service, ARS-149 U.S. Govt. Printing Office, Washington, D.C. 34 p.

Sharpley, A.N., J.J. Meisinger, A. Breeuwsma, J.T. Sims, T.C. Daniel, and J.S. Schepers. 1998. Impacts of animal manure management on ground and surface water quality. p173-242. In: Hatfield, J.L., and B.A. Stewart (eds.) Animal Waste Utilization: Effective use of manure as a soil resource. Sleeping Bear Press Inc., Boca Raton, FL.

Sims, J.T., A.C. Edwards, O.F. Schoumans, and R.R. Simard. 2000. Integrating soil phosphorus testing into environmentally based agricultural management practices. J.Environ. Qual. 29:60-71.

Smeltzer, E. 1992. Reducing phosphorus levels in Lake Champlain. P. 9-21 *In* A clean lake for tommorow: proceedings. Lake Champlain Committee, Burlington, VT.

Uusi-Kämppä, J, B. Braskerud, H. Jansson, N. Syverson, and R. Uusitalo. 2000. Buffer zones and constructed wetlands as filters for agricultural phosphorus. J. Environ. Qual. 29:151-158.

Valk, H., J.A. Metcalf, and P.J.A. Withers. 2000. Prospects for minimizing phosphorus excretion in ruminants by dietary manipulation. J. Environ. Qual. 29:28-36.

Vermont Dept. of Agric., Food and Markets (VDAFM). 1996. Vermont agricultural nonpoint source pollution reduction program law and regulations. Title 6, Chapter. 215. VDAFM, Montpelier, VT.

Vermont Department of Environmental Conservation (DEC) and New York State Department of Environmental Conservation (DEC). 1997. A phosphorus budget, model, and load reduction strategy for Lake Champlain. Dept. Environ. Conserv., Water Quality Div., Waterbury, VT.

Vermont RCWP Coordinating Committee. 1991. St. Albans Bay Rural Clean Water Program, Final Report, 1991. Vermont Water Resources Research Center, University of Vermont, Burlington, VT.

Watzin, M.C. 1992. State of knowledge about Lake Champlain. P. 33-62 *In* A research and monitoring agenda for Lake Champlain. Proceedings of a workshop, Dec. 17-19, 1991, Burlington, VT. Tech. Report No. 1. Lake Champlain Basin Program, Grand Isle, VT.

Weld, J.L., R.L. Parsons, D.B. Beegle, A.N. Sharpley, W.J. Gburek, and W.R. Clouser. 2002. Evaluation of phosphorus-based nutrient management strategies in Pennsylvania. J. Soil Water Conserv. (*in press*).

Wu, Z., L.D. Satter, and R. Sojo. 2000. Milk production, reproductive performance, and fecal excretion of phosphorus by dairy cows fed three amounts of phosphorus. J. Dairy Sci. 83:1028-1041.

ZEBRA MUSSEL MONITORING IN THE RICHELIEU RIVER: A FRUITFUL INTERACTION BETWEEN SCIENTIFIC RESEARCH AND PUBLIC INVOLVEMENT

Yves de Lafontaine and Marcel Comiré*

ABSTRACT

As part of a collaboration between scientists from Environment Canada and COVABAR (a non-governmental organization), a 4-year study was conducted to monitor the presence, spread and potential impacts of zebra mussels (*Dreissena polymorpha*) along the Richelieu River. Taking advantage of riverbed cleanups conducted every September, a sampling program was set up to determine the annual colonization and spatial distribution of zebra mussels along the river between 1997 and 2000. A protocol for effective sampling coordination and scientific data acquisition was developed to allow that objects removed from the water be examined and the zebra mussels attached to them be counted and sampled. Results confirmed that the first massive invasion by zebra mussel occurred upstream of the Chambly Basin in 1996 and two years later at downstream sites. Size distribution of zebra mussels varied between years and sites which suggested strong variability in recruitment patterns along the river. This variability could be due to inter-annual variation in river discharge during summertime. Colonization rate of unionids by zebra mussels was still low in these early years of colonization in the river. It is recommended that a long term sampling program coupled with specific ecological studies be conducted in order to assess the abundance and impacts of zebra mussels in the Richelieu River. Sampling responsibility for such monitoring should be transferred to non-governmental organization in order to increase public awareness to the problem of exotic species and biodiversity.

1. INTRODUCTION

As a direct result of the worldwide trade of goods and merchandise, the introduction of non-indigenous species has become a major environmental issue threat to the

* Yves de Lafontaine, Environment Canada, St. Lawrence Centre, 105 McGill, Montréal, QC, Canada H2Y 2E7. Marcel Comiré, COVABAR, 806 chemin Richelieu, Beloeil, QC, Canada J3G 4P6

biological integrity of aquatic ecosystems. At least 160 exotic species have invaded the Great Lakes during the past 200 years, 85 in the St. Lawrence River, and 113 in the Hudson River (Mills et al., 1996, 2000; de Lafontaine and Costan, 2002). Of these, the zebra mussel (*Dreissena polymorpha*) introduced into the Great Lakes in the mid-1980s has rapidly spread through many water bodies in eastern North America and certainly typifies the undesirable invasive pest in aquatic systems (Claudi and Leach, 2000). The significant impact of zebra mussel at both ecosystemic and socio-economic levels (Pimentel et al., 2001) has promoted concern and action by governmental agencies, media and general public in United States and Canada.

The invasion and rapid dispersal of zebra mussel over so vast a territory was mainly attributed to its relatively long planktonic larval stage (15–25 days) (Claudi and Mackie, 1993), which means it can be carried long distances by currents. Rivers and canals are thus major vectors of transportation of zebra mussels from one region to another and their subsequent colonization of other ecosystems. Spread of the zebra mussel from the Great Lakes watershed into the Hudson River occurred presumably via the Erie Canal and the Mohawk River in the state of New York (Mills et al., 2000). Once in the Hudson, the species then invaded southern Lake Champlain in 1993 (Eliopoulos and Stangel, 1999), and its larval stages were first reported in the Richelieu River at the northern end of the lake in 1996 (de Lafontaine and Cusson, 1997). This example illustrates the key role of rivers and other waterways in colonization by zebra mussel populations and in the eventual homogenization of aquatic biodiversity as a result of the introduction of a non-native species (Gido and Brown, 1999).

The discovery and present northward extent of zebra mussels within Lake Champlain renders the Richelieu River at high risk of invasion by this mollusk. It has been shown that the colonization of riverine environments by zebra mussels depends largely on the production of larvae in upstream lakes (Borcherding and de Ruyter van Steveninck, 1992). Given the speed at which the zebra mussel propagates in the bodies of water that it infests, a sampling initiative to monitor the invasion and potential impact of zebra mussels in the Richelieu River was felt necessary. During the initial stages of colonization, it is important that sampling be conducted every year to determine the extent and speed of species spread. Annual monitoring of the colonization and spatial distribution of zebra mussels in the Richelieu was initiated as part of a riverbed cleanups organized and conducted every September since 1995 by the Conseil Régional de l'Environnement de la Montérégie (CREM—Montérégie Regional Environment Council) and the Comité de Conservation et de Valorisation du Bassin de la Rivière Richelieu (COVABAR). As zebra mussels attach themselves to a variety of substrates (Kilgour and Mackie, 1993 ; Marsden and Lansky, 2000), trash and anthropogenic debris afford opportunities for settlement and attachment, and serve as sampling units to detect and monitor the presence of zebra mussel populations along the Richelieu River.

This paper presents the results of the zebra mussel monitoring surveys conducted as part of the annual riverbed cleanups between 1997 and 2000. Our objective was two fold: 1) to determine the abundance and spatial distribution of zebra mussel in the Richelieu River, and 2) to assess the validity of using the annual cleanup as a monitoring survey for gathering information on population dynamics of zebra mussels in the river. Specifically, the data were examined to (1) assess changes in degree of zebra mussel colonization of the waterway since 1997, (2) determine the validity of abundance estimates made by examining objects in comparison with a quantitative assessment by quadrats and (3) determine the extent and potential impact of zebra mussel colonization of unionids. Specimens gathered in 1997 and 1998 were used to determine the phenotypic variability and the length-weight relationship of individuals in the Richelieu River population and to compare it with other populations. The length-weight relationship may also serve for assessing the potential impact of zebra mussels, using weight-based metabolic processes (i.e. filtration rate, respiration rate). Finally, we examined the role and mutual benefits of cleanup activities for scientific data acquisition and public education.

2. MATERIALS AND METHODS

2.1 Sampling

Sampling was carried out on September 14, 1997, September 20, 1998, September 19, 1999, and September 17, 2000, as part of the annual cleanup of the Richelieu riverbed organized by CREM, the Comité de Mise en Valeur de la Vallée du Richelieu (CMVVR—Richelieu Valley Enhancement Committee) and the Comité de Conservation et de Valorisation du Bassin de la Rivière Richelieu (COVABAR—Richelieu River Watershed Conservation and Enhancement Committee). In 1997, sampling was conducted at 12 different sites over a distance of 85 km between Saint-Paul de l'Ile-aux-Noix et Saint-Ours (Figure 1, Table 1). Four sites were located upstream of the Chambly Basin (near the town of Chambly) and 6 sites were downstream. In 1998, work was done at 5 of the 12 sites examined in 1997 (Table 1). One site (Saint-Jean-sur-Richelieu) was in the upper Richelieu (upstream of the Chambly Basin) and the other four were in the lower Richelieu. In 1999, the only two sites selected for clean-up were downstream (Saint-Marc-sur-Richelieu and Saint-Ours), while in 2000, five sites, all downstream of the Chambly Basin, were selected. Only the Saint-Marc-sur-Richelieu site was sampled each year. The sites were selected by the cleanup organizers on the basis of their needs and objectives and not with regard to a zebra mussel sampling strategy. The sampling sites in a given municipality varied each year so that unexplored adjacent areas (within 1 km) could be cleaned up and thus more objects would be picked up.

Sampling consisted of removing anthropogenic objects and debris from the water by divers and in examining them by volunteers, who counted the number of attached zebra mussels. The divers were also asked to pick up unionids (native freshwater mussels) when they found them. The diving and cleanup took an average of four hours at each site and was concentrated within a strip of less than 15 m from the shore. The total area inspected by the divers varied, depending on the number of objects and the number of divers at each site. When there were too many objects at some sites to examine them all exhaustively and in detail, objects were randomly selected from the batches brought up by the divers. The information noted was type of object examined (i.e., bottle or tire), and in the case of unionids, whether they were alive or dead; type of substrate (i.e., metal, wood, glass, etc.);

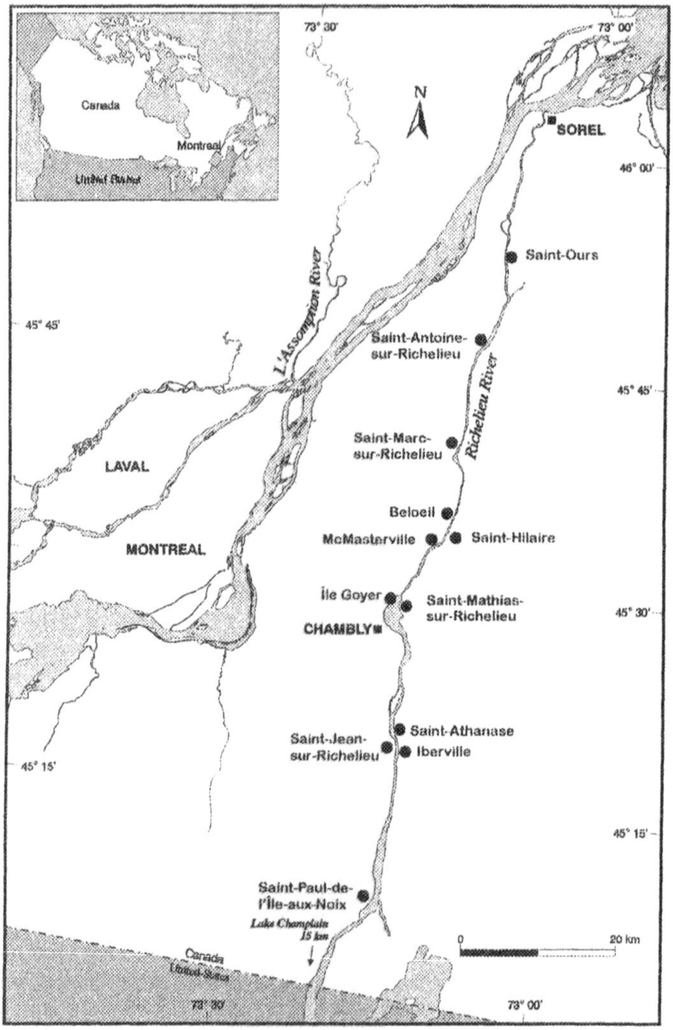

Figure 1. Location of sampling sites on the Richelieu River

size (length, width, height) in centimetres; colonizable surface (i.e., percentage of surface of object that was not buried in sediment at the bottom of the river); and number of zebra mussels attached. The zebra mussels were removed and placed in a plastic bag identified with a code corresponding to the site and object number. When there were a great many zebra mussels or it was impossible to examine the entire object, only a fraction of the total surface of the object was examined. In this case, the percentage of the surface area actually examined was calculated. Data on quantity and type of trash and debris removed were also compiled for years when zebra mussels were not sampled (1995, 1996, and 2001).

In 1998, quantitative sampling was also done at each of the five clean-up sites as well as next to the Chambly wharf and marina, in the Chambly Basin (Figure 1). Two divers would place a quadrat measuring one square metre randomly on the riverbed, in an area that had not yet been cleaned up, so that they would not be disturbed by higher turbidity. Three or four quadrats were sampled at each site. All the zebra mussels as well as the unionids in each quadrat were collected and placed in a net bag, then taken to the surface. The zebra mussels were placed in a plastic bag identified with a code corresponding to the site and quadrat number.

At each site, only two or three specimens of various species of unionids were kept for identification purposes. The other specimens, once examined, were placed in a basin of water and then later put back on the riverbed by the divers. All the samples of zebra mussels were placed in a freezer (-20°C) at the end of the day, while the unionids were cleaned and dried or preserved in a solution of 70% (v/v) denatured ethanol.

2.2 Laboratory Analyses

Zebra mussels from all samples were identified and counted in the St. Lawrence Centre laboratory. Zebra mussels (*Dreissena polymorpha*) and quagga mussels (*Dreissena bugensis*) were distinguished on the basis of morphological characteristics (Domm et al., 1993; Claxton et al., 1997). The shell length of each of up to 300 mussels per site was measured using digital vernier calipers (± 0.01 mm). The phenotype of individual mussels collected in 1997 was also determined based on descriptions of color and band patterns provided by Smirnova et al. (1993) and Marsden et al. (1996). All the zebra mussels gathered during the quantitative sampling (by quadrat) were identified and measured (length, width, height) using digital vernier calipers (± 0.01 mm). Each mussel was then dissected to separate the soft tissue from the shell. The soft tissue and the shell were oven dried at 105°C for 24 h then weighed (dry weight) using a Mettler electronic scale (± 0.0001 g).

The objects gathered and examined were divided into two main categories: anthropogenic objects and unionids. The anthropogenic objects were grouped by 12 main types of substrate: glass, metal, plastic, rubber, cement, wood, cloth, fibreglass, paper, organic matter, polypropylene and pottery/sandstone. The glass objects were chiefly bottles, the rubber objects chiefly tires and the metal objects chiefly aluminum beer or soft-drink cans or steel car or bicycle frames. The unionids were identified as live or dead (empty shells) in most cases. The surface area of each object was calculated from measurements of length, width and height, taking into account the object's geometric form. The intensity of colonization at a given site or for a particular type of substrate was determined by the percent of objects having zebra mussels attached. The abundance of zebra mussels was estimated as the number of mussels per object or per m^2 of surface area of an object or quadrat. Unless otherwise noted, descriptive statistics presented in this paper were arithmetic mean and standard deviation. The statistical treatment of data was done using SAS software and all tests were applied with a level of significance set at 0.05.

3. RESULTS

3.1 Sampling Effort and Debris Removal

Since 1995, cleanups were performed at different sites between Saint-Paul-de-l'Ile-aux-Noix near the outflow of Lake Champlain and Saint-Ours in the Richelieu. The number of sites varied between years (Figure 2) and sites located downstream of the Chambly Basin were more frequently selected, especially during the 4 years when zebra mussels sampling was carried out. Seven sites have been visited only once while Saint-Marc-sur-Richelieu and Beloeil were selected seven and six times respectively since 1995. A total of 1170 volunteers participated to these operations and the average number of volunteers per site tended to increase over time (Figure 2). A total of 590 divers were involved, averaging 7 to 19 divers per site to collect objects. The mean number of helpers on shore varied between 6 and 22 per site to examine objects and debris and record information relative to trash materials and zebra mussels presence. This reveals a sustained interest by the general public to these annual events.

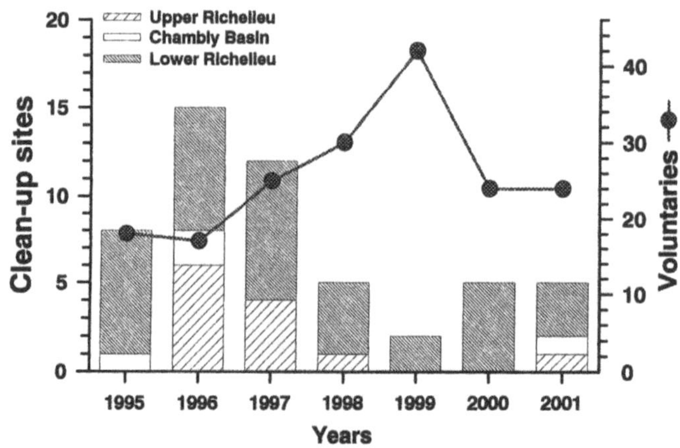

Figure 2. Number of sites and mean number of volunteers per site during cleanups between 1995 and 2001 in the Richelieu River

Over the years, approximately 57 tons of trash and debris were removed from the various locations. The total quantity taken each year varied between 3 and 16 tons depending on the number of sites visited. The average quantity per site varied between 0.64 and 2.2 tons and did not exhibit any trend over time. On average, one diver removed ~ 0.1 ton each year (0.05 - 0.129 ton per diver between 1995 and 2001). The average weight of debris was 0.7 ton per 100 objects and there was no sign of any declining trend over time (Figure 3a). Cans and bottles consisted of the majority of objects (in numbers) followed by construction parts and household items (Figure 3b). The relative proportion of the various classes of objects remained relatively stable over time, indicating little variation in the type of substrates used as sampling units for zebra mussels monitoring.

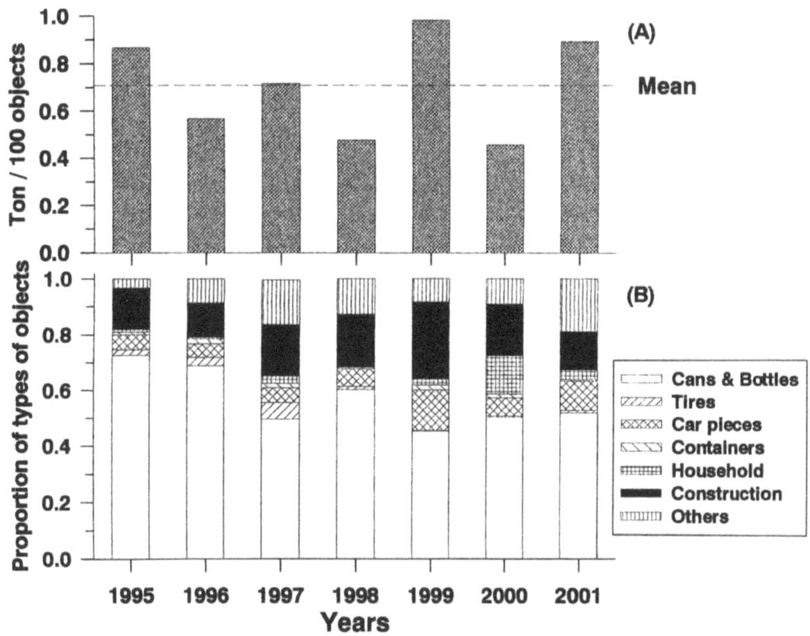

Figure 3. Annual variation in (a) average weight (tons per 100 objects) and in (b) relative proportion of different classes of debris and objects removed during cleanups

3.2 Zebra Mussels Distribution and Abundance

A total of 7,097 objects, including 1,616 unionids, were examined, and 7,190 zebra mussels were gathered at all sampling sites between 1997 and 2000 (Table 1). All specimens collected were of the species *Dreissena polymorpha*; no quagga mussel (*Dreissena bugensis*) has yet been found in samples from the Richelieu.

The presence of zebra mussels was confirmed at all sites from Saint-Paul de l'Île-aux- Noix to Saint-Ours (Figure 1, Table 1). As the number of objects examined varied greatly from site to site and from year to year, the percentage of objects colonized by the zebra mussels was calculated to compare the relative intensity of colonization (Figures 4 and 5). In 1997, the colonization rate at the four upstream sites varied between 33.3% and 88.9% but declined sharply and did not exceed 6.5% at downstream sites (Figure 4). Overall, 98.7% of all zebra mussels were collected at the four upstream sites and only 1.3% were collected at the 6 sites in the lower Richelieu. The mean number of mussels per object varied between 10 and 15 at the three most upstream sites, dropped to 0.77 at Saint-Athanase and was below 0.10 at all downstream sites.

Table 1. Number of objects examined (including unionids) and zebra mussel abundance at twelve sites sampled between 1997 and 2000

Municipality	*Site*	*Year*	*No. of objects examined*	*No.of objects colonized*	*No. of zebra mussels gathered*	*Mean no. of zebra mussels per object*	*Mean no. of zebra mussels per colonized object*
Upstream							
Saint-Paul de l' Île-aux-Noix	1	1997	18	16	186	10.33	11.63
Saint-Jean-sur-Richelieu	2	1997	145	92	1,844	12.72	20.04
		1998	250	118	3,101	12.40	26.28
Iberville	3	1997	81	72	1200	14.81	16.67
Saint-Athanase	4	1997	180	60	139	0.77	2.32
Downstream							
Ile Goyer	5	1997	299	3	3	0.01	1.00
Saint-Mathias	6	1997	218	3	7	0.03	2.33
		1998	442	10	18	0.04	1.80
		2000	178	3	3	0.02	1.00
McMasterville	7	1997	289	1	1	0.003	1.00
Saint-Hilaire	8	1997	183	0	0	0	0
		1998	560	0	0	0	0
		2000	208	21	46	0.22	2.19
Beloeil	9	1997	285	3	3	0.01	1.00
		1998	866	1	1	0.001	1.00
		2000	242	20	36	0.15	1.80
Saint-Marc-sur-Richelieu	10	1997	77	5	5	0.06	1.00
		1998	542	9	9	0.02	1.00
		1999	300	42	57	0.19	1.36
		2000	254	21	43	0.17	2.05
Saint-Antoine	11	1997	227	2	12	0.05	6.00
		2000	374	27	37	0.10	1.37
Saint-Ours	12	1997	664	13	14	0.02	1.08
		1999	216	76	425	1.97	5.59

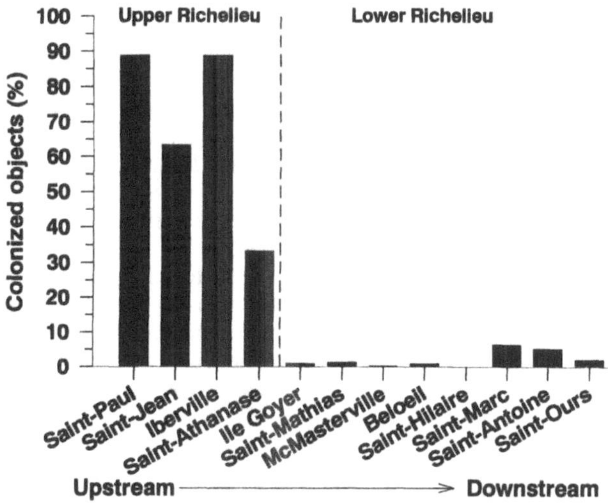

Figure 4. Zebra mussel colonization rates at 12 sites along the Richelieu River in 1997

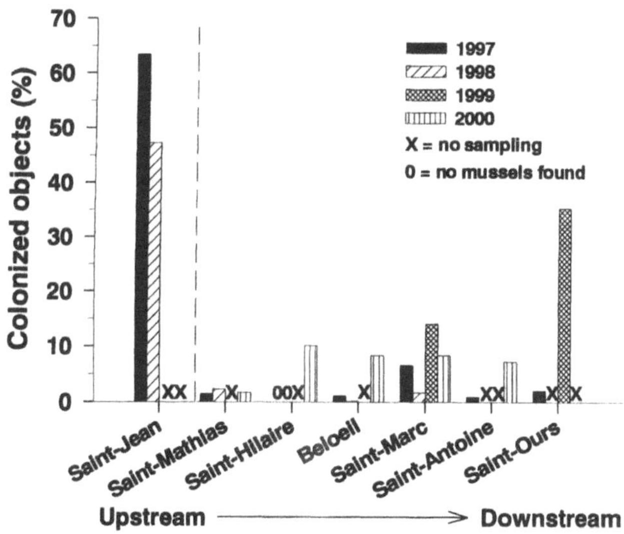

Figure 5. Zebra mussel colonization rate at sites samples between 1997 and 2000 in the Richelieu River

The colonization rate at Saint-Jean-sur-Richelieu (upstream of Chambly Basin) was 63.4% in 1997 and 47.2% in 1998—much higher than the overall downstream rate of 2.4% (Figure 5). In 1998, more than 3,000 zebra mussels (99.1% of the total number of mussels taken) were gathered at Saint-Jean-sur-Richelieu, while the numbers were quite low (between 0 and 18 mussels per site) at the four sites downstream of the Chambly Basin (Saint-Marc-sur-Richelieu, Beloeil, Saint-Hilaire and Saint-Mathias). Although the abundance of mussels was relatively low at downstream sites in 1999 and 2000, colonization has gradually increased over the years (Table 1, Figure 5). The abundance index remained relatively stable upstream (12 to 13 mussels per object) but increased tenfold between 1997 and 2000 downstream and the greatest progression was recorded at the Saint-Ours site between 1997 and 1999 (Table 1).

Quantitative sampling by 1 m^2 quadrats in 1998 found zebra mussels only at Saint-Jean-sur-Richelieu, where densities varied between 15 and 161 mussels/m^2 ($n = 4$, mean = 88, SD = 59). No mussels were found in the quadrats at the four sites in the lower Richelieu or at Chambly, where underwater exploration by two diver biologists turned up no sign of mussels, either on the natural riverbed or on the walls of the wharf and lock.

3.3 Effect of Substrate Type

The level of colonization (percent of objects colonized) on various types of substrate indicated that zebra mussels have no significant preference for any particular type of substrate (G test likelihood ratio = 8.213, $p = 0.314$), particularly at the upstream sites where there was greater colonization. When the analysis was restricted to glass, metal and plastic substrates, for which high enough numbers of objects were examined ($n > 50$), colonization of glass substrates was found to be slightly lower than for metal or plastic, but not significantly (G test = 3.9, $p = 0.14$). Nonetheless, this tendency to lesser colonization of glass objects was more noticeable using the mean number of mussels per object (Figure 6). Zebra mussel abundance on glass objects (9.9 mussels per object ± 2.4 in 1997; 9.3 mussels ± 1.9 in 1998) was on average half to a quarter of that on metal objects (22.8 mussels ± 5.6 in 1997; 32.8 mussels ± 10.8 in 1998) or plastic (21.3 mussels ± 6.4 in 1997; 35.0 mussels ± 19.2 in 1998). The surface area of the objects examined varied between 0.004 and 1 m^2, but for the most part between 0.01 and 0.1 m^2, and no significant relationship was found between zebra mussel abundance and object surface area (Pearson correlation test, $p > 0.05$). It appeared pointless to use a correction factor that would take into account the size of the objects in calculating the colonization differences associated with the type of substrate.

The zebra mussels colonized various species of unionids. Fourteen different species of unionids were identified during cleanup operations between 1997 and 1999 (see species list in Appendix 1). ***Lampsilis radiata*** and ***Elliptio complanata*** were the two most common species and were collected at all sites. The degree of colonization of unionids was not significantly different from that of anthropogenic objects as a whole (all sites and

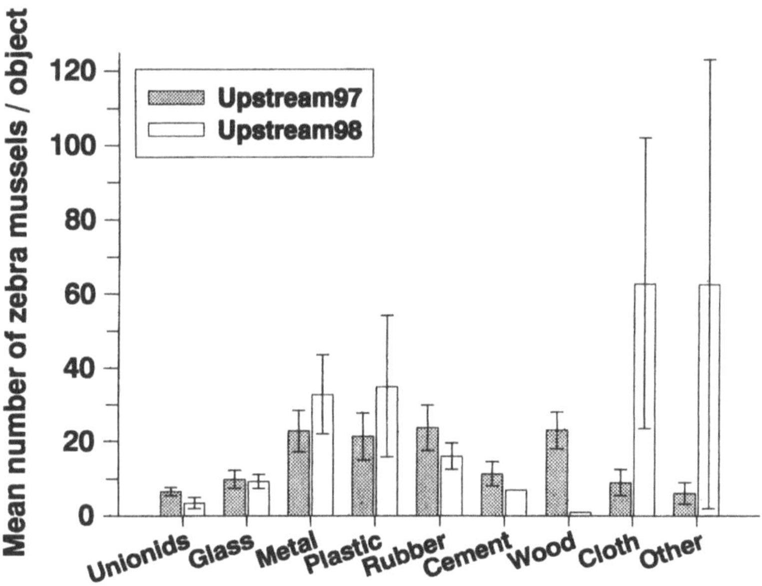

Figure 6. Mean number (lines indicate standard deviation) of zebra mussels on different types of substrate gathered at upstream sites along the Richelieu River in 1997 (4 sites) and 1998 (1 site)

years combined) (t test = -0.0799; $p = 0.94$, $n = 23$) (Table 2). Despite the huge differences between the number of unionids and the number of objects examined, the intensity of colonization of the two types of substrates was very similar for each sector and year. The steep gradient in colonization between the upstream and downstream sector of Chambly Basin was very clear both for inert objects and unionids (Table 2). The colonization rate of live and dead unionids by zebra mussels was very similar for a given sector in a particular year (Table 3). The difference between upstream and downstream colonization rates and the disproportionate number of live versus dead unionids in the two sectors and from year to year made statistical analysis of the results impossible. Although the cumulative results suggest a zebra mussel colonization rate three times higher for live unionids than dead (10.7% vs. 3.5%), the findings in the downstream stretch of the river, combined for all four years, show no significant difference in colonization rates of live and dead unionids (Table 3). The mean colonization rate was less than 10 zebra mussels per unionid at upstream locations in 1997 and 1998 (Figure 6).

Table 2. Zebra mussel colonization of inert objects and unionids in the Richelieu River between 1997 and 2000

	Inert objects			*Unionids*		
Sector/ Year/ Number of sites	*Examined*	*Colonized*	*Percent colonized (%)*	*Examined*	*Colonized*	*Percent colonized (%)*
Upstream 1997 – 4	293	175	59.7	131	76	58.0
Upstream 1998 – 1	246	116	47.2	4	2	50.0
Downstream 1997 – 7	1,399	25	1.8	843	5	0.6
Downstream 1998 – 4	2,000	18	0.9	410	4	1.0
Downstream 1999 – 2	426	98	23.0	90	20	22.2
Downstream 2000 – 5	1,117	81	7.2	139	11	7.9

Table 3. Zebra mussel colonization of live and dead unionids at upstream and downstream sites between 1997 and 2000

	Live unionids			*Dead unionids*		
Sector/ Year	*Examined*	*Colonized*	*Percent Colonized (%)*	*Examined*	*Colonized*	*Percent Colonized (%)*
Upstream 1997	125	71	56.8	6	5	83.3
Upstream 1998	4	2	50.0	—	—	—
Downstream 1997	458	4	0.9	385	1	0.26
Downstream 1998	167	2	1.2	227	2	0.88
Downstream 1999	31	8	25.8	59	12	20.3
Downstream 2000	72	5	6.9	63	6	9.5
Downstream Total	728	19	2.6	734	21	2.8
Total	857	92	10.7	740	26	3.5

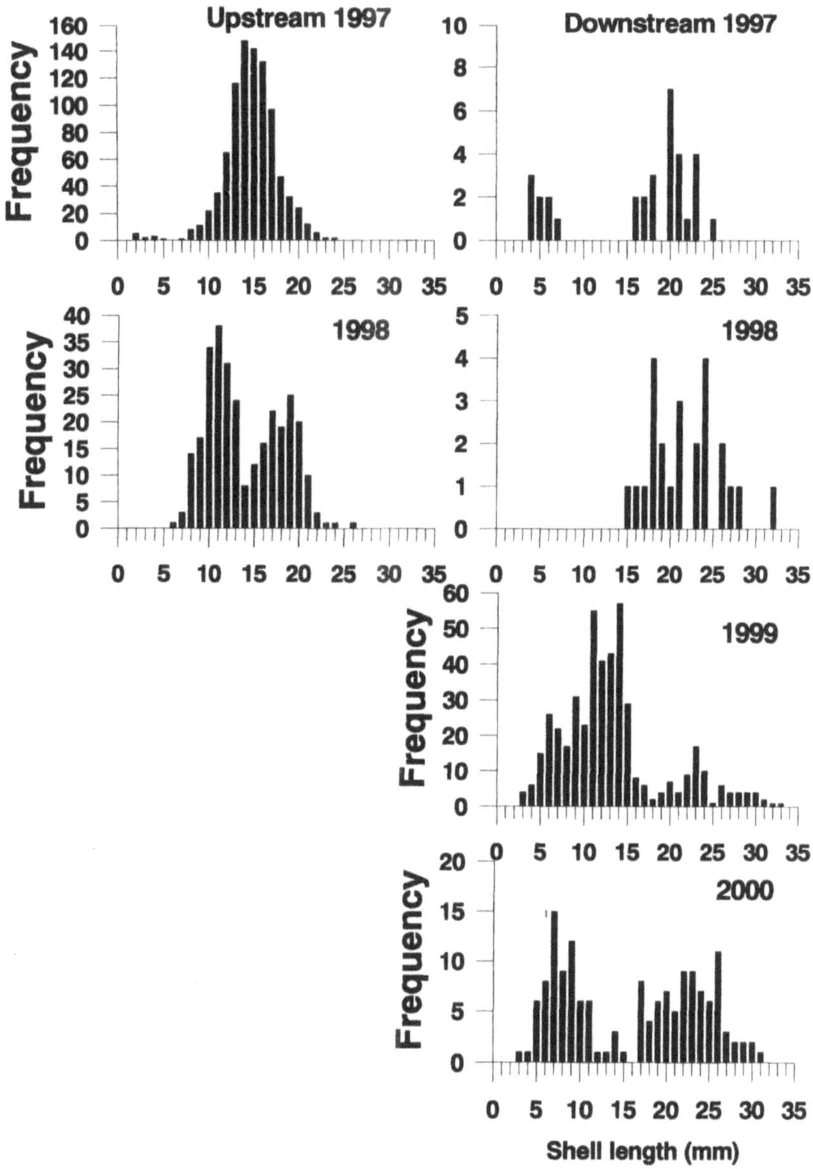

Figure 7. Size distribution of zebra mussels in the Richelieu River upstream and downstream of the Chambly Basin between 1997 and 2000

3.4 Size of Zebra Mussels and Cohort Analysis

The shell length of zebra mussels collected during the 4 years varied from 2.1 to 33.3 mm ($n = 1{,}307$). The size distribution and mean length of the mussels differed significantly between the upstream and downstream sites and from year to year (Figure 7) (Kolmogorov-Smirnov D test, $p < 0.0001$; ANOVA: $F = 12.27$, $p < 0.001$). Fisher's least significant difference test (multiple comparison of pairs) indicated significant differences in mussel length between 1997 and 1998 for the upstream sites, and between 1999 and 2000 for the downstream sites.

At upstream sites, the 1997 zebra mussel population varied from 7.4 to 20.3 mm in size with a single mode at 14 mm (Figure 7). In 1998, the size distribution ranged from 6.8 to 26.6 mm and was characterized by two distinct cohorts with modal values of 11 and 19 mm. Specimens under 15 mm in length (threshold value separating the two cohorts) represented 56.7% of all zebra mussels measured in 1998.

The very low numbers of zebra mussels gathered at the downstream sites in 1997 ($n = 32$) and 1998 ($n = 24$) produce truncated size distributions and cannot be used to determine with any accuracy the number of cohorts that make up the populations. All the mussels were longer than 15 mm, except for eight specimens gathered at a site in 1997, which were between 4 and 7 mm in length. Mussels between 7 and 15 mm were virtually absent in downstream samples in these 2 years. Despite the low numbers, the size distributions of mussels downstream were significantly different from those upstream in 1997 (K-S: $D = 0.588$, $p < 0.001$) and in 1998 (K-S: $D = 0.643$, $p < 0.001$). In 1999 and 2000, the size distribution changed radically and was clearly multimodal, indicating that there were two (or even three) distinct zebra mussel cohorts. The proportion of individuals under 15 mm was 73.4% in 1999 and 45.4% in 2000. The size distributions in those two years differed significantly (K-S: $D = 0.354$, $p < 0.001$).

In 1999, the size distribution was characterized by a principal mode between 11 and 14 mm and secondary modes of 6 mm and 23 mm. In 2000, mussels smaller than 10 mm dominated largely, while the proportion of mussels between 12 and 16 mm was very small, in contrast with earlier observations both upstream and downstream. Despite slight variations in the numbers of mussels, multimodal size distribution was observed among the mussels gathered at the various sites.

Founded on the premise that each cohort corresponds to a different annual recruitment peak, cohort analysis allows us to estimate the strength of annual colonization. Attribution of cohorts to years of larval production depends, however, on the presumed growth rate of the mussels. Given the lack of information on individual growth of zebra mussels in the Richelieu River, our analysis was based on the results of population growth in other bodies of water. In their review of the literature, Jantz and Neumann (1992) stated that the mean length of zebra mussels at the end of the first growth season (0+ cohort) varied between 4 and 7 mm, while the mean length of one year old mussels (1+ cohort) was 13 to 17 mm. Bigger specimens were older. After examining samples gathered in three consecutive years from the Hudson River, Strayer et al. (1996) reported multimodal size distributions in which each mode corresponded to a different production year. The size of newly settled juveniles (settled during the current year) varied from 4 to 8 mm in September, while a second mode of about 15 mm corresponded to mussels produced the preceding year (Strayer et al., 1996). Two-year-old mussels measured more than 20 mm. Nalepa et al. (1995) also reported cohorts of very similar size for zebra mussels beginning to colonize Lake Huron. In northern Lake

Champlain near the Richelieu, Eliopoulos and Stangel (1999) observed that the mean length of juvenile mussels attached to artificial substrates varied between 3 and 7 mm (mean 5 mm) in mid-October, a month later than our sampling in the Richelieu.

In light of this information, it would seem plausible to consider that most zebra mussels > 6 mm collected in mid-September in the Richelieu would be at least one year old and older : those under 15 to 17 mm would be one year old (1+ cohort) and the larger ones would be two or more years old (2+ cohort). Zebra mussels produced in the current year (0+ cohort) in the Richelieu would be no longer than 5 or 6 mm and these appeared to be few and far between in our collections. The relative abundance of annual recruitment of zebra mussels in the Richelieu was estimated as the product of the relative proportion of mussels in each cohort and estimated mussel abundance (Table 4). Upstream, the 1997 recruitment (7.02) represented 57% of that calculated for 1996 (12.34). Downstream, recruitment was very low in 1996 and 1997. On the other hand, 1998 recruitment at the downstream sites appeared to be much higher than that of 1999 and 1997 (Table 4) and largely contributed to the first significant colonization by zebra mussels in the lower Richelieu.

Table 4. Relative proportion of various cohorts defined by length classes of zebra mussels sampled at upstream and downstream sites along the Richelieu River

		1+ cohort		*2+ cohort*	
Sector/Year	*Abundance*	*%*	*Abundance*	*%*	*Abundance*
Upstream 1997	12.72	97.0	12.34	3.0	0.38
Upstream 1998	12.40	56.6	7.02	43.3	5.38
Downstream 1997	0.025	12.5	0.003	62.5	0.016
Downstream 1998	0.012	4.2	0.0005	95.8	0.0115
Downstream 1999	0.88	81.4	0.716	18.5	0.163
Downstream 2000	0.12	46.1	0.055	53.9	0.065

3. 5 Allometric Relationships and Phenotypic Pattern

The relationship between the length and height of zebra mussel shells at the Saint-Jean-sur-Richelieu site did not vary between years (ANCOVA, $p > 0.05$) (Table 5). The relationship between total dry weight and shell length of the zebra mussel was typically allometric (Figure 8), with an exponent close to 3, which is generally to be expected for this type of biological relationship (Peters, 1984). The relationship between dry weight of soft tissue and length is also allometric, with a coefficient of 2.88. The predictive model for this relationship explains 84.7% of the variance. A coefficient between 2.6 and 2.9 was also observed between the weight of soft tissue and shell length in earlier studies (Roe and MacIsaac, 1997). On the other hand, the dry tissue weight varied linearly or proportionally with the dry shell weight (Figure 8, Table 5).

Zebra mussels of six different phenotypes were collected in the Richelieu River and the majority being either of phenotype CC (77.3%) or AA (17.2%) (Table 6). The relative proportion of the six phenotypes in zebra mussels from the Richelieu differed significantly (*G* test likelihood ratio = 131.8, $p<0.001$) from that reported in the St. Lawrence River in 1993 (Lapierre and de Lafontaine, 1996) but was close to that calculated for the mussel populations from the Ponto-Caspian region and central Russia (Smirnova et al., 1993) (Table 6). This would provide further support to the hypothesis that zebra mussels in the Richelieu River were issued from Lake Champlain populations and not from the St.Lawrence River (de Lafontaine and Cusson, 1997).

Table 5. Empirical relationships between morphometric measurements of zebra mussels from the Richelieu River in 1997 and 1998

Site	*Year*	*n*	*Relationship*	r^2
Richelieu (12 sites)	1997	942	ln (height) = -0.400 + 0.924 ln (length)	0.940
Saint-Jean-	1997	336	ln (height) = -0.441 + 0.938 ln (length)	0.887
sur-Richelieu	1998	157	ln (height) = -0.3196 + 0.886 ln (length)	0.890
	1998	157	ln (total dry weight) = -3.145 + 3.016 ln (length)	0.951
	1998	157	ln (dry tissue weight) = -5.899 + 2.884 ln (length)	0.847
	1998	157	ln (dry shell weight) = -3.225 + 3.0028 ln (length)	0.952
	1998	157	dry tissue weight = 0.0407 + 1.187 dry shell weight	0.780

Table 6. Relative proportion of various phenotypes in zebra mussels from the Richelieu River and other locations

		Phenotypes					
Region	*n*	AA	CC	DD	MM	OO	RR
Richelieu River	969	0.172	0.773	0.013	0.001	0.033	0.008
St. Lawrence River[1]	341	0.267	0.508	0	0	0.210	0.015
Lake Erie[2]		0.12	0.380	0.03	0.190	0.170	0.110
Ponto-Caspian area[2]		0.094	0.814	0.022	0.006	0.056	0.008
Central Russia[2]		0.170	0.708	0	0.010	0.110	0.002
Baltic Sea[2]		0.080	0.899	0	0.001	0.015	0.005

[1] Lapierre and de Lafontaine, 1996; [2] Smirnova et al., 1993

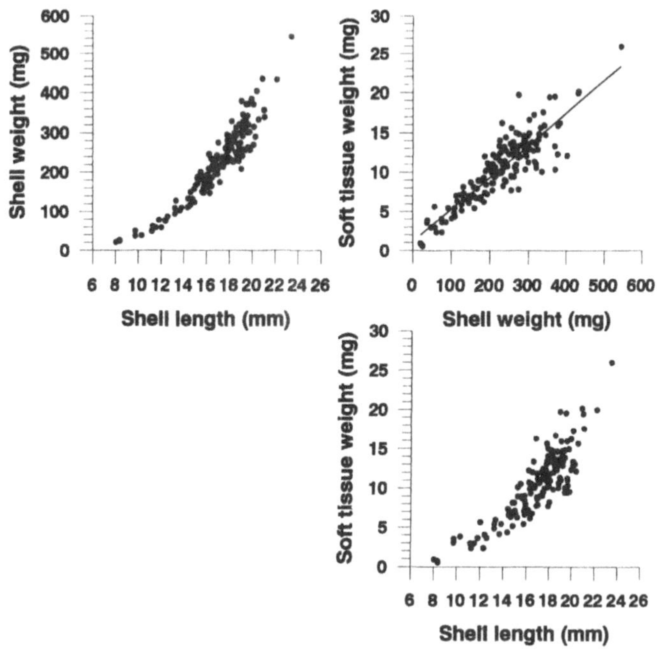

Figure 8. Relationship between weight and shell length of zebra mussels from the Richelieu River in September 1998

4. DISCUSSION

4.1 Colonization Rate and Spatial Distribution

The annual monitoring performed as part of Richelieu River cleanup campaigns from 1997 to 2000 confirmed the occurrence of zebra mussels at all the sampling sites between Saint-Paul de l'Île aux Noix and Saint-Ours, indicating that the species is now present over almost the entire length of the river. The spatial distribution of mussels was not uniform, but characterized by a steep gradient in abundance between the areas upstream and downstream of the Chambly Basin. The upper Richelieu (upstream of the Chambly Basin) was clearly more colonized than the lower Richelieu, both in the intensity of colonization and the relative abundance of mussels per object (Table 1). Downstream, mussel abundance and distribution gradually increased between 1997 and 2000 (Table 1, Figure 3), which suggests that zebra mussel populations are still in the process of invading and establishing colonies in this area. These findings suggest that zebra mussel colonization of the Richelieu follows a pattern similar to the downstream march model defined by Horvath et al. (1996). According to this model, zebra mussel colonization along a river mainly depends on the larval production of populations located upstream. Invasion of the entire length of a river is conditional upon the establishment of stable, abundant breeding populations all along the waterway. In all likelihood, zebra mussels have propagated in the same way in the Hudson (Strayer et al., 1996), the Illinois (Stoeckel et al., 1997) and the Rhine (Borcherding and de Ruyter van Steveninck, 1992). In the Rideau River (Ontario), the colonization by zebra mussels of the lower course of the river extended rapidly from an upstream source point that provided the inoculum for the production of larvae that drift downstream, where they settle (Martel, 1995).

The intensity of colonization and index of abundance revealed strong upstream-downstream gradient in zebra mussel colonization patterns in the Richelieu River. Altough it could be argued that sampling zebra mussels from objects did not allow us to obtain very precise density estimates, quantitative index of abundance could be obtained by estimating the densities of objects per unit of searching area or searching distance. Given that the colonization rate was 12.4 mussels per object at upstream sites in 1998 (Table 1) and that most objects had a surface area ranging from 0.01 to 0.1 m^2, this would yield estimates of densities varying from 100 to 1000 mussels per m^2, if the riverbed was covered with objects. This not being the case, mussel densities were certainly much lower, more probably in the range of 1 to 50 mussels per m^2, assuming that debris coverage was ~1-5%. These values agree reasonably with those of 15 to 160 mussels/m^2 derived from quadrat samples from the same location. Both of these estimates were lower than those (500 to 7,500 mussels/m^2) reported for populations already established for some time in a lake or river (Ramcharan et al., 1992; Nalepa et al., 1995; Martel, 1995; Strayer et al., 1996). Densities in the Hudson River fluctuated between 1,000 and 10,000 mussels/m^2 barely three years after the initial invasion (Strayer et al., 1996), while those in the Rideau were as high as 3,500 mussels/m^2 four years after they were discovered. Since colonization by zebra mussels is still at an early stage in the Richelieu River, it is therefore recommended that monitoring as part of riverbed cleanups be maintained and that quantitative sampling be developed and carried out to estimate relative densities and quantify the impact of zebra mussels.

The Chambly Basin seems to play a key role in determining the observed difference between upstream and downstream sites in terms of colonization rates and relative abundance of zebra mussels. The physical and chemical characteristics of the waters of the Richelieu are significantly different downstream of the basin: greater flow rate, turbidity, conductivity, major nutrients (N, P) for primary producers and agricultural contaminant loads (Piché and Simoneau, 1998). These changes are associated with two tributaries (the L'Acadie and Des Hurons rivers) that run through farmland and empty into the Richelieu at the Chambly Basin. However, with the exception of flow rate, those factors known to influence zebra mussels colonization, vary only to a limited degree between upstream and downstream sections. Levels of calcium, which are crucial to zebra mussel growth (Claudi and Mackie, 1993), do not vary much between upstream and downstream, and were between 16.34 and 20.82 mg/L in 1998 (Y. de Lafontaine, unpublished data). These values, though relatively low and not optimal for zebra mussels, are still above 11 mg/L, the level deemed critical for survival, growth and reproduction (McMahon, 1996). According to Mellina and Rasmussen (1994), high zebra mussel densities generally develop in water with calcium levels above 21 mg/L. Eliopoulos and Stangel (1999) have shown, however, that ambient concentrations of between 14 and 20 mg/L had no negative effect on zebra mussels in Lake Champlain. We believe that this conclusion also applies to Richelieu River populations. Current speed in the lower Richelieu could act as a limiting factor on zebra mussel colonization, although once again, observed values (0.6–1.0 m/sec, see Piché and Simoneau, 1998) are below the critical levels reported in the literature (Claudi and Mackie, 1993). Thus, while the possibility cannot be totally excluded, it seems unlikely that the physical-chemical conditions and quality of the water of the Richelieu are the cause of the pronounced gradient in zebra mussel colonization between the stretches upstream and downstream of the Chambly Basin.

The low abundance of zebra mussels downstream of Chambly Basin would appear to be attributable, rather, to weak recruitment stemming from translocation of larvae. Zebra mussel colonization of the upper Richelieu (upstream of Chambly Basin) seems to result chiefly from larval drift from Lake Champlain (de Lafontaine and Cusson, 1997). Sampling in 1996 and 1997 showed that the density of veliger larvae dropped by more than 90% between the upper and lower Richelieu (de Lafontaine and Cusson, 1997; Y. de Lafontaine, unpublished data), which would mean lower colonization potential downstream. Recent studies have shown that zebra mussels in rivers do not usually reach abundance levels comparable to those of upstream lakes and do not generally dominate lotic habitats (Horvath et al., 1999; Toczylowski et al., 1999). The situation in the Richelieu, with its upstream-downstream gradient of zebra mussel densities, thus seems typical of the dynamics of this species in rivers. Our current findings, however, do not allow us to conclude whether the present spatial gradient is stable or whether it will attenuate over time. According to the downstream march model (see above, Horvath et al., 1996), colonization and abundance of zebra mussels in the lower Richelieu could increase over time, once upper Richelieu populations have become well established and achieved high enough densities to ensure sustained and abundant larval production. Long-term monitoring, coupled with specific studies focussing on larval production and drift and on colonization by juveniles, is required to shed light on the dynamics of colonization and dispersal of zebra mussels from the upper to the lower Richelieu.

The variability of the size structure of mussels in the two parts of the river and from year to year reveals both spatial and temporal variability in recruitment of zebra mussels in the Richelieu (Table 4). Significant colonization of the upper Richelieu probably occurred in 1996, with important colonization of the lower Richelieu not taking place until two years later, in 1998. This is supported by the absence of very large (>25 mm) and older mussels at upstream sites in 1997 and by the fact that the presence of zebra mussels was not noticed during riverbed cleanups in 1995 and 1996 (Figure 2). A number of findings suggest that recruitment and colonization in 1997 were relatively weaker than in 1996 or 1998. First, mussel abundance per object upstream of the Chambly Basin remained stable (~12 mussels/object) between 1997 and 1998 (Table 1), while the size distribution clearly indicated the presence of two cohorts in 1998 as opposed to just one in 1997 (Figure 7). Second, the absence of mussels smaller than 15 mm downstream (though there were some upstream) in 1998 also suggests very low recruitment in 1997. Similarly, the low proportion of mussels in the 2+ cohort (> 15 to 17 mm) in 1999 may well be indicative of low recruitment in 1997 and earlier. Yet each year there was a very small proportion of mussels in the 0+ cohort. This situation may be due to (1) late colonization of the Richelieu by zebra mussels or (2) volunteers' difficulty for detecting very small mussels (< 3 mm) on objects. It seems unlikely that our attribution of various size groups of mussels to production years might be erroneous and that mussels between 7 and 15 mm might belong to the 0+ cohort, as that would mean a growth rate far superior to that observed in populations in Lake Champlain and other sites at the same latitude. In our opinion, the environmental conditions of the Richelieu River, which are relatively similar to those in Lake Champlain and in the St.Lawrence River, could not support such a difference in growth rate.

The mean flow rates of the Richelieu in summer (July and August) have been highly variable in the last few years, depending on the rainfall (Table 7). The highest mean monthly flow for July, which is the peak period for veliger larvae in the river (de Lafontaine and Cusson, 1997), was recorded in 1998 (661 m^3/s). This flow, two to five times higher than in other years, could have transported more larvae downstream and been responsible for the greater colonization of the lower Richelieu in 1998, as evidenced by the size of the 1+ cohort observed during the clean-ups in 1999 (Table 4). It is also interesting to note that the mean monthly flow was also very high in July 1996 (504 m^3/s) at the time when the initial colonization of the upper Richelieu probably occurred. On the other hand, the very low flows recorded in the summer of 1999 (Table 7) probably led to less downstream colonization than the year before (Table 4). We therefore hypothesize a causal link between river flow fluctuations and strength of recruitment and colonization of zebra mussels in the Richelieu. Although many earlier studies have noted the importance of the larval supply on zebra mussel recruitment in lotic systems (Borcherding and de Ruyter van Steveninck, 1992; Martel, 1995; Nichols, 1996; Stoeckel et al., 1997), the effect of fluctuations in flow on controlling recruitment was still very speculative. Confirmation of this hypothesis and the development of an empirical model of the relationship between summer flow and zebra mussel recruitment in the Richelieu would require long-term monitoring of mussel colonization along the river. Monitoring of this type must include an analysis of size and annual growth of mussels in order to determine the structure of zebra mussel cohorts.

Table 7. Mean flow (m^3/s) of the Richelieu River for the months of June, July and August, between 1995 and 2000

Year	*June* *Mean (SD) Min-Max*	*July* *Mean (SD) Min-Max*	*August* *Mean (SD) Min-Max*
1995	241 (51) 155–321	125 (15) 96–157	171 (29) 104–214
1996	729 (106) 573–940	504 (24) 458–563	383 (67) 284–495
1997	465 (104) 275–632	269 (24) 224–311	204 (14) 175–225
1998	469 (62) 384–631	661 (63) 508–741	451 (36) 396–523
1999	226 (59) 139–328	122 (15) 102–164	89 (9) 63–103
2000	677 (101) 505–873	390 (52) 309–504	273 (31) 224–334

According to Ackerman et al. (1994), recruitment of zebra mussels is a two-stage process: (1) attachment of pediveliger larvae followed by metamorphosis into juveniles and (2) movement of juvenile or adult mussels (also called translocators) to new areas. It has already been demonstrated that juveniles can be translocated when the aquatic plants to which they are attached are pulled up and carried away by the current (or by boats) (Lewandowski, 1982; Johnson and Carlton, 1996). While the first mechanism seems to have largely contributed to the colonization of the upper Richelieu through the transportation of larvae from Lake Champlain (de Lafontaine and Cusson, 1997), the second process might explain the presence of large mussels (> 20 mm) downstream in the first two years (1997 and 1998). Large mussels were rarely found at the upstream sites, but were more abundant downstream. The lack of a continuum of relative proportions of the various mussel cohorts along the Richelieu may suggest that translocation is at work. It is nevertheless impossible to specify whether the transfer is due to the drifting of plants or other floating objects from Lake Champlain or to overland transportation (i.e., by boat trailers, Johnson and Carlton, 1996). This method of spreading zebra mussel populations in the Richelieu appears, however, to be less significant than colonization by larval drift.

So far, it has been argued that zebra mussels invasion into the Richelieu River was a northward extension of the Lake Champlain populations. The absence of quagga mussels (which are present in the St. Lawrence River but still absent in Lake Champlain) and the different phenotypic composition of mussels between the Richelieu and the St. Lawrence River (Table 6) would support the above suggestion and indicate that possible transfer of zebra mussels from the St. Lawrence River into the Richelieu was very limited and negligible.

4.2 Potential Impact on Unionids

Although the impacts of zebra mussels on the economic and recreational use of the Richelieu River remain to be demonstrated, one of the most quickly observable direct biotic effects is predicted to be the decline in native unionid populations (Ricciardi et al., 1995; Nalepa et al., 1996). The impact on unionids is highly variable, however, depending on the degree of infestation by zebra mussels, which in turn depends on environmental conditions and habitat type (Nichols and Amberg, 1999; Horvath et al., 1999). It has recently been shown that unionids were not colonized by zebra mussels any more than were various other solid substrates and that zebra mussel densities on unionids were similar to those measured on rigid substrates in the same area (Toczylowski and Hunter, 1997; Toczylowski et al., 1999). Our findings tend to confirm this, as they reveal no significant difference in degree of colonization of unionids in comparison to inert objects in the Richelieu (Table 2, Figure 6). This suggests that in the early stages of colonization, zebra mussel attachment is essentially related to the availability of a rigid substrate, without preference for unionids. In our study, the mean level of infestation was under 10 zebra mussels per unionid, which is still much lower than the levels reported in other rivers strongly colonized by zebra mussels (> 100 mussels/unionid, Tucker, 1994; Ricciardi et al., 1995). According to observations by Hart et al. (2001), the degree of colonization of unionids observed in the Richelieu in 1998 was too low to have a significant negative effect on unionid survival rates. Ricciardi et al. (1995) predicted that high unionid mortality rates are to be expected when colonization reaches about 100 zebra mussels per unionid. Considering that the infestation rate on Unionidae went from 10 to 100 zebra mussels per unionid in four years in the Mississippi (Hart et al., 2001), it appears imperative to reassess the degree of colonization of unionids in the upper Richelieu as soon as possible. The ratio of the number of live and dead unionids could provide an indication of the impact of zebra mussel colonization on unionids in the Richelieu and be compared with data accumulated during the last four cleanup operations.

4.3 Monitoring and Cleanup Operations

The results presented above provided tangible evidence that fruitful collaborative partnerships may yield interesting and relevant information from both scientific and educational perspective. Aside from the benefit of removing debris and polluting objects, riverbed cleanups do also provide numerous interesting opportunities for public education on several environmental issues (i.e. pollution, exotic species, biodiversity) and for better understanding and management of aquatic systems. As shown here, riverbed cleanups can also help acquiring relevant scientific information when used for synoptic spatial biomonitoring.

From a scientific standpoint, riverbed cleanups present powerful tools for both biological monitoring and assessment of aquatic systems. The overall findings from the clean-up campaigns conducted in the Richelieu River each September have allowed us to confirm and document the spread of zebra mussels all along the river since their first invasion in 1996. Four years later, the spatial distribution of the zebra mussel remained very heterogeneous and characterized by a steep abundance gradient between the stretches of the river upstream and downstream of the Chambly Basin. As a means of determining the presence of zebra mussels at the various sites along the Richelieu River, the intensive collection done during the cleanup operations turned out to be more efficient than the sampling of quadrats. The monitoring by the cleanup crews was indeed very effective in confirming the presence of zebra mussels at a given site and comparing degrees of colonization at different sites. In 1998, zebra mussels were reported in the quadrats at only one highly colonized site (Saint-Jean-sur-Richelieu) and no mussels were found in the quadrats at downstream sites, despite their presence on objects at the same sites. This is mainly explained by the very large search effort deployed by the high number of divers and volunteers participating to each cleanup. Such large sampling effort which would be almost impossible to implement otherwise, and surely represents the best way to «discover» and confirm the presence of new introduced species while they are still at low densities and therefore rare. The cleanups also facilitated the development of relevant information on the diversity of unionids in the Richelieu River. Severe decline in populations abundance of unionids have been noted throughout the Great Lakes and in North American waters and many species are now under threat and endangered (Metcalfe-Smith et al., 1998). Information on the distribution and abundance of these species is desperately needed and could well become integrated in future cleanup activities. Here again, the cleanups introduce the possibility to accurately estimate and monitor the impact of zebra mussels on unionids.

From an educational point of view, riverbed cleanups in the Richelieu remained of large interest to the public as shown by the sustained numbers of volunteers and helpers since 1995 (Figure 2). The quantity of trash and debris removed from the water annually (Figure 3) indicates that these activities are still necessary. As demonstrated here, the addition of a scientific monitoring component to the cleanup operations is easily feasible and can help ensuring public involvement in river assessement and management and we advocate that sampling responsibility for monitoring purposes be taken up by local public organizations.

To date, the Richelieu is the only tributary of the St. Lawrence that has been invaded and colonized by zebra mussels. The impact of any future translocation of mussels from the Richelieu to the St. Lawrence has not been assessed, but should be fairly moderate and confined mainly to the south shore of Lake St. Pierre. It is important to keep in mind, however, that colonization of the Richelieu by zebra mussels increases the risk of the species' spreading to other bodies of water in Quebec. The Richelieu is heavily used by pleasure craft, and some of them may frequent other inland waters. In this regard, the implementation and application of a prevention program should be a high priority in order to reduce the risk of propagation of the zebra mussel and other undesirable species. This can best be achieved through solid involvement and mutually beneficial interactions of public groups, scientists and decision-makers.

5. REFERENCES

Ackerman, J.D., Sim, B., and R. Claudi, R., 1994, A review of the early life history of zebra mussels (*Dreissena polymorpha*): comparisons with marine bivalves, *Can. J. Zool.* **72**: 1169-1179.

Borcherding, J., and de Ruyter van Steveninck, E.D., 1992, Abundance and growth of *Dreissena polymorpha* larvae in the water column of the River Rhine during downstream transportation, in: *The Zebra Mussel* Dreissena polymorpha*: Ecology, Biological Monitoring and First Applications in the Water Quality Management,* D. Neumann and H.A. Jenner, eds., Gustav Fischer, Stuttgart, pp. 29-44.

Claudi, R., and Leach, J.H., 2000, *Nonindigenous Freshwater Organisms: Vectors, Biology, and Impacts,* CRC Press Inc., Lewis Publishers, Boca Raton, Florida, 464 p.

Claudi, R., and Mackie, G.L., 1993, *Practical Manual for Zebra Mussel Monitoring and Control,* CRC Press Inc., Lewis Publishers, Boca Raton, Florida, 227 p.

Claxton, W.T., Martel, A., Dermott, R.M., and Boulding, E.G., 1997, Discrimination of field-collected juveniles of two introduced dreissenids (*Dreissena polymorpha* and *Dreissena bugensis*) using mitochondrial DNA and shell morphology, *Can. J. Fish. Aquat. Sci.* **54**: 1280-1288.

de Lafontaine, Y., and Costan, G., 2002, Introduction and transfer of alien aquatic species in the Great Lakes - St. Lawrence River drainage basin, in: *Alien invaders in Canada's Waters, Wetlands, and Forests,* R. Claudi, P. Nantel, and E. Muckle-Jeffs, eds., Department of Natural Resources Canada, Ottawa, pp. 73-92.

de Lafontaine, Y., and Cusson, B., 1997, Veligers of zebra mussels in the Richelieu River: an intrusion from Lake Champlain?, *Proceedings of the Second Northeast Conference on Nonindigenous Aquatic Nuisance Species,* April 18-19, 1997, Burlington, Vermont.

Domm, S., McCauley, R.W., Kott, E., and Ackerman, J.D., 1993, Physiological and taxonomic separation of two Dreissenid mussels in the Laurentian Great Lakes, *Can. J. Fish. Aquat. Sci.* **50**: 2294-2297.

Eliopoulos, C., and Stangel, P., 1999, *Lake Champlain 1998 Zebra mussel monitoring program,* Vermont Department of Environmental Conservation, Waterbury, Vermont, USA, 32 p.

Gido, K.B., and Brown, J.H., 1999, Invasion of North American drainages by alien fish species, *Freshw. Biol.* **42**: 387-399.

Hart, R.A., Davis, M., Grier, J.W., and Miller, A.C., 2001, Survival of unionids following removal of attached zebra mussels, *J. Freshw. Ecol.* *16*: 29-34.

Horvath, T.G., Martin, K.M., and Lamberti, G.A., 1999, Effect of zebra mussels, *Dreissena polymorpha,* on macroinvertebrates in a lake-outlet stream, *Am. Midl. Nat.* **142**: 340-347.

Horvath, T.G., Lamberti, G.A., Lodge, D.M., and Perry, W.L., 1996, Zebra mussel dispersal in lake-stream systems: source-sink dynamics?, *J.N. Am. Benthol. Soc.* **15**: 564-575.

Jantz, B., and Neumann, D., 1992, Shell growth and aspects of the population dynamics of *Dreissena polymorpha* in the River Rhine, in: *The Zebra Mussel* Dreissena polymorpha*: Ecology, Biological Monitoring and First Applications in the Water Quality Management,* D. Neumann and H.A. Jenner, eds., Gustav Fischer, Stuttgart, pp. 48-66.

Johnson, L.E., and Carlton, J.T., 1996, Post-establishment spread in large-scale invasions: dispersal mechanisms of the zebra mussel *Dreissena polymorpha, Ecology 77:* 1686-1690.

Kilgour, B.W., and Mackie, G.L., 1993, Colonization of different construction materials by the zebra mussel (*Dreissena polymorpha*), in: *Zebra Mussels: Biology, Impacts, and Control,* T.F. Nalepa and D.W. Schloesser, eds., Lewis Publishers, Boca Raton, Florida, pp. 167-173.

Lapierre, L., and de Lafontaine, Y., 1996, *Colonisation de l'épave du Lady Sherbrooke par les Moules zébrées*, Environnement Canada, Conservation de l'Environnement, Centre Saint-Laurent, Scientific and Technical Report Series ST-48, 17 p.

Lewandowski, K., 1982, The role of early developmental stages in the dynamics of *Dreissena polymorpha* (Pall.) (Bivalvia) populations in lakes. II. Settling of larvae and the dynamics of settled individuals, *Ekol. Pol.* **30**: 223-286.

Marsden, J.E., Spidle, A.P., and May, B., 1996, Review of genetic studies of Dreissena spp., *Amer. Zool.* **36**: 259-270.

Marsden, J.E., and Lansky, D.M., 2000, Substrate selection by settling zebra mussels, *Dreissena polymorpha*, relative to material, texture, orientation, and sunlight, *Can. J. Zool.* **78**: 787-793

Martel, A., 1995, Demography and growth of the exotic zebra mussel (*Dreissena polymorpha*) in the Rideau River (Ontario), *Can. J. Zool.* **73**: 2244-2250.

McMahon, R.F., 1996, The physiological ecology of the zebra mussel, *Dreissena polymorpha*, in North America and Europe, *Amer. Zool.* **36**: 339-363.

Mellina, E., and Rasmussen, J.B., 1994, Patterns in the distribution and abundance of zebra mussel (*Dreissena polymorpha*) in rivers and lakes in relation to substrate and other physicochemical factors, *Can. J. Fish. Aquat. Sci.* **51**: 1024-1036.

Metcalfe-Smith, J.L., Staton, S.K., Mackie, G.L., and Lane, N.M., 1998, Changes in the biodiversity of freshwater mussels in the Canadian waters of the lower Great Lakes drainage basin over the past 140 years, *J. Great Lakes Res.* **24**: 845-858.

Mills, E.L., Strayer, D.L., Scheuerell, M.D., and Carlton, J.T., 1996, Exotic species in the Hudson River basin : a history of invasions and introductions, *Estuaries* **19**: 814-823.

Mills, E.L., Chrisman, J.R., and Holeck, K.T., 2000, The role of canals in the spread of nonindigenous species in North America, in: *Nonindigenous Freshwater Organisms: Vectors, Biology and Impacts*, R. Claudi and J.H. Leach, eds., CRC Press Inc., Lewis Publishers, Boca Raton, Florida, pp. 347-380.

Nalepa, T.F., Wojcik, J.A., Fanslow, D.L., and Lang, G.A., 1995, Initial colonization of the zebra mussel (*Dreissena polymorpha*) in Saginaw Bay, Lake Huron: population recruitment, density, and size structure, *J. Great Lakes Res.* **21**: 417-434.

Nalepa, T.F., Hartson, D.J., Gostenik, G.W., Fanslow, D.L., and Lang, G.A., 1996, Changes in the freshwater mussel community of Lake St. Clair: from Unionidae to *Dreissena polymorpha* in eight years, *J. Great Lakes Res.* **22**: 354-369.

Nichols, S.J., 1996, Variations in the reproductive cycle of *Dreissena polymorpha* in Europe, Russia, and North America, *Am. Zool.* **36**: 311-325.

Nichols, S.J., and Amberg, J., 1999, Co-existence of zebra mussels and freshwater unionids: population dynamics of *Leptodea fragilis* in a coastal wetland infested with zebra mussels, *Can. J. Zool.* **77**: 423-432.

Peters, R.H., 1984, *The ecological implications of body size*, Cambridge University Press, Cambridge, U.K., 329 p.

Piché, I., and Simoneau, M., 1998, Le bassin de la rivière Richelieu: Profil géographique, sources de pollution, intervention et assainissement et qualité des eaux, in: *Le bassin versant de la rivière Richelieu: l'état de l'écosystème aquatique — 1995*, Ministère de l'Environnement et de la Faune, Direction des écosystèmes aquatiques, Québec, Envirodoq No. EN980604, rapport EA-13, section 1.

Pimentel, D., McNair, S., Janecka, J., Wightman, J., Simmonds, C., O'Connell, C., Wong, E., Russel, L., Zern, J., Aquino, T., and Tsomodo, T., 2001, Economic and environmental threats of alien plant, animal, and microbe invasions, *Agricult. Ecosyst. Environm.* **84**: 1-20.

Ramcharan, C.W., Padilla, D.K., and Dodson, S.I., 1992, A multivariate model for predicting population fluctuations of *Dreissena polymorpha* in North American lakes, *Can. J. Fish. Aquat. Sci.* **49**: 150-158.

Ricciardi, A., Whoriskey, F.G., and Rasmussen, J.B., 1995, Predicting the intensity and impact of *Dreissena* infestation on native unionid bivalves from *Dreissena* field density, *Can. J. Fish. Aquat. Sci.* **52**: 1449-1461.

Roe, S.L., and MacIsaac, H.J., 1997, Deepwater population structure and reproductive state of quagga mussels (*Dreissena bugensis*) in Lake Erie, *Can. J. Fish. Aquat. Sci.* **54**: 2428-2433.

Smirnova, N.F., Biochino, G.I., and Vinogradov, G.A., 1993, Some aspects of the Zebra mussel (*Dreissena polymorpha*) in the former European USSR with morphological comparisons to Lake Erie, in: *Zebra Mussels : Biology, Impacts and Control*, T.F. Nalepa and D.W. Schloesser, eds.,. Lewis Publ., Boca Raton, Florida, pp. 217-226.

Stanczykowska, A., and Lewandowski, K., 1993, Thirty years of *Dreissena polymorpha* ecology in Mazurian Lakes of northeastern Poland, in: *Zebra Mussels : Biology, Impacts and Control*, T.F. Nalepa and D.W. Schloesser, eds.,. Lewis Publ., Boca Raton, Florida, pp. 3-33.

Stoeckel, J.A., Schneider, D.W., Soeken, L.A., Blodgett, K.D., and Sparks, R.E., 1997, Larval dynamics of a riverine metapopulation: implications for zebra mussel recruitment, dispersal, and control in a large-river system, *J. N. Am. Benthol. Soc.* **16**: 586-601.

Strayer D.L., Powell, J., Ambrose, P., Smith, L.C., Pace, M.L., and Fischer, D.T., 1996, Arrival, spread, and early dynamics of a zebra mussel (*Dreissena polymorpha*) population in the Hudson River estuary, *Can. J. Fish. Aquat. Sci.* **53**: 1143-1149.

Toczylowski, S.A., and Hunter, R.D., 1997, Do zebra mussels preferentially settle on unionids and/or adults conspecifics?, in: *Zebra Mussels and Aquatic Nuisance Species,* F. D'Itri, ed., Ann Arbor Press Inc., Chelsea, Michigan, pp. 125-140.

Toczylowski, S.A., Hunter, R.D., and Armes, L.M., 1999, The role of substratum stability in determining zebra mussel load on unionids, *Malacologia* **41**:151-162.

Tucker, J.K., 1994, Colonization of unionid bivalves by the zebra mussel, *Dreissena polymorpha*, in Pool 26 of the Mississippi River, *J. Freshw. Ecol.* **9**: 129-134.

Appendix 1. Unionid species gathered in the Richelieu River during cleanup operations

Site (Years)	*Species*
Saint-Paul de l'Île-aux-Noix (1997)	*Lampsilis radiata* *Elliptio complanata*
Iberville (1997)	*Lampsilis radiata* *Lampsilis ovata* *Ligumia recta* *Pisidium compressum* *Pyganodon cataracta* *Elliptio sp.*
Saint-Jean sur Richelieu (1997)	*Lampsilis radiata* *Elliptio sp.* *Elliptio complanata*
Saint-Athanase (1997)	*Lampsilis radiata* *Lampsilis ventricosa* *Ligumia recta* *Elliptio complanata* *Elliptio sp.* *Pyganodon cataracta*
Saint-Mathias (1997)	*Lasmigona costata* *Lampsilis radiata* *Leptodea fragilis* *Ligumia recta* *Elliptio complanata*
Île Goyer (1997)	*Lampsilis radiata* *Leptodea fragilis* *Elliptio complanata*
McMasterville (1997)	*Lampsilis radiata* *Lampsilis ventricosa* *Lasmigona costata* *Leptodea fragilis* *Ligumia recta* *Elliptio dilatata* *Elliptio complanata* *Alasmidonta undulate*
Saint-Hilaire (1997)	*Lampsilis radiata*

Site (Years)	*Species*
	Lampsilis ovata
	Lampsilis ventricosa
	Leptodea fragilis
	Ligumia recta
	Elliptio complanata
	Elliptio sp.
Beloeil (1997)	*Lampsilis radiata*
	Leptodea fragilis
Saint-Marc sur Richelieu (1997)	*Lampsilis radiata*
	Lampsilis ventricosa
	Elliptio complanata
	Elliptio crassidens
	Elliptio sp.
Saint-Antoine sur Richelieu (1997)	*Lampsilis ovata*
	Lampsilis radiata
	Lampsilis ventricosa
	Leptodea fragilis
	Elliptio complanata
Saint-Ours (1997)	*Lampsilis ovata*
	Lampsilis radiata
	Lampsilis ventricosa
	Alasmidonta undulata
	Ligumia recta
	Elliptio complanata
	Elliptio crassidens
	Elliptio dilatata
	Pisidium compressum
	Pyganodon cataracta
Saint-Mathias (1998)	*Lampsilis radiata*
	Lampsilis ventricosa
	Leptodea fragilis
	Ligumia recta
	Elliptio complanata
	Elliptio dilatata
	Elliptio sp.
	Pyganodon cataracta
	Pyganodon sp.
Saint-Hilaire (1998)	*Lampsilis radiata*
	Lampsilis ventricosa
	Ligumia recta
	Elliptio complanata

Site (Years)	*Species*
	Elliptio dilatata
	Alasmidonta undulata
Beloeil (1998)	*Lampsilis radiata*
	Lampsilis ventricosa
	Pyganodon grandis
	Elliptio complanata
	Elliptio dilatata
	Alasmidonta undulata
Saint-Marc sur Richelieu (1998)	*Lampsilis radiata*
	Lampsilis ventricosa
	Leptodea fragilis
	Ligumia recta
	Pyganodon cataracta
	Elliptio crassidens
	Elliptio sp.
Saint-Marc sur Richelieu (1999)	*Lampsilis radiata*
	Lampsilis ventricosa
	Leptodea fragilis
	Ligumia recta
	Pyganodon cataracta
	Elliptio complanata
Saint-Ours (1999)	*Lampsilis radiata*
	Lampsilis ventricosa
	Ligumia recta
	Elliptio complanata
	Elliptio crassidens
	Elliptio dilatata

WATER QUALITY IMPROVEMENTS FOLLOWING RIPARIAN RESTORATION IN TWO VERMONT AGRICULTURAL WATERSHEDS

Donald W. Meals*

ABSTRACT

Achievement of management goals for Lake Champlain will require significant reductions of phosphorus (P) loads from agriculture, the dominant nonpoint source (nps) in the Basin. Cost-effective nonpoint source reduction strategies must be based on reliable treatment techniques beyond the basic erosion control and animal waste storage practices promoted in the past. *The Lake Champlain Basin Agricultural Watersheds National Monitoring Program (NMP) Project* evaluated the effectiveness of low-cost livestock exclusion, streambank protection, and riparian restoration practices in reducing concentrations and loads of nps pollutants from grazing land at the watershed level. Treatment and control watersheds in the Missisquoi River Basin in northwestern Vermont were monitored from 1994 through 2000 in a paired-watershed design. Monitoring included stream discharge recording, flow-proportional sampling for nutrients and suspended solids, intensive grab sampling for indicator bacteria, and tracking of land use and agricultural management in the watersheds. Treatment (riparian fencing, alternative water supplies, protected stream crossings, and streambank bioengineering) was completed in 1997 at a cost of ~$40,000. Landowner participation in land treatment was entirely voluntary and all treatments were 100% cost-shared by the project and cooperators. Post-treatment data documented significant reductions in nutrient and suspended solids concentrations and loads (~20-50%) and bacteria counts (~40-60%) from both treated watersheds. Although water quality in one treated watershed was seriously compromised by mismanagement of one large farm in the final year, the project clearly demonstrated the potential for inexpensive riparian zone restoration to reduce nps pollution from grazing lands in Vermont.

keywords: bacteria, nonpoint source pollution, paired-watershed, phosphorus, riparian zone, Vermont

* New England Interstate Water Pollution Control Commission/VT Dept. of Environmental Conservation, Waterbury, VT 05671. dmeals@wcvt.com

1. INTRODUCTION

Lake Champlain, the sixth largest freshwater lake in the U.S., is undergoing cultural eutrophication due to excessive phosphorus (P) load from its 20,800km^2 drainage basin, which spans portions of Vermont, New York, and Quebec, Canada. About 71% of the average 647 t/yr P load comes from nonpoint sources (nps) (VT Dept. of Environmental Conservation and NY Dept. of Environmental Conservation 1997) and the majority of the nps P load to Lake Champlain has been attributed to agricultural land in the basin (Meals and Budd 1998, Hegman et al. 1999). The management strategy for Lake Champlain adopted by the States of Vermont and New York and the Province of Quebec calls for reductions of P loading from both point and nonpoint sources (Lake Champlain Basin Program 1996). Vermont's nps management strategy for Lake Champlain and its tributaries will rely on implementation of controls of known effectiveness, particularly for agriculture, to reduce P loads to meet state water quality standards established for the Lake.

Early efforts to reduce agricultural nps pollution in Vermont since 1980 have focused on structural animal waste management measures in the state's predominantly dairy agriculture. Grazing impacts on water quality have not been addressed, although dairy cows in Vermont traditionally spend half of the year on pasture with free access to streams. Direct deposition of waste into streams, destruction of riparian vegetation, and trampling of streambanks and streambeds are all potential problems associated with unrestricted livestock grazing. Livestock grazing is widely recognized to potentially contribute increased nutrients, sediment, and microorganisms to surface waters (Duda and Finan 1983, Correll et al. 1995, Belsky et al. 1999).

The *Lake Champlain Basin Agricultural Watersheds Section 319 NMP Project* sought to quantify the effectiveness of livestock exclusion, streambank protection, and riparian restoration practices in reducing sediment, nutrient, and bacteria runoff from agricultural land to surface waters. Treatment effectiveness was evaluated through water quality monitoring at watershed outlets using a paired-watershed design. The project was one of twenty-three special nps control projects in the National Monitoring Program, funded in part by the U.S. EPA under Section 319 of the Clean Water Act. Project design and results have been described in detail elsewhere (e.g., Meals 2001a and 2001b).

2. METHODS

2.1 Study area

The project was conducted within the Missisquoi River drainage of the Lake Champlain Basin in Vermont in the towns of Berkshire and Richford, VT (Fig. 1). The Missisquoi River drains the most intensively agricultural region of the Basin, contributing the greatest nps P load (~82 t/yr) to Lake Champlain among its tributaries (VT Dept. of Environmental Conservation and NY Dept. of Environmental Conservation 1997). Typical of the region, water quality in the study streams is impaired by nutrients, bacteria, and organic matter originating from animal wastes generated from dairy farms, crop production, and livestock activity within streams and riparian areas (VT Agency of

Natural Resources 1996). The macroinvertebrate and fish communities in all three study streams are impaired, reflecting degradation from sediment and nutrient enrichment, and generally fail to comply with Vermont biological water quality standards. The 690 ha Samsonville Brook watershed (WS1) and the 1,422 ha Godin Brook watershed (WS 2) received riparian restoration treatment. The 954 ha Berry Brook watershed (WS 3) was the control.

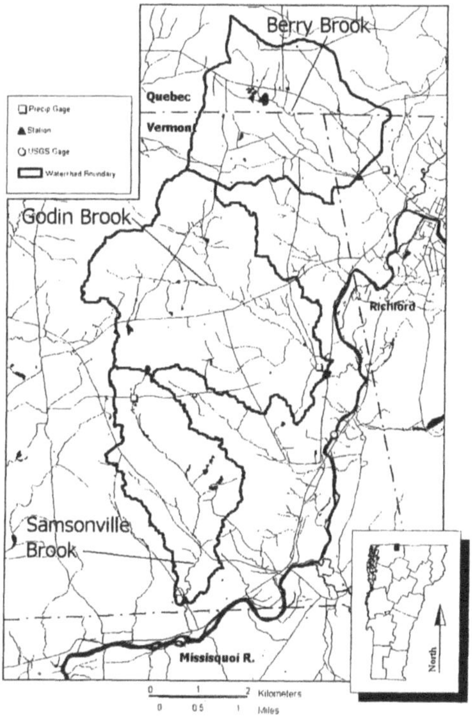

Figure 1. Project location map

The study watersheds occupy part of a broad swath of land that rises gently from the Champlain lowlands in the west through the foothills of the Green Mountains to the east. Maximum relief in the study area is ~250 m, from a low of ~150 m above mean sea level to maximum elevations of 400 m in WS 3. Most watershed soils formed in unstratified glacial drift of various metamorphic minerals, although each of the streams enters areas of deep alluvial and lacustrine soils near its confluence with the Missisquoi River.

The climate of the area is of the cool, continental type with cold winters, warm summers, a short growing season, and pronounced seasonal variations in temperature and precipitation. Mean annual temperature is 6.3 °C. Annual precipitation averages 1,049 mm; average annual snowfall is 291 cm. About 60% of watershed land area is forested; just 2-3% of each watershed is in residential use. One-third of watershed area is in agriculture, predominantly dairy. Most watershed farmers rely on grass production for livestock feed. More than 25% of each watershed is in hay or mixed hay/pasture, with only 1-3% of the land devoted to silage corn production. In 2000, farms in the study watersheds reported 864, 2306, and 379 A.U. (Animal Units, 1 A.U.=~450 kg) in WS 1, WS 2, and WS 3, respectively.

2.2 Study design

To account for year-to-year variations in climate/hydrologic inputs, the project followed a paired watershed design (USEPA 1993), using two watersheds – treatment and control - and two periods of monitoring - calibration and post-treatment. During the *Calibration Period*, neither watershed receives treatment and water quality data are collected similarly at each watershed outlet. The treatment watershed has a change in management implemented at the end of the calibration period, whereas management in the control watershed remains unchanged. During the *Treatment Period*, management changes are active in the treatment watershed, while the control watershed remains under original conditions. Monitoring continues according to the original design throughout the treatment period. The basis of this approach is that a quantifiable relationship (i.e., by linear regression) exists between the treatment and control watershed for paired water quality data (e.g., TP concentration), and that this relationship holds until a major change is made in one of the watersheds. At that time, a new relationship may exist.

In this study, the effects of treatment were evaluated on the basis of statistically significant changes in the relationships among the study watersheds for monitored water quality variables (USEPA 1993, USEPA 1997). The principal hypothesis tested was that treatment in WS 1 and WS 2 would yield significant improvements in water quality compared to the control. Two treatment watersheds (WS 1 and WS 2) were monitored to assess different levels of treatment

2.3 Stream Monitoring

Streamflow and water quality were monitored at the outlet of each watershed for three years preceding land treatment in order to achieve satisfactory calibration between the two watersheds. Precipitation was recorded continuously by gages located in each watershed. Secure, heated monitoring stations were located at each watershed outlet, where streamflow was recorded continuously by a bubbler-type flow meter. At each station, a refrigerated automatic sampler was paced by signals from the flow meter; for each volume increment of stream flow, a fixed volume of sample was collected and stored in carboys within the sampler. The resulting flow-proportional composite water samples were retrieved weekly for analysis for total phosphorus (TP), total Kjeldahl nitrogen (TKN), and total suspended solids (TSS) . Grab samples were collected twice weekly for indicator bacteria (*E. coli*, fecal coliform, and fecal streptococci), dissolved oxygen, conductance, and water temperature. Samples were analyzed at the Vermont Department of Environmental Conservation (DEC) Laboratory by USEPA or APHA methods (USEPA 1983, USEPA 1985, American Public Health Association 1995).

Vermont DEC biologists conducted annual assessments of macroinvertebrate and fish communities in each of the monitored streams and in ecoregion control streams representing biological reference conditions. Methods used to collect aquatic macroinvertebrates and fish are documented in the Vermont Department of

Environmental Conservation Field Methods Manual (VT DEC, 1989); these methods follow a protocol comparable to the riffle/run sampling portion of Rapid Bioassessment Protocol III (RBPIII) as described in Plafkin et al. (1989). Standard metrics were used to assess community status, including the EPT Index and BioIndex for macroinvertebrates and the IBI for fish.

2.4 Land use and agricultural management monitoring

Changes in land use/land cover (i.e., forest, agriculture, residential, open, idle) and agricultural activity (i.e., corn, hay, pasture) were monitored and updated annually through farmer record-keeping, interviews with each owner and/or operator of agricultural land, and review of US Department of Agriculture (USDA) aerial photography. Land ownership and land use/land cover were compiled and mapped in a geographic information system database.

2.5 Land treatment plan

During the calibration period, critical areas needing treatment in WS 1 and WS 2 were identified through baseline farm inventories, inspection of stream corridors, and interpretation of aerial video imagery. Treatment plans were developed for individual farms in cooperation with landowners, who participated voluntarily. Landowners signed agreements to ensure operation and preservation of the treatments for the duration of the project and providing for joint maintenance of treatments by the project and the landowner.

Land treatment consisted of a suite of practices designed to treat and protect streams and riparian zones, including exclusion of livestock from selected areas of streams, creation of protected riparian zones, improvement or elimination of heavily used livestock stream crossings, and revegetation of degraded streambanks. The treatment required fencing, watering systems, reducing the number and size of livestock crossing areas, bridging or armoring necessary crossing areas, and streambank erosion control through bioengineering techniques. Personnel from the USDA Natural Resources Conservation Service (NRCS) and from the U.S. Fish and Wildlife Service provided technical assistance in treatment design. The project budget, other conservation programs, the landowners themselves, and local volunteer groups provided funds and labor for treatment implementation.

3. RESULTS

3.1 Calibration period water quality

Selected water quality data over the calibration period (May, 1994 - September, 1997) are summarized in Table 1. On the average, the three watersheds received similar precipitation, although there were variations among individual storm events. Stream discharge followed typical regional patterns of low flow in summer and winter and peak flows during spring snowmelt. Except for small year-to-year differences due to variations in precipitation and stream discharge, water quality was fairly consistent through the calibration period in each of the monitored streams. Concentrations of TP,

TKN, and TSS were typical of streams draining agricultural watersheds in Vermont (Meals 1990, VT Rural Clean Water Program Coordinating Committee 1991), and the three watersheds exhibited generally similar concentrations over the calibration period. Indicator bacteria counts were high and varied seasonally by five orders of magnitude. *E. coli* counts routinely exceeded the Vermont water quality criterion for recreation of 77 organisms/100 ml. Nutrient and sediment export tended to be low during base flow, punctuated by major flux during periods of high discharge, particularly during spring runoff. Nutrient and sediment export rates tended to exceed those reported from other U.S. agricultural watersheds (e.g., Frink 1991), but were comparable to values reported in other small Vermont agricultural watersheds (Meals 1990, VT Rural Clean Water Program Coordinating Committee 1991).

3.2 Treatment

Land treatment was fully implemented in WS 1 and WS 2 from May - November, 1997. Landowner participation in land treatment was entirely voluntary and all treatments were 100% cost-shared by the project and cooperators. Eight landowners signed treatment agreements; additional work was done on four other properties in the watersheds. In WS 1, treatment included fencing of ~730 m of Samsonville Brook, creating a 2 – 5 m wide protected riparian zone and eliminating three livestock crossing areas. Treatment along Godin Brook in WS 2 included protection of both sides of 2300 m of stream and wetland with livestock exclosure fencing, a livestock bridge, a 300 m stabilized livestock travel lane; and three culvert and two armored stream crossings for livestock. The width of the fenced riparian zones created varied from ~2 - 8 m, depending on topography and the amount of land the landowner was willing to set aside. Livestock watering systems were installed on two farms to replace stream access as a drinking water supply.

Within all protected riparian zones, streambank stabilization was done through bioengineering techniques, including planting willows as cuttings, live stakes, and fascines and installation of tree revetments and brushrolls to protect banks from the erosive force of water. These devices also help capture and settle sediment and promote revegetation of streambanks. No deliberate plantings were made in riparian zones; natural regrowth of native grasses, shrubs, and other vegetation following removal of grazing pressure was ample. Treatment installation was done by project staff, landowners, volunteers from a local river basin association, and a Vermont Youth Conservation Corps work crew. Total treatment installation costs (materials plus labor) were: $3,789 in WS 1 and $34,565 in WS 2.

The length of stream treated represented less than 10% of the total stream length in each of the treated watersheds. In WS 1, just 7% of the 10,400 m of total stream length was treated; about 9% of the 24,800 m of stream length in WS 2 received treatment. The stream length protected in WS 1 represented 49% of the ~1,400 m of stream adjacent to grazed land; this treatment affected >90% of the livestock population grazing riparian pastures. In WS 2, ~28% of the 8,150 m of pasture-adjacent stream were protected, accounting for ~20% of livestock with access to streams. More livestock stream crossings were eliminated or protected in WS 2 than in WS 1.

Table 1. Median values and (interquartile range) for water quality and stream discharge, and range of annual nutrient and sediment export over Calibration Period (May 1994 - May 1997) and Treatment Period. (October 1997 – November 2000).

VARIABLE	CALIBRATION PERIOD			TREATMENT PERIOD		
	Treatment		Control	Treatment		Control
	WS 1	WS 2	WS 3	WS 1	WS 2	WS 3
Weekly total precipitation (cm)	1.14 (0.28-2.39)	1.30 (0.46-2.62)	1.27 (0.46-2.54)	1.33 (0.47-2.30)	1.59 (0.55-2.49)	1.36 (0.50-2.17)
Weekly mean Discharge (m^3/s)	0.06 (0.02-0.21)	0.16 (0.08-0.38)	0.19 (0.10-0.36)	0.12 (0.04-0.29)	0.16 (0.08-0.39)	0.21 (0.11-0.43)
Weekly mean [TP] (mg/L)	0.088 (0.052-0.166)	0.102 (0.039-0.199)	0.084 (0.052-0.174)	0.072 (0.42-0.125)	0.106 (0.061-0.172)	0.074 (0.040-0.125)
Weekly mean [TKN] (mg/L)	0.50 (0.35-0.77)	0.50 (0.35-0.93)	0.50 (0.3-0.65)	0.53 (0.38-0.64)	0.62 (0.43-0.85)	0.44 (0.31-0.58)
Weekly mean [TSS] (mg/L)	27.8 (12.2-65.1)	18.6 (5.1-43.4)	16.6 (6.7-35.5)	15.7 (7.0-34.4)	19.1 (7.9-31.7)	10.2 (4.6-24.2)
Weekly mean *E. coli* (#/100 ml)	153 (24-472)	700 (57-4600)	560 (49-5250)	70 (10-235)	251 (21-1730)	335 (32-3500)
Weekly mean conductance (μ S/cm)	88 (70-113)	105 (87-135)	107 (87-125)	84 (64-105)	116 (91-142)	114 (92-130)
Weekly mean water temperature (°C)	7.7 (0.7-16.5)	7.9 (0.7-16.5)	8.0 (0.9-16.1)	6.9 (1.0-15.4)	7.8 (1.1-16.8)	7.8 (1.6-16.0)
TP export (t/yr)	0.99 - 1.99	0.93 - 3.51	0.93 - 2.88	0.57 - 1.06	1.62 - 3.02	1.19 - 7.15
TP export (kg/ha/yr)	1.4 – 2.9	0.6 – 2.5	1.0 – 3.0	0.8 – 1.5	1.1 – 2.5	1.2 – 7.5
TKN export (t/yr)	6.7[1]	11.9[1]	8.3[1]	3.0 – 3.8	6.4 – 9.2	5.8 – 9.1
TKN export (kg/ha/yr)	9.7[1]	8.4[1]	8.7[1]	4.4 – 5.7	4.5 – 6.5	6.1 – 9.5
TSS export (t/yr)	289 - 963	198 - 1324	195 - 966	191 - 481	619 - 969	370 – 1235
TSS export (kg/ha/yr)	419 - 1396	139 - 931	204 - 1013	277 - 698	435 - 681	388 - 1295

[1] Due to laboratory error, only one full year of Calibration Period TKN data exist.

3.3 Post-treatment water quality

Selected water quality data over the Treatment Period (October 1997 – November 2000) are summarized in Table 1. Despite the occurrence of several unusually extreme discharge events during the first post-treatment year and a regional drought in the second year, post-treatment median values and ranges for most water quality variables were comparable to those observed during the calibration period. An exception were *E. coli* bacteria levels; both median and the range of *E. coli* counts observed appeared to decline markedly in all three watersheds for the post-treatment period. In 1999, at end of the second post-treatment year, erosion and runoff from a major land clearing project and poor riparian management associated with expansion of a large farm in WS 2 seriously degraded water quality in Godin Brook for the remainder of the study.

Before proceeding with the paired watershed analysis, linear regressions between WS 3 (independent variable) and WS 1 (dependent variable) and similarly between WS 3 and WS 2 were performed on log-transformed data to determine if conditions for acceptable calibration between WS 1 and WS 3 and between WS 2 and WS 3 had been met. These procedures are fully documented elsewhere (Meals 1998). Comparison of residual variance for the regressions with the smallest worthwhile difference (USEPA 1993) showed that changes as small as $\pm$ 10-20% in pollutant concentration and export could be detected with statistical significance. Thus, the calibration period was adequate to detect reasonable changes following treatment.

Changes in water quality in response to treatment were tested using Analysis of Covariance (ANCOVA). ANCOVA compares two groups of data (e.g., calibration and treatment) while adjusting for changes in explanatory variables using a combination of analysis of variance and regression. In the paired watershed design, data from the control watershed are used to account for the effects of year-to-year climate variations on water quality variables of interest. For a given variable of interest in a treatment watershed, the explanatory variable (covariate) is the same variable observed concurrently in the control watershed. Calibration and Treatment Period regressions between treatment and control watersheds were tested for significant differences in slopes, intercepts, or both by analysis of variance of regression coefficients; differences between Calibration and Treatment water quality were deemed to be significant if slope and/or intercept of the regressions differed at $P \leq 0.10$ (Grabow et al. 1999). All analyses were performed in the statistical software *JMP 4.0* (SAS Institute 2000) according to the approach outlined by Grabow et al. (1999). Results of this analysis are presented in detail elsewhere (Meals 2001b); following is a summary of the main results.

Three full years of post-treatment data were evaluated for evidence of response to treatment. Direct comparisons of calibration and post-treatment data by Student's t-Test indicated that TP and TSS concentrations and *E. coli* counts declined significantly in WS 1 following treatment, but few significant differences between calibration and treatment periods were observed in WS 2 or WS 3 . However, the utility of the paired-watershed design is in the ability to detect change by accounting for background variability by including data from the control watershed in Analysis of Covariance (ANCOVA).

Significant response to treatment was detected based on changes in both treated watersheds relative to the control watershed for most physical/chemical water quality variables. Examples of paired regression plots are shown in Figures 2 and 3.

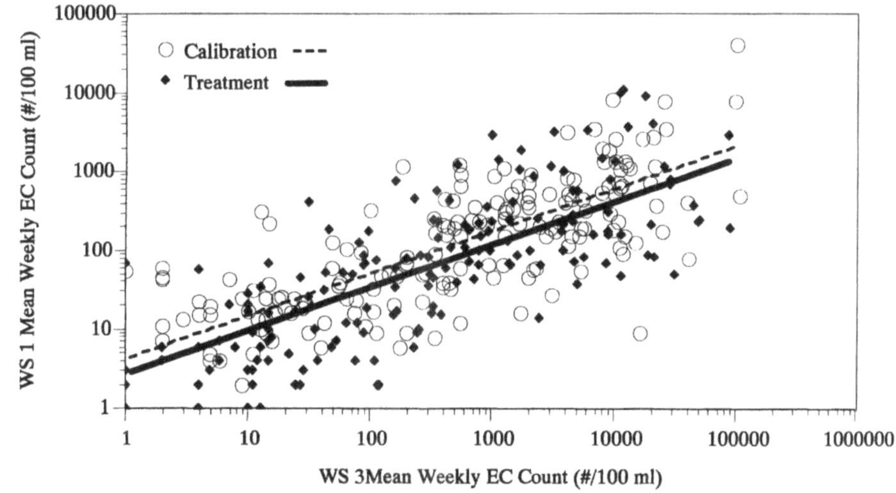

Figure 2. Paired regression plot for weekly mean *E. coli* bacteria counts, WS 1 vs. WS 3. Points and line labeled "Calibration" represent data pairs from pre-treatment years, May 1994 - May 1997; "Treatment" points and line represent data pairs from post-treatment years, October 1997 – September 2000. Intercepts of regression lines differ significantly, P=0.004.

The parallel lines in Figure 2 indicate a significant decrease in *E. coli* counts over the full range of conditions in WS 1 following treatment, compared to counts in the control watershed (WS 3) for the same period. The crossed lines in Figure 3 suggest post-treatment reductions in TP export from WS 1 compared to export from WS 3 to the right of the intersection of the two regression lines, i.e., in the upper range of export conditions, above ~30 kg TP/wk. Similar crossed calibration and treatment paired regression lines for other variables (not shown) indicate that in WS 1, TP, TKN, and TSS concentrations in the mid- to high-range characteristic of active runoff decreased significantly compared to the control in the Treatment period. Specific conductance in WS 1 was slightly, but significantly, lower in the Treatment period than during Calibration. Post-treatment water temperatures were slightly higher in cold weather and slightly cooler in hot weather compared to those in the Calibration period. Low-flow stream discharge in WS 1increased, but high flows decreased in the Treatment period, possibly responding to a similar pattern in precipitation compared to the control watershed. Export of TP, TKN, and TSS from WS 1 all decreased significantly relative to the control over the mid- to high-range following treatment.

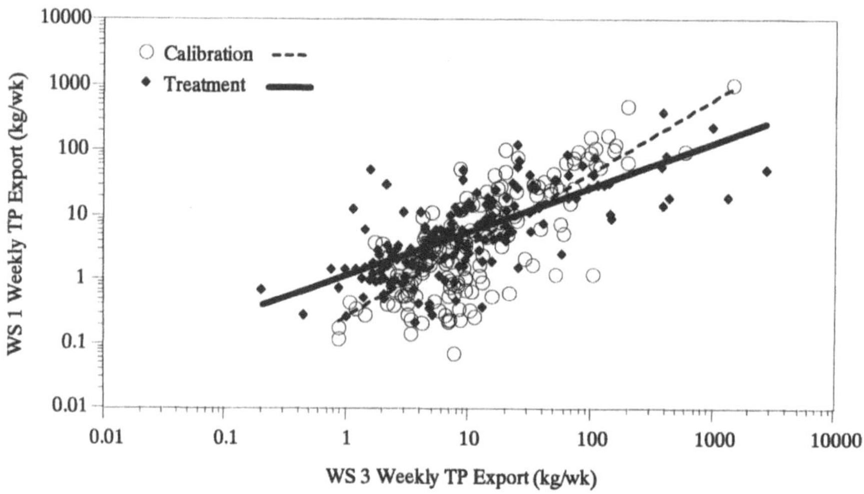

Figure 3. Paired regression plot for weekly TP export, WS 1 vs. WS 3. Points and line labeled "Calibration" represent data pairs from pre-treatment years, May 1994 - May 1997; "Treatment" points and line represent data pairs from post-treatment years, October 1997 – September 2000. Slopes and intercepts of regression lines differ significantly, P<0.001.

For WS 2/WS 3, calibration and post-treatment regressions differed significantly for all measured variables except dissolved oxygen and TP export. TP, TKN, and TSS concentrations in WS 2 decreased significantly in the Treatment period relative to the control over the mid- to high-range of concentrations. Significant drops in post-treatment *E. coli* counts were observed in WS 2 over the entire range of bacteria counts. No change in TP export from WS 2 and only very small decreases in the extreme high range of TKN and TSS export were documented. This was a reversal of larger reductions in export noted in the first two post-treatment years (Meals and Hopkins 2002).

4. DISCUSSION

The average treatment response for each water quality variable, estimated by comparing values predicted by regression relationships for each treatment watershed for the Calibration and Treatment periods at the average value for the Control watershed, are shown in Table 2.

Results suggest that the effect of treatment on water quality in WS 1 was substantial, but that in some cases, average water quality did not improve in WS 2. The increased export indicated in average response is largely an artifact of the crossed lines shown, for example, in Fig. 2, where post-treatment reductions occurred only in the upper range of export.

Table 2. Estimates of average response of monitored variables to treatment. Percentage changes were derived by comparing predicted values from Calibration and Treatment regressions at the mean value observed in the control watershed during the Treatment period.

Variable	WS 1 % change in mean	WS 2 % change in mean
[TP]	-15%	+18%
[TKN]	-12%	0
[TSS]	-34%	+40%
E. coli	-29%	-44%
Fecal coliform	-38%	-46%
Fecal strep.	-40%	-20% (n.s.)[1]
Conductance	-11%	+1%
Water temperature	-6%	+3%
TP export	+47%	+12%
TKN export	+125%	+16%
TSS export	+40%	+32%

[1] n.s. = nonsignificant: Calibration and Treatment regressions did not differ significantly, $P \leq 0.10$

It should be noted that the water quality changes in the treated watersheds took place while essentially no change in water quality occurred in the control watershed during the Treatment Period (as indicated by results of t-Test of Calibration and Treatment period means). The only significant difference in WS 3 water quality between Calibration and Treatment periods was a decrease in TSS concentration. Thus, it appears that the changes in the treatment watersheds were probably not due to deterioration in control watershed water quality alone.

Because the majority of annual nps export tends to occur during a few high-magnitude events, rather than at the average condition, a more appropriate assessment of response can be made with respect to export over the full range of conditions. All Treatment period values from the control watershed were entered into the Calibration regressions to yield predictions of WS 1 or WS 2 export under pre-treatment management conditions but post-treatment hydrologic conditions. The summed differences between observed and predicted export indicate the net change in export (Table 3). For WS 1, a reduction of 49% in TP load or ~827 kg TP/yr is estimated by this model, confirming the greater treatment effect at high export levels that was illustrated earlier in Fig. 3. Similarly, TKN and TSS loads show substantial reductions following treatment. For WS 2, on the other hand, major increases in export are indicated. These increases are in marked contrast to reductions observed in WS 2 after the first two years of treatment (Meals 2000). The deterioration of water quality in Godin Brook was undoubtedly due to repeated episodes of concentrated field runoff from an expanding farm in the middle of WS 2 that began in 1999 and continued through the end of the project. Destruction of a natural forested riparian zone, direct concentrated runoff of excessive manure applications, severe soil erosion, and deposit of soil masses directly into the stream combined to overwhelm the effects of riparian restoration elsewhere in the watershed.

Table 3. Estimates of changes in watershed pollutant export over Treatment period. Estimates are derived by summing differences between observed and predicted weekly Treatment period loads.

	WS 1		WS 2	
	% Change	Mass change	% Change	Mass change
TP export	-49%	-827 kg/yr	+80%	+1,181 kg/yr
TKN export	-38%	-2,171 kg/yr	+44%	+ 23,513 kg/yr
TSS export	- 28%	-114,893 kg/yr	+104%	+392,202 kg/yr

The positive water quality effects observed in this study are generally consistent with those reported in the literature. Williamson et al. (1996) reported 85% reduction in sediment load, 27% reduction in particulate P load, and 40% reduction in particulate N load from grazed watersheds in New Zealand following implementation of improved riparian management. Daniels and Gilliam (1996) documented 50 - 80% reductions in sediment and P loads from agricultural cropland in North Carolina where grass and riparian filters were implemented. Robinson et al. (1996) reported 70% and 85% sediment reduction in runoff through 3 m and 9.1 m grass vegetated filter strips, respectively. Even very narrow grass buffers less than 1 m wide have been shown to be capable of reducing concentrations of TP and ammonia-N by 40% and 60%, respectively, in cropland runoff in Texas (Eghball et al. 2000). Tate et al. (2000) reported that non-grazed, vegetated buffer strips reduced TSS concentration and loads and TP loads from irrigated pastures in California. Clausen et al. (2000) showed that restoration of a riparian buffer adjacent to Connecticut corn land decreased overland flow concentrations of TP by 73%, TKN by 70%, and TSS by 92%.

Research on the use of vegetated buffers for bacteria removal has produced conflicting results. High initial removals (≥90%) have been reported for runoff containing high bacterial densities (>10^5 organisms/100 ml) have been reported, but bacteria numbers in runoff leaving buffer areas appear to level off at about 10^4 – 10^5 organisms/100 ml (Crane et al. 1983, Moore et al. 1988). Increases in fecal coliform densities in overland flow through grass buffers has been observed during summer months and attributed to regrowth of bacteria within the buffers (Moore et al. 1988). Coyne et al. (1998) reported fecal coliform numbers in cropland runoff in Kentucky were generally reduced by ~one order of magnitude after passage through grassed strips of different widths but by the end of a rainfall event, bacteria counts leaving the filter strips exceeded counts entering the strips. The filter strip appeared to become a reservoir for sediment-bound organisms trapped from surface runoff. It is worth noting that even the substantial reductions of *E. coli* counts observed here did not result in complete compliance with Vermont water quality standards in the monitored streams.

The specific mechanisms responsible for the observed pollutant removal are uncertain and beyond the scope of this study. In general, three processes may have occurred: (1) reduction of direct deposit of nutrients, bacteria, and organic matter from livestock in or near the stream; (2) filtration of overland flow by the restored riparian zone (i.e., as a vegetated filter strip); and (3) reduction of streambank erosion. It was impossible to sort out the importance of these within the study design. It should be cautioned that direct comparison of this project's results on a watershed scale with other studies on the plot or field scale are not appropriate. The performance of the riparian zones as "vegetated filter strips," for example, was probably less effective than that observed in plot or field studies because no attempt was made to promote sheet flow or prevent concentrated overland flow through the protected riparian zones. Furthermore, as noted earlier, treatment was implemented on a small proportion of each of the treated watersheds.

Complete results of macroinvertebrate and fish monitoring are presented and discussed elsewhere (Meals 2001b); principal findings are summarized briefly here. The macroinvertebrate community in WS 1 responded significantly to treatment, based on comparisons of several biometrics against ecoregion controls and between Calibration and Treatment periods using the nonparametric rank sum Dunn's test. Significant increases in EPT and BioIndex values indicated improvements in biological integrity; in the second and third years after treatment, BioIndex values for WS 1 met Vermont Class B water quality biocriteria. Macroinvertebrate biometrics indicated significant shifts in the percent composition of several major orders of aquatic insects and the percent composition of the functional groups within the treatment streams, suggesting that a shift in the food base was most responsible for improved biological integrity. Improvements noted in the macroinvertebrate community in WS 2 (Godin Brook) after two years of treatment were reversed in the final year due to catastrophic sedimentation events from upstream disturbances.

The fish assemblages showed no significant changes in structure compared to the ecoregion control or between the Calibration and Treatment Periods for either of the two treatment watersheds, based on Mann-Whitney rank sum tests of density, IBI, and species richness. Insufficient treatment or a lag in response of the fish community may explain the lack of response. In addition, some fish species may not have been available to migrate into the streams from the Missisquoi River which continues to be impaired by high sediment and nutrient loads. Although no observable changes in fish assemblage structure were recorded, some physical habitat improvements were noted in both Samsonville and Godin Brooks after treatment.

5. CONCLUSIONS

The project was successful in implementing riparian restoration treatments that addressed a significant portion of grazing-related water quality problems in two small Vermont agricultural watersheds. The simple livestock exclusion and streambank bioengineering measures were inexpensive to install and maintain; the ~$40,000 expended in treating the two watersheds might easily be consumed in installing structural practices on a single farm. Data collected through intensive monitoring documented significant reductions in phosphorus, nitrogen, suspended solids, and indicator bacteria in response to treatment. Riparian zone treatments had a positive effect on stream biota, although community composition continued to indicate impacts of moderate organic enrichment. Improvements in both water quality and stream biota in Godin Brook (WS 2) after two years of treatment were reversed in the final year due to catastrophic pollutant loading from farm mismanagement upstream, demonstrating the potential for a single acute problem to offset management improvements elsewhere in the watershed. Overall, project results indicate that riparian zone protection/restoration is a cost-effective tool for reducing nonpoint source pollutant concentrations and loads from livestock grazing lands in the Lake Champlain Basin.

6. ACKNOWLEDGMENTS

This project was funded through the U.S. EPA Section 319 National Monitoring Program, with substantial cost share from the Vermont Department of Environmental Conservation and the University of Vermont. Rick Hopkins from the VT DEC provided administrative and field support; biomonitoring was conducted by biologists from the VT DEC. The financial and technical support of the U.S. Fish and Wildlife Service and the USDA-Natural Resources Conservation Service is gratefully acknowledged. Volunteers from the Missisquoi River Basin Association and Ben & Jerry's Homemade, Inc. and members of the Vermont Youth Conservation Corps helped with streambank restoration. Thanks are also due to participating farmers, landowners and watershed residents.

7. REFERENCES

American Public Health Association. 1995. *Standard Methods for the Examination of Water and Wastewater* 19th edn, American Public Health Association/American Water Works Association/Water Environment Federation, Washington, DC.

Belsky A.J., A. Matzke and S. Uselman. 1999. Survey of livestock influences on stream and riparian ecosystems in the western United States. *J. Soil and Water Cons.* 54:419-431.

Clausen, J.C., K. Guillard, C.M. Sigmund, and K. M. Dors. 2000. Water quality changes from riparian buffer restoration in Connecticut. *J. Environ. Qual.* 29:1751-1761.

Correll D.A., T.E. Jordan, and D.E. Weller. 1995. Livestock and pasture land effects on the water quality of Chesapeake Bay watershed streams. in: *Animal Waste and the Land –Water Interface*, K. Steele (ed.), Lewis Publishers, New York.

Coyne, M.S., R.A. Gilfillen, A. Villalba, Z. Zhang, R. Rhodes, L. Dunn, and R.L. Blevins. 1998. Fecal bacteria trapping by grass filter strips during simulated rain. *J. Soil and Water Cons.* 53(2):140-145.

Crane, S.R., J.A. Moore, M.E. Grismer, and J.R. Miner. 1983. Bacterial pollution from agricultural sources: a review. *Trans. ASAE* 26:858-866.

Daniels, R.B. and J.W. Gilliam. 1996. Sediment and chemical load reduction by grass and riparian filters. *Soil Sci. Soc. Am. J.* 60:246-251.

Duda A.M. and D.S. Finan. 1983. Influence of livestock on nonpoint source nutrient levels of streams. *Trans. ASAE* 26:1710-1716.

Eghball B., J.E. Gilley, L.A. Kramer and, T.B. Moorman. 2000. Narrow grass hedge effects on Phosphorus and Nitrogen in runoff following manure and fertilizer application. *J. Soil and Water Cons.* 55:172-176.

Frink C.R. Estimating nutrient exports to estuaries. 1991. *J. Environ. Quality* 20:717-724.

Hegman, W., D. Wang, and C. Borer. 1999. *Estimation of Lake Champlain Basinwide Nonpoint Source Phosphorus Export.* Technical Report No. 3, Lake Champlain Basin Program, Grand Isle, VT.

Grabow, G.L., J. Spooner, L.A. Lombardo, and D.E. Line. 1999. Detecting Water Quality Changes Before and After BMP Implementation: Use of SAS for statistical analysis. *NWQEP Notes*, No. 93, North Carolina State Water Quality Group Newsletter, NCSU, Raleigh, NC.

Lake Champlain Basin Program. 1996. *Opportunities for Action.* Lake Champlain Basin Program, Grand Isle, VT.

Meals, D.W. *LaPlatte River Watershed Water Quality Monitoring and Analysis Program Comprehensive Final Report 1979-1989.* 1990. Program Report No. 12, Vermont Water Resources Research Center, University of Vermont, Burlington, VT.

Meals, D.W. 1998. *Lake Champlain Basin Agricultural Watersheds Section 319 National Monitoring Program Project, Comprehensive Calibration Period Report.* Vermont Dept. Environ. Conserv., Water Quality Div., Waterbury, VT.

Meals, D.W. 2000. *Lake Champlain Basin Agricultural Watersheds Section 319 National Monitoring Program Project, Year 6 Annual Report.* VT Dept. Environ. Conserv., Water Quality Div., Waterbury, VT.

Meals, D.W. 2001a. Water quality response to riparian restoration in an agricultural watershed in Vermont, USA. *Water Sci. & Technol.* 43:175-182.

Meals, D.W. 2001b. *Lake Champlain Basin Agricultural Watersheds Section 319 National Monitoring Program Project,Final Project Report May, 1994 – November, 2000.* Vermont Dept. Environmental Conservation, Waterbury, VT.

Meals, D.W. and L.F. Budd. 1998. Lake Champlain Basin nonpoint source phosphorus assessment. *J.American Water Resources Assoc.* 34(2):251-265.

Meals, D.W. and R.B. Hopkins. 2002. Phosphorus reductions following riparian restoration in two agricultural watersheds in Vermont, USA. *Water Science & Technology* 45(9):51-60.

Moore, J.A., J. Smyth, S. Baker, and J.R. Miller. 1988. *Evaluating Coliform concentrations in runoff from various animal waste management systems.* Special Report 817, Agricultural Experiment Stations, Oregon State University, Corvallis, OR.

Plafkin, J.L., M.T. Barbour, K.D. Porter, S.K. Gross, and R.M. Hughes. 1989. *Rapid bioassessment protocols for use in streams and rivers: benthic macroinvertebrates and fish.* EPA/444/4-89-001. U.S. Environ. Prot. Agency,Wash., DC.

Robinson, C.A., M. Ghaffarzadeh, and R.M. Cruse. 1996. Vegetative filter strip effects on sediment concentrations in cropland runoff. *J. Soil and Water Cons.* 50(3):227-230.

SAS Institute. 2000. *JMP* Version 4.0.0. SAS Institute Inc., SAS Campus Drive, Cary, NC.

Tate, K.W., G.A. Nader, D.J. Lewis, E.R. Atwill, and J.M. Connor. 2000. Evaluation of buffers to improve thequality of runoff from irrigated pastures. *J. Soil and Water Cons.* 55(4):473-479.

U.S. Environmental Protection Agency 1983. *Methods for Chemical Analysis of Water and Waste,* EPA-600/4-79-020, Washington, DC.

U.S. Environmental Protection Agency. 1985. *Test Methods for Escherichia Coli and Enterococci in Water by the Membrane Filter Procedure,* PA 600/4-85/076, Environmental Monitoring and Support Laboratory, Cincinnati, Ohio.

U.S. Environmental Protection Agency. 1993. *Paired Watershed Study Design,* EPA 841-F-93-009, Office of Water, Washington, DC.

U.S. Environmental Protection Agency. 1997. *Linear Regression for Nonpoint Source Pollution Analysis,* EPA-841-B-97-007. Office of Water, Washington, DC.

Vermont Agency of Natural Resources. 1996. *State of Vermont 1996 Water Quality Assessment, 305(b) Report,* Vermont Dept. Environ. Conserv., Water Quality Div., Waterbury, VT.

Vermont Department of Environmental Conservation. 1989. *Field Methods Manual.* VT DEC Laboratory, Waterbury, VT.

Vermont Department of Environmental Conservation and New York State Department of Environmental Conservation. 1997. *A Phosphorus Budget, Model, and Load Reduction Strategy for Lake Champlain,* Dept. Environ. Conserv., Water Quality Div., Waterbury, VT.

Vermont Rural Clean Water Program Coordinating Committee. 1991. *St. Albans Bay Rural Clean Water Program Final Report, 1980-1990.* Vermont Water Resources Research Center, University of Vermont, Burlington, VT.

Williamson R.B., C.M. Smith, and A.B Cooper. 1996. Watershed riparian management and its benefits to a eutrophic lake. *J. Water Resour. Planning and Manage.* 122(1):24-32.

TEMPORAL AND SPATIAL VARIABILITY IN NON-POINT SOURCE PHOSPHORUS IN RELATION TO AGRICULTURAL PRODUCTION AND TERRESTRIAL INDICATORS: THE BEAVER BROOK CASE STUDY, PIKE RIVER BASIN, QUÉBEC

Aubert R. Michaud*, Richard Lauzier and Marc R. Laverdière

ABTRACT

The phosphorus (P) transfer system within the Beaver Brook watershed, a small agricultural watershed (11 km^2) tributary to the Pike River, was described through outlet monitoring and spatially discrete sampling of upstream subwatersheds. Spatial variability of P exports was related to the type of agricultural production system and to the attributes of the landscape. Over the three years of study, mean annual exports of dissolved reactive soluble (DRP), bioavailable (BioP) and total phosphorus (TP) from the watershed were 0.57, 0.93 and 1.54 kg-P ha^{-1}, respectively. These export loads were temporally episodic in nature: 75% of total P exports occurred within 6% of the monitoring period. These episodes occurred during peak streamflow events and under late-winter/early-spring conditions when overland runoff was generated through snowmelt or precipitation on thawing or saturated soils. Under base flow conditions, the flux-stream discharge relationship indicated that TP point sources contributed about 5% of annual exports. During elevated streamflow conditions, high bioavailability and particulate ratio of P exports indicated in-stream P storage and transformation mechanisms, as well as a significant contribution of subsurface P transport to the main stream channel. The majority of spatial P concentration patterns could be explained on the basis of the runoff curve number, and the localisation of the subwatershed within the landscape. Under comparable levels of hydrological activity, significant differences in subwatershed water quality were explained by typical agronomic (source) indicators,

* Aubert R. Michaud, Institut de recherche et développement en agroenvironnement, Québec, (Québec) Canada, G1P 3W8

namely P budget. The implications for farm and land management are that for a tangible reduction in P exports to occur, the importance of considering both runoff and nutrient management must be stressed.

1. INTRODUCTION

Following investment in control of water contamination from industrial and urban point sources, the current focus on water quality problems in Lake Champlain's Missisquoi Bay has largely shifted to pollution originating from non-point sources. Over 79% of the yearly P load received by the Bay (143 Mg-P yr^{-1}) has been attributed to agricultural land use, which accounts for 26% of Missisquoi watersheds by area (Hegman et al., 1999). This non-point source water quality issue is not exclusive to Missisquoi Bay or Lake Champlain. In fact, the overwhelming majority of water quality problems in southern Quebec and the Northeastern United States result from non-point pollution (Gangbazo and Babin, 2000; Hamilton and Miller, 2002).

The availability of agroenvironmental indicators based on comprehensive sets of relevant factors can enhance strategic planning and the efficient implementation of non-point P export-prevention measures at the farm-, subwatershed- or territory-scale. Based on such a premise, the concept of the P-index, ranking non-point P export risks according to a range of source and transport factors (Lemunyon and Gilbert, 1983: Gburek et al., 2000a), was developed by researchers and advisory personnel.

Source factors represent P potentially exportable through overland and subsurface flows, ranking a site's vulnerability according to soil test P as well as the origin, placement and timing of P inputs. Several studies have linked P in surface runoff to soil test P. However, a threshold soil test P has been judged too limited to be the sole criterion to guide P management and P applications, unless used in conjunction with an estimate of a site's potential for runoff and soil erosion (Wolf et al., 2000).

Transport factors determine whether potential P contributing areas will translate into P loss. Surface runoff is generally considered the primary hydrological pathway for P flux. Given the distribution of precipitation intensities inherent to weather patterns over subwatersheds of Quebec and the Northeastern United States, saturation-excess runoff is the most common type of runoff to occur. Consequently, the occurrence of overland flow is generally controlled by cumulative rainfall and snowmelt events rather than by rainfall intensity (Lapp, 1996). Under these climatic conditions, saturation-excess runoff is largely a function of landscape, rises in watertable and stream level, impermeable subsoils, slope breaks and convergent subsurface flows. Given their landscape-wide distribution, these processes reveal hydrologically-active variable-source areas (VSA) (Beven and Wood, 1983), the extent of which varies both seasonally and within precipitation events.

Subsurface throughflow by way of preferential (macropore) and artificial drain flow has also been identified as a significant pathway for P fluxes. Subsurface P exports are of particular concern where manures are applied to drained cracking-clay soils (Preedy et al., 2001; Beauchemin et al., 1998). Also worthy of concern, when rainfall occurs soon after slurry application, is rapid incidental P transfer via overland flow (Sharpley and Withers, 1994; Gangbazo, 1991; Grando, 1996). Given the wide distribution of marine clay soils and artificially drained lands in the study area and throughout the Saint-Lawrence valley region, subsurface preferential and artificial flow conditions are favoured. Typically most subsurface flow occurs in early spring, when partitioning of surface and subsurface flow is influenced by soil frost depth, snow cover and timing of low-intensity precipitation. Another important hydrologic pathway, typical of Quebec's long and narrow parcels, is lateral flow along plow pans oriented along the topographical gradient (Brunelle and Savoie, 2000).

Though effective control of P flux relies on the management of both source and transport factors, little work has described how these factors interact at a watershed scale. The current watershed study sought to (i) document the mechanisms of P transfer at work in the watershed, and (ii) link P sources and transport mechanisms to the spatial variability of P exports in order to support farm and watershed-scale planning and management of non-point P.

2. MATERIALS AND METHODS

This study of the Beaver Brook watershed was conducted in support of a cooperative agricultural rehabilitation initiative (Lauzier and Gasser, 2002). A diagnostic study of aquatic ecosystems had established the basin's vulnerability to P exports (Caumartin and Vincent, 1994). A simulated rainfall plot study quantifying P emissions by overland flow on typical soil series of the watershed (Michaud and Laverdière, 2003) provided additional support to the conclusions drawn from the watershed study.

A continuous three-year monitoring campaign (Oct. 1997-Sept. 2000) measured stream flow and water quality at the watershed outlet. Over this same period additional flow measurements and grab samples were obtained on a frequent basis at selected locations throughout the watershed. A geographic information system (GIS) was used to integrate these data with cropping systems information, including: P mass balance analysis, erosion potential, and soil test P. ANOVA and ANCOVA procedures served to estimate and relate P exports to cropping system characteristics, soil test P, and physical attributes of the watershed landscape.

2.1 Study Area Description

The Beaver Brook drains into the Pike River, a major tributary of Lake Champlain's Missisquoi Bay, situated in the southern Quebec near the Canada/US border (Figure 1). The 6.4 km main channel of the Beaver Brook drains the 11 km^2 Beaver Brook watershed, of which 88% by area is devoted to agricultural pursuits. The tributary drainage channels entering the Beaver Brook have all been engineered to accommodate an intensive system of artificial sub-surface drainage that covers roughly half the cropland area in the watershed. The watershed is subject to the warm spring and summer growing periods and cold winters typical of southeastern Quebec. The long-term mean annual precipitation is about 1057 mm (Environment Canada, 2003).

The western-most portion of the watershed is flat while the eastern portion consists of rolling hills. The elevation rises from about 30 m above MSL at the watershed outlet to

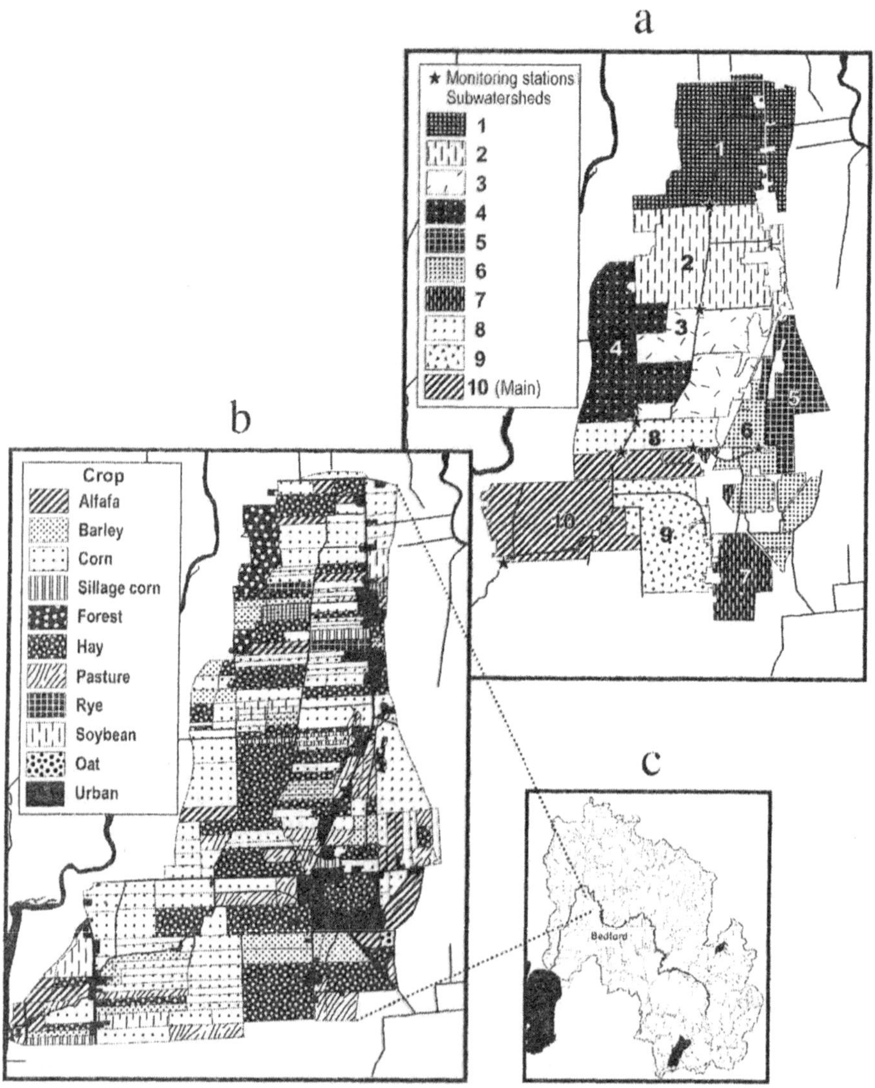

Figure 1. Beaver brook subwatersheds and monitoring stations (a), land use for crop year 1999 (b) and location within Pike River watershed (c).

about 60 m above MSL in the eastern uplands (Figure 2). Poorly-drained lacustrine and marine clays (predominately Sainte-Rosalie series: Orthic Humic Gleysol; Very fine clayey, mixed, nonacid, mesic, Typic Humaquepts) occupy the flat western portions of the watershed and line the main stream channel. Glacial calcareous tills, including the poorly-drained Bedford (Orthic Humic Gleysol; Fine-loamy, mixed, calcareous, frigid, Typic Humaquepts) and Sainte-Brigide series (Orthic Humic Gleysol; Coarse-loamy, mixed, calcareous, mesic, Typic Humaquepts) occupy the higher elevations, with the well-drained Sainte-Sebastien series (Gleyed sombric Brunisol; Loamy-squeletal, mixed, nonacid, frigid, Aquic Dystrochrepts) occupying the highest and eastern-most ridge of the watershed (Figure 3). The original soils data of Can et al. (1948) were validated by early spring multi-spectral images and auger samplings of six transects across the study area.

The cropping systems of the 24 farms on the watershed reflected a predominately dairy and cash crop production agriculture, but some hog and poultry operations were also present. Grains, corn (Zea mays L.), perennial forages, and soybeans (Glycine max Merr.) are widely cultivated in rotation. Most fields are long and narrow, typical of Quebec's French heritage. Land use and cropping patterns for the 1998 growing-season are shown in Figure 1.

2.2 Cropping system descriptors

Over the period of the study, systematic descriptions were made of the cropping systems implemented on the farms situated within the watershed. Farm fields in the watershed were aggregated into 267 soil management units (SMU) based on physical boundaries or homogeneity of management. The soils in all SMU's were sampled and tested for Mehlich-3 soil P et al. levels (Mehlich, 1984). Phosphorus saturation levels were then calculated for each SMU based on the Mehlich-3 P/Al ratio used as a P desorption indicator for Quebec soils (Tran and Giroux, 1990; Giroux and Tran, 1996). All 24 farm managers in the watershed provided detailed agronomic data for each SMU for the 1998 and 1999 growing seasons, including the quality, quantity, timing, and application method of mineral fertilisers and manures. These agronomic data were balanced with regional crop P uptake rates (Beaudet, 2002) to generate annual P mass balance surpluses or deficits for each SMU. Manure management indices were computed following a method adapted from Lemunyon and Gilbert (1993) and Bolinder et al. (1998). Soil test P, agronomic data, and P mass balance of each SMU were entered into a GIS database (MAPINFO, 1996).

Soil loss was estimated according to the Revised Universal Soil Loss Equation (RUSLE) (Renard et al., 1993). Runoff coefficients were estimated using the curve number (CN) method (SCS, 1972) as adapted to Quebec conditions by Monfet et al. (1979). The hydrologic group and soil erodibility index of each SMU was based on cropping, tillage, and subsurface drainage conditions and recent soil survey data (Grenon et al., 1999; Tabi et al., 1990). For each SMU these parameters, along with slope length and gradients (LS), and estimates of the RUSLE crop management factor (C) were entered into the GIS database, allowing soil loss and runoff coefficient estimates to be computed for each SMU/crop-year combination.

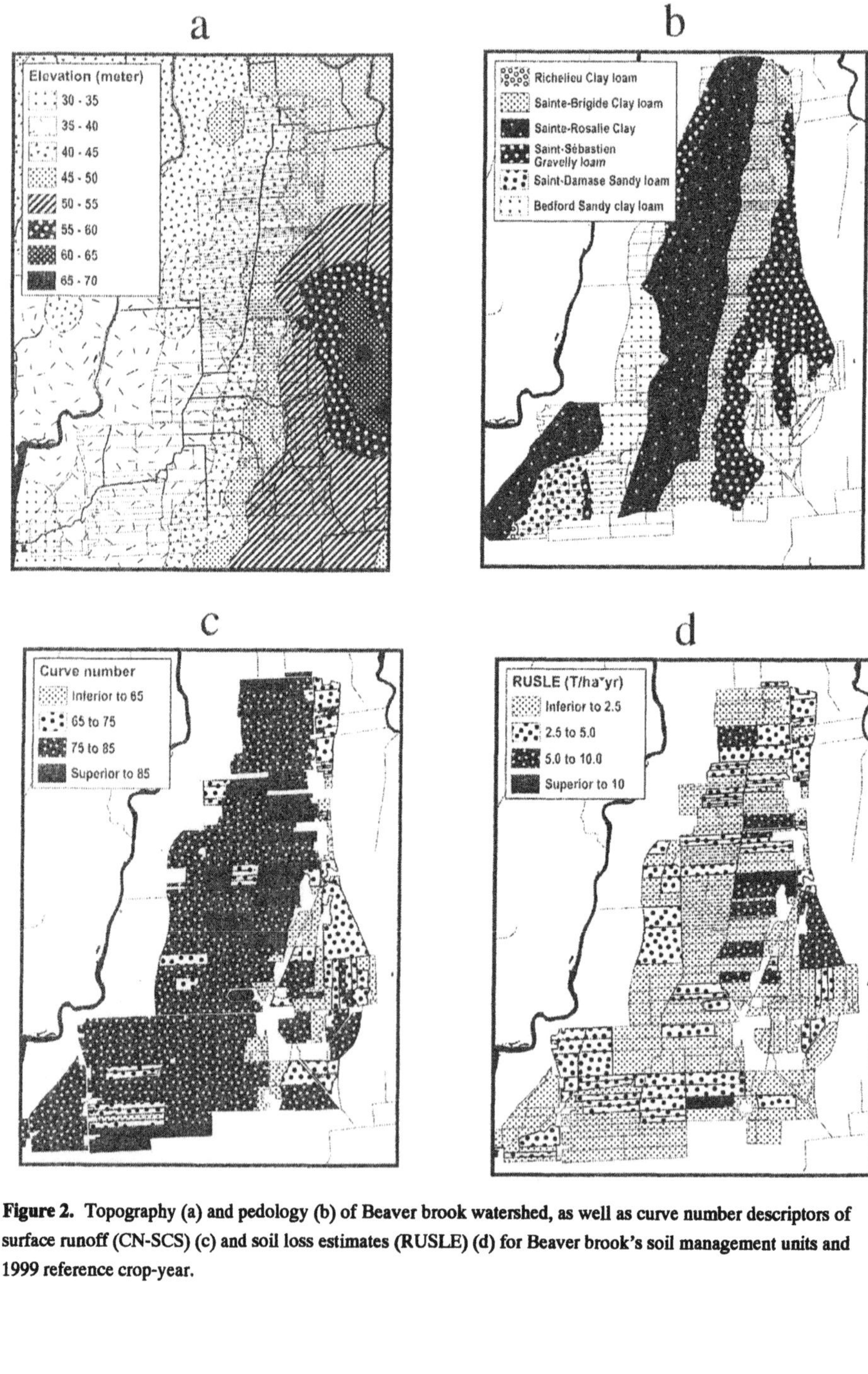

Figure 2. Topography (a) and pedology (b) of Beaver brook watershed, as well as curve number descriptors of surface runoff (CN-SCS) (c) and soil loss estimates (RUSLE) (d) for Beaver brook's soil management units and 1999 reference crop-year.

2.3 Hydrologic and Water Quality Monitoring in the Study Area

Precipitation was measured at 15 min intervals over the entire three-year study period. A main rain-gauge was located near the centre of the watershed and three secondary gauges were positioned around the periphery of the watershed. Snow precipitation data were obtained from nearby meteorological station in Phillipsburg, QC. and Farnham, QC.

The Beaver Brook watershed was divided into 10 subwatersheds and the flow and water quality emanating from each was monitored (Figure 1a). The Main Station (№ 10), located at the watershed outlet, provided a continuous record of flow thanks to a V-notch flume equipped with a bubbler type stage recorder (Hydrologic, 2003). The stage-discharge relationship for this station was developed from propeller flow-meter measurements using the BDH cubic spline rating curve model (CEHQ, 2000). A January 1998 ice storm obstructed the channel and damaged the equipment at the Main Station, resulting in the recording of flow data being interrupted from January through late March 1998. Stream monitoring also included 9 upstream sampling stations comprising the Spatial Sampling Network (subwatersheds № 1-9) each equipped with a lymnimetric scale and passive peak water level gauge. On four occasions over the study period, discharge under peak flow conditions was measured at each of the upstream sampling stations.

A total of 169 grab samples were collected at the Main Station over the period of study, using a protocol biased for elevated and peak flows. In addition, a total of 38 sampling campaigns were carried out throughout the Spatial Sampling Network: 30 during periods of elevated and peak flow in the late fall, winter, and early spring, and 8 during base flow. Two simultaneous upstream/downstream routes allowed all 10 stations to be sampled within 90 minutes. These sampling campaigns were complemented by field observations of overland runoff, surface ponding, sub-surface drain activity, and channel flow conditions.

All water samples were kept at 4°C until analysis. Total suspended solids (TSS) were determined by filtration through a 0.45 μm filter (Greenberg et al., 1992a). Dissolved reactive phosphorus (DRP) was determined colorimetrically on filtered samples (< 0.45 μm) using the molybdenum-blue method of Murphy and Riley (1964). Bioavailable phosphorus (BioP) was determined using the 0.1 N NaOH extraction method of Sharpley et al. (1991), while total phosphorus (TP) concentration was measured using the persulphate digestion technique (Greenberg et al., 1992b).

2.4 Data Analysis

Daily sediment fluxes and P export from the watershed (Main Station) were computed using the FLUX 5.0 software (Walker, 1998). The four flow-strata regression was based on a log concentration/flow (log C/Q) relationship of the mean daily flow (Q) and water quality concentrations (C) from the 169 grab samples. The coefficients of variation of TP, BioP, DRP, and TSS export estimates, calculated using the jack-knife procedure of Walker (1987), were 0.12, 0.14, 0.07, and 0.20, respectively, well within the acceptable range of Walker (1998). The concentration and load residuals provided by

FLUX 5.0 were independent of flow, date, season, concentration, and export, with no outliers being detected at a confidence level of $\alpha = 0.05$.

Spatial and temporal variability and relationships among the subwatersheds, cropping systems data and the water quality and flow data collected throughout the spatial sampling network were analysed with SAS (SAS, 2000), using the Analysis of Variance (ANOVA) and Analysis of Covariance (ANCOVA) procedures. All water quality data were log-transformed prior to analysis. The analysis showed normality of residuals, homogeneity of variance, independence from subwatershed size, and the absence of outliers ($\alpha = 0.05$ CL).

The ANOVA analyses considered three sources of variability: (i) Sampling event effect - reflecting both inter- and intra-event sources of temporal variability within the data set, (ii) Sampling station effect - reflecting sources of spatial variability in the data, and (iii) Crop-year effect - reflecting both spatial and temporal variability since sampling occurred over two distinct growing seasons whereas crop rotation and nutrient inputs for a given SMU changed yearly according to the rotation. This effect also reflects yearly differences in precipitation, which, in turn, affect nutrient and sediment export.

Differences in water quality patterns among the subwatersheds with respect to the main outlet monitoring station were tested by ANCOVA (Grabow et al., 1999), using watershed outlet discharge as the covariate. Conceptually, this procedure enabled a simultaneous comparison of water quality patterns of all subwatersheds, while considering the varying hydrological conditions that prevailed during spatial sampling runs. Flow dependency of subwatershed water quality parameter concentrations was best described through reduced-form ANCOVA models of the form:

$$\log[SW_i] = b_o + b_1(\log Q) + b_{sw} \qquad (1)$$

where:

$[SW_i]$	is the concentration of the water quality parameter for the ith subwatershed;
b_o	is the model intercept;
b_1	is the model slope;
Q	is the outlet discharge;
b_{sw}	is the subwatershed regression parameter (intercept) estimate.

3. RESULTS AND DISCUSSION

3.1. Runoff and Soil Loss Descriptors

Runoff coefficients, CN, computed for individual SMU's, ranged from a low of 58 for hay production on permeable soils (hydrological group I) with subsurface drainage, to a high value of 89 for row crop production on bedded, low permeability soils, without subsurface drainage (Figure 2c). Longitudinal spatial topographic gradients, soil parent-material permeability and extent of subsurface drainage systems within the Beaver Brook watershed were responsible for contrasting runoff coefficients among subwatersheds. Curve number estimates for upland subwatersheds (№ 5, 6 and 7) were lower than those of the lowland subwatersheds (№ 1, 4 and 9) or of the main channel subwatersheds (№ 2, 3 and 8) (Table 1).

Mean estimated annual soil loss (RUSLE) from individual watershed SMUs was 2.25 Mg ha^{-1} (Table 3). The maximum soil loss estimate for any single SMU was 14 Mg ha^{-1} for corn production on rolling terrain within subwatershed № 5 (Figure 2d). Subwatershed-scale estimates of soil loss, weighed for individual SMU surface areas (Table 1), were found to be relatively homogeneous within watersheds, averaging 2.5 Mg ha^{-1}. Higher soil loss estimates for subwatershed № 5 (3.17-3.27 Mg ha^{-1}) were related to steeper gradients among its SMU's.

Table 1. Selected hydrologic descriptors of Beaver brook subwatersheds derived from individual soil management units (SMU) data weighed for surface area.

Sub-watershed No.	Type [(1)]	**Cumulative Surface area** Total (ha)	Cropped (%)	**Elevation** Min (m)	Range (m)	**Flow** Lenght (km)	**Sub.** Drain (%)	**Tilled** Area (%) 1998	1999	**Runoff Curve** Number-NCRS 1998	1999	**Soil Loss-RUSLE** (T ha^{-1}) 1998	1999
1	M, T	171	84	42.0	6.7	1.4	50	81	74	82	83	2.85	2.94
2	M	329	82	40.6	16.4	2.4	58	84	81	82	82	2.58	2.54
3	M	443	84	38.1	18.9	3.9	49	77	71	82	83	2.59	2.51
4	T	105	99	38.2	5.1	1.8	40	66	73	84	84	2.41	2.69
5	T	75	96	52.1	13.4	1.3	84	65	72	68	68	3.27	3.17
6	T	169	84	46.1	19.3	1.9	67	52	50	71	71	2.12	2.09
7	T	47	71	44.6	8.6	1.3	83	83	83	73	70	1.47	0.87
8	M	863	82	38.0	9.9	4.3	53	70	67	80	80	2.36	2.35
9	T	99	97	37.1	15.9	1.8	43	43	53	81	81	2.22	1.60
10	M	1058	88	35.0	30.4	6.4	55	69	67	80	81	2.28	2.23

[(1)] M: Main stem; T: Tributary

3.2 Terrestrial Phosphorus Descriptors

3.2.1. Phosphorus Budget

The soil surface P mass balance for the Beaver Brook watershed (Σ[mineral + manure inputs]-Σ [crop uptake]) indicated that manure and mineral P inputs exceeded crop uptake by 11.30 and 8.61 Mg P, respectively, for the 1998 and 1999 cropping seasons. Table 2 details P mass balance data for the entire watershed (№ 10) as well as for individual subwatersheds (№ 1-9) of the spatial sampling network. Mineral fertiliser was the dominant P input in the watershed for both crop years, with watershed area-weighted averages of 16.0 kg P ha^{-1} in 1998, and down to 13.5 kg P ha^{-1} in 1999. Manure

Table 2. Soil phosphorus saturation, phosphorus budgets at soil surface and manure management descriptors of Beaver brook subwatersheds derived from individual soil management units (SMU) data weighed for surface area.

Catchments No.	Type [(1)]	Soil P [(2)] Saturation (%)	1998 P budget Manure Inputs (kg/ha)	Mineral Inputs (kg/ha)	Crop Uptake (kg/ha)	Balance[(3)] (kg/ha)	Manured Land (%)	Manure Mgt Index [(4)]
1	M, T	7.91	19.6	15.4	18.3	16.6	36	0.67
2	M	8.23	15.9	16.7	18.7	14.0	28	0.79
3	M	8.53	13.7	17.0	17.7	12.9	30	1.11
4	T	7.70	5.8	18.5	16.2	8.2	28	0.62
5	T	19.38	7.0	15.3	17.7	4.6	11	0.63
6	T	16.05	9.6	12.8	15.8	6.6	23	0.95
7	T	18.31	44.9	9.63	17.3	37.2	41	1.65
8	M	10.19	14.1	15.6	17.0	12.7	30	1.14
9	T	10.99	25.8	4.8	14.1	16.6	24	0.95
10	M	10.23	13.1	16.0	16.9	12.2	25	0.96

Catchments No.	Type [(1)]	1999 P budget Manure Inputs (kg/ha)	Mineral Inputs (kg/ha)	Crop Uptake (kg/ha)	Balance[(3)] (kg/ha)	Manured Land (%)	Manure Mgt Index [(4)]
1	M,T	7.6	14.3	18.4	3.6	9	0.19
2	M	7.6	14.3	18.4	3.6	9	0.19
3	M	7.5	14.1	17.5	4.1	9	0.25
4	T	17.1	17.1	16.9	17.4	11	0.42
5	T	14.0	14.0	17.0	11.0	19	1.20
6	T	9.3	10.5	15.7	4.1	13	0.82
7	T	44.9	0.0	12.7	32.2	41	1.65
8	M	13.2	13.0	16.8	9.5	13	0.47
9	T	25.8	2.2	13.1	14.9	24	0.95
10	M	12.5	13.5	16.7	9.3	12	0.45

[(1)] M: Main stem; T: Tributary [(2)] Mehlich-3 extractible P/Al. [(3)] Mineral + manure inputs - Crops uptake

[(4)] Manure management index:

SMU Index	Risk Class	Type, Period, Mode Manure management [a]
0	None	
1	V. low	CG, In
2	Low	Sp, In
4	Moder.	Sp-Slu-Ov/CG-Sma-Ov/PH-In
8	High	Sp-SMa-Ov/CG-Slu-Ov
16	V. high	PH-Ov

[a]Where: Sma:Solid manure
Slu: Liquid manure
Sp:Spring
CG: Crop growth
PH: Post harvest
In: Incorporated
Ov: Overland (no incorporation)

P inputs were relatively constant from year to year, with watershed area-weighted means of 13.1 kg P ha^{-1} and 12.5 kg P ha^{-1}, respectively, for 1998 and 1999. However, as a result of management constraints related to an exceptionally wet fall in 1999, manured acreage was nearly half of that in 1998. From computations of P budget for individual subwatersheds, area-weighted P surplus resulting from manure inputs were consistently higher within subwatershed № 7 in 1998 (37 kg P ha^{-1}) and 1999 (32 kg P ha^{-1}), than in any other subwatershed. An inter-annual variation in area-weighted P balance was also documented for subwatersheds № 1, 2, 3,4 and 5 (Table 2).

From a management perspective, manure inputs, rather than mineral inputs, generally accounted for the surplus P balance of individual SMU's. Phosphorus inputs typically reflected manuring rates based on the manure's nitrogen, rather than P. Figure 3a presents the distribution in P balance for individual SMU's and both crop years. Phosphorus balances remained generally below 25 kg-P ha^{-1} under sole mineral fertilisation, while manured SMU's reached up to 150 kg-P ha^{-1}. Application methods of mineral fertilizers and manure were found to be relatively homogeneous among watershed SMU's for both crop years, and were representative of relatively low environmental pressures. Mineral fertilisers were systematically incorporated through secondary tillage or band-applied in springtime. Manure inputs were generally incorporated through fall primary tillage, spring secondary tillage or broadcast on hay fields during summertime. The resulting manure management indices computed for individual SMU's indicate, for both years, a relatively higher risk of manure source P transfer within subwatershed № 7 than within any other subwatershed (Table 2).

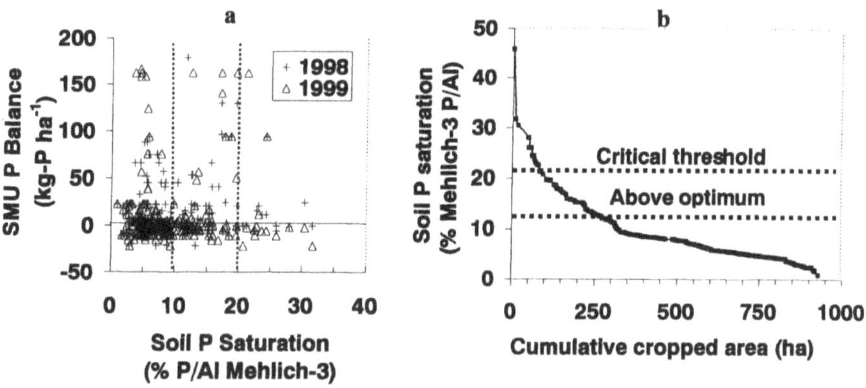

Figure 3. Terrestrial P budget at soil surface in relation to soil P saturation (Melich-3 P/Al) (a) and distribution of soil P saturation levels (b) for Castors brook's soil management units.

3.1.2. *Soil Phosphorus*

The distribution of soil P saturation values for the SMU's in the Beaver Brook watershed indicates that 34% of the watershed's cropped area had reached the supraoptimal soil P saturation ratio value (Mehlich-3 P/Al) of 10% (Giroux et Tran, 1996), while an additional 10% of the landbase had a P saturation ratio above the critical agroenvironmental threshold of 20% (Figure 3b). This distribution in soil P saturation values is consistent with local data from a study on the Pike River watershed (Deslandes et al., 2003; in this issue) and regional observations (Beaudet, 2002). For both 1998 and

1999, the soil P saturation levels of individual SMUs showed no significant correlation with their P balances (Figure 3a), indicating that historically, manure applications were rotated across the entire cropped area of the watershed. From a subwatershed-scale perspective, area-weighted soil P saturation values indicate a spatial gradient in soil P enrichment favouring upland subwatersheds № 5, 6 and 7 (Table 2). The relatively high soil test P levels in this part of the watershed reflect elevated farm-scale P mass balances, limitations in land base available for manure disposal and correspondingly higher rates of manure P inputs.

Given the annual surplus in the P balance of the Beaver Brook watershed (9-12 kg ha^{-1}) and typical soil P enrichment dynamics documented for Quebec soils (Giroux et al., 1996), one can deduce a mean annual increase in Mehlich-3 soil test P of 3 kg ha^{-1}. From a management perspective, this trend in soil P enrichment highlights the importance of considering soil P saturation ratios in the development of nutrient management plans within the study area. Up to 23% of the watershed's land base presents P saturation values below 5% (Figure 3b) and can still benefit from soil P enrichment through timely manure inputs. However, cropped land with a P saturation level of over 10% (roughly 200 kg-P ha^{-1} by Mehlich 3) or above the critical threshold of 20% (likely over 400 kg-P ha^{-1} by Mehlich 3) calls for careful P input management.

3.3 Phosphorus in the aquatic ecosystem

3.3.1 Phosphorus Concentrations and Loading

Throughout the monitoring period P and TSS concentrations in the Beaver Brook were strongly dependent upon stream flow. The relationships of TP concentration and quantity of TP exported through discharge at the watershed outlet are presented in Figures 4a and 4b, respectively. Under FLUX modelling, following flow-stratification of data, all P fractions and TSS concentrations showed significant linear relationships with stream discharge ($P < 0.01$). An examination of the C/Q relationship at the Beaver Brook outlet, coupled with field observations on surface and subsurface flow activity during sampling runs, led to the identification of three stream flow regimes, namely base flow ($Q<0.06$ m^3 s^{-1}), elevated flow ($0.06<Q<0.35$ m^3 s^{-1}) and peak flow ($Q>0.35$ m^3 s^{-1}). The inverse C/Q relationship (Figure 4a) under the base flow regime is indicative of accumulation of P stock within the main channel of the Beaver Brook. Increases in stream discharge typically result from groundwater flow and promote the dilution of P concentrations in the water column. The relatively steep C/Q slope during the elevated flow regime is primarily associated with the resuspension, in response to active interflow (subsurface runoff) activity within the watershed, of particulate P stocked within the hydrological network. Under peak flow conditions, P concentrations in the Beaver Brook were at their maximum in response to overland runoff activity on significant, but variable portions of the landscape. Since most intense surface runoff events occurred in late winter and early spring and were typically associated with frozen soil conditions, a flat to negative C/Q slope (Figure 4a) suggests that the erosive potential of runoff was likely triggered by consolidation of soil aggregates. Similar observations were reported by Dorioz and Ferhi (1993) from the Mercube watershed study in Haute-Savoie, France.

According to FLUX modelling, mean annual DRP, BioP, TP and TSS loads were 0.57, 0.93, 1.54 and 820 kg-P ha^{-1}, respectively. A daily time-series of precipitation, streamflow, TP and TSS fluxes throughout the three-year monitoring program showed that the hydrologic years 1997-1998 and 1998-1999 were relatively dry, with precipitation below the long term annual mean (1057 mm), whereas the spring of 2000 was particularly wet (Figure 5), causing severe soil and crop management problems for the local farming community. Considering the three hydrologic years under study, over 75% of the TP load lost from the Beaver Brook watershed was concentrated into only 6% of the entire monitoring period. This flashy temporal distribution of loadings reflects the hydraulic efficiency of the watershed's surface and subsurface drainage system.

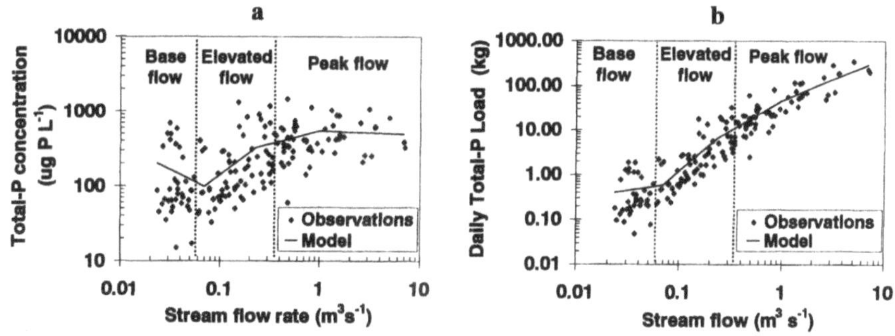

Figure 4. Concentration model (a) and load model (b) of total P (TP) for Castors brook during 1997-2000 monitoring period.

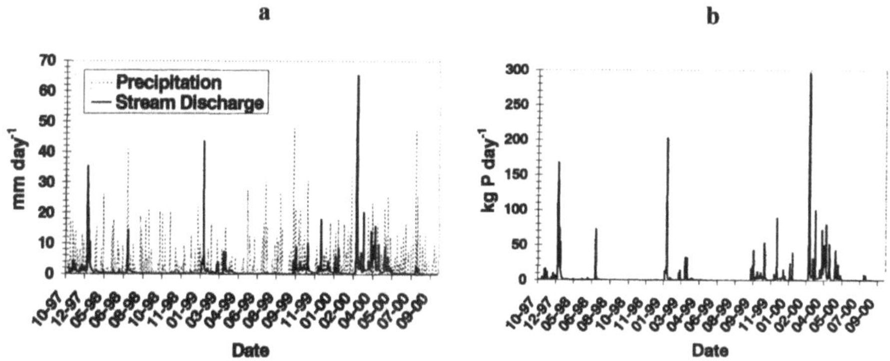

Figure 5. Daily time series of precipitation, stream discharge (a) and total P fluxes (b) modeled for Beaver brook 1997-2000 monitoring period.

Estimated TP loading agreed with model estimates of 1.4 kg-TP ha^{-1} for the Pike River basin (Hegman et al., 1999) based on the diagnostic-feasibility study of 1991 (VDEC and NYDC, 1997) and updated land use data. Estimates of TP loading for the

Beaver Brook confirmed the relatively high vulnerability of the lower Pike River's tributary watersheds to non-point source P export, compared to that of Lake Champlain's main tributaries (Budd and Meals, 1994; Hegman et al., 1999).

From a mass balance perspective, annual TP loadings (1.54 kg-P ha^{-1}) accounted for 6% of total annual P inputs into the Beaver watershed, implying that 85% of the yearly mineral and manure P inputs applied in excess of crop uptake is stocked within the watershed. This 15% TP export/balance ratio for the Beaver Brook watershed is consistent with the 16% reported for the entire Pike River watershed, based on TP loading of 1.4 kg-P ha^{-1} (Hegman et al., 1999) and a 8.6 kg-P ha^{-1} surplus in terrestrial P mass balance (Deslandes et al., 2003; in this issue). Phosphorus export from the Beaver Brook watershed, as it relates to terrestrial P balance, is also consistent with the range of published values for North America. Gangbazo et al. (2002) reported ratios similar to those of the Beaver Brook for the North Boyer River, a 30 km^2 agricultural watershed in Quebec's Bellechasse region. They estimated a two-year mean annual loading of 2.03 Mg-P ha^{-1} yr^{-1}, corresponded to 15% of the surplus P balance (14 Mg-P ha^{-1} yr^{-1}), or 10% of total manure and mineral inputs estimated at 21 kg ha^{-1} (Laflamme et al., 1997). Baker and Richards (2002) reported similar export/input ratios of 6.4 and 7.0%, respectively, for the Maumee and Sandusky watersheds of Northern Ohio.

3.3.2. Export Typology

In order to better describe the influence of hydrologic conditions on phosphorus loadings, P and TSS fluxes time series have been partitioned according to the phosphorus export typology concept (Dorioz et al., 1998), which is based on a description of phosphorus export regimes under unique conditions of hydrology, P inputs and storage in the river system. Table 3 partitioned overall phosphorus and sediment loadings according to the three previously mentioned stream flow regimes, as well as with respect to crop growth (May to September) and recharge periods (October to April).

*3.3.2a. **Base Flow Regime***. Under base flow conditions, P loadings essentially originate from point sources in the watershed, related to low-density households (approximately 15 persons per km^2), two unconfined solid manure pits (herd size below 50 animal units) and farmstead point sources. Typically, stream P was highly reactive under base flow conditions (Table 3). The C/Q negative slope from the concentration model (Figure 4a) suggest that TP inputs exceeded TP export under the base flow regime, leading to an accumulation of point source P within the watershed. Applying the Dorioz and Blanc (2001) approach in determining point P loadings from the C/Q relationship, the annual point source flux is about 200 kg-P ha^{-1} yr^{-1} for the Castor Brook, for a corresponding streamflow of 0.063 m^3 s^{-1}. Based on this estimate, P point sources accounted for approximately 5% of the annual phosphorus loadings exported from the watershed, a value bracketed by values of 3% and 8%, respectively, from the Lake Champlain diagnostic-feasibility study for the entire Pike River watershed (VDEC and NYDEC, 1992), and from Hegman et al. (1999) for the downstream hydrologic units of the Pike River.

3.3.2b. Peak Flow Regime. Peak flow events were responsible for most of the annual TP exports (71%) from the Beaver Brook watershed. Particulate P accounted for close to two-thirds of the total phosphorus loading. Total P showed relatively low bioavailability (50%), compared to other flow regimes. This likely reflects the detachment and transport of coarser sediment out of field micro-depressions, as well as channel erosion within ditches and of Beaver Brook stream banks. Peak flows predominantly occurred in late winter and early spring in response to coincidental favourable conditions including a rapid melting of a deep snowpack (March 1998), rain on top of melting snow and frozen soils (January 1999) or an exceptional amount and duration of precipitation on saturated soils (April 2000).

Table 3. Partitioning of precipitation, stream discharge, total suspended solids (TSS) flux, total phosphorus (TP) flux, bioavailable phosphorus (BioP) flux and dissolved reactive phosphorus (DRP) flux according to stream flow regime and hydrologic periods for the 1997-2000 monitoring period of Castors brook.

Flow [1] Regime (m3/s)	Flow [2] Period	Time Frame (% Total)	Precipitation (mm)	Precipitation (% Total)	Stream Discharge (mm)	Stream Discharge (% Total)	TSS Annual Flux (kg ha^{-1})	TSS Annual Flux (% Total)
Peak	Recharge	2.9	55	7	153	39	411	50
(>0.35)	Crop growth	2.1	81	10	55	14	140	17
Elevated	Recharge	22.4	155	19	107	27	153	19
(0.06-0.35)	Crop growth	8.0	91	11	40	10	57	7
Base	Recharge	24.7	143	17	18	5	28	3
(<0.06)	Crop growth	39.8	297	36	23	6	31	4
TOTAL		100	822	100	395	100	819	100

Flow [1] Regime (m3/s)	Flow [2] Period	TP Annual Flux (kg ha^{-1})	TP Annual Flux (% Total)	BioP Annual Flux (kg ha^{-1})	BioP Annual Flux (% Total)	DRP Annual Flux (kg ha^{-1})	DRP Annual Flux (% Total)
Peak	Recharge	0.80	52	0.41	44	0.28	49
(>0.35)	Crop growth	0.29	19	0.14	16	0.10	18
Elevated	Recharge	0.28	18	0.25	27	0.12	21
(0.06-0.35)	Crop growth	0.10	7	0.09	10	0.05	8
Base	Recharge	0.03	2	0.02	2	0.01	2
(<0.06)	Crop growth	0.03	2	0.02	2	0.01	2
TOTAL		1.54	100	0.93	100	0.57	100

(1) Stream monitoring period extends from Oct.1997 to Sept. 2000, excluding period of December 13th to March 26th 1998 (276 mm of precipitation).

(2) Recharge period: October to April; crop growth period: May to September.

During the crop growth period, cumulative precipitation, rather than precipitation intensities largely controlled streamflow and P export. Five-day antecedent rainfall of 25 to 75 mm was required to yield a daily streamflow of 3.5 mm, typically associated with the onset of overland flow conditions on the Beaver Brook. Within the crop growth period, in the early fall of 1999 and in May 2000, peak flows were observed in response to significant five-day antecedent rainfall. Over the three-year monitoring period, only one peak flow event occurred during the summer months (July 1998). The watershed received 90 mm of rain over a storm event which was intensively monitored through discrete sampling (Figure 6).

3.3.2c. Elevated Flow Regime. Depending on antecedent precipitation and the hydrological state of watershed, elevated streamflows ($0.06\ m^3s^{-1} < Q < 0.35\ m^3s^{-1}$) occurred either: (i) in receding peak flow events, generally occurring during the recharge period, or (ii) in the rising limb of the hydrograph, either preceding peak flow events or in response to moderate precipitation. Under an elevated streamflow regime, subsurface runoff and in-stream resuspension are considered the dominant hydrologic pathways of P transfer to the watershed's outlet. Hydrograph separation indicated that a $0.35\ m^3s^{-1}$ stream flow rate generally correspond to the point of inflection on the falling limb of hydrographs, indicating a transition from dominant surface to dominant subsurface flow conditions. Using this method, Lapp et al. (1998) estimated that interflow contributed from 9.5 to 21.3% to annual orthophosphate loads in the Saint-Esprit experimental watershed, a 26 km^2 rural basin of the Lanaudière region of Quebec. Monitoring data for the Beaver Brook indicated that conditions of elevated flow regime accounted for 25% of annual TP loadings. Under such conditions, P export was largely in the particulate form (56%) and presented a relatively high degree of P bioavailability (88%), contrasting markedly with the 50% P bioavailability of peak stream flows loadings.

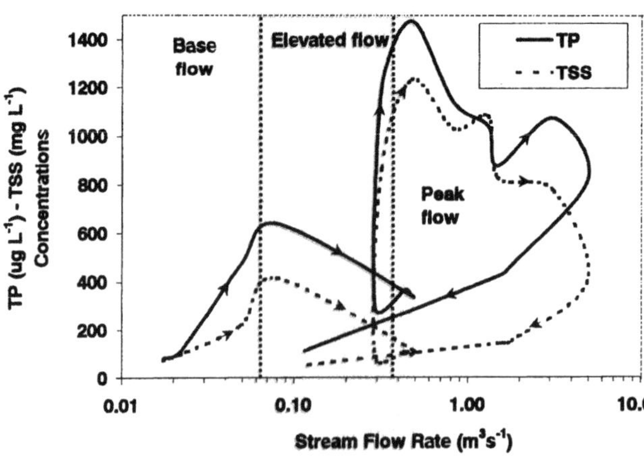

Figure 6. Clockwise hysteresis curves for total phosphorus (TP) and total suspended solids (TSS) versus streamflow rate monitored at Beaver brook outlet gauging station, in response to a 90 mm precipitation event between the 7th and 11th of july 1998.

During the flow recession stage of major runoff events, artificial drainage, implemented on 52% of the basin cropped area, presumably acted as the main pathway of P transport to streams. On long, narrow fields, lateral flow over the plow pan formed during fall tillage along the slope gradient, as well as non-erosive surface runoff within the immediate saturated surroundings of the low-lying, main channel streambanks, presumably account for a portion of the P export under the elevated flow regime.

Relatively high PP and BioP levels in the Beaver Brook, under conditions conducive to subsurface P transport processes, were consistent with field-scale surface and subsurface monitoring of P loading under physiographic conditions typical of the Beaver Brook watershed. Phosphorus loading of subsurface drains beneath low-lying clay soil sites neighbouring the Beaver Brook watershed rose in proportion to drainflow, and accounted for as much as 80% of the TP load (Jamieson et al., 2002). In the Beauce/Appalache region of Eastern Quebec, Giroux and Enright (2002) reported that on a LeBras silt loam the particulate fraction of P accounted for roughly half of the annual drain flow loadings. However, Coté (2002) reported much lower particulate fractions (18%) in drain flow loadings from experimental plots on the same soil series, but under conservation tillage. Subsurface drain monitoring on several cracking clay soils in Quebec,also showed that particulate P accounted for roughly half of TP exports (Beauchemin et al., 1998).

Elevated stream flows just above the base flow transition point, also typically exhibited high PP and BioP levels early in the rising limb of hydrographs, indicating significant resuspension of in-stream stocked P. The twin-peaked C/Q time-series observed during the summer storm event of July 7 to 11, 1998 illustrates the resuspension phenomenon particularly well. Very early in the event, TP and TSS concentrations rose sharply when stream discharge reached an elevated flow regime (>0.06 $m^3 s^{-1}$), and were much higher than those observed about 12 hours later, during the first peak of stream discharge (0.50 $m^3 s^{-1}$). Flow then decreased slightly (0.30 $m^3 s^{-1}$), with corresponding decreases in TP and TSS concentrations. In response to a second, sharp increase in stream flow rate, TP and TSS concentrations peaked at 1.47 mg L^{-1} and 1.23 g L^{-1}, respectively, three hours before peak stream discharge (5 $m^3 s^{-1}$). Such time lags between peak P concentrations and peak streamflows are typically interpreted as in-stream resuspension of sediments and PP previously stocked within the hydrological network (Dorioz and Blanc, 2001). During the final decline in stream flow, TSS and TP concentrations exhibited a second hysterisis loop, indicating the exhaustion of in-stream sediment and P stocks, as well as the diminishing detachment and transport capacity of surface runoff. This phosphorus retention phenomenon within the Beaver Brook watershed was also apparent from the spatial variability in subwatershed P export, as discussed below.

3.4 Subwatershed-scale Variations

The main objective of the spatial sampling protocol was to detect significant differences in P concentration patterns in the main channel and tributary subwatersheds of the Beaver Brook watershed, under contrasting physiographic conditions and management. Spatial variability of P transfer was supported by 38 discrete samplings from 10 stations under rapidly evolving and widely ranging hydrological conditions.

3.4.1. Spatial vs Temporal Variability

Analysis of variance (ANOVA) performed on the data collected in the Spatial Sampling Protocol (namely 38 grab samples taken at each of 10 stations in the watershed) suggested that not only did TP concentrations vary spatially and temporally (i.e. according to flow) within the watershed, but so did BioP, DRP, and TSS concentrations. The ANOVA model considered (i) crop year effect (combines variation due to annual differences in both weather and field management), (ii) sampling station effect (spatial variability), and (iii) sampling event effect, nested within the crop year effect (temporal variability related to the timing of the grab samples).

Table 4. Summary of analysis of variance of crop-year effect, sampling event effect, and sampling station effect on total phosphorus (TP), bioavailable phosphorus (BioP), dissolved reactive phosphorus (DRP) and total suspended solids (TSS) concentrations from spatial sampling program of Castors brook subwatersheds.

Source of variation (1)	% of Total Variation from ANOVA (2)			
	TP	BioP	DRP	TSS
Crop year effect	8*	ns	7*	8**
Sampling event effect (3)	47***	59***	52***	43***
Sampling station effect	16***	16***	14***	8***
Crop year * Station interaction	ns	ns	ns	ns
Model R^2	0.76	0.77	0.75	0.60

(1): Variance analysis conducted on log-transformed data.

(2): *, **, *** : Resprectively associated to F values significant at 0.05, 0.01 and 0.001 level.

(3): Nested within Crop year effect.

Statistical model : $Y_{ijk} = \mu + C_i + E_{j(i)} + S_k + \varepsilon_{ijk}$

Where : Y_{ijk} = Water quality dependent variable for the ith year, j^{th} sampling event and k^{th} sampling station;

μ = Overall mean;

C_i = Crop year effect;

$E_{j(i)}$ = Sampling event effect (nested within year);

S_k = Sampling station effect;

ε_{ijk} = Residual error.

The $E_{j(i)}$ and ε_{ijk} are assumed to be independently and identically distributed, and $E_{j(i)}$ is assumed to be independent of ε_{ijk}.

About 47% of the variation in TP concentration was accounted for by the timing of the sampling events (Sample Event Effect) while spatial variability among the subwatersheds (Sampling Station Effect) and Crop Year Effect accounted for only 16% and 8%, respectively, of the variability (Table 4). While the percentage of variability explained by each type of effect was significant, clearly the temporal effect (the effect of stream flow) had the predominant impact on TP concentration. A similar accounting of variation among the three effects was apparent for BioP, DRP, and TSS concentrations (Table 4). When an additional ANOVA was applied to only those samplings carried out during peak flow and elevated flow conditions, the Sampling Station Effect (spatial variability) accounted for 30% of the overall variability in TP concentration, while the Sampling Event Effect (temporal variation) was reduced to 29%. Thus, there were significant spatial and temporal patterns in TP concentration within the Beaver Brook watershed.

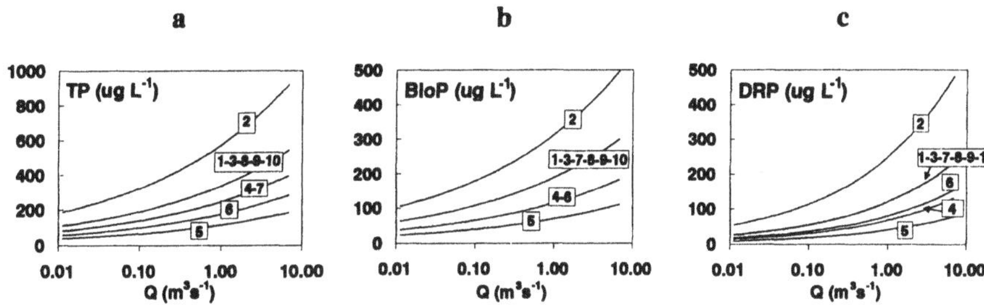

Figure 7. ANCOVA models for total P (a), bioavailable P (b) and dissolved reactive P (c) for Beaver brook subwatersheds using stream flow rate at outlet gauging station as covariate.

Analysis of covariance (ANCOVA) was also applied to these spatial sampling data to confirm the dependency of TP, BioP, and DRP concentration on flow at the watershed outlet. A reduced model ANCOVA (differences in intercept) was applied to test differences in linear regression estimates using the discharge at the watershed outlet as covariate. These ANCOVA models for TP, BioP, and DRP are shown in Figure 7. The regression intercept estimators were significantly different, indicating that the spatial pattern in TP, BioP, and DRP concentration was relatively independent of stream flow.

3.4.2. Relating Spatial Variability to Terrestrial Indicators

Conceptually, the spatial pattern in P transfer reflects the relative influences of terrestrial P sources and hydrologic pathways of P transport within subwatersheds. From a management perspective, determining how these source and transport factors influence and interact on P transfer is of strategic importance. However, the influence of P sources and hydrological processes operate at different temporal and spatial scales. One way to discriminate the relative influence of P sources and transport processes is to compare water quality profiles of individual subwatersheds with contrasting P sources or hydrologic descriptors. The spatial arrangement of sampling stations within the Beaver Brook watershed enabled comparisons among discrete samples collected along the main stream channel, as well as among independent subwatersheds of the hydrologic network. Comparisons among main stream channel stations provided indications on in-stream processes such as P inputs, dilution, retention and transformation processes along the stream reaches. Sampling stations on lateral branches and upstream reaches of the main channel of the Beaver Brook enabled comparisons among subwatersheds with distinctive management or physiographic properties. Considering the basin landforms described earlier, tributary subwatersheds № 1,4, and 9 were representative of lowland conditions, while stations № 5, 6 and 7 represent upland conditions.

3.4.2a. Upland vs lowland: landscape driven variability. Water quality data from the Spatial sampling Network indicated that hydrologic controls exerted a dominant influence over soil P levels in explaining spatial variability in water quality within the Beaver Brook data. ANOVA-adjusted means of TP concentrations for individual subwatersheds and crop-year combinations (N=20) were found to be significantly correlated with subwatershed area-weighted runoff coefficients (CN) for peak flow ($r=0.68$; $P < 0.01$) as well as elevated flow conditions ($r=0.61$; $P < 0.01$).

The uplands subwatersheds in the Beaver Brook watershed generally had a lower CN than did the lowland subwatersheds (Table 1). On the upland watersheds subsurface drainage was implemented on over 75% of the cropped area, whereas less than 50 % of the lowland cropped areas are drained (Table 1). The manner in which TP was lost from fields in upstream subwatershed № 6 vs. downstream subwatershed №1 was compared. Both subwatersheds had an area of 143 ha, but subwatershed № 6 had a soil P saturation ratio of 16%, while that of subwatershed №1 was 8% (Table 2). Even with this difference in soil P saturation ratio, during peak flow samplings the concentration of BioP in runoff from subwatershed № 1 was consistently higher than that from subwatershed № 6 (Fig 9). For equivalent TSS concentrations, the concentration of BioP remained below 200 $\mu g\ L^{-1}$ for subwatershed № 6 and consistently between 160 and 600 $\mu g\ L^{-1}$ for subwatershed № 1. These findings appear to conflict with the fieldwork of several other researchers describing a soil P control on runoff P (Sharpley et al., 1996; Pote et al., 1999; Kleinman et al., 2000) and with observations from runoff plot studies on Beaver Brook benchmark soils (Michaud and Laverdiere, 2003).

However, field observations indicated that subwatershed № 6 produced only limited surface runoff during peak flow sampling events, while the low-lying lowland subwatershed became saturated on extended near-stream surfaces in response to elevated stream and water table levels. The landform and surface drainage patterns in subwatershed № 1 also created an intimate connectivity with the main channel of the Beaver Brook. Figure 8a also shows that concentrations of BioP in runoff from lowland subwatershed № 1 varied seasonally, with average concentrations during the crop growth period being, at equivalent flow and TSS concentration, consistently nearly two-fold those during the non-growth or recharge period. This is attributed to in-stream P stock releases during early fall storm flows, following an extended period of low-flow period.

3.4.2b. Source factors: nothing is gained, nothing is lost. Upland subwatersheds №5 and №7 have similar cropping patterns, soil P saturation levels, and CN (Tables 1 and 2). However, the P balance for subwatershed №7 was dominated in 1998 and 1999 by 37 and 32 $kg\ ha^{-1}\ yr^{-1}$ manure applications, respectively, whereas subwatershed №5 only received 5 and 11 $kg\ ha^{-1}\ yr^{-1}$, respectively. Erosion estimates (RUSLE) averaged about 1 $Mg\ ha^{-1}\ yr^{-1}$ for subwatershed №7 and somewhat over 3 $Mg\ ha^{-1}\ yr^{-1}$ for subwatershed №5. During peak flows exceeding 2 $m^3\ s^{-1}$, BioP concentrations in the runoff from subwatershed №7 exceeded those of subwatershed №5 (Figure 8b) because the onset of surface runoff mobilized the relatively manure-rich topsoil in subwatershed №7.

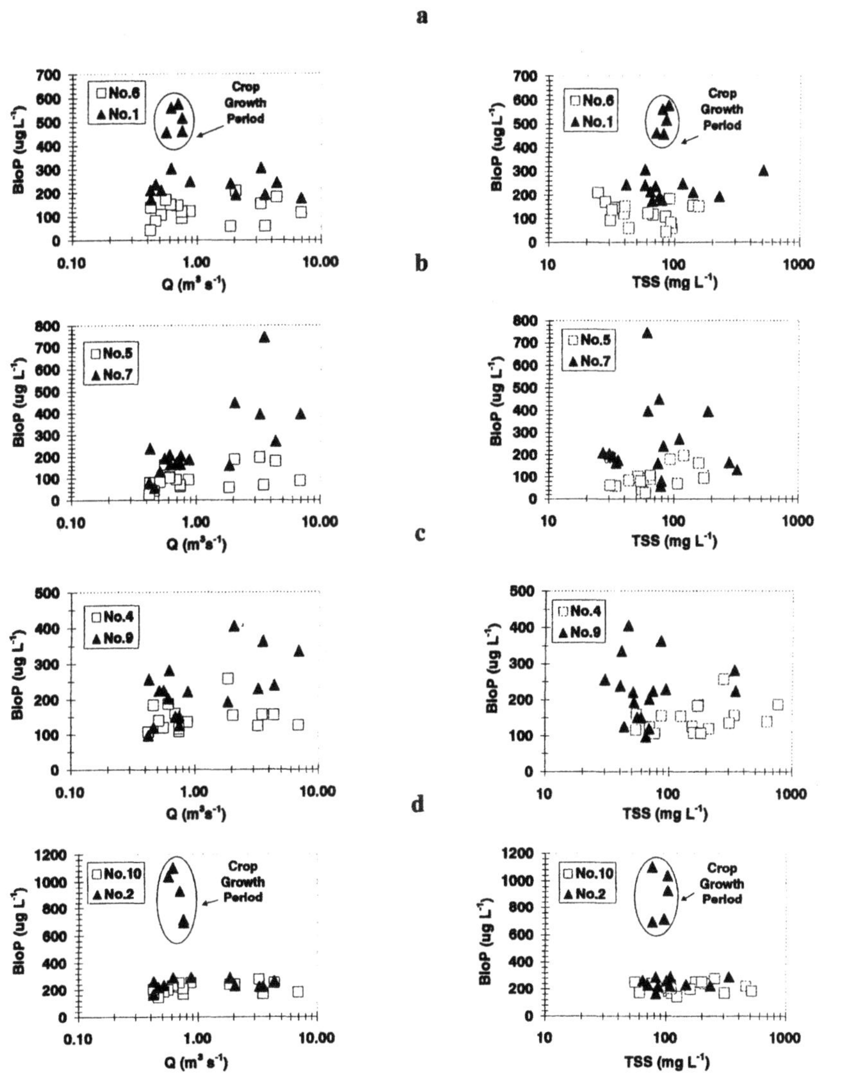

Figure 8. Bioavailable P (BioP) concentration versus stream discharge and total suspended solids (TSS) concentration of discrete water samples collected during peak flow regime from Beaver Brook subwatersheds contrasting in P sources (b-c) and position within landscape (a-d). Under ANOVA procedure, all pairs of subwatersheds have significantly different bioavailable phosphorus concentration model-adjusted means ($p<0.05$) for peak flow regime, using Tukey-Kramer's adjustment for multiple comparisons.

The BioP concentrations in runoff from lowland subwatersheds № 4 and № 9 can also be compared. These two subwatersheds had similar CNs, subsurface drainage areas, and P balances. The soil P saturation and manure inputs were greater for subwatershed № 9 than №4, while the RUSLE erosion was greater for subwatershed № 4 than № 9. The BioP concentrations in the runoff from subwatershed № 9 exceeded those of subwatershed №4 during peak flows (Figure 8c). As with upland subwatersheds it is believed this pattern is explained by mobilisation of the relatively manure-rich topsoil in subwatershed № 9.

An enhanced mobility of P from manure sources is consistent with findings reported in the literature. Increased P export from recently manured plots under simulated rainfall conditions was reported by Michaud and Lavaliere (2003). Gangbazo (1991) and Grando (1996) also working in Quebec, reported similar results after the application of liquid manures. Haygarth and Jarvis (1999) concluded that manure additions have a greater effect on soil P availability and leaching than do mineral fertilizers. Sharpley et al. (2001) documented higher P levels in runoff from soil receiving manures than mineral fertilizers.

4. CONCLUSIONS

Spatial discrete water sampling and continuous flow gauging allowed a description of the temporal and spatial patterns in phosphorus exports from the Beaver Brook watershed. Most of the annual load in TP was associated with flashy peak flow events in late winter and early spring period, when saturation-excess overland runoff developed over extended field surfaces. Examination of the C/Q relationship for downstream stations supported the identification of distinct flow regimes, which proved to be useful in describing spatial variability of water quality descriptors among upstream subwatersheds. Analysis of variance of water quality data, based on flow regime subsets, as well as analysis of covariance using stream flow as a covariate, provided contrasting spatial patterns among the watershed's subwatersheds. Flow-dependent water quality data and landscape descriptors indicated that contrasting terrestrial flow paths prevailed among subwatersheds during peak stream flow regime. Monitoring data at an elevated flow stage indicated that subsurface flow inputs of P to the main channel as well as in-stream storage and release of highly bioavailable particulate P were important mechanisms of P transfer in the watershed.

Agroenvironmental indicators combining cropping systems and physical landscape attributes provided a coherent explanation of observed spatial patterns in P fractions and concentrations. Globally, the spatial pattern in water quality highlighted a landscape-driven hydrologic control on P export, despite an opposing gradient in terrestrial soil P enrichment. Spatial patterns in subwatershed P concentrations were best revealed by the surface runoff descriptor (CN) and subwatershed elevation within landscape. If landscape hydrology was shown to exert a dominant influence on P exports, on a watershed scale, manure P management was also shown to influence P exports, particularly amongst the most hydrologically active lowland subwatersheds, as well as for upland subwatersheds.

This research demonstrates that planning and implementation of measures to prevent agricultural non-point P exports must document landscape hydrological activity and address both nutrient and runoff management. Long term control of topsoil P enrichment remains of critical importance, particularly for lowland units of the landscape, subject to subsurface preferential P transfer and overland runoff. Punctual, excessive manure P inputs translated into P export to aquatic ecosystems. These observations support the concept of critical source areas (CSA), where most P exports originate from a relatively small fraction of the watershed's land, combining high-P sources and surface runoff activity (Gburek et al., 2000b). A practical implication of the CSA concept is to adapt cropping systems to the hydrological functioning of subwatersheds, particularly with respect to rate and timing of manure inputs.

Acknowledments. This publication is a result of work sponsored by the Ministère de l'Agriculture, des pêcheries et de l'alimentation du Québec. We wish to acknowledge the collaboration of the Centre d'Expertise hydrique du Québec for their support in hydrometric monitoring. We also acknowledge the 24 farm managers of the Beaver Brook watershed and the Coopérative de Solidarité de la Rivière-aux-Brochets. We also thank Pierre Audesse and the personnel from IRDA laboratory. Special appreciation also goes to Alan Cassell for inspiring comments and continuing support in preparing this publication.

REFERENCES

Baker, B. and R.P. Richards. 2002. Phosphorus budgets and riverine phosphorus export in northwestern Ohio watersheds. J. Environ. Qual. **31** :96-108.

Beauchemin, S., R.R. Simard et D. Cluis. 1998. Forms and concentration of phosphorus in drainage water of twenty-seven tile-drained soils. J. Environ. Qual. **27** :721-728.

Beaudet, P. 2002. La fertilisation et la gestion du risque agroenvironnemental. In Ordre des agronomes du Québec (ed. Cd-rom) Colloque sur le phosphore : une gestion éclairée. 20 p.

Bevin, K. and E.F. Wood. 1983. Subwatershed geomorphology and the dynamics of runoff contributing areas. J. Hydrol., **65**: 139-158.

Bolinder, M.A., Simard, R.R., Beauchemin, S., and K.B. MacDonald. 1998. Indicator of risk of water contamination : methodology for the phosphorus component. Agriculture et agro-alimentaire Canada. Rapport interne № 24, 30p.

Brunelle A. and V. Savoie. 2000. Guide des pratiques de conservation en grandes cultures. Module 7-B: Diagnostic et correction des problèmes de drainage. Conseil des productions végétales du Québec. 32 p.

Budd, L.F. and D.W. Meals. 1994. Lake Champlain non-point source pollution assessment. Lake Champlain basin program Technical Report 6A and 6B. Grand Isle, Vermont.

Cann, D.B., P. Lajoie and P.C. Stobbe. 1948. Soil survey of Shefford, Brome and Missisquoi counties in the province of Quebec. King's printer, Ottawa, Ont.

Caumartin, J. and R. Vincent. 1994. Diagnostic environnemental de la Rivière-aux-Brochets. Ministère de l'environnement et de la faune du Québec. Québec, Qc.

Centre d'expertises hydriques du Québec. 2000. Logiciel de traitement de données hydrométriques BDH, version 1.0. Québec, Qc.

Côté, D. 2002. When Optimized Manure Management Leads to Fertile Soil, Healthy Air and Pure Water. Lake Champlain Research Consortium, Symposium 2002: Lake Champlain in the new millennium. Saint-Jean-sur-Richelieu, Québec.

Deslandes, J., A.R. Michaud, and F. Bonn. 2003. Use of GIS and remote sensing to develop indicators of phosphorus non-point source pollution in the Pike river basin. In: T. Manley and P. Manley, ed., Lake Champlain in the new millennium. Kluwer academic/plenum pub. New York NY.

Dorioz, J.M., E.A. Cassell, A. Orand and K.G. Eisenman. 1998. Phosphorus storage, transport and export dynamics in the Foron river watershed. Hydr. Process. **12(2)**: 285-309.

Dorioz, J.M., E. Pilleboue and A. Ferhi. 1989. Dynamique du phosphore dans les bassins versants: importance des phénomènes de retention dans les sediments. Water Res., 23(2), 147-158.

Dorioz, J.M. and A. Ferhi. 1993. Non-point pollution and management of agricultural areas : phosphorus and nitrogen transfer in an agricultural watershed. Wat. Res. 28(2):395-410.

Dorioz, J-M. and P. Blanc. 2001. Maîtrise de la charge externe en phosphore des plans d'eau et fonctionnement des basins versants. In : L'eau dans l'espace rural, vie et milieu aquatiques. INRA editions, 113-132.

Gangbazo, G. 1991. Effets des évènements hydrologiques sur la perte d'azote et de phosphore suite à l'épandage de lisier de porc. PhD Thesis, Institut National de la Recherche Scintifique (INRS-EAU). 324 p.

Gangbazo, G. et F. Babin. 2000. Pollution de l'eau des rivières dans les bassins versants agricoles. Vecteur environnement. 33-4 : 47-57.

Gangbazo, G., D. Cluis and E. Buon. 2002. Suspended sediments and phosphorus transport in an agricultural watershed. Vecteur Environnement 35: 44-53.

Gburek, W.J., A.N. Sharpley, L. Heathwaite and G.J. Folmar. 2000a. Phosphorus management at watershed scale: a modification of the phosphorus index. J. Env. Qual. 29: 130-144.

Gburek, W.J., A.N. Sharpley and G.J. Folmar. 2000b. Critical areas of phosphorus export from agricultural watersheds, In A.N. Sharpley, Ed., Agriculture and phosphorus management: the Chesapeake bay, Lewis Publishers, Boca Raton, Florida.

Giroux , M., and T.S. Tran. 1996. Critères agronomiques et environnementaux liés à la disponibilité, la solubilité et la saturation en phosphore des sols agricoles du Québec. Agrosol 9(2) :51-57.

Giroux, M., D. Carrier and P. Beaudet. 1996. Problématique et méthode de gestion des charges de phosphore appliquées aux sols agricoles en provenance des engrais de ferme. Agrosol 9(1) :36-45.

Giroux, M. and P. Enright. 2002. Concentrations et charges d'azote et de phosphore perdues dans les drains souterrains selon les cultures et les modes de fertilisation. Site de Saint-Lambert-de-Lauzon. Cahiers de l'observatoire de la qualité des sols du Québec, Institut de recherche et de développement en agroenvironnement Inc.

Grabow, G.L., J. Spooner, L.A. Lombardo, and D.E. Lione. 1999. Detecting water quality changes before and after BMP implementation : Use of SAS for statistical analyses. NEQEP Notes, № 93, North Carolina State Water Quality Group Newsletter, NCSU, Raleigh, NC.

Grando, S. 1996. Effets de deux modes d'épandage de lisier de porc sur la qualité de l'eau de ruissellement. Mémoire de fin d'études, ENITA de Bordeaux. 64 p.

Greenberg, E.A., L.S. Clesceri and A.D. Eaton. 1992a. Total Suspended Solids Dried at 103-105 °C. Standard method for examination of waste and waster water, 18th edition, pp 2-56.

Greenberg, E.A., L.S. Clesceri and A.D. Eaton. 1992b. Persulfate Digestion Method. Standard method for examination of waste and waster water, 18th edition, pp 4-112.

Grenon, L., J,-M. Cossette, M. Deschênes and L. Lamontagne. 1999. Etude pédologique du comté de Rouville, Québec. Agriculture et Agroalimentaire Canada, Direction générale de la recherche, Centre de recherche et de développement sur les sols et les grands cultures. 264 pages.

Haygarth, P.M. and S.C. Jarvis. 1999. Transfer of phosphorus from agricultural soils. Adv. Agron. 66: 195-249.

Hamilton, P.A. and T.L. Miller. 2002. Lessons from the National Water-Quality Assesment. J. Soil Water Conserv. 57:164-A-24A.

Hegman, W., D. Wang et X.C. Borer. 1999. Estimation of Lake Champlain basin wide non-point source phosphorus export. Technical report № 31. Lake Champlain Basin Program publication series.

Hydrologic, 2003. Levelmeter an recorder LPN 8. Hydrologic Co. Saint-Martin d'Heres, France.

Jamieson, A., C. Madramootoo and P. Enright. 2002. Evaluating Phosphorus Losses in surface and subsurface runoff from two agricultural fields in Quebec : Lake Champlain Research Consortium, Symposium 2002: Lake Champlain in the new millennium. Saint-Jean-sur-Richelieu, Québec.

Kleinman, P.J.A., R.B. Bryant, W.S. Reid, A.N. Sharpley, and D. Pimentel. 2000. Using phosphorus behavior to identify environmental thresholds. Soil Science 165: 943-950.

Laflamme D., I. Piché, A.R. Michaud, Y. Bédard, G. Trencia, R. Laroche, L. Champagne and J.-M. Gouin. 1997. Situation environnementale du bassin de la rivièere Boyer. MENV, MAPAQ, SLV 2000, Groupe d'Intervention de la rivièere Boyer, Qc, Canada.

Lapp, P. 1996. The hydrology and water quality of an intensive agricultural watershed in Québec. M.Sc. Thesis. Department of Agricultural Engineering, McGill University, Montreal, Canada.

Lapp, P., C.A. Madramootoo, P. Enright amd J. Perrone. 1998. Water quality on an intensive agricultural watershed in Québec. J. Amer. Water Res. Assoc. 34(2): 427-437.

Lauzier, R. and E. Gasser. 2002. The Coopérative de solidarité de la rivière aux Brochets : concrete measures to reduce phosphorus loadings. Lake Champlain Research Consortium, Symposium 2002: Lake Champlain in the new millennium. Saint-Jean-sur-Richelieu, Québec.

Lemunyon, J.L. and R.G. Gilbert. 1993. The concept and need for a phosphorus assessment tool, J. Prod. Agric., 6:483-496.

MAPINFO, 1996. Mapinfo Corporation. Troy, NY.

Michaud, A. R. and M. R. Laverdière. 2003. Effects of cropping, soil type and manure application on phosphorus export and bioavaibility. Can. J. Soil Sci. (under review).

Mehlich, A., 1984. Mehlich-3 soil extractant: A modification to Mehlich-2 extractant. Commun. Soil Sci. Plant Anal. 15:1049-1416.

Monfet, J. 1979. Évaluation du coefficient de ruissellement à l'aide de la méthode SCS modifiée. Service de l'hydrométrie, Ministère de l'Environnement du Québec, Publication № HP-51, Québec.

Murphy, J., and J.R. Riley. 1962. A modified single solution method for the determination of phosphates in surface waters. Anal. Acta 27:31-36.

Pote, D.H., T.C. Danial, D.J. Nichols, A.N. Sharpley, P.A. Moore, Jr., D.M. Miller and D.R. Edwards. 1999. Relationship between phosphorus levels in three Ultisols and phosphorus concentration in runoff. J. Env. Qual. 28: 170-175.

Preedy, N., K. McTiernan, R. Matthews, L. Heathwaite and P. Haygarth, 2001), Rapid incidental phosphorus transfers from grassland, J. Environn. Qual. 30:2105-2112.

Renard, K.G., G.R. Foster, G.A. Weesies, D.K. McCooland, and D.C. Yoder. 1993. Predicting erosion by water : A guide to conservation planning with the revised universal soil loss equation (RUSLE). USDA-ARS agricultural Handbook.

Sharpley, A.N., W.W. Troeger and S.J. Smith. 1991. The measurement of bioavailable phosphorus un agricultural runoff. J. Env. Qual. 20:235-238.

Sharpley A.N. and Withers, P.A. 1994. The environnementally sound management of agricultural phosphorus, Fertilizer Res., 39: 133-146.

Sharpley, A.N., T.C. Daniels, J.T. Sims, and D.H. Pote. 1996. Determining environmentally sound soil phosphorus levels. J. Soil Water Conserv. 51: 160-166.

Sharpley A.N., R.W. McDowell, J.L. Weld and P.J.A. Kleinman. 2001. Assesing site vulnerability to phosphorus loss in an agricultural watershed. J. Environ. Qual. 30 : 2026-2036.

Soil Conservation Service, U.S. Department of Agriculture. 1972. Hydrology. In : National Engineering Handbook. Washington D.C. pp.10.5-10.6.

Statistical analysis system. 2000. SAS 8.0 for Windows. SAS Institute Inc. Cary, NC.

Tabi, M., L. Tardif, D. Carrier, G. Laflamme and M. Rompré. 1990. Inventaire des problématiques de dégradation des sols du Québec : région agricole du Richelieu-Saint-Hyacinthe. Ministère de l'Agriculture, des pêcheries et de l'alimentation du Québec, Entente auxiliaire Canada-Québec sur le Développement agroalimentaire. 119 p.

Tran, T.S. et M.Giroux. 1990. Relation entre les propriétés du sol et la disponibilité du P à la plante. Agrosol 3(1) : 7-12.

Vermont Department of Environmental Conservation (VDEC) and New York Department of Environmental Conservation (NYDEC). 1992. Lake Champlain Diagnostic-Feasibility Study. Interim progress report. Waterbury, VT and Albany, NY.

Walker, W.W. 1987. Empirical methods for predicting eutrophication in impoundments. Report 4: Applications manual Tech. Rep. E-81-9. Prep. For U.S. Army Corps Eng. Waterways experiment station, Vicksburg, MS.

Walker, W. 1998. Flux, Stream loads computations, version 5.0. Environmental laboratory USAE Waterways Experiment Station. Vichsburg, Misisipi, USA.

Wolf, A., A.N. Sharpley, L. Ressler, T. Esser, and Les E. Lanyon. 2000. Concerns and recommendations of the workgroups. In A.N. Sharpley, Ed., Agriculture and phosphorus management: the Chesapeake bay, Lewis Publishers, Boca Raton, Florida.

A BENEFIT-COST FRAMEWORK TO EVALUATE THE IMPACT OF LEGISLATION SUPPORTING REDUCTION OF AGRICULTURAL POLLUTION AT THE WATERSHED LEVEL

Esther Salvano[1], Alain N. Rousseau[1,2], Guy Debailleul[3], and Jean-Pierre Villeneuve[1]

ABSTRACT

This paper introduces the development of a methodology to conduct benefit-cost analyses of management scenarios that aim at reducing agricultural pollution at the watershed level. The proposed methodology, which relies on benefit transfer, was integrated into the distributed hydrological system GIBSI. A case study focusing on evaluation of water quality benefits (*i.e.*, aesthetic and visual uses) of mandatory nutrient management plans in the municipality of St-Patrice-de-Beaurivage in the Chaudière River basin, Quebec, Canada, was used to demonstrate the potential of the methodology. Improved water quality provided for an average annual gain of 29,107 days adding up to $1,135,160 in total benefits. Preliminary results were representative of the 1982-1985 meteorological series used for the simulations. Application of the nutrient management plan required treatment of 98,346 m^3 of liquid manure for a total cost of $1,180,150. Since benefits and costs were of the same order of magnitude, these preliminary results showed that this environmental legislation has the potential to be economically feasible.

1. INTRODUCTION

In Quebec, Canada, land applications of manure and inorganic fertilizers in excess of crop requirements have contributed in part to high concentrations of fertilizer elements (*e.g.*, nitrate and phosphate) and pathogens (*e.g.*, *Cryptosporidium, Escherichia coli*) in groundwater and surface waters [Gangbazo *et al.*, 1999]. In some agricultural watersheds, such application rates have caused eutrophication (*e.g.*, Yamaska river [Delisle, *et al.*,

[1] Université du Québec, Institut national de la recherche scientifique, Centre Eau, Terre et Environnement (INRS-ETE), 2800 Einstein, Sainte-Foy, QC, Canada, G1V 4C7
[2] Contact author: phone (418)654-2621; fax (418)654-2600; email : *alain_rousseau@inrs-ete.uquebec.ca*
[3] Université Laval, Département d'Économie agroalimentaire, Sainte-Foy, QC, Canada, G1V 7P4

1998]) and contamination of surface waters (*e.g.*, Boyer river [Laflamme *et al.*, 1997]). As a result, cropland runoff has emerged as the major source of nonpoint sources (NPSs) of pollution responsible for degradation of aquatic ecosystems [Baril, 1998].

In recent years, governments and stakeholders have used the sustainable development paradigm, as defined by the Brundtland Commission [WCED, 1990], to plan and support agricultural development [Smith and McDonald, 1998]. Under this paradigm, NPSs of pollution from cropland represents a threat to sustainable agriculture, that is: the production of food and other services by management practices that are economically feasible as well as environmentally and socially acceptable [Legg, 2000; Bonny, 1994].

All those concepts and observations have compelled governments to establish measures and environmental regulations to curtail agricultural NPSs of pollution [Debailleul, 1998]. In 2002, the Quebec government introduced a new legislation supporting the reduction of agricultural pollution (*Règlement sur les exploitations agricole - REA*). This legislation is going to be progressively applied to high-risk farms to protect soil and water from adverse agricultural activities such as fertilisation in excess of crop requirements and inadequate manure storage practices [MENV, 2002a]. Since agricultural contaminants are transported downstream and high-risk farms are spatially distributed in the landscape, there is a need to assess the efficiency of on-farm best management practices (BMPs) at the watershed level.

Distributed hydrological modeling and benefit-cost analyses provide valuable frameworks to determine the water quality impact of environmental legislations [Rousseau *et al.*, 2000a; Gangbazo et Painchaud, 1999; Villeneuve *et al.*, 1998a]. There exist integrated modeling systems (IMSs) that can be used to study water quantity and quality problems at the watershed level [Rousseau *et al.*, 2003]. Among them, there is GIBSI (*Gestion Intégrée des Bassins versants à l'aide d'un Système Informatisé*), an IMS designed to evaluate impacts of various point and nonpoint discharges on the water quality and yield of a watershed [Mailhot *et al.*, 1997; Villeneuve *et al.*, 1998b; Rousseau *et al.*, 2000a,b]. The integration of a benefit-cost analysis module to GIBSI will provide a framework to study the economic efficiency of agricultural BMPs to maintain or restore water uses within a watershed. According to Faucheux and Noël [1995] and Boardman *et al.* [1996], benefit-cost analyses provide a means to weigh the costs and benefits of environmental legislation and policies, and, ultimately, to value water resources in terms of spending related to water uses and willingness-to-pay for the conservation and improvement of the quality of the resource. Since most environmental resources are not traded on any market, it is difficult to determine their monetary values. With the willingness-to-pay concept, it is possible to create a fictitious market to value those resources. In this article, the non-monetary values will not be described. To our knowledge, benefit-cost studies focusing on water quality improvement or restoration of designated use of water through environmental legislation at watershed level have not been performed in Quebec so far.

The primary objectives of this paper are to present the benefit-cost framework provided by GIBSI and illustrate, using a case study on the Chaudière River basin (Quebec, Canada), the economic feasibility of the *REA* legislation. This paper is organized in three sections. The first section presents an overview of GIBSI. The second section describes the benefit-cost framework integrated in GIBSI. Finally, the third section weighs the costs and benefits associated with a nutrient management plan resulting from the application of the *REA* legislation in one of the most prominent agricultural sub-basins of the Chaudière River basin, the Beaurivage River sub-basin.

2. OVERVIEW OF GIBSI

GIBSI is an IMS comprised of a database (including spatial and attribute data), simulation models (hydrology, soil erosion, agricultural-chemical transport, and water quality), management modules (land use, point source, agricultural production systems, and reservoir management modules), a geographic information system, and a relational database management system. For the decision maker, GIBSI offers the ability, using a scenario approach, to predict impacts of industrial, forestry, urban, and agricultural BMPs with respect to targeted water quality and yield standards for various uses of water (*i.e.*, agricultural, fish and wildlife, industrial, recreational). The development process of GIBSI was preceded by a need analysis based on BMP scenarios that involved potential users and the Quebec Ministry of Environment (QMENV). To ensure a successful implementation at the QMENV, a water management professional was directly involved in the development process. A description of system development and application can be found in Mailhot *et al.* [1997], Villeneuve *et al.* [1998b], and Rousseau *et al.* [2000a].

2.1. Computational Unit and Management Scale of Agricultural NPS Scenarios

The computational unit (*i.e.*, spatial simulation unit or SSU) in GIBSI consists of an elementary sub-basin comprised of one river segment and three soil layers. This computational unit may be viewed as a relatively homogeneous hydrological unit [Fortin *et al.*, 2001]. The computational time step is the day. Basin discretization is done using a digital elevation model and digitized river and lake network [Turcotte *et al.*, 2000]. Interpolation of meteorological data, potential evapotranspiration and vertical water budget are performed for each SSU. Surface runoff is routed through all the cells of the digital elevation model of a SSU. Soil erosion and agricultural contaminant transport processes are simulated for each agricultural land use unit of a SSU (*i.e.*, corn, grain and hay/pasture fields). Snowpack dynamic is computed for three classes of land use unit of a SSU (*i.e.*, deciduous forests, coniferous forests and open areas). Water quality is simulated for each segment of the river network.

From an agricultural NPS management point of view, each SSU is characterized by attribute data such as livestock (*i.e.*, animal units), crop rotation system, manure application practices, tillage, soil conservation, fertilization and pest management practices, and agronomic balances for nitrogen and phosphorus. Figure 1 presents the management scales of agricultural management scenarios (*e.g.*, implementation of a nutrient management plan): the SSU, sub-basin or administrative unit level (*e.g.*, municipality). Finally, a delineation tool may be used to select the drainage basin, made of SSUs, upstream any given river segment.

2.2. Probability of Exceeding Water Quality Standards

Simulation results are used to evaluate, for a given time period (*e.g.*, summer), probabilities of exceeding water quality standards (WQSs) designating the water use (*e.g.*, swimming or fishing) of each river segment [Rousseau *et al.*, 2002a,b]. These probabilities may be expressed in terms of normalized values (*i.e.*, relative values) or days (*i.e.*, absolute values) as follows:

$$P([X] > WQS) = 1 - F_{[X]}([WQS]) \qquad (1)$$

where $[X]$ is the concentration of the studied contaminant and $F_{[X]}(WQS)$ is the cumulative of the probability density function evaluated at a WQS. The number of days where $[X]$ is greater than the WQS is obtained by multiplying the right hand side of Eq. (1) by the total number of days within the studied period[4].

3. METHODOLOGY

The monetary valuation of damages or benefits constitutes an essential part of environmental economics [Barde, 1992]. Valuation methods focus on determining the monetary gain or loss of an individual's well-being following the improvement or degradation of a service rendered by a natural resource. The proposed methodology represents an adaptation of the framework developed by the English Foundation for Water Research [FWR, 1996]. The purpose of the FWR framework is to value benefits of improving water quality and perform benefit-cost analyses of improving coastal, estuary and river waters. The FWR framework is based on benefit transfer which reduces the costs associated with economic studies. Benefit transfer is the application of monetary values obtained from a particular non-market goods analysis to an alternative or secondary policy decision setting [Brookshire and Neil, 1992]. This involves the use of a data set developed under well-defined conditions to a quite distinct alternative application.

From this perspective, the FWR framework may be viewed as a pragmatic way of conducting benefit-cost analyses [Brouwer, 2000]. Benefit transfers are often the only possible means of valuing environmental benefits when few data sets are available or financial resources too scarce to conduct large scale studies [Barbier *et al.*, 1997]. For this study, available data on the total economic value of nature for Canadians have made possible the adoption of a methodology based on benefit transfer [Environnement Canada, 2000; 1999]. Benefit transfer estimates have been used in legal and governmental investigations for several years now [Boyle and Bergstrom, 1992; Brookshire and Neill, 1992; Smith, 1992; Bergstrom, 1996; Brouwer and Spaninks, 1999; Brouwer, 2000; Podar et al., 2000]. At this point, it is important to raise two cautionary notes: (i) it is difficult to clearly assess the economic value of natural resources using willingness-to-pay values and benefit transfer and (ii) the quality of benefit transfer data is subject to the uncertainty associated with the original studies [Desaigues et Point, 1993]. Figure 2 illustrates the proposed five-step methodology.

3.1. First Step: Problem Definition

The first step focuses on the identification of the river segment of interest where current conditions preclude attainment of the designated use of water, the delineation of the neighbouring region from which estimates of potential users and benefits can be drawn, and the identification of the sought-after objectives used to elaborate alternative management scenarios to be simulated with GIBSI. For this study, designated uses of water were classified into four categories as in Simoneau *et al.* [1998]: primary contact uses (swimming, kayaking); secondary contact uses (fishing, canoeing), aesthetic or visual uses (camping, hiking, cycling, picnicking), and potable water uses. WQSs are used to govern each one of these categories. For example, primary contact uses are characterized

[4]. Note that for some water quality parameters, an on-site assessment of the aesthetic status is required (*e.g.*, visual and olfactory assessments associated with the presence of aquatic plants)

using aesthetic, bacteriological, and organoleptic proprieties of water. The QMENV aesthetic WQS defining recreational use of water requiring primary (*i.e.*, direct and prolonged) contact is for prevention of river eutrophication, that is phosphorus concentrations ([P]) should not exceed 0.03 mg P/l [Ministère de l'Environnement du Québec, 1990, rev. 1992]. For the QMENV [2002b], meeting WQSs means that restoration or protection of any current or potential use of water is achieved.

3.2. Second Step: Water Quality Simulations

The second step involves water quality simulations using GIBSI. Comparison between values of simulated water quality parameters defining WQSs for a base-case scenario (*i.e.*, reference state) and any alternative management scenario allows for determining whether or not there are any water quality improvements. Simulation results are used to evaluate, for a given time period (*e.g.*, summer), probabilities of exceeding WQS designating the water use of interest of the studied river segment.

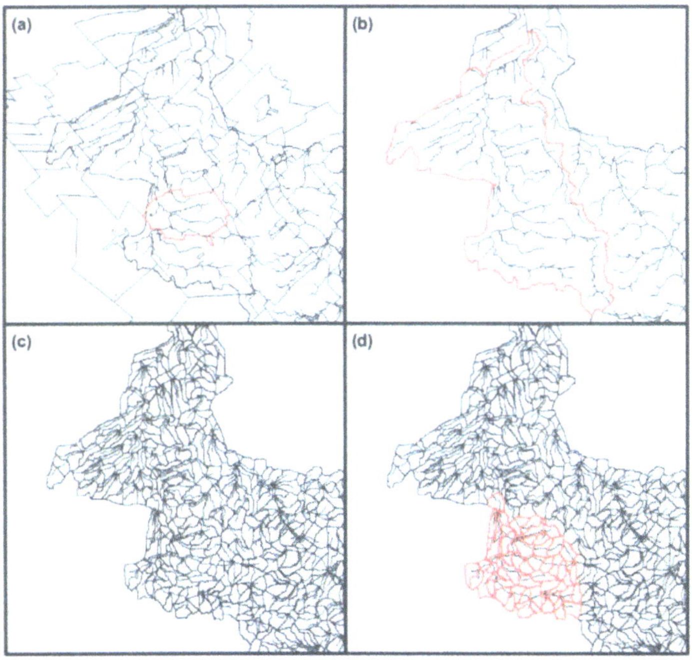

Figure 1. Management scales of agricultural BMPs management scenarios: (i) municipality, (b) sub-basin, (c) SSU, and (d) sub-basin made of SSUs located upstream a studied river segment

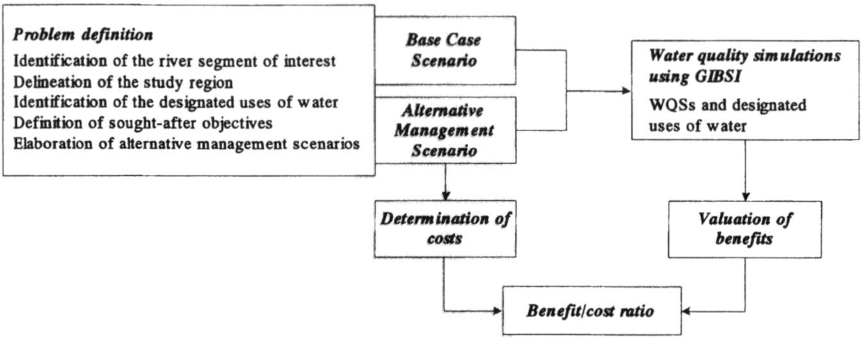

Figure 2. Illustration of the proposed five-step methodology

3.3. Third Step: Valuing Benefits

The third step focuses on valuing the benefits associated with: (i) improving water quality (*i.e*, environmental benefits) and (ii) restoring or improving a designated water use. The first possibility is for cases where there is a water quality improvement but not sufficient enough to allow for a designated water use. The second possibility is for cases where there is an actual restoration or improvement of a designated water use. In this case, the benefits associated with the improvement are evaluated. For this study, it is important to stress that socio-economical benefits are not included in the estimation of benefits. Thus, benefits represent the preferences of individuals, customarily shown by their behaviour in markets and backed up by their willingness-to-pay [Hanley and Spash, 1994]. It is noteworthy that the economic value gives an order of magnitude of the benefits of improved water quality. Table 1 introduces the algorithms used to estimate the monetary benefits associated with the use of water.

The first computation involves estimation of a daily monetary value per recreational activity. This value comes from the data set compiled by Environment Canada [Environnement Canada, 2000; 1999] on the total economic value of nature for Canadians and more precisely for the province of Quebec. This value (willingness-to-pay) is given by the sum of the actual expenditures per activity (Expenditures_$) and the direct benefits (Consumer_Surplus_$) for gaining access to that activity. The benefits are nothing but the sum of daily spending (including lodging, transportation, food, equipment and other miscellaneous items) and direct benefits for practicing the activity. Table 2 presents the expenditures and consumer surplus values used in this study [Environnement Canada, 1999]. It is important to mention that this methodology can under or over evaluate the willingness-to-pay of the targeted population. This is one of the limits of the methodology.

The population size of potential users contained within the neighbouring region of the studied river segment (Population_Zone) may be obtained using two approaches: (i) a predetermined value or (ii) estimation of the number of potential users within a 5- or 10-km radius depending on the use of water. In the latter, the number of potential users is

The population size of potential users contained within the neighbouring region of the studied river segment (Population_Zone) may be obtained using two approaches: (i) a predetermined value or (ii) estimation of the number of potential users within a 5- or 10-km radius depending on the use of water. In the latter, the number of potential users is given by the product of the population and its percentage practicing the targeted activity (Population_Activity). The total number of days of activity is obtained by multiplying Population_Activity by the number of days a user will practice the activity (Days_Usage). Table 3 presents the participation rate for each activity identified in this study. The total number of potential days of practicing the targeted activity within a given period of time (*e.g.*, summer season) is computed using the probability of exceeding the WQSs designating the studied water use given by GIBSI. To determine the net benefits of improving water quality, this computational procedure is performed for the base-case scenario and any alternative management scenario.

Table 1. Algorithms used to estimate monetary benefits associated with the use of water

Computational Algorithms
• Daily Monetary Value per Activity WTP_$ = Expenditures_$ + Consumer_Surplus_$
• Population in the Studied Region (i) Population_Zone = Population_Study (ii) Population_Zone = Population_Area * Study_Area
• Population Practicing the Activity Population_Activity = Population_Zone * %_Population
• Total Number of Days for the Activity Total_Days_Activity = Population_Activity * Days_Activity
• Total Number of Potential Days of Practicing the Activity Days_Usage = Days_Total_Activity * (1 – Probability_Exceedence)
• Monetary Benefits for Designated Water Use $_Benefits_Simulation = Days_Usage * WTP_$

Table 2. Expenditures and consumer surplus values for outdoor activities and sport fishing in Quebec [Environnement Canada, 1999]

Category of Activities	Expenditures [$]		Consumer Surplus [$]	
	Annual*	Daily*	Annual*	Daily*
Outdoor Activities	531	34	77.7	5
Sport Fishing	378	29	66.1	7.6

* *per participant*

Table 3. Participation rate for outdoor activities and sport fishing in Quebec [Environnement Canada, 1999]

Activity	Participation Rate [%]
Swimming	16.4
Hiking	29.3
Picnicking	22.2
Camping	12.4

Hiking & Tourism	17.1
Canoeing & Kayaking	8.6
Cycling	10.8
Photography	11.6
Sport Fishing	17.6

3.4. Fourth Step: Determining Costs

The fourth step aims at determining costs associated with agricultural BMP scenarios simulated by GIBSI. At first, a drainage basin made up of the SSUs, upstream the studied river segment of the discretized basin, is delineated to compile the following attribute data: livestock production, crop production, fertilization practices, and soil conservation practices. Table 4 introduces the algorithms used to estimate costs associated with agricultural management scenarios.

Table 4. Algorithms used to estimate costs associated with agricultural management scenarios

Computational Algorithms
Animal Production
• Land Application Technique
Land_Application_$ = Land_Application_Technique_$ * Area
• Livestock
Benefits_Livestock_$ = Revenue_Livestock_$ - Costs_Livestock_$
Crop Production and Cultivating Practices
• Crop Production
Benefits_Crop_$ = Revenue_Crop_$ - Costs_Crop_$
• Cultivating Practices
Cultivating_Practice_$ = Costs_Practice_$ * Area
Fertilization and Soil Conservation Practices
• Fertilization Practices
Land_Application_Fertilizer_$ = Fertilizer_$ * Rate_Fertilizer * Area
Stocking_Surplus_$ = Volume_Surplus * Stocking_$
Treatment_Surplus_$ = Volume_Surplus * Treatment_$
• Soil Conservation Practices
Pratice_Conservation_$ = Costs_Conservation_$ * Area

3.4.1. Animal Production

There exist two possible management scenarios regarding animal production, the first one dealing with manure application practices (solid and/or liquid manure application techniques), the other with livestock modifications (*i.e.*, increasing or decreasing the number of animal units). For manure application practices, unit costs (*e.g.*, $/ha) depend on the selected application techniques. For livestock production, production costs and revenues must be evaluated to determine and compare benefits for the base-case scenario and the alternative management scenario. A decrease in benefits is considered as an opportunity cost which refers to foregone benefits [Revéret, 1990] and, thus, it represents the payment or decrease in income required to adopt the alternative management scenario [Eisen-Hecht and Kramer, 2002]. This cost is added to the other costs.

3.4.2. Crop Production and Cultivating Practices

There exist two possible management scenarios regarding crop production, the first one focusing on crop rotation systems (forage crops; "corn silage and grain corn"; grain cereals) the other on crop management practices (tillage, seeding and harvesting calendar). As for livestock production, production costs and revenues must be evaluated to determine and compare benefits for the base-case scenario and the alternative management scenario. A reduction in benefits is considered as an opportunity cost that is added to other scenario costs. Total cost is obtained by multiplying unit costs by the total cropland area within the region where scenarios apply.

3.4.3. Fertilization and Soil Conservation Practices

Nine possible fertilization practices are used to implement a nutrient management plan with respect to nitrogen (N) and phosphorus (P). On-farm or animal manure (organic and inorganic) and commercial (inorganic) sources of N and P are considered in this study. Surpluses of solid and/or liquid manure may be stocked or treated for commercial use. Meanwhile, there are two possible ways of cultivating that relate to soil conservation practices (cross-slope and along-the-slope cultivating). Fertilization costs are estimated using commercial fertilizer costs, application costs considering application rates and cropland surface area. Storage and treatment costs depend on surplus volumes. For this study, note that treatment only applies to liquid manure produced by swine production systems. Surpluses arise when agronomic balances for N and P are positive, that is on-farm N and P supplies exceed crop requirements. The sum of the costs associated with management of animal manure and application of commercial fertilizers represents the total cost for fertilization practices. Meanwhile, the total cost for soil conservation practices depends on cultivating practices and cropland surface area. Finally, total cost for fertilization and soil conservation practices correspond to the sum of the two previous total costs.

3.5. Fifth Step: Present Value and Benefit-Cost Ratio of Alternative Scenarios

The last step of this five-step methodology consists of selecting a discount rate and computing a benefit-cost ratio. Any delay in reaching an environmental benefit or restoring a designated water use diminishes the value of an alternative scenario. One way to account for time preferences is to calculate the present value of an environmental benefit in t years from now using a discount rate [Barde, 1992; Desaigues and Point, 1993]. This allows for a fair comparison of various alternatives. Since the attractiveness of alternative scenarios depends on the discount rate, selection of appropriate rate of discount represents an important aspect of any benefit-cost analysis. It reflects the extent to which decision makers value the present over the future. A large discount rate signifies a large depreciation of future worth.

4. CASE STUDY

4.1. Problem Definition

Figure 3 depicts the main sub-basins of the Chaudière River basin. Current water quality and agricultural conditions within the basin have allowed us to narrow down on the intensively farmed Saint-Patrice-de-Beaurivage municipality within the Beaurivage River sub-basin. Over the years, repeated land applications of manure and inorganic fertilizers in excess of crop requirements have led to high phosphorus concentrations in surface waters and soils [Simard *et al.*, 1995; Simoneau *et al.*, 1998; Bédard *et al.*, 1998]. In this sub-basin 86 % of the phosphorus load to surface waters results from agricultural activities [Simoneau *et al.*, 1998]. Certain river segments of the Beaurivage sub-basin offer a good potential for restoring recreational activities such as hiking and camping. Table 5 presents land use, livestock, and cropland characteristics of the aforementioned municipality and sub-basin. Note that nearly 24 % of the sub-basin swine production takes place in the municipality of Saint-Patrice-de-Beaurivage. Figure 4 depicts the location of current and potential recreational uses of water within the sub-basin. Overall, there are nine sites for primary contact uses (swimming, kayaking), eight sites for secondary contact uses (fishing, canoeing), four sites for aesthetic or visual uses, and one site for potable water use.

Implementation of the *REA* legislation as a management scenario was validated with local stakeholders. As previously mentioned, this provincial legislation supporting the reduction of agricultural pollution [MENV, 2000a] is going to be progressively applied to high-risk farms to protect soil and water from adverse agricultural activities. The *REA* legislation provides for: (i) distance requirements for manure application near wells and waterways, (ii) specific conditions for land application of manure, (iii) appropriate manure storage structures, (iv) specific practices for manure application, and (v) treatment or elimination of animal manure. Also, this legislation provides for a mandatory nutrient management plan (referred to as *PAEF* for *Plan agroenvironnementale de fertilisation*) for all farms located within a basin characterized by manure N and P in excess of plant requirements. To prevent over fertilization, implementation of a *PAEF* requires accounting for soil and crop characteristics. In the Chaudière River basin, *PAEFs* have been drawn since 1999. *PAEF* represents the key *REA* management tool.

Table 6 displays the *REA* specifications that can be simulated using GIBSI. Distance requirements for manure application near wells and waterways represent the only standard that cannot be implemented through the agricultural scenario builder of GIBSI. The time of year to apply manure can be specified through the crop management calendar. Land application techniques recognized by the QMENV, such as incorporation of liquid manure within the first 10 cm of soils, can be implemented through the animal production scenario builder. Mandatory nutrient management plans can be constructed at the SSU, sub-basin, or municipal level, and thus, to an ensemble of high-risk farms located within the boundaries of these various management scales.

For this study, the alternative management scenario considered the implementation of nutrient management plans for all farms located within the St-Patrice-de-Beaurivage municipality. Under these plans, solid manure, and if necessary liquid manure are applied to meet crop requirements for N and P. The surplus of liquid manure is treated. The base-case scenario assumes that all solid and liquid manure produced within the municipality are applied on cropland.

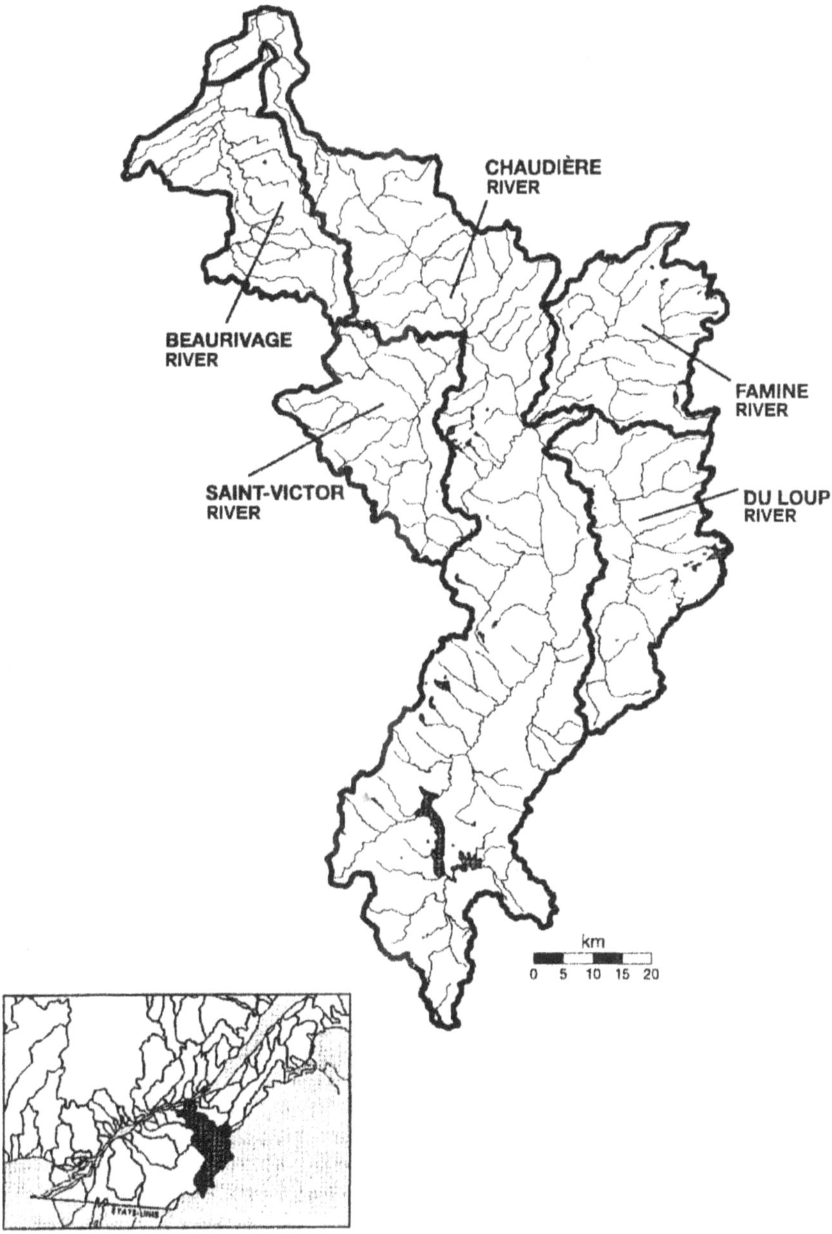

Figure 3. Principal sub-basins of the Chaudière River basin

Table 5. Land use, livestock, and cropland characteristics of the Saint-Patrice-de-Beaurivage municipality and the Beaurivage River sub-basin [Villeneuve et al., 1998b]

Characteristics	Beaurivage River Sub-Basin	Municipality of Saint-Patrice-de-Beaurivage
Area [ha]	71,820	8,575 (11.9)*
Agricultural Land [%]	24.5	29.4 (14.0)
Pasture	17.8	20.6 (13.8)
Cereal	6.6	8.5 (15.3)
Corn	0.1	0.1 (5.1)
Population	20,000	1,100 (5.5)
Livestock [A.U.]		
Cattle	19,128.2	3,533.8 (18.5)
Swine	30,782	7,253.3 (23.6)
Poultry	1,977.5	78.6 (4.0)
Others	362.2	20.4 (5.6)
Cultivated Area [ha]		
Forage	12,814	1,768 (13.8)
Cereal	4,743	727 (15.3)
Corn	524	27 (5.2)

* *Proportion of the Beaurivage River Sub-Basin (%)*

4.2. Water Quality Simulations

The May 15 through September 15 period (123 days) corresponded to the time period considered for computation of the probability of exceeding the WQS of each scenario for the designated use of water. Because the WQS for prevention of river eutrophication is usually designed using a ${}_{30}Q_5$ (30-day low flow with a 5-year return period) [Ministère de l'Environnement du Québec, 1990, rev. 1992] and that this design stream flow occurred in 1982 (July 10 through August 8), based on frequency analysis of observed flows at the Saint-Lambert gauge station near the basin outlet [Villeneuve *et al.*, 1998b], water quality simulations were performed using four years of meteorological data (1982-1985). It is noteworthy that those years were simulated independently instead of continuously. At this point, we are not interested by the cumulative effect. We assumed that a ${}_{30}Q_5$ design flow occurred at the studied river segment during the same period. Ideally, exceeding probabilities should be evaluated using longer meteorological records (*e.g.*, 30 years). A statistical analysis of the 1964-1995 precipitation record indicated that years 1982 (1,006 mm) and 1985 (1,079 mm) were considered dry years. The average precipitation for the Chaudière River basin is 1,210 mm. Whereas, years 1983 and 1984 were estimated to be an average year and a wet year, respectively. Note that the driest year registered 964 mm of precipitation and the wettest year 1,500 mm. Figure 5 shows the location of the studied river segment with respect to the Saint-Patrice-de-Beaurivage municipality. Comparison between probabilities of exceeding the WQS for prevention of river eutrophication for the base-case scenario and the *REA* scenario indicates a 34-day gain. Note that Figure 5 is the exceeding probability map for the base-case scenario. The two curves (a) and (b) depicted on the graph shown in Figure 6 represent the complementary cumulative curves of the probability density function for the base-case scenario and the *REA* scenario, respectively. The intersection of the WQS line (horizontal line) with those two curves indicates that the probabilities of exceedence for the base-case scenario and the *REA* scenario were 0.38 and 0.12, respectively.

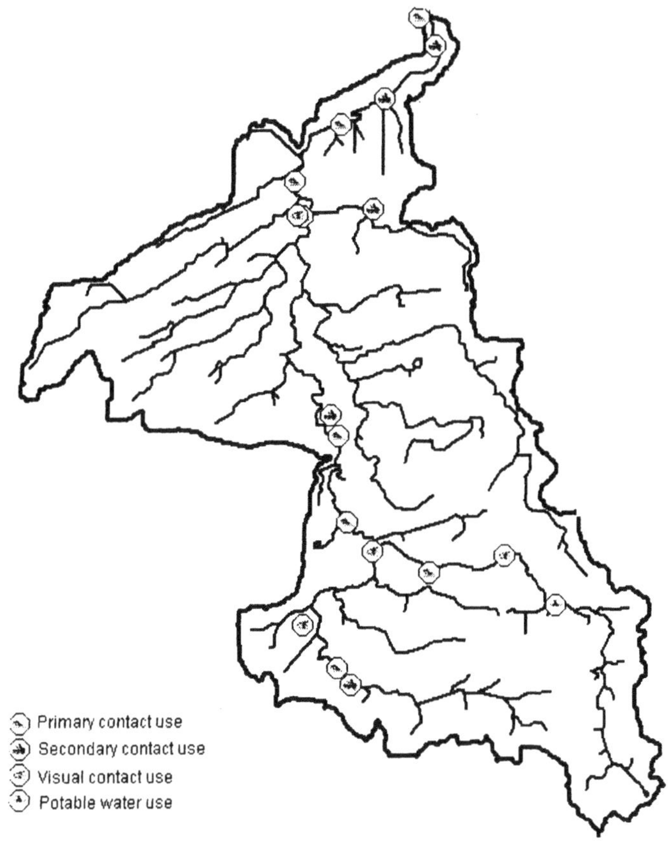

Figure 4. Location of current and potential recreational uses of water within the Beaurivage River sub-basin

Table 6. *REA* specifications simulated by GIBSI

REA Specifications	Description *	GIBSI
Distance Requirements	Spring, Well, Water Course, Ditch, Wetland : 15 m	No
Manure Storage	Solid Manure:	
	- Distance Requirements (15 m)	Yes
	- Vegetative Cover	Yes
	- Containment of Liquid Manure Runoff	Yes
	Liquid Manure : appropriate manure storage structures	Yes
	Containment of Liquid Manure Runoff	Yes
Manure Disposal	Land Application or treatment or transformation	Yes
	With respect to Environmental Regulation	Yes
Land Application	Forbidden	
	- spreading between 01/10 and 13/03	Yes
	- spreading on frozen and snow covered soils	Yes
	- spreading on land without a riparian zone	No
	- distance requirements for spreading (3 m)	No
	- spreading waste at distances of over 25 m	No
	According to an Agro-Environmental Fertilization Plan	Yes
Manure Treatment	Manure transformation in useful products	Yes
N Application Rate	Inorganic Fertilizer Only	Yes
(< Crop Requirement)	Combination of Commercial and Animal Manure	Yes
	Animal Manure	Yes
P Application Rate	Inorganic Fertilizer Only	Yes
(< Crop Requirement)	Combination of Commercial and Animal Manure	Yes
	Animal Manure	Yes

* *MENV [2002]*

4.3. Valuing Benefits

For this study, expenditures and consumer surplus values for aesthetic and visual uses (namely, camping and hiking) of water amounted to $34/day/person and $5/day/person, respectively, for a total of $39/day/person [Environnement Canada, 1999]. The population included within a 10-km radius of the studied river segment was 13500. Benefits generated by the population outside this 10-km radius are not accounted for in this analysis. Assuming that 24.2 % (3,267 persons) of the population will use the surrounding area for hiking and 29.3 % (3,955 persons) camping the total number of days for these two activities came up to 69,408 for the base-case scenario and 98,515 for the *REA* scenario. Hence, as shown in Table 7, total benefits for the base-case scenario were $2,706,921 and $3,842,081 for the *REA* scenario. Therefore, the difference between these scenarios represents a net benefit of $1,135,160. It is noteworthy that these results are representative of the 1982-1985 meteorological series.

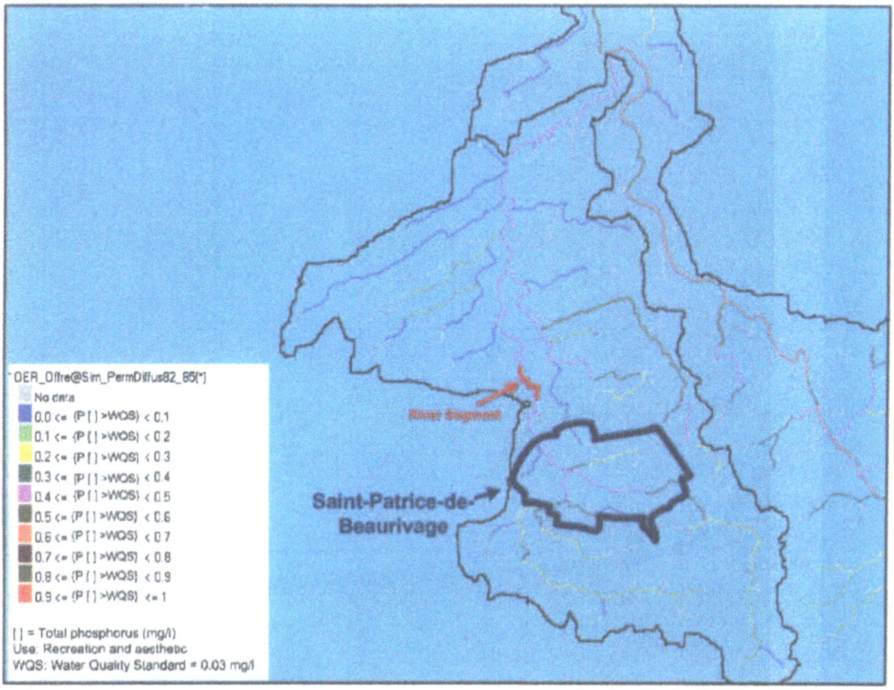

Figure 5 Probabilities of exceeding the WQS for prevention of river eutrophication of each river segment for the base-case scenario ([P] < 0.03 P mg/l)

4.4. Determining Costs

Table 8 presents the agronomic balances for N and P for the municipality of Saint-Patrice-de-Beaurivage. Surpluses for N and P for this municipality were 134.3 kg/ha and 86.2 kg/ha, respectively, which corresponds to 98,346 m^3 of liquid manure. Treatment cost was assumed to be \$12/m^3 [FPPQ, 2001], for a total cost of \$1,180,150. The unit cost accounted for operational (manpower, energy and inputs) and immobilization (acquisition and installation depreciated over the life expectancy of the technology, and capital) costs.

Table 7. Benefits for the base-case scenario and the *REA* scenario

Activity	Population	Number of Days		Monetary Benefits [$]	
		Base-case scenario	*REA* Scenario	Base-case scenario	*REA* Scenario
Hiking	3,267	31,396	44,562	1,224,439	1,737,913
Camping	3,955	38,012	53,953	1,482,482	2,104,168

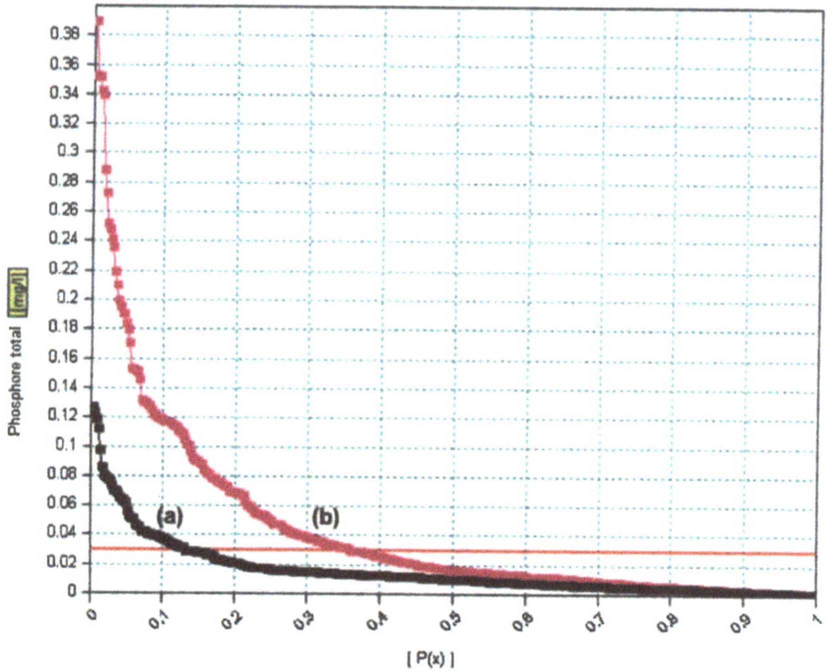

Figure 6. Complementary cumulative curve of the probability density function for: (a) the base-case scenario and (b) the *REA* scenario. Intersection of the horizontal line with those two curves indicates probabilities of exceeding the WQS for prevention of river eutrophication

4.5. Benefit-Cost Ratio

First, it is noteworthy to restate that socio-economic benefits were not considered as part of the *REA* scenario (*e.g.*, economic downfalls of setting up a liquid manure treatment plant; improved air quality, improved public health). The benefit-cost ratio of the *REA* scenario is 0.96 assuming a 0 % discount rate. This ratio indicates that benefits and costs are of the same order of magnitude despite the inherent uncertainties associated with benefit-cost analyses. The small difference between the costs and benefits leaves some room for discussion among decision makers and stakeholders. For example, if the general population is keen on improving water quality, it may be politically feasible to distribute the costs to all citizens regardless of their intent to use the studied river segment for recreational activities such a hiking and camping. The use of a higher discount rate makes the initial cost weigh much more heavily in the calculation, leading to a less important benefit-cost ratio [Tienteberg, 2000]. Finally, cheaper treatment technologies may eventually increase the benefit-cost ratio.

Table 8. Agronomic balances for N and P for the municipality of Saint-Patrice-de-Beaurivage

	Nitrogen (kg/ha)	Phosphorus (kg/ha)
Offer	185.82	117.38
Crop Requirement	51.51	31.19
Balance	134.31	86.19

5. SUMMARY AND CONCLUSION

This paper introduced, using a case study, the development of a methodology to conduct benefit-cost analyses of various alternative management scenarios focusing on supporting the reduction of agricultural pollution at the watershed level. The proposed methodology is primarily based on benefit transfer and water quality simulations and accounts for potential users contained within the neighbouring region of the studied river segment. This methodology has the potential to be quite useful for assessing the economic feasibility of environmental policies or legislations supporting sustainable agricultural development. Its integration within a hydrological modeling framework such as GIBSI represents a cost-effective way of estimating the economic feasibility and simulating and the water quality impacts of such policies or legislations at the watershed level. The availability of data on the total economic value of nature for Canadians allowed for the development of this framework based on benefit transfer. The methodology may be summarized or viewed as a five-step procedure. The first step focuses on problem definition, the second step on valuing benefits, the third step on water quality simulations, the fourth step on determining costs, and the fifth step on evaluating a benefit-cost ratio.

The case study presented in this paper focused on evaluating the water quality benefits of the *REA* legislation in terms of improving aesthetic and visual uses (namely, camping and hicking) of water on a river segment downstream the municipality of St-Patrice-de-Beaurivage in the Chaudière River basin, Quebec. Improved water quality provided for an average annual gain of 29,107 days adding up to $1,135,160 in total benefits. These results were representative of the 1982-1985 meteorological series used for the water quality simulations. Application of nutrient management plans within the aforementioned municipality required the treatment of 98,346 m^3 of liquid manure for a total cost of $1,180,150. Since benefits and costs were of the same order of magnitude, it somehow proved that the *REA* legislation is economically feasible. The use of a higher discount rate would make the initial cost much higher resulting in a smaller benefit-cost ratio. It is noteworthy that the economic value only represented an order of magnitude of the benefit of water quality improvements. On another front, these results have to be considered preliminary as ongoing research focuses on assessing model uncertainty. This uncertainty is a combination of incomplete understanding of modeled systems, and inadequate measurements of model parameters and input variables.

Finally, as stated by Eisen-Hecht and Kramer [2002], use of a water quality model allows decision makers and stakeholders flexibility to compare costs and benefits for different alternatives and tests the effectiveness of a proposed environmental management policy or legislation.

ACKNOWLEDGMENTS

The authors wish to acknowledge the financial support of the GIBSI I and GIBSI-II projects by the following programs, "*Fonds de recherche et de développement technologique en environnement, Projet de recherche exploratoire en environnement (FRDT-E, PREE)*" and "*Programme d'aide à la recherche et au développement en environnement (PARDE)*", respectively, which are administered by the Quebec Ministry of Environment.

REFERENCES

Barde, J.P., Économie et politique de l'environnement. Presses Universitaires de France, Paris. 2e Édition. 379 p. 1992.

Barbier, E.B., M.C., Acreman and D. Knowler, Évaluation économique des zones humides : Guide à l'usage des décideurs publics et planificateurs. Bureau de la Convention de Ramsar, Gland, Suisse. 1997. [On line]. http://www.ramsar.org/lib_val_f_intro.htm, date of access November 6, 2000.

Baril, P., La prévention : les apports diffus. Dans Symposium sur la Gestion de l'Eau au Québec, Volume 1 : Recueil de Textes des Conférenciers. Édité par : Villeneuve, J.-P., A. N. Rousseau, et S. Duchesne. Université du Québec, Institut National de la Recherche Scientifique, INRS-Eau, Québec, Canada., pp. 97-101. 1998.

Bédard, Y, S. Gariépy and F. Delisle, Bassin versant de la rivière Chaudière : l'activité agricole et ses effets sur la qualité de l'eau, Ministère de l'Environnement et de la Faune du Québec et Saint-Laurent Vision 2000, 116 pages, 1998.

Bergstrom, J.C., Current Status of Benefits Transfer in the U.S. : A Review. European-American University of Georgia, Department of Agricultural and Applied Economics, Faculty Series 96-09, April 1996.

Boardman, A.E., D.H. Greenberg, A.R. Vining and D.L. Weimer, Cost-Benefit Analysis: Concepts and Practice. Prentice Hall, Upper Saddle River, New Jersey, 1996.

Bonny, S., Les possibilités d'un modèle de développement durable en agriculture. Le cas de la France. *Courrier de l'environnement no 23*, pp.5-15., 1994.

Boyle, K.J. and J.C. Bergstrom, Benefit Transfer Studies: Myths, Pragmatism, and Idealism. *Water Resources Research*, 28 (3): 657-663, 1992.

Brookshire, D.S. and H.R. Neill, Benefit Transfers : Conceptual and Empirical Issues. *Water Resources Research*, 28 (3): 651-655, 1992.

Brouwer, R., Environmental value transfer : state of the art and future prospects. *Ecological Economics*, 32 (1): 137-152, 2000.

Brouwer, R. and F.A. Spaninks, The Validity of Environmental Benefits Transfer : Further Empirical Testing. *Environmental and Resource Economics*, 14: 95-117, 1999.

Debailleul, G., Le processus d'intensification de l'agriculture québécoise et ses impacts environnementaux : une rétrospective à méditer. *Vecteur Environnement*, 31 (2) : 49-54, 1998.

Delisle, F., S. Gariépy and Y. Bédard, *Bassin versant de la rivière Yamaska : l'activité agricole et ses effets sur la qualité de l'eau.* Ministère de l'Environnement et de la Faune en collaboration avec Saint-Laurent 2000. 124 pp., 1998.

Desaigues, B. et P. Point, *Économie du patrimoine naturel – La valorisation des bénéfices de protection de l'environnement.* Economica, Paris, 317 pp., 1993.

Eisen-Hecht, J.I. and R.A. Kramer, A cost-benefit analysis of water quality protection in the Catawba basin. *Journal of the American Water Resources Association*, 38 (2):453-465, 2002.

Environnement Canada, *L'importance de la nature pour les Canadiens: Les avantages économiques des activités reliées à la nature.* Groupe de travail fédéral-provincial-territorial sur l'importance de la nature pour les Canadiens. 49 pp., 2000.

Environnement Canada, *L'importance de la nature pour les Canadiens: Rapport sommaire de l'Enquête.* Groupe de travail fédéral-provincial-territorial sur l'importance de la nature pour les Canadiens, 1999.

Faucheux, S. and J.F. Noël, *Économie des ressources naturelles et de l'environnement.* Armand Colin Éditeur, Paris, 1995.

Fortin, J.P., R. Turcotte, S. Massicotte, R. Moussa, J. Fitzback and J.P. Villeneuve, A distributed watershed model compatible with remote sensing and GIS data. Part 1: description of the model. *Journal of Hydrologic Engineering*, 6(2), 91-99, 2001.

FPPQ. Rapport d'évaluation des technologies de gestion et de traitement du lisier de porc. Rapport du groupe de travail « Transfert technologique » du Plan agroenvironnemental de la production porcine pour la Fédération des producteurs de porcs du Québec, 2001.

FWR, Assessing the benefits of surface water quality improvements Manual. FWR/CL 0005. Foundation for Water Research, Marlow, 1996.

Gangbazo, G., D. Cluis and C. Bernard, Connaissances acquises en pollution diffuse agricole au Québec – 1993–1998 : Analyse et perspectives. *Vecteur Environnement*, section scientifique, juillet 1999, pp. 36-45, 1999.

Gangbazo, G. and J. Painchaud, Incidence des politiques et programmes d'assainissement agricole sur la qualité de l'eau de six rivières - 1988 -1995. *Vecteur Environnement*, 32 (1): 29-36, 1999.

Hanley, N. and C. L. Spash, Cost-Benefit Analysis and the Environment. Edward Elgar, 1993.

Laflamme, D., I. Piché, A. Michaud, Y. Bédard, G. Trencia, R. Laroche, L. Champagne and J.M. Gouin, Situation environnementale du bassin de la rivière Boyer. Ministères de l'Environnement et de la Faune, de l'Agriculture, des Pêcheries et de l'Alimentation en collaboration avec Saint-Laurent 2000 et le Groupe d'intervention pour la restauration de la rivière Boyer, 1997.

Legg, W., Sustainable agriculture : an economic perspective. Conference in the UK on Sustainable Agriculture. [On line]. http://www.ocde.org/agr/News/cont-8.htm, date of access June 2, 2000.

Mailhot A., A.N. Rousseau, S. Massicotte, J. Dupont and J.P. Villeneuve, A Watershed-Based System for the Integrated Management of Surface Water Quality: The GIBSI System. *Water Science Technology*, 36 (5): 381-387, 1997.

Ministère de l'Environnement du Québec (MENV). Règlements sur les exploitations agricole – En bref. 39 pages, 2002a.

Ministère de l'Environnement (MENV), IMAGE : Information sur les milieux aquatiques pour la gestion de l'eau. [On line] http://www.menv.gouv.qc.ca/eau/sys-image, date of access, January 31, 2002b.

Ministère de l'Environnement du Québec, Critères de qualité de l'eau (1990; rév. 1992). Service de l'évaluation des rejets toxiques et Direction de la qualité des cours d'eau, ministère de l'Environnement du Québec, Québec. 1992.

Revéret, J.-P., L'analyse coûts-avantages – Un outil contesté dans le domaine de l'environnement. Bureau d'audiences publiques sur l'environnement, Gouvernement du Québec. 29 p, 1984.

Podar, M., V. K. Smith, G. L. Van Houtven, S. Pattanayak, T. H. Bingham, *Improving the Practice of Benefit Transfer : A Preference Calibration Approach.* Préparé pour la U.S. Environmental Protection Agency, Office of Water and Office of Policy, Economics, and Innovation, 68-C6-0021. 2000.

Rousseau, A.N., A. Mailhot and J.-P. Villeneuve. Information technologies in the wider perspective: integrating management functions across the urban-rural interface. (Submitted for publication to *Environmental Modelling and Software*, 02/03), 2003

Rousseau, A. N., A. Mailhot, S. Gariépy, E. Salvano and J.P. Villeneuve, Calcul de probabilités de dépassement d'objectifs environnementaux de rejets de sources ponctuelle et diffuse à l'aide du système de modélisation intégrée GIBSI. *Revue des Sciences de l'Eau*, vol. 15, numéro spécial, pp. 121-148, 2002a.

Rousseau, A. N., A. Mailhot and J.P. Villeneuve, Development of a risk-based TMDL assessment approach using the integrated modelling system GIBSI. *Water Science and Technology*, vol. 45, number 9, pp. 317-324, 2002b.

Rousseau, A. N, A. Mailhot and J.P. Villeneuve, La gestion intégrée de l'eau par bassin versant : Problématique, éléments de solution et exemples d'application du système informatisé GIBSI. *Vecteur environnement*, vol. 33, numéro 5, pp. 27-30 et 51-54, 2000a.

Rousseau, A. N, A. Mailhot, R. Turcotte, M. Duchemin, C. Blanchette, M. Roux, N. Etong, J. Dupont and J.P. Villeneuve, GIBSI – An integrated modelling system prototype for river baisin management. *Hydrobiologia* 422/423, pp. 465-475, 2000b.

Simard, R.R., D. Cluis and G. Gangbazo, Animal unit density and P loads in soils of the Beaurivage Watershed. *Revue Canadienne de la Science du sol*, vol 73, no. 4, page 653, 1995.

Simoneau, M., L. Pelletier and N. Martel, Le bassin de la rivière Chaudière : profil géographique, sources de pollution et interventions d'assainissement. Direction des écosystèmes aquatiques, ministère de l'Environnement et de la Faune, Québec. Envirodoq no EN980022, pp. 1.1 à 1.34, 1998.

Smith, V. K., On Separating Defensible Benefit Transfers From « Smoke and Mirrors ». *Water Resources Research*, vol. 28, no. 3, pp. 685-694, 1992.

Smith, C.S. and G.T. McDonald, Assessing the sustainability of agriculture at the planning stage. *Journal of Environmental Management*, 52, pp. 15-37, 1998.

Tietenberg, T.H, Environmental and Natural Resource Economics. Fith Edition Addison-Wesley Longman inc. 630 pages. 2000.

Turcotte R., J.P. Fortin, A.N. Rousseau, S. Massicotte and J.P. Villeneuve, Determination of the drainage structure of a watershed using a digital elevation model and a digital river and lake network. *Journal of Hydrology*, 240: 225-242, 2001.

Villeneuve, J.P., P. Hubert, A. Mailhot and A.N. Rousseau, La modélisation hydrologique et la gestion de l'eau. *Revue des Sciences de l'eau*, vol. 11, pp. 19-40, 1998a.

Villeneuve, J.P., C. Blanchette, M. Duchemin, J.F. Gagnon, A.N. Rousseau, M. Roux, J.F. Tremblay and R. Turcotte, Rapport Final du Projet GIBSI, Mars 1998, Rapport No. R-462, INRS-Eau, Sainte-Foy, Québec, 1998b.

WCED, Our Common Future. Australian edition. Melbourne: Oxford University Press, 1990.

IMPORTANCE OF HYDRIC SOILS AND NEAR-LAKE AREAS AS PHOSPHORUS SOURCE AREAS IN THE LAKE CHAMPLAIN BASIN: EVIDENCE FROM A LANDSCAPE-LEVEL MODEL

Nicole Seltzer and Deane Wang*

1. INTRODUCTION

The passage of the Clean Water Act in 1972 made maintaining and improving surface water quality a national goal. Much of the initial effort was levied against point sources of pollution (Puckett, 1995); however, controlling non-point pollution has become a priority as control of the remaining point source problems becomes increasingly less cost-effective. Non-point phosphorus (P) pollution, in particular, has received frequent public attention as it threatens many of our nation's streams and rivers. Eutrophication, typically caused by high P levels, has been identified as the number one water quality problem in many regions (NYC DEP, 1999; LCBP, 1994).

In light of the above, the identification of non-point sources is critical to minimizing phosphorus' impact on surface waters. Progress in this arena has been hindered by limited resources, as water quality assessment often requires continuous monitoring (Heidtke and Auer, 1993). To aid in the implementation of water quality programs, managers have turned to using a variety of models to both predict P levels (e.g., Auer et al., 1997) and to identify land areas most responsible for elevated P levels (e.g., Zollweg et al., 1995). The latter is an important management tool, as targeting source areas for clean up allows the implementation of cost-effective remedial strategies (Sharpley, 1995).

Within the literature, three general categories of non-point P models exist to predict P at varying spatial scales: complex mechanistic models (often developed for the plot or field scale), empirical models (used at any scale), and exploratory regression models (often employed at the landscape level). Mechanistic models such as Chemical Runoff

* School of Natural Resources, University of Vermont, Burlington, Vermont, 05405, Deane Wang is the corresponding author.

and Erosion from Agricultural Management Systems (CREAMS) (Knisel, 1980) and the Areal Nonpoint Source Watershed Environment Response Simulation (ANSWERS) (Beasley et al., 1982, 1985) incorporate both physical and chemical processes in order to predict nutrient and sediment loading. Some of these models can rely upon process-related equations that are uncertain and often inaccurate (Novotny and Chesters, 1989; Levine et al., 1993; Harris, 1998), do not utilize a resolution fine enough to capture the resolution of the parameters that drive the model (Baun, 1986), and are limited to use by large agencies or research groups that have the funds and computer technology to run them (Levine et al., 1993; Johnes and Hodgkinson, 1998).

Empirical models generally use a single equation to express the relationship between an easily measured parameter (i.e., land use) and non-point P. Export models (e.g., Johnes ,1996; Johnes and Hodgkinson, 1998; Heidtke and Auer, 1993) fall into this category. These models typically have low data requirements, can reflect a number of spatial scales, and serve to obtain a rough estimate of average long-term nutrient totals (Heidtke and Auer, 1993). Drawbacks include the inability to model episodic events and incorporate spatial and temporal information (Harris, 1998) and a lack of reliable export coefficients in the literature (Johnes and Hodgkinson, 1998; Levine et al., 1993).

Exploratory regression models are able to go beyond the single watershed scale. Model construction usually involves the application of spatial data to look for relationships between nutrient levels and various landscape parameters (e.g. Lamon, 1995; Robertson, 1997; Tufford et al., 1998). These models are an essential tool to study large-scale system behavior and provide the opportunity to integrate many types of data (e.g. SPARROW model used for the Chesapeake Bay and Gulf of Mexico, Smith et al., 1997, Alexander et al., 2000). A helpful feature of regression models is that they can combine spatial variations in landscape features such as geology, land use and biological communities (Johnes and Hodgkinson, 1998). In this way, they can predict overall patterns in dependent variables, permitting the identification of outlier watersheds.

The choice of model type and spatial scale is an important one, as it has bearing on the results generated. The processes affecting P transfer through the environment are different at varied spatial scales, causing wide variation in the results found in studies conducted at different scales (Wiley et al., 1997; Levin, 1992). Ultimately, researchers must consider the suitability of each spatial scale in solving the perceived problem. Recent discussion in the literature has stressed the importance of the integration of spatial scales in modeling, and ideas abound on the best way to do it (Wiley et al., 1997; Johnes and Hodgkinson, 1998; Levine et al., 1993).

An important first step in the integration of spatial scales is understanding which P transfer processes are most important at each scale. Within the Lake Champlain Basin, phosphorus research using mechanistic (Cassell et al., 1998) and empirical (Budd and Meals, 1994; Weller et al., 1996; Meals and Budd, 1998; Hegman et al., 1999) model has been conducted at a variety of scales. However, exploratory regression models at the whole basin scale are lacking. Landscape level studies of the basin are important because they can provide insight into the links between large scale compartments of the P cycle, which may not be apparent at a smaller scale.

For this study, a GIS database for the Vermont side of the Lake Champlain Basin was constructed to include a broad variety of landscape variables including land use, hydrology, geology, and location. Multiple regression was then used to explore those variables and their relationship to measured P levels in stream outlets in the studied portion of the Basin. The relationships that emerge, while only "correlations" and not proof of causation, can be used to infer P transfer processes that may be of use to watershed managers.

2. METHODS

2.1. Study Area

The Lake Champlain Basin stretches from the Adirondack Mountains of New York in the west, to the Green Mountains of Vermont in the east. Its northern border encompasses a small area of the plains of Quebec, Canada (Figure 1). The basin is 21,326 km^2, 56% of which lies in Vermont, 37% in New York, and 7% in Canada.

Phosphorus levels within Lake Champlain have been a recent topic of interest in all three regions. In 1993, a Water Quality Agreement between Vermont, New York and Quebec was signed, establishing in-lake P criteria. As a result of this agreement, a variety of research has been done on in-lake P levels (VT and NY DEC, 1997), the role of land use on P levels (Hegman and others 1999, Budd and Meals, 1993; Weller and others 1996), and the effect of different agricultural management practices on P levels (Clausen and Meals, 1989).

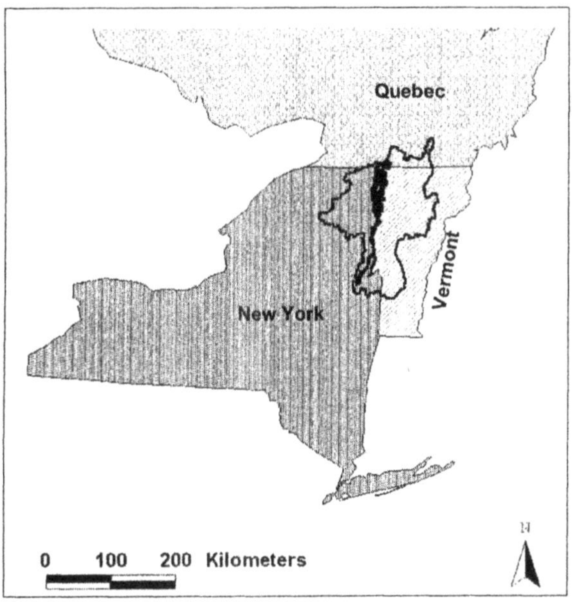

Figure 1. Location of the Lake Champlain Basin relative to Quebec, Canada; New York; and Vermont.

The broadest set of landscape data in Vermont are available for those 15 watersheds that drain into the lake from the Vermont side of the basin (Figure 2). Therefore, only those 15 watersheds were included in the study area. The land draining directly into the lake without an identifiable first order stream was not included in the study area, as no P data were available for it. The 15 study area watersheds, which range in size from 3,000 to 300,000 hectares, were hand digitized from 1:24,000 USGS topographical maps. The land use in the study area (1993 coverage) was 21% agriculture, 72% forest and 7% urban. The soils are glacial till in many areas, with large clay deposits near the lake.

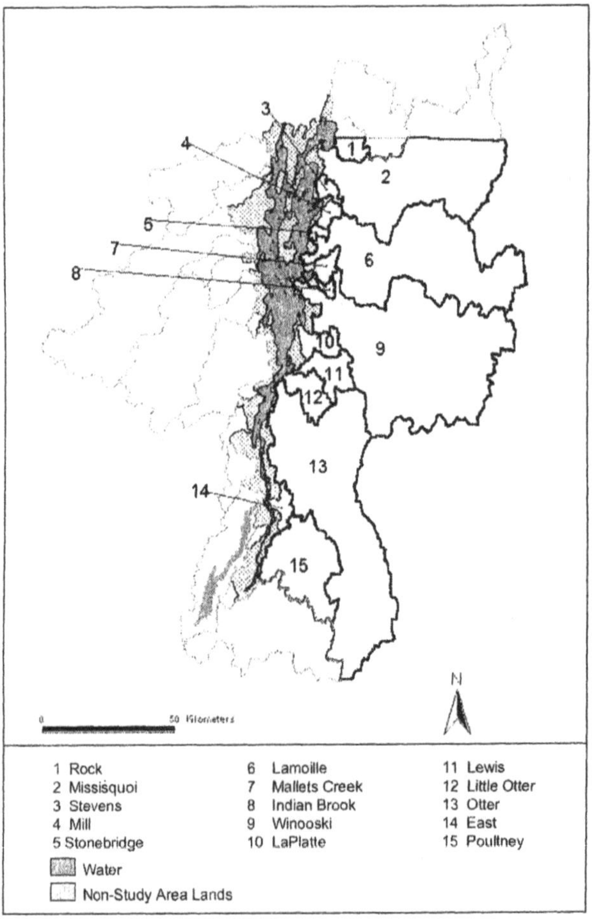

Figure 2. The 15 watersheds in the Lake Champlain Basin included in the study area.

2.2. Water Quality

Total Phosphorus (TP) concentration was measured at 15 stations during the period March 1990 to April 1992 by the Vermont and New York Departments of Environmental Conservation (VT DEC and NY DEC, 1997). Sample stations were located at the outlet of each of the 15 watersheds, upstream from any backwater influence from the lake. The number of samples taken at each site ranged from 103 to 82 in low, moderate and high flows, with 9%-35% representing high flows. TP samples were analyzed in VT DEC labs using EPA's Standard Method 4500-PF.

The plant operations records for all permitted municipal and industrial wastewater P discharges to surface waters were examined by the DEC, and from this an estimate was made of the point source P load to the lake. In this manner, P inputs to the lake could be partitioned between point and non-point sources.

The focus of this study was on both the non-point P concentrations in streams and rivers and the mass of non-point P export to Lake Champlain. For the purposes of this study, the P concentration used is a volume-weighted, concentration (mg P liter^{-1}) for a year period, while P export is an mass (kg P watershed^{-1} year^{-1}). According to Osborne and Wiley (1988) it is important to examine the impact of pollutants in terms of both concentration (as used in loading coefficients), and export (area-based export coefficients). In this study, the export regression models used independent variables based upon total watershed area (i.e., total hectares of agriculture in each watershed), while the concentration regression model used independent variables based upon percentage of watershed area (i.e., percent of the watershed that is agriculture). If study watersheds vary greatly in size, explained variation (r^2) in export models will then include a component of size because larger watersheds will export more P. Concentration models do not have this issue in interpretation of results.

Two of the 15 watersheds included in the study, the Rock and Missisquoi, crossed the boundary between the United States and Canada. Because of a lack of spatially explicit data for the Canadian portion of the basin at the time of this study, the P export for the Canadian portion of these two watersheds was reduced based upon a previous model constructed for the Lake Champlain Basin by Hegman et al. (1999). To reduce the P export, regionally derived land use export coefficients were applied to the area of agricultural, urban and forested land within the Canadian portions of the watersheds. The calculated P export was then subtracted from the original export given in the DEC report (VT DEC and NY DEC, 1997) to obtain an estimate of the P exported from Vermont lands. Note that only the P export measurements were adjusted; P concentration was not reduced for these two watersheds because concentration is not necessarily dependent on the size of the watershed.

2.3. GIS Data Layers

A variety of GIS coverages were used to quantify several aspects of landscape structure. The individual variables (Table 1) were placed into one of four datasets based upon whether they were related to land use, hydrology, geology, or location.

2.3.1. Land Use

Sixteen land use categories for the study area were obtained from classification of LANDSAT Thematic Mapper satellite data (VCGI, 1997A). Thirteen of the sixteen individual variables were used in analyses, as well as three aggregated variables (Table 1). A digital line graph coverage of major roads was obtained (VCGI, 1993) in order to estimate the density of paved roads in each watershed. Data on the number of farm animals in each watershed (Hegman et al., 1999) were also incorporated into analyses.

Table 1. Information on variables used in analyses. Variables shown are export model variables; variables used in the concentration models contain similar information, but are ratios of the variable to total watershed area (i.e., percentage rather than total square hectares of total agriculture). * indicates that totals include areas classified as wetland; wetland total includes agricultural areas classified as wetland.

Dataset	Indep. Variable	GIS Data Coverage Used	Coverage Source	Resolution
Borders		Lake and Watersheds	Hegman et al., 1999	1:24,000
Land Use	Barren Land	LANDSAT Land Use	VCGI, 1997A	1:250,000
	Total Agriculture*	"	"	"
	Brush	"	"	"
	Orchards	"	"	"
	Other Ag.	"	"	"
	Row Crops	"	"	"
	Hay / Pasture	"	"	"
	Total Forest*	"	"	"
	Decid. Forest	"	"	"
	Conifer. Forest	"	"	"
	Mixed Forest	"	"	"
	Total Urban*	"	"	"
	Residential	"	"	"
	Commercial	"	"	"
	Industrial	"	"	"
	Other Urban	"	"	"
	Roads	RDSCLx	VCGI, 1993	1:100,000
	Animal Units	VT & Quebec Animal Unit	Hegman et al., 1999	N/A
Hydrology	Surface Waters	National Hydrography dat.	USGS, 1999	1:100,000
	Total Wetland*	VSWInn; LANDSAT Land Use	VCGI, 1996; VGCI, 1997A	1:24,000; 1:250,000
	For. Wetland	"	"	"
	Non-for. Wetland	"	"	"
	Farmwater	Agricultural Streams	Seymour, 1998	1:5,000
	Avg. Precipitation	Precipitation Surface	created for this study	1:250,000
Geology	Clay	STATSGO	NRCS, 1994	1:250,000
	K-Factor > 3.0	"	"	"
	Hydric	"	"	"
	Slope > 25%	DEMnnnn	VCGI, 1997B	1:24,000
Locational	Riparian Agric.	National Hydrography dat.; LANDSAT Land Use	USGS, 1999; VCGI, 1997B	1:100,000; 1:250,000
	Riparian Forest	"	"	"
	Riparian Urban	"	"	
	Riparian Wetland	National Hydrography dat.; LANDSAT LU; VSWInn	USGS, 1999; VCGI 1997B; VCGI, 1996	1:100,000; 1:250,000; 1:24,000
	Riparian Clay	National Hydrography Dataset; STATSGO	USGS, 1999; NRCS, 1994	1:100,000; 1:250,000
	Riparian Hydric	"	"	"
	Near-lake Agriculture	Lake Champlain; LANDSAT Land Use	Hegman et al., 1999; VCGI, 1997B	1:24,000; 1:250,000
	Near-lake Forest	"	"	"
	Near-lake Urban	"	"	"
	Near-lake Wetland	Lake Champlain; VSWInn; LANDSAT Land Use	Hegman et al., 1999; VCGI, 1996	1:24,000; 1:24,000; 1:250,000
	Near-lake Hydric	Lake Champlain and Watersheds; STATSGO	Hegman and others, 1999; NRCS, 1994	1:24,000; 1:250,000

2.3.2. Hydrology

Digital hydrology data, including both streams and lakes, were obtained from the USGS (USGS, 1999) in order to quantify the amount of surface water in each watershed. Wetlands data were gained from two sources: the LANDSAT digital land use data described previously, and the Vermont Significant Wetlands Inventory (VSWI) (VCGI, 1996). Wetlands were classified as either forested or non-forested. The *Farmwater* variable was created from The Vermont Housing and Conservation Board's detailed coverage of streams, brooks and rivers that flow over agricultural lands within the Lake Champlain Basin (Seymour, 1998).

Precipitation data were obtained for 57 stations in and around the basin for the years 1951 - 1998 for the United States, and 1951 - 1991 for Canada (NCDC, various years; Environment Canada, 1994). These data were then averaged to create an annual value, and ArcView GIS 3.2 (ESRI, 1999) was used to create a precipitation surface for the entire basin following the geospatial interpolation methods of Hegman and others (1999).

2.3.3. Geology

Surficial geology data were obtained from the State Soil Geographic (STATSGO) database (NRCS, 1994), at a resolution of 1:250,000. The STATSGO database was used, in conjunction with its accompanying relational tables, to extract information on the presence or absence of hydric and clay soils. The structure of the STATSGO relational tables was such that a threshold had to be established in order to classify soils as hydric or clay. A soil polygon in the GIS soil coverage is predominately one soil type, but there can be upwards of 15 other types present. If a soil polygon consisted of greater than 10% hydric soils, it was classified as hydric in this study. Similarly, a threshold of 40% clay was established to categorize a soil polygon as clay for this study. Those soils, with a K-Factor greater than 3.0, represented by the variable *K-Factor > 3*, are considered highly erodible (Brady, 1990), and so were also included in the study. Defining these soil polygons using these % thresholds was done empirically, as the regressions were stronger than when a few different %'s were used. Testing of threshold %'s was not comprehensive.

A 7.5 minute Digital Elevation Model (DEM) (VCGI, 1997B) was used to extract information on slope within each of the watersheds. The variable *Slope* represents the amount of land in each watershed with greater than 25% slope.

2.3..4 Location

Two sets of locational variables were created, riparian variables and near-lake variables. The buffer utility included in ArcView 3.2 (ESRI, 1999) was used to extract land use, wetland and geology information from a 500 meter distance around all streams identified in the USGS digital stream water coverage (VCGI, 1997A). The resulting seven riparian variables (Table 1) were created in order to allow comparison of landscape variables within river corridors to data derived from the whole watershed (Osborne and Wiley, 1988; Johnson et al., 1997; Tufford et al., 1998).

The second set of locational variables, the near-lake variables, were also created with the ArcView buffer utility (Table 1). Information on land use, geology and wetlands was extracted from an empirically derived 5.6 km buffer zone around the lake's shoreline. The resulting five near-lake variables were included in the model building process to test the hypothesis that characteristics of land closer to the lake have a significant influence on P levels in the lake.

2.4. Statistical Methods

Statistical analyses were run using both P concentration (mg P l^{-1}) and P export (kg P watershed^{-1} year^{-1}) as dependent variables. The distribution of all dependent and independent variables was assessed using visual observation of histograms as well as the Shapiro-Wilk statistic (SAS Institute, 1988). Most variables violated the assumption of normality, tending instead to be log-normally distributed. Therefore, all independent variables for the export model were transformed to log_{10} *(100 + x)* prior to all analyses, while the independent variables for the concentration model were transformed to log_{10} *(0.5 + x)*. These transformations gave all variables a normal distribution. Dependent variables for the two models were only log-transformed.

Principal components analysis (PCA) on the covariance matrix was used to reduce the initial number of independent variables to a smaller set that still explained the majority of the dataset variation (Johnson and Gage, 1997). Separate PCA's were conducted for each model; one on each of the four datasets (land use, hydrology, geology, or location). For those principal components that explained greater than one percent of the total dataset variation, the relationships between the principal components and each variable, as well as between the individual variables, were examined using the correlation matrices. Variables were kept or discarded based upon two criteria: 1) variables could not be highly correlated within the dataset, and 2) variables kept must significantly correlate with at least one of the retained principal components. As this initial variable reduction procedure was not the final variable selection, variables were retained if there was any question of their contribution to the dataset.

Using the new, smaller set of variables, forward-selection linear regression models with an initial probability for entry of 0.25 were constructed with non-point P concentration (mg P l^{-1}) or export (kg P watershed^{-1} year^{-1}) as the dependent variables. Variables significant at this level were then subjected to a more rigorous examination of their partial F-test, contribution to the model's r^2 and adjusted r^2, and minimization of the model's variance inflation factor, in order to identify that variable or combination of variables that best explain P export or concentration. Outliers were analyzed using the SAS system's PROC REG / INFLUENCE statistics (SAS Institute, 1988). The residuals of the final models were analyzed for normality and homogeneity of variance.

3. RESULTS

The principal components analysis reduced the number of independent variables from 39 to 10 in the export model datasets, and from 39 to 15 in the concentration model datasets. Many of the variables were highly inter-correlated, a common feature among landscape and other spatially distributed data (Johnson and Gage 1997), and so were deleted from consideration in the final model. In both models, a few highly correlated variables had similar, strong relationships with P. In the export model, ***Total Agriculture***, ***Row Crops*** and ***Hay/Pasture*** had highly significant correlations ($r > 0.9$) with ***P (kg watershed^{-1} year^{-1})***. These three independent variables are also highly inter-correlated, with $r > 0.95$. A similar situation occurred in the concentration model, with ***%Other Agriculture***, ***%Coniferous Forest*** and ***%Deciduous Forest*** all having significant relationships with ***P (mg l^{-1})*** ($r > 0.7$) as well as being inter-correlated ($r > 0.55$). In these cases, the variable that created the final model with the highest adjusted r^2 was kept, and all others deleted. Those variables kept for consideration in the final model are shown in Table 2.

Table 2. Final datasets used in the loading and export models. (Data were obtained from land cover information using ARC GIS.)

Export Model	Loading Model
Total Agriculture	%Total Agriculture
Other Ag/Mixed	%Orchards
Industrial	%Industrial
Other Urban	%Total Urban
Clay	%Clay
Hydric	%Hydric
Farmwater	%Farmwater
Total Wetland	%Total Wetland
Near-lake Agriculture	%Near-lake Agriculture
Near-lake Hydric	%Near-lake Hydric
Commercial	%K-factor > 3
	%slope
	%Brush
	%Surface Water
	%Total Forest
	%Barren

Forward selection linear regression using those export model variables shown in Table 2 as independent variables and *P (kg watershed^{-1} year^{-1})* as the dependent variable, showed that the combination of the variables *Hydric*, *Near-lake Agriculture* and *Near-lake Hydric* best explained P variation. The final export model, *P (kg watershed^{-1} year^{-1}) = -1.568 + 1.042(Hydric) + 1.091(Near-lake Hydric) - 0.878(Near-lake Agriculture)*, is plotted in Figure 3a. The model has an adjusted r^2 of 0.93 and an F-value with Prob>F of 0.001. The coefficients for *Hydric*, *Near-lake Hydric*, and *Near-lake Agriculture* have a Prob>F of 0.001, 0.011, and 0.022, respectively.

Using the concentration model dataset (Table 2), forward selection regression showed that the combination of *%Total Agriculture*, *%Near-lake Agriculture* and *%Near-lake Hydric* best explained *P (mg l^{-1})* variation. The final concentration model, *P (mg l^{-1})* = -3.1 + 1.396(*%Total Agriculture) + 0.84(%Near-lake Hydric) – 1.04(%Near-lake Agriculture)* is plotted in Figure 4. The model has an adjusted r^2 of 0.686 and an F-value with Prob>F of 0.001. The coefficients for *%Total Agriculture*, *%Near-lake Hydric*, and *%Near-lake Agriculture* have a Prob>F of 0.001, 0.03, and 0.021, respectively. The residuals of both models show homogeneity of variance and are normally distributed.

4. DISCUSSION

4.1. Significance of Soil Property Parameters

The inclusion of the variable *Hydric* in the export model may denote the influence of soil properties on P transport. The variable *Hydric* represents the hectares of soil in each watershed that is predominately hydric in character, according to NRCS databases. These soils are either close enough to the groundwater table or have poor enough drainage to be seasonally wet. These properties have an effect on the way that both water and P are transmitted.

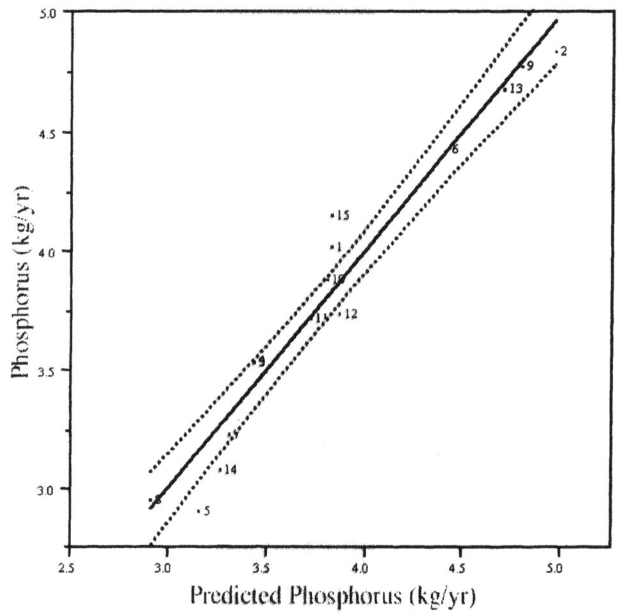

Figure 3. The final export regression model showing the regression line and the 95% confidence interval.

Proponents of the Variable Source Area (VSA) concept contend that soils located near surface waters produce most of the surface runoff in a watershed, and that the area of these contributing soils increases with rainfall intensity and duration (Pionke et al., 1990; Troendle, 1985; Gburek, 1990, Gburek and Sharpley, 1998). As the moisture content of these soils rises, the ability of water to percolate through the soil column decreases. As rains continue, the soil's infiltration capacity is exceeded and surface runoff occurs (Brady, 1990). Thus, as the soil in areas further away from the stream become saturated, they also begin to contribute to surface runoff, expanding the source area. This means that those soils in the watershed with greater infiltration capacities provide little surface runoff, with the processes of infiltration and groundwater recharge being dominant. As the soils classified as hydric in this study have properties that minimize their infiltration capacities, they may be an identifiable source of surface runoff in the basin. While all of the soils classified as hydric may not be located within a defined Variable Source Area, their ability to transmit surface runoff makes them a potential P source area.

Surface runoff has the energy to erode large amounts of soil and the capacity to carry particulates rich in P, making it an important mechanism in P transfer (Sharpley et al., 1994). Relying upon the premise of the VSA concept, if "contributing" soils are correctly classified, it should be possible to identify those areas of a watershed that are likely to be a large source of P to surfacee waters (Zollweg et al., 1995). A spatial overlap between high P-producing land uses and a high proportion of hydric soils should

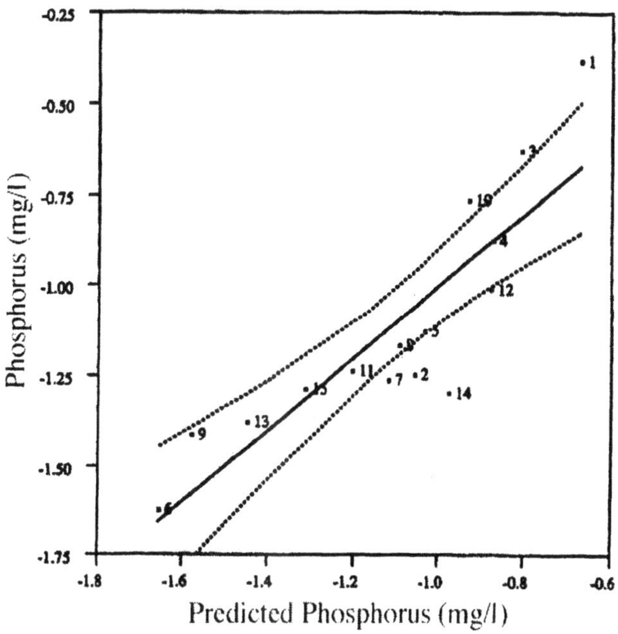

Figure 4. The final concentration regression model showing the regression line and the 95% confidence interval.

exist; however, for these source areas to exist. Analysis of the GIS database shows that an average of 82% of the study area's agriculture is located on areas classified here as hydric. As such, the significance of the variable *Hydric* may denote the importance of variable source area-related P transfer processes. The variables *Hydric* and *Total Agriculture* are unfortunately somewhat inter-correlated, making inferences about process more difficult.

The other "geological" variable did not help explain the variation in either export or concentration models due to either a lack of relationship or the high inter-correlation between land use and geologic parameters, which is common in these types of studies (Richards et al., 1996; Beaulac and Reckhow, 1982; Hill, 1981). Many recent studies have found significant relationships between P and land use (Robertson, 1997; Thierfelder, 1998; Tufford et al., 1998), which has reinforce the inclusion of only land use information in P models (Tufford et al., 1998; Driver and Troutman, 1989; Johnes and Heathwaite, 1997; Jordan et al., 1997; Meals and Budd, 1998; Hegman et al., 1999). However, this study, as well as other research that has found significant correlations between P and geology (Dillon and Kirchner, 1975; Robertson, 1997; Vaithiyanathan and Correll, 1992), suggests that focusing on the phosphorus-land use link may limit our future understanding of P dynamics in the landscape.

4.2. Significance of Locational Parameters

The addition of the near-lake variables into both the export and concentration models suggest that the location of landscape elements is important. One process-related mechanism that may explain the significance of these variables is in-stream nutrient retention. If in-stream retention is considerable, P entering a stream in the upland areas of a watershed would not subsequently reach the watershed outlet. Land near the lake may therefore have a larger influence on P levels in the lake than land in the watershed's headwaters (Thierfelder, 1998).

In-stream processes have the potential to influence the chemical and physical forms of P in a stream, which in turn affect how P is delivered downstream (Hill 1997; Meyer and Likens 1979; Minshall and others 1983). In a review of the transport mechanisms of P in streams and wetlands, Reddy and others (1999) cited many examples of studies in which streams retained significant amounts of P. A P transport model for the Lake Okeechobee Basin (Zhang and others 1996) found that, on average, one-third of the P reaching the stream would be assimilated upstream of the lake. House and Warwick (1998) concluded that both soluble dissolved P and total dissolved P levels decrease with distance downstream, probably due to uptake by sediments, macrophytes and benthic algae. These studies are in contrast to research that finds either no attenuation, or short-term attenuation followed by a large export (Hill, 1997; Wang et al., 1999; Meyer and Likens, 1979; Dorioz et al., 1989).

Given the variation in results of in-stream studies, a regional perspective provided by studies such as ours can be a source of information on the importance of this process. Isolated studies at a small scale are useful for exploring the mechanisms of in-stream retention and whether it is significant in specific stream environments. In order to view the management implications of significant in-stream nutrient retention however, a more regional perspective should be taken.

4.3. Negative Model Coefficients

Note that the coefficients on the variables *Near-lake Agriculture* and *%Near-lake Agriculture* are negative in both models. The negative coefficient, when taken literally, means that as the amount (or percentage) of agriculture in the near-shore zone increases, the P-export (or concentration) to the lake should decrease, a non-intuitive relationship. The possible source of this negative relationship becomes evident when the ratio of *Near-lake Hydric* : *Near-lake Agriculture* is examined. When the two variables are plotted separately against the ratio, it becomes clear that watersheds with a small amount of agriculture in the near-lake zone also have small amounts of hydric soil there. As the amount of agriculture increases, however, the relative amount of hydric soil in the near-lake zone decreases. As previously discussed, the relative amount of hydric soil versus agriculture in a watershed may have a large influence on P transfer, as agriculture on hydric soils may exports larger amounts of P than agriculture on non-hydric soils. The lower relative amount of near-lake hydric soil in watersheds with large amounts of near-lake agriculture would therefore mean lower overall P export. This relationship may account for the negative coefficient on the *Near-lake Agriculture* and *%Near-lake Agriculture* variables.

5. CONCLUSIONS

The results of this study exemplify the use of a landscape-level analysis to suggest P transfer processes are important at a large scale. It was conducted in order to develop landscape level research in the Lake Champlain Basin, as field scale studies, or even studies at the single watershed level, do not always allow us to identify the connections that exist among distinct P transfer processes (Schlosser and Karr, 1981). This study's use of a different research perspective illuminated two potentially important P transfer processes, hydric soil export and potential in-stream nutrient retention, that have been largely unexplored in the Lake Champlain region.

Landscape-level analyses have benefits that reach beyond their ease of use and simple data requirements. The information gained through this type of research raises questions about the assumptions we rely upon in modeling P in our landscape. The significance of both hydric soils and near-lake lands in this study point to complexities within the P transfer system that are not reflected in simply land use-based export coefficient modeling. Also beneficial is the landscape level model's focus on variability in P output between watersheds, making the recognition of outlier watersheds straightforward. If followed by a search for variables to bring in the outliers, such as the near-lake lands in this study, landscape level models can be useful in providing information that may point to why one watershed area produces more P than another. These important benefits help investigators identify new research questions, making the landscape level model a useful tool in setting research priorities for a region.

It is important to note that landscape scale models complement, but cannot replace the information gained from smaller scale studies. This study suggested the importance of both P source areas and in-stream nutrient retention. To substantiate this, a series of field experiments at a smaller scale should be conducted.

Identification of the landscape areas where P is exported or retained has important implications for both the scientific and management communities. Researchers can incorporate this knowledge into the development of spatially integrated models. These models can then be used by managers to narrow the scope of their remedial efforts by targeting discrete portions of the landscape, allowing for cost-effective management strategies. Incorporating various spatial scales into our research agendas results in a more complete understanding of the P transfer system, which in turn improves our decision-making abilities.

6. ACKNOWLEDGMENTS

The authors would like to thank three anonymous reviewers, who provided insightful comments and detailed editorial assistance that greatly improved the quality of the manuscript.

7. REFERENCES

Alexander, R.B., Smith, R.A., and Schwarz, G.E., 2000, Effect of stream channel size on the delivery of nitrogen to the Gulf of Mexico, Nature **403**: 758-761.

Auer, M.T., S. Doerr, S. Effler, and E. Owens, 1997, A Zero Degree of Freedom Total Phosphorus Model: 1. Development for Onondaga Lake, New York, *Journal of Lake and Reservoir Management* **13**(2):118-130.

Baun, K . 1986. *Application of the ANSWERS model in a nonpoint source program: final report.* Wisconsin: Wisconsin Department of Natural Resources, Nonpoint Source and Land Management Section.

Beasley, D.B., L.F. Huggins, and E.J. Monke. 1982. Modeling sediment yields from agricultural watersheds. *Journal of Soil and Water Conservation* **37**:113-117.

Beasley, D.B., E.J. Monke, E.R. Miller, and L.F. Huggins. 1985. Using simulation to assess the impacts of conservation tillage on movement of sediment and phosphorus into Lake Erie. *Journal of Soil and Water Conservation* **40**: 233-237.

Beaulac, M., and K. Reckhow. 1982. An Examination of land Use – Nutrient Export Relationships. *Water Resources Bulletin,* **18**:1013-1024.

Brady, Nyle C. 1990. *The Nature and Property of Soils* (10th ed.) . Macmillan, New York.

Budd, L.F., and D.W. Meals. 1994. *Lake Champlain nonpoint source pollution assessment. Lake Champlain Basin Program Technical Report 6A and 6B.* Grand Isle, Vermont.

Cassell, E.A., J.M Dorioz, and D.C. Braun. 1998. Modeling Phosphorus Dynamics in Ecosystems: Mass Balance and Dynamic Simulation Approaches. *Journal of Environmental Quality* **27** (2) : 293-307.

Clausen, J.C., and D.W. Meals. 1989. Water quality achievable with agricultural best management practices. *Journal of Soil and Water Conservation* **44**(6): 594-596.

Dillon, P., and W. Kirchner. 1975. The Effects of Geology and Land Use on the Export of Phosphorous from Watersheds. *Water Research* **9**:135-148.

Dorioz, J.M., E. Pilleboue, and A. Ferhi. 1989. Phosphorus dynamics in watersheds: role of trapping processes in sediments. *Water Research* **23** (2) : 147-158.

Driver, N., and B. Troutman. 1989. Regression models for estimating urban storm-runoff quality and quantity in the United States. *Journal of Hydrology* **109**:221-236.

Environment Canada. 1994. Monthly Climate Data and 1961-1990 Normals on CD-ROM. Canadian Meteorological Service, Dorval, Quebec, Canada.

Environmental Systems Research Institute (ESRI). 1999. *ArcView 3.2* On-line software help. Redlands, CA.

Gburek, William J. 1990. Initial contributing area of a small watershed. *Journal of Hydrology,* **118**: 387-403.

Gburek, W.J., and A.N. Sharpley. 1998. Hydrologic controls on phosphorus losses from upland agricultural watersheds. *Journal of Environmental Quality* **27**:267-277.

Harris, G.P. 1998. Predictive models in spatially and temporally variable freshwater systems. *Australian Journal of Ecology* **23**:80-94.

Heidtke, T.M., and M.T. Auer. 1993. Application of a GIS-based nonpoint source nutrient loading model for assessment of land development scenarios and water quality in Owasco Lake, New York. *Water Science and Technology* **28**(3-5): 595-

Hegman, W., D. Wang, and C. Borer. 1999. *Estimation of Basinwide Nonpoint Phosphorus Export – Revised Final Draft Report.* Lake Champlain Basin Technical Report #31. Grand Isle, Vermont.

Hill, Alan R. 1981. Stream Phosphorous Exports from Watersheds with Contrasting Land Uses in Southern Ontario. *Water Resources Bulletin,* **17**:627-634.

Hill, A.R. 1997. The potential role of in-stream and hyporheic environments as buffer zones. Pages 115-127 in *N.E. Haycock, T.P. Burn, K.W.T. Goulding, and G. Pinay (eds.)* Buffer Zones: Their Processes and Potential in Water Protection. Quest Environmental, U.K.

House, W.A., and M.S. Warwick. 1998. A mass-balance approach to quantifying the importance of in-stream processes during nutrient transport in a large river catchment. *The Science of the Total Environment* **210/211**: 139-152.

Johnes, P.J. 1996. Evaluation and management of the impact of land use change on the nitrogen and phosphorus load delivered to surface waters: the export coefficient modelling approach. *Journal of Hydrology* **183**: 323-349.

Johnes, P.J., and A.L. Heathwaite. 1997. Modelling the impact of land use change on water quality in agricultural catchments. *Hydrological Processes* **11**:269-286.

Johnes, P.J., and R.A. Hodgkinson. 1998. Phosphorus loss from agricultural catchments: pathways and implications for management. *Soil Use and Management* **14**:175-185.

Johnson, L., C. Richards, G. Host, and J. Arthur. 1997. Landscape influences on water chemistry in Midwestern stream ecosystems. *Freshwater Biology* **37**:193-208.

Johnson, L., and S. Gage. 1997. Landscape approaches to the analysis of aquatic ecosystems. *Freshwater Biology* **37**: 113-132.

Jordan, T.E., D.L. Correll, and D.E. Weller. 1997. Nonpoint source discharges of nutrients from Piedmont watersheds of Chesapeake Bay. *Journal of the American Water Resources Association* **33**(3): 631-645.

Knisel, W.G. Jr. (ed.). 1980. *A Field Scale Model for Chemicals, Runoff and Erosion from Agricultural Management Systems,* Vol. 3, USDA Conservation Research Report 26, US Government Printing Office, Washington D.C.

Lake Champlain Basin Program. 1994. *Opportunities for Action: An evolving plan for the future of the Lake Champlain Basin.* Lake Champlain Basin Program Technical Lamon, E.C. 1995. A Regression Model for the Prediction of Chlorophyll a in Lake Okeechobee, Florida. *Lake and Reservoir Management* **11**(4): 283-290.

Lamon, E. C., 1995, A regression model for the prediction of chlorophyll a in Lake Okeechobee, *Lake and Reservoir Management* **11**(4):283-290.

Levin, S.A. 1992. The Problem of Pattern and Scale in Ecology. *Ecology* **73**(6): 1943-1967.

Levine, D., C. Huntsaker, S. Timmins, and J. Beauchamp. 1993. *A Geographic Information System Approach to Modeling Nutrient and Sediment Transport.* Oak Ridge National Laboratory, Environmental Sciences Division. Pub No. 3993. Oak Ridge, TN.

Meals, D.W. and Budd, L.F., 1998, Lake Champlain Basin nonpoint source phosphorus assessment. *J. Am. Water Res. Assoc.* **34**(2): 251-265.

Meyer, J., and G. Likens. 1979. Transport and Transformation of Phosphorus in a Forest Stream Ecosystem. *Ecology* **60** (6) : 1255-1269.

Minshall, G.W., R. Petersen, K. Cummins, T. Bott, J. Sedell, C. Cushing, and R. Vannote. 1983. Interbiome comparison of stream ecosystem dynamics. *Ecological Monographs* **53**(1): 1-25.

National Climatic Data Center (NCDC). Various years. Climatological Data. Asheville, North Carolina.

Natural Resources Conservation Service (NRCS). 1994. *State Soil Geographic (STATSGO) Data Base Data Use Information.* Misc. pub. No. 1492. Fort Worth, Texas.

New York City Department of Environmental Protection. 1999. Development of a water quality guidance value for Phase II Total Maximum Daily Loads (TMDLs) in the New York City Reservoirs. *New York City Department of Environmental Protection Report.* Valhalla, NY.

Novotny, V., and G. Chesters. 1989. Delivery of sediment and pollutants from nonpoint sources: A water quality perspective. *Journal of Soil and Water Conservation* **44**(6): 568-576.

Osborne, L., and M. Wiley. 1988. Empirical relationships between land use/cover and stream water quality in an agricultural watershed. *Journal of Environmental Management* **26**:9-27.

Pionke, H.B., W.J. Gburek, A.N. Sharpley, and R.R. Schnabel. 1990. Flow and nutrient export patterns for an agricultural hill-land watershed. *Water Resources Research*, **32**: 1795-1804.

Puckett, Larry J. 1995. Identifying the Major Sources of Nutrient Water Pollution. *Environmental Science and Technology*, **29**:408A-414A.

Reddy, K.R., R.H. Kadlec, E. Flaig, and P.M. Gale. 1999. Phosphorus Retention in Streams and Wetlands: A Review. *Critical Reviews in Environmental Science and Technology* **29**(1):83-146.

Richards, C., L. Johnson, and G. Host. 1996. Landscape scale influences on stream habitats and biota. *Canadian Journal of Fish and Aquatic Science*, **53**(Suppl. 1):295-311.

Robertson, Dale M. 1997. Regionalized Loads of Sediment and Phosphorous to Lakes Michigan and Superior—High Flow and Long Term Average. *Journal of Great Lakes Research*, **23**:416-439.

SAS Institute, Inc. 1988. *SAS / STAT Users Guide* (Release 6.03 ed.). Pp. 837-843. Cary, NC.

Schlosser, I.J., and J.R. Karr. 1981. Water quality in agricultural watersheds: impact of riparian vegetation during base flow. *Water Resources Bulletin* **17**(2): 233-240.

Seymour, Susan. 1998. *Streams, brooks, and rivers that flow over farmland properties Metadata.* Cartographic Technologies, Inc., Brattleboro, VT.

Sharpley, A. 1995. Identifying sites vulnerable to phosphorus loss in agricultural runoff. *Journal of Environmental Quality* **24**(5):947-951.

Sharpley, A.N., S.C. Chapra, R. Wedepohl, J.T. Sims, T.C. Daniel, and K.R. Reddy. 1994. Managing agricultural phosphorus for protection of surface waters: Issues and options. *Journal of Environmental Quality* **23**:437-451.

Smith, R.A., Schwarz, G.E., and Alexander, R.B., 1997, Regional interpretation of water-quality monitoring data, *Water Resources Res.* **33**:2781-2798.

Thierfelder, Thomas. 1998. The morphology of landscape elements as predictors of water quality in glacial/boreal lakes. *Journal of Hydrology* **207**:189-203.

Troendle, C.A. 1985. Variable Source Area Models. pp. 347-407. IN Anderson, M.G. and T.P. Burt (eds.), *Hydrological Forecasting.*,Wiley, New York.

Tufford, D., H. McKellar, and J. Hussey. 1998. In-stream nonpoint source nutrient prediction with land-use proximity and seasonality. *Journal of Environmental Quality* **27**:100-111.

United States Geological Survey. 1999. *National Hydrography Dataset Catalog Unit (CU) Nos. 02010005, 02010003, 02010007, 02010001 and 02010002.* http://nhd.usgs.gov/data.html.

Vaithiyanathan, P., and D. Correll. 1992. The Rhode River Watershed: Phosphorous Distribution and Export in Forest and Agricultural Soils. *Journal of Environmental Quality*, **21**:280-288.

Vermont Center for Geographic Information, Inc. 1993. *RDSCLx Metadata.* Burlington, Vermont.

Vermont Center for Geographic Information, Inc. 1996. *VSWInn Metadata.* Burlington, Vermont.

Vermont Center for Geographic Information, Inc. 1997A. *Final Report: Development of Land Cover/Land Use Geographic Information System Data Layer for the Lake Champlain Basin and Vermont.* Burlington, Vermont.

Vermont Center for Geographic Information, Inc. 1997B. *DEMnnnn Metadata*. Burlington, Vermont.

Vermont Department of Environmental Conservation (DEC) and New York DEC 1997. *A phosphorus budget, model, and load reduction strategy for Lake Champlain: The Lake Champlain diagnostic-feasibility report*. Waterbury, Vermont and Albany, New York.

Wang, D., S.N. Levine, D.W. Meals, J. Hoffman, J. Drake, and E.A. Cassell. 1999. *Importance of Instream Nutrient Storage to P Export from a Rural, Eutrophic River in Vermont, USA. In Lake Champlain in Transition : From Research to Restoration* (pp. 205-223). The American Geophysical Union.

Weller, C., M. Watzin, and D. Wang. 1996. Role of Wetlands in Reducing Phosphorus Loading to Surface Water in Eight Watersheds in the Lake Champlain Basin. *Environmental Management*, **20**:731-739.

Wiley, M.J., K.L. Steven, and P.W. Seelbach. 1997. Reconciling landscape and local views of aquatic communities: lessons from Michigan trout streams. *Freshwater Biology* **37**: 133-148.

Zhang, J., T.S. Tisdale, and R.A. Wagner. 1996. A Basin Scale Phosphorus Transport Model for South Florida. *Applied Engineering in Agriculture* **12**(3): 321-327.

Zollweg, J.A., W.J. Gburek, H.B. Pionke, and A.W. Sharpley. 1995. GIS-based delineation of source areas of phosphorus within agricultural watersheds of the northeastern USA. *Modelling and Management of Sustainable Basin-scale Water Resource Systems (Proceedings of a Boulder Symposium, July 1995)*. IAHS Publ. No 231, 1995. Boulder, Colorado.

USING A LANDSCAPE APPROACH TO INTERPRET DIFFUSE PHOSPHORUS POLLUTION AND ASSIST WITH WATER QUALITY MANAGEMENT IN THE BASINS OF LAKE CHAMPLAIN (VERMONT) AND LAC LÉMAN (FRANCE)

Deane Wang[1], Jean-Marcel Dorioz[2], Dominique Trevisan[2], David C. Braun[1], Lisa J. Windhausen[1], and Jean-Yves Vansteelant[2]

Keywords: diffuse pollution, nonpoint, phosphorus, landscape, multiple regression, Lake Leman, Lake Champlain

ABSTRACT

Diffuse pollution should be recognized as a landscape-level phenomenon. As such, it requires an observational approach consistent with the complex structure and function of the landscape system. We developed a landscape-level approach to study the transfer of phosphorus in rural areas of the Lake Champlain and Lac Léman basins. We began by developing a concept of P dynamics that captured some of the diversity and complexity of P movement through the land (transfer system). Given this initial concept of the diffuse pollution in the landscape, we adopted a synoptic watershed sampling strategy to begin the quantitative description of diffuse P pollution. Data from these types of studies were then analyzed using multiple regression to infer connections between activities on the land and phosphorus flux to surface waters. Our inferences include: 1) land cover determines phosphorus flux during high flow but not during low flows periods, 2) during high flow events, natural wetlands are a significant sink for diffuse phosphorus in surface waters, 3) fluxes and concentrations are higher when the basins are intensively plowed, 4) in the context of plowed areas, agricultural practices as opposed to land cover is a more important determinant of phosphorus flux in watersheds, and 5) the position of elements in the landscape is an important factor controlling diffuse phosphorus pollution. The method and basis for arriving at these conclusions are discussed. We suggest that synoptic sampling of water quality over extensive areas in a landscape, coupled with multiple regression to analyze relationships among P fluxes and landscape variables, is

[1] School of Natural Resources - University of Vermont, Burlington, Vermont, USA
[2] Station d'Hydrobiology - Institut National de Recherche Agronomique, Thonon, France

an appropriate tool for determining driving factors, analyzing the diversity of processes, and finding generality in complex landscape systems.

1. INTRODUCTION

Human activity in the landscape has a substantial power to alter the quality of surface waters and ecosystems dependent on that quality. In the last two decades, as control of point sources has progressed, diffuse sources of pollution have been recognized as a major cause of this water quality degradation (U.S. Environmental Protection Agency, 1995; Pucket, 1995; CIPEL, 1988; Novotny and Olem, 1994). In temperate regions in freshwaters, phosphorus is generally the water quality parameter of greatest threat relative to eutrophication because it limits aquatic plant productivity (Vollenveider, 1968; Barroin, 1980; Hecky and Kilham, 1988). However, progress in controlling diffuse P pollution of surface waters has been slow (Pucket, 1995). In the watersheds of many large lakes, diffuse phosphorus has been identified as an important pollutant to control because of the threat of continued P loading further driving eutrophication (Lake Champlain Management Conference, 1996; CIPEL, 1984).

Understanding of the mechanisms of diffuse pollution is a critical step in achieving reductions. Developing this understanding at a landscape scale is key (e.g., Whigham et al., 1988; Johnston et al., 1990; Kling et al., 2000; Jones et al., 2001), because the transfer of diffuse P is a landscape-level phenomena primarily involving 1) erosion and runoff from a complex diversity of landscape surfaces differentiated by a variety of factors including soil, topography, land cover, land use practices (Sharpley and Halvorson, 1994), and 2) transfer and attenuation through ditches, stream, rivers, and wetlands.

Initial approaches for describing the export of diffuse phosphorus pollution in rural regions have excluded much of the complexity and diversity inherent in landscapes. Data from experimental work at the field-scale are very useful for elucidating processes of P export from land surfaces (Koro and Bernard, 1995; McIsaac et al., 1995). The watershed unit has also been used in many scientific studies to study the hydrological factors and to evaluate the role of agriculture in exporting phosphorus to surface water (Dorioz and Ferhi, 1994; Jordan-Meille et al., 1998; McDowell et al., 2001). However, because these studies are representative of only a small portion of actual field conditions, results can not be accurately extrapolated to the diversity of situations in a typical landscape.

Because of this landscape diversity, the challenge in managing diffuse P pollution is to understand P loading and transport dynamics at the appropriate scales. Mechanistic models (e.g., CREAMS-WT, Heatwole et al., 1988; AGNPS, Young et al., 1989; Lenzi and Di Luzio, 1997) require extensive data, which are not easily available, and their results are often disappointing at the scale of large and complex watersheds (Leite, 1990). This contrasts the relative success of empirical models, including P loading coefficients and export functions, for general land cover (e.g., Vollenveider, 1968; Bovay, 1980; Beaulac and Reckhow, 1982; Meals and Budd, 1994). These empirical averaging approaches are useful at a broad scale to 1) characterize the difference in diffuse pollutant export among regions (Dillon and Kichner, 1975; Omernik et al., 1976; Wendt and Corey, 1980), or 2) capture a long term trend (Cassell et al., 1998), but they neglect the diversity and complexity of the landscape and oversimplify the interaction of hydrology, land cover, management practices (e.g., Romkens et al., 1973; Gaynor and Findlay, 1995), biogeochemistry of the site (Sharpley et al., 1993), and the spatial relationship among landscape elements (Whigham et al., 1988). In addition, within broad classes of land cover, the variability of agricultural practices, which can control the intensity of erosion and surface runoff (Boiffin et al., 1988; Sharpley et al., 1994;Vansteelant et al., 1997), may play an important role in determining P fluxes. Thus, neither mechanistic or simple empirical models can adequately handle the complexity of P processes in the landscape.

Given the difficulty in describing P dynamics at the scale of the landscape, some researchers have begun to develop more complex, multi-scale, GIS-based, hybrid models to predict P load (or more broadly -- water quality) based on land cover and other landscape variables such as proximity to streams or riparian areas, presence of buffers, connectivity, and source-sink concepts (e.g., Tim et al., 1992; Soranno et al., 1996; Johnson et al., 1997; Tufford et al., 1998). We feel that a landscape approach is essential, and that it should consolidate concepts to include study of the 1) diversity field conditions that mobilizes phosphorus, 2) variety of routes phosphorus can take in moving from the field into the hydrologic network, 3) possibilities of attenuation in riparian areas and wetlands, and 4) hierarchy of transformation and storage mechanisms in first, second, third, etc. order streams. Furthermore, management of diffuse pollution requires an understanding of the human factors causing the pollution and creating, disturbing, or modifying the routes of P through the land. These human factors transcend the scale of the field and the farm, extending to the community and the region, and hence also dictate a landscape approach.

In this paper we review and discuss a series of our initial studies employing an empirical landscape approach to understand diffuse phosphorus pollution (including Weller et al., 1996; Windhausen et al., 2003; Vansteelant et al., 1997; Trévisan et al., 1995; Dorioz and Trévisan, 2001). The general question that forms the background to this work was how do the activities on and organization of the landscape contribute to the eutrophication of lakes (Lac Léman and Lake Champlain). We use broad-scale observations of the spatial and temporal patterns in surface water chemistry of watersheds in conjunction with multivariate analyses to evaluate the extent to which diffuse phosphorus fluxes are a function of 1) the content of the landscape (the distribution of land cover), 2) attenuating landscape elements (wetlands, riparian zones), 3) the practices that take place within each land cover type, and 4) the spatial position of the landscape elements. Because we feel that our approach deviates from the norms of experimental science in water quality research, we discuss the foundations of our inferences about diffuse phosphorus pollution.

In presenting syntheses of our previous work, we hope to demonstrate that our general approach using broad-scale observations is useful for understanding diffuse phosphorus pollution because it complements "classical" watershed research by emphasizing the diversity and position of landscape elements within the watershed. This additional knowledge can help form a basis for making management decisions to reduce P flux to surface water. The initial studies we present do no adequately describe the spatial and temporal patterns of nutrient fluxes in our landscapes, but only suggest use of a landscape approach to begin this difficult task.

2. AN APPROACH TO STUDYING DIFFUSE POLLUTION IN THE LANDSCAPE

We began by developing a concept that first, recognizes diffuse phosphorus transfer as a landscape-level phenomenon and second, captures some of the diversity and complexity of phosphorus movement through a heterogeneous landscape.

2.1. The Phosphorus Transfer System

Because there is no important gaseous phase for phosphorus in our landscapes, phosphorus can be considered to be a conservative element, accounted for using a mass balance. Occurring in inorganic, organic, soluble and particulate forms, phosphorus emission from land surfaces includes a combination of forms that can undergo a variety of transformations during transport (Dorioz et al., 1998). Total phosphorus from diffuse sources is dominated by the flux of particulate phosphorus (PP), primarily a surface phenomena closely tied to runoff and erosion (Ryden et al., 1973; Sharpley et al., 1993). This movement of total phosphorus is primarily a discontinuous process, in motion during surface runoff events (Verhoff et al., 1982). The potential for this runoff begins with any process reducing the infiltration rate of the soil surface. Impermeable surfaces (e.g., roads, houses, compacted areas) and undisturbed vegetated surfaces (e.g., forests, grasslands) comprise the extremes of a permeability gradient. Cultivated areas vary spatially and temporally in permeability, depending on soil type, soil cover (e.g., vegetation, mulches) and the degree of soil "crusting." "Crusting" is a precipitation-induced degradation of the surface porosity and concurrent evolution of a surface crust of fine clays (Boiffin et al., 1988; Auzet et al., 1990). Most of the cultivated soils in the Lac Léman area are sensitive to this crusting, which enhances surface runoff and associated phosphorus emission. Runoff can occur under very low rainfall intensities (Vansteelant et al., 1997).

Starting as a series of small rills at the field scale, surface runoff generally must connect with the network of temporary depressions and ditches in order to carry its load of phosphorus to streams and then higher order rivers (Jordan-Meille et al., 1998). Subsurface flow carries much less phosphorus due to fixation in the subsoil (van Riemsdjick et al., 1987; Dorioz and Ferhi, 1994; Gilliam, 1994), except in some specific conditions including sandy soils, long-term over-fertilization, and agricultural drainage (Sims et al., 1998). Highly permeable buffer strips of permanent vegetation (e.g., grass, shrubs, hedgerows) and riparian buffers can attenuate the P flux to the hydrologic network (Gilliam, 1994; Uusi-Kamppa et al., 2000). While in the network, particulate P can settle and soluble P can sorb onto the network surfaces or be taken up by biota (Wang et al., 1999). Resuspension and partial desorption from previously stored P can occur generally during stormflows (Dorioz et al., 1998). In addition, major obstacles like wetlands and lakes can store and/or transform the phosphorus emitted from upstream fields (Johnston 1991, Detenbeck et al., 1993).

Thus a set of interacting processes move P through the landscape in a "transfer system," that includes emission from sources (mainly soils or sediments) and transport (including transformation and attenuation), This transfer system comprise a highly diverse and complex set of landscape structures arrayed in a complex and diverse hydrologic hierarchy - and interacting with some of the human activities, especially agricultural practices and management of the hydrologic network. The notion of hierarchy and complexity are characteristic of the landscape science literature (e.g., Allen and Starr, 1982; Urban et al., 1987; Turner et al., 1995). Given this complex transfer system, the challenge is to design an observational approach that can capture enough of both the fundamental processes of phosphorus dynamics and the diversity of responses of the various landscape components (e.g., fields, buffers, streams, wetlands).

2.2. Data Collection in a Heterogeneous Landscape

Conceptual models and associated observation strategies are molded by the environments in which they are formed. The study areas are representative of rural areas in the Lake Champlain and Lac Léman Basins, which both cover a wide geographic area with considerable landscape diversity (Table 1). Each basin contains mountains and valleys with a rural-urban mix of land uses and land cover. Despite a large number of inhabitants, the basins are largely rural because the residents are mostly concentrated in urban areas. There are many hundreds of active farms (mainly dairy) and forestry operations arrayed in a complex and diverse mosaic pattern including many thousands of individual parcels.

With typically complex bedrock geology due to periods of uplift and weathering, both regions are further complicated by a glacial surficial geology. Glacially compacted subsoils are fairly impermeable leading to an abundance of streams, which drain much of the precipitation (>60 cm/yr) from the landscape (Hamid et al. 1989). This extensive and diverse land cover creates a considerable challenge for measurement and sampling. We focused on predominantly rural areas with no point sources of pollution. The land cover included in the studies cover the range of dominant agricultural practices.

2.3. A Phosphorus Observation Strategy Employing Watershed Networks

To observe phosphorus movement in the landscape, we chose small watersheds as the basic unit of study. Watersheds were minimally sized to provide a measurable signal (discharge and concentration) that can be related to variables describing the state of the basin. The complexity of the P transfer system led us to emphasize collection of data at broad-scales using networks of watersheds. While it would be ideal to know the chemical flux for many points in the landscape over a period of time (preferably replicate observations from homogeneous landscape elements, e.g., multiple corn fields with a particular slope, soil, agronomic practice combination), this detail and quality of data is prohibitively expensive to collect. An alternative, continuous monitoring of water quality from one watershed, while providing detailed and accurate estimates of phosphorus fluxes and mechanisms of transfer, would not capture the diversity of P dynamics at the landscape scale. Thus extrapolation of this site intensive information would be difficult to support. The compromises between intensive sampling of one (or a few) watersheds and a few periodic observations from a network of watersheds depend

Table 1. Parameters of the study areas.

Parameter	LaPlatte Watershed	Chablais	Eight eastern watersheds
Basin context	Study 1	Study 2	Study 3
Location	Lake Champlain Basin - Vermont, New York, USA; Quebec, Canada.	Lac Léman Basin - Haute Savoie, France; Geneva, Vaud, Valais, Switzerland.	see study 1
Basin size	21,326 km^2	7,393 km^2	see study 1
Approx. population		1.5 million	see study 1
Maximum elevation		4,634 m	see study 1
Trophic state of lake	mesotrophic	mesotrophic	see study 1
Study area			
Dependent variables	TP, SRP, TSS, concen-tration and g ha^{-1} day^{-1}	TP, PP, TSS concentration	TP, kg ha^{-1} yr^{-1}
Independent variables: type	general land use, details on wetland types	general land use, cultural practices by field, aggregate indices	general land use, details on wetland types
Independ. variables: number	15	30	17
Spatial dimension: number of watershed units	15	14	8
Temporal dimension: number of samples or intervals	18	7	1
Time frame	annual	winter: January - March	annual
Sampling regime	events, base flow	seasonal high flows	periodic over the year
Range of watershed sizes	149 to 1396 ha	15 to 244 ha	3058 to 21,005 ha

on the structure of the system (Figure 1). If the landscape diversity is not high, then extrapolation of intensive studies is justified. If the chemical quality of stream water is fairly constant, then infrequent but more spatially extensive sampling is not rendered useless by high temporal variation.

Given the enormous landscape variability of the lake basins in our studies, and what we knew about the pattern of temporal variations (Dorioz et al., 1991), we developed a phosphorus sampling strategy with the following characteristics:

(a) sample extensively:
- many watersheds (10-100 ha)
- watersheds with varying land cover and agricultural practices, include a diversity of cover types and their proportional representation

(b) sample infrequently, but differentiate:
- wet period events
- seasonal base flows

(c) use grab samples of surface water:
- collections are "near" synoptic, describing one point in time
- during declining limb of events to avoid the high variability in the beginning of the chemograph
- focus on chemical concentration, obtain flux if possible
- collect approximate flow information, water budget indices, etc.

We considered a variety of indicators of phosphorus movement including total phosphorus (TP), soluble reactive phosphorus (SRP), total suspended solids (TSS), and particulate phosphorus (PP), analyzed via Standard Methods (APHA 1985). The speciation of P is of great relevance to understanding P dynamics as well as P impact on receiving waters (Bostrom et al., 1988; Logan et al., 1979). For each measure of water quality, an appropriate unit needs to be selected (Table 2). The social question revolves around eutrophication of receiving surface water, thus implicating P flux (e.g., kg yr^{-1} or kg ha^{-1} yr^{-1}) as the unit of interest. In order to calculate these numbers, intensive monitoring of stream water using automated gauging stations is necessary. In Lake Champlain these data were sometimes available but generally we sampled water quality (mg l^{-1}) along with concurrent estimation of Q (l sec^{-1}) to yield an instantaneous estimate of P flux (kg min^{-1}, extrapolated to kg day^{-1}). Without a good estimate of Q, phosphorus must be expressed as simple concentration.

The relevance of each unit of measure (kg min^{-1} or mg l^{-1}) is dependent on 1) the question being asked, and 2) the water yield from precipitation across the sampled watersheds. If the percentage of precipitation reaching the stream in each watershed is similar, then concentration is a good surrogate measure of instantaneous flux when expressed on a unit area basis (kg ha^{-1} min^{-1}). In addition, knowing the P flux at the time of sampling is really the primary question if you want to know when in the year the highest risk of P flux exists. Knowing only the concentration of P in a grab sample is not nearly so useful as knowing flux, yet it can still provide information about the relative risk of P export from the land.

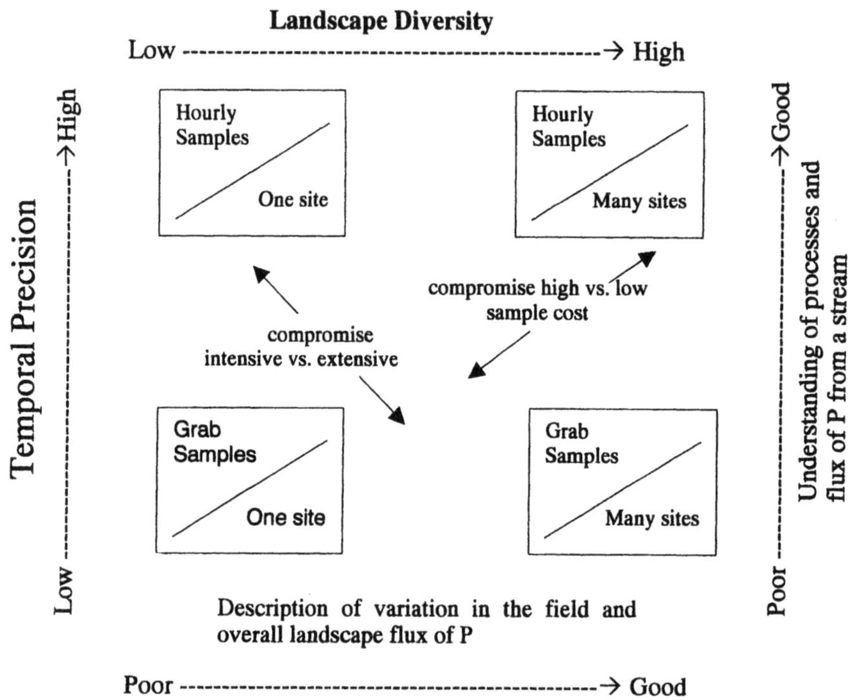

Figure 1. Trade-off of intensive vs. extensive observations, an example from study of the structure of the P transfer system.

Table 2. Measures of P Flux - Compromises between extensive and intensive observations

Observation	Unit	Comment
Annual flux	kg yr^{-1} or kg ha^{-1} yr^{-1}	sample intensive, expensive to collect, best measure for estimating P flux from the landscape
Instantaneous flux	kg min^{-1} or kg ha^{-1} min^{-1}	extensive sampling possible, good measure of flux at a point in time, annual calculation not possible
Volume-weighted concentration	mg l^{-1} for the year/period	extensive sampling possible, suggestive of a time-cumulative impact
Average concentration	mg l^{-1} for the year/period	extensive sampling possible, some reduction in sample variance, difficult to relate to absolute flux, allows relative comparisons among watersheds
Concentration	mg l^{-1}	maximum extensive sampling possible, allows relative comparisons among watersheds

2.4. A Landscape Observation Strategy

In our studies, observations were made 1) directly, and land condition noted on a map, or 2) remotely, employing satellite imagery, air photographs, and GIS to help classify the landscape. Our classification of the land surfaces began with simple cover types (crop, forest, residential developed, etc.). However, the cover type is not the only characteristic important for the potential emission of P, and other classifications were tried. Specific practices leading to that land cover may be a critical factor in diffuse flux. Other land characteristics such as a particular conservation tillage of a crop cover may be a more important determinant of surficial emission of P via runoff than whether the crop is corn or cereal, or whether the soil P concentration is high or low (Vansteelant et al., 1997). In addition, soil "crusting," a precipitation-induced degradation of the surface porosity and concurrent evolution of a surface crust of fine clays, leads directly to runoff and associated phosphorus emission. Thus crusting could also provide a good predictor of P flux. In a similar way, impermeable surfaces generate large volumes of runoff during rainfall events and thus cause resuspension of stream sediment and attendant P fluxes. These agricultural practices and watershed characteristics were described using a survey of farms and fields.

We used a variety of approaches to quantify the amount of landscape that is potentially emitting or attenuating phosphorus (Table 3). Each has relevance to different expressions of the phosphorus observed and often has a different conceptual basis. The simplest and most direct measure is area, but this is not necessarily the unit best suited to expressing the landscape to phosphorus relationship. For example, the emergent vegetation at the periphery of an open-water wetland may be the active unit of P processing, thus suggesting the use of perimeter as an appropriate measure. Furthermore, all positions in the landscape may not be equally as important. For example, riparian areas may be more functional quantities to consider for both P emission and attenuation.

On the contrary, surfaces without nearby connection to the hydrologic network may not participate in the landscape dynamics of phosphorus.

Table 3. Some possible measures of landscape

Observation	Unit	Measure of
area	ha	a land cover, or land use/practice, of a soil with a certain P concentration or in a certain physical condition
area	%, of total watershed	land cover, physical condition, etc. normalized as a percentage of the total area of a watershed
perimeter	km	the landscape element, e.g. wetlands
number	count	landscape elements, most useful if the landscape elements are of the same size
active area	ha, within bounds	land cover, etc. that contributes to an elemental flux, e.g., within 10 m of a flowing stream.
weighted area	ha, adjusted for effectiveness	land cover adjusted for its potential contributions, e.g., land within 10 m of a stream may contribute 2x the P of land 20 m from the stream

Each landscape metric was used only with compatible units of phosphorus. For example, area as hectares is the appropriate unit where the unit of P flux is kg of P per watershed. The percentage of the total watershed area of a contributing landscape element is compatible with P expressed as concentration or specific flux (kg ha^{-1} min^{-1}).

Given these considerations discussed above, our landscape observation strategy had the following characteristics:

(a) cover a diversity of landscape elements
- diversity of surfaces
- diversity of geo/topo/bio variables
- include attenuating and "sink" features

(b) consider surface conditions and diversity of agricultural practices
- reduction of permeability due to compaction and "crusting"
- cultural practices such a method of plowing, timing, etc.

(c) include relative spatial information
- connection or proximity to the hydrologic network
- stream order in the hydrologic network

This variety of ways of expressing and quantifying the different landscape surfaces resulted in many non-mutually exclusive sets of landscape descriptors that could be used to relate to phosphorus fluxes.

2.5. Analyzing Observations Of Phosphorus In The Landscape

In its simplest form, the analysis of the relation between P flux or concentration and the landscape requires a matrix with one dependent variable representing P from each watershed and several corresponding independent variables describing landscape attributes of each watershed. Our observations of phosphorus and the landscape generally resulted in a 3-dimensional data matrix of various chemical constituents expressing phosphorus or TSS, many times of sampling, and many different landscape attributes for each watershed. We interpreted the patterns within these spatial and temporal observations of chemical information to infer function at the scale of the landscape.

To evaluate our matrices, we generally relied on multiple regression to relate patterns of phosphorus (Table 2) to the structure of the landscape as measured by our various spatial representations (Table 3). Empirical models are of the form P = bo +b1(X1)+ b2(X2) + b3(X3); where Xi's are variables in each land cover category employed. We limited models to three land cover variables. Given the different possible expressions of phosphorus (e.g., TP, SRP, PP, kg ha^{-1}, mg l^{-1}, annual or seasonal average, volume-weighted mean, annual maximum, individual sample date) and the many possible ways of representing the landscape variables, choosing the analyses to conduct was not straight forward. The number of possible regressions, to relate P to the landscape, runs into the thousands. To reduce the combinations of regressions, clustering was used to group sample dates. Clustering was accomplished using a Ward's clustering algorithm (JMP 3.1, SAS Institute) on a date-by-watershed matrix of phosphorus values.

To further complicate choosing the specific regressions to run, statistical considerations create further options, which increases the number of possible regressions. For example, lack of homoscedasticity requires a data transformation in order to meet requirements for hypothesis testing. Some variables needed this transformation, others did not. We chose to use a process-based, intuitive selection of independent landscape variables, rather than attempt all possible combinations of dependent and independent variables. This choice has some ramifications for estimation of the significance of any relationship that we might discover and is discussed in section 4. Step-wise regression was also used to help with the selection of variables.

The results of our multiple regressions provide no direct information about causes and effects, only indications of statistical relationship, which could indicate spurious, indirect, or causal relationships. Despite the ambiguity of the method, we found consistent and statistically significant relationships (high and significant r^2), which we interpreted as evidence of a functional relationship between phosphorus flux and certain attributes of the landscape.

2.6 Brief Description Of Methods Employed

Methods are described in greater detail in the original sources (Weller et al., 1996; Trévisan et al., 1995; Vansteelant et al., 1997; Dorioz and Trévisan, 2001; Windhausen et al., 2003). A brief overview of study design, analytical, and statistical methods are provided here.

In Study 1, SRP, TP and TSS were observed in 15 subwatersheds in the LaPlatte River watershed from October 1994 to October 1995. Subwatersheds ranged in size from 149 to1396 ha (Table 1). Fluxes were estimated via synoptic grab samples and estimates of flow using a velocity-integrating flow meter according to the conventional mid-section method (Rantz, 1982). Discharge rating curves were developed for each sampling site according to Kennedy (1984). Subwatershed boundaries were delineated on U.S.G.S. 7.5 minute topographic maps and verified in the field, particularly in areas of low relief. GIS coverages were created by digitizing land cover and subwatershed boundaries using ARC/INFO based on 1988 1:5000 panchromatic orthophotographs, updated with field surveys in 1990. The overall accuracy of the linework was reported as +/- 25 m (Appleton, 1993). The land use coverage followed a classification system that was adapted for use in Vermont (Vermont Center for Geographic Information System, 1992) from the Anderson land use and land cover system (Anderson et al., 1976). Row cropland in the study area was mapped onto 1:5000 orthophotographs during summer field surveys in 1995 and then transferred to the GIS layers. Wetland classification (Cowardin et al., 1979) was based on interpretation of National High Altitude Photograph (NHAP) color infrared imagery, enlarged to a scale of 1:20,000 from an original 1:58,000.10% of the wetlands were field checked. Wetland boundaries were rectified using 1988 1:5000 orthophotographs. Hydric soils were mapped using criteria in the Federal Manual for Identifying and Delineating Wetlands (Federal Interagency Committee for Wetland Delineation, 1989). The area of these drainage classes in each subwatershed was determined from an existing soils coverage of the LaPlatte River basin (U.S.D.A. Natural Resource Conservation Service, 1995). TP was analyzed colorimetrically on a spectrophotometer using the molybdenum blue method following digestion in a sulfuric acid-persulfate solution (U.S. Environmental Protection Agency, 1983). SRP was measured on 0.45 μm membrane filtered samples colorimetrically by the molybdenum blue method within 48 hours of sample collection. Simple and multiple regression was used to assess relationships among dependent and independent variables. All combinations of explanatory variables were evaluated for multicollinearity using the variance inflation factor (VIF, Neter et al., 1989) and tested for normality (Shapiro-Wilks, $p < 0.05$).

Study 2, conducted in the Chablais, was part of a larger set of studies to understand sources of diffuse pollution in the Lac Léman Basin (CIPEL, 1984; Dorioz and Ferhi, 1994; Pilleboue and Dorioz, 1986). In the Chablais study, stream water was collected from 14 watersheds on seven dates in the generally high flow period from 19 January to 6 March 1995 (Table 1). This time period had the highest risk of runoff due to the condition of the fields after harvest in winter (little vegetative cover, low EVT, high water table, see Dorioz et al., 1998). For independent variables, data on 17 specific land cover and/or agricultural practices were collected using low altitude photography coupled with field surverys to identify planted crops, agricultural practice in use, area of roads, residential area, and natural cover in forest. Percentage of "crusted" soil (precipitation-induced degradation of the surface porosity and concurrent evolution of a surface crust of fine clays, Boiffin et al., 1988) in the watershed was determined by visual evaluation. The concentration of P in soils was analyzed by standard methods. All area units for the independent variables were converted to percent of total watershed area. In addition to the individual sample dates, a simple numerical average for each of the water quality parameters was calculated. Multiple regression was used to evaluate the relationship between dependent and independent variables.

In study 3 (Weller et al., 1996), annual TP flux was estimated for eight watersheds ranging in size from 3,058 to 21,005 ha. (Smeltzer and Quinn 1996, VT DEC & NY DEC 1997, unpublished report). Landscape and wetland variables were obtained from a variety of sources, but were predominantly interpreted from 1988 Landsat Thematic Mapper (TM) images. Primary land cover information was modified using a geographic information system (GIS). Again simple and multivariate analyses were performed.

3. RESULTS - THE RELATIONSHIP BETWEEN DIFFUSE PHOSPHORUS POLLUTION AND LANDSCAPE STRUCTURE

Using the approach described, we started with the simplest concept about P and the landscape (P flux is a function of land cover in the landscape) and then progressively modified that concept to include greater complexity (also a function of landscape attenuators of P transfer, and of landscape position).

3.1. Phosphorus Flux Is Related To The Area Of Different Land Cover

The notion that phosphorus flux is determined by the surface area of certain land cover within a watershed is supported by many watershed studies (Johnson et al. 1976, Beaulac and Reckhow, 1982; Meals and Budd, 1994). Does our broad scale approach of using synoptic grab samples across a diversity of watersheds support these other studies?

We first used the LaPlatte River watershed in Vermont to evaluate this question (Windhausen et al., 2003, this volume). For each water quality measure for each sample date, we ran regressions with simple land cover variables (total agriculture, row crops, forest, residential, wetlands, etc.). None of the more than 60 regression run for total suspended solids (TSS) concentration indicated that a relationship existed and these values are not reported. Soluble reactive phosphorus (SRP) concentration followed the same pattern as total phosphorus (TP) and on certain dates the two correlated very closely (up to $r^2 = 0.99$, and very often over 0.6, see Table 4). After obtaining this result, TP was analyzed more comprehensively than SRP.

To reduce the amount of data and sample variation, we calculated volume-weighted mean concentration for clusters of dates corresponding to different TP flux regimes (Table 5). The clusters consisted of four groups of dates that resembled each other in having similar TP fluxes values for a similar set of watersheds. In addition, a volume-weighted mean for the entire year of samples was calculated. These different ways of deriving a dependent variable related to TP resulted in a complex matrix of r^2's. Some of these values are included in Table 5.

Using the samples means, a very simple land cover model taking into account just the area of three very general categories of land cover (set 1 = % of agric land, developed land, and wetland) can "explain" 60 % of the variability in the dependent variable (flux or concentration). This increases to about 80% (set 2 or set 3) using a more precise descriptors of agricultural land. As a result of the similarity between SRP and TP samples, the r^2's observed for the relationship of SRP and TP export to land use also followed each other fairly closely ($r^2 = 0.74$) suggesting that overall, the two forms of phosphorus flux are dominated by a similar process in the landscape: runoff.

In addition, although the r^2's are generally lower for TP concentration, concentration seems to tell the same story as TP flux, thus suggesting that concentration captures much of the relationship between landscape surface condition and the flux of diffuse phosphorus to surface waters. Using the empirical model with set 3 (X1 = croplands ; X2 = grasslands ; X3 = wetlands), produces an r^2 of 0.58 for concentration of P. This is an important result as it mirrors finding in study 2 (discussed below), where the landscape in

Table 4. R^2 values for (a) the relationship of SRP and TP concentration for 15 watersheds on 15 dates, (b) the relationship of SRP flux to land use variable set 1 (cropland, wetland, cows), and (c) the relationship of TP flux to land use variable set 1.

Date	(a) R^2 of [SRP] to [TP	(b) R^2 for SRP flux	(c) R^2 for TP flux
11/01/94	0.99***	0.47	0.56
11/29/94	0.82***	0.55	0.44
01/22/95	0.87***	0.31	0.11
02/25/95	0.45*** (0.93 excl. #8)[1]	0.63	0.34
03/08/95	0.52***	0.79	0.87
03/14/95	0.09 (0.61excl. #6)	0.65	0.73
04/13/95	0.72***	0.51	0.44
04/28/95	0.60***	0.31	0.34
05/18/95	0.40** (0.75 excl. #11)	0.19	0.05
06/12/95	0.83***	0.03	0.23
08/04/95	0.27* (0.97 excl.#5)	0.12	0.14
08/06/95	0.96***	0.43	0.21
08/12/95	0.80***	0.22	0.28
10/06/95	0.94***	0.42	0.40
10/15/95	0.76***	0.25	0.26

[1] watershed #8 produced a large amount of sediment on this sample date resulting in an outlier in the [SRP] to [TP] relationship; excluding this value the r^2 is much higher, 0.93.

the Lake Léman basin is very similar to that of Lake Champlain. These results validate our broad scale approach: using synoptic grab samples across a diversity of watersheds is able to differentiate land cover types and their role in P export.

However, there is additional complexity inherent in this type of approach. For individual sample dates, many of the regressions were not statistically significant for any combination of independent land cover variables tried. The r^2 for most dates was generally low, except for on several occasions when streamflow was high due to heavy rainfall on wet or frozen soils in the spring. In these cases, agricultural land, cropland, the presence of cows in the stream, and wetland, generally appeared as good indicators of P flux. In addition, the r^2's for means for the spring high flow period and the whole year were significant and ranged as high as 0.95. Thus temporal variability in the relationship between land cover and P flux is high, probably due to large changes in the mechanics and ecology of P emission and transport under different hydrologic and seasonal conditions.

Table 5. R^2 values for regressions of TP concentration (mg/l) and TP flux (g ha^{-1} day^{-1}) vs. five land use variable sets for individual sample dates and means for clusters of dates. Volume weighted means are shown for concentration data. Models are of the form: $Y=\beta 0+\beta 1(X1)+\beta 2(X2)+\beta 3(X3)+\ldots$ where the X variables in each land use variable set are described below. Clustering was based on temporal patterns of TP flux across the study watersheds.

Sample	Land Use Var. Set 1		Land Use Var. Set 2		Land Use Var. Set 3		Land Use Var. Set 4		Land Use Var. Set 5	
	[TP]	TP flux	[TP]	TP flux	[TP]	TP flux	[TP]	TP flux	[TP]	TP flux
low flow* mean	0.55[1]	0.45[1]	0.54	0.35	0.52[1]	0.38	0.51	0.43	0.51	0.31
02/25/95	0.28	0.19	0.26	0.37	0.32	0.32	0.28	0.20	0.25	0.34
05/18/95	0.12	0.03	0.07	0.02	0.12	0.04	0.16	0.09	0.14	0.05
06/12/95	0.12	0.30	0.22	0.30	0.22	0.21	0.21	0.28	0.31	0.23
wint/summer mean	0.36	0.15	.036	0.29	0.42	0.29	0.37	0.17	0.39	0.30
11/01/94	0.26	0.45[1]	0.12	0.25	0.26	0.47[1]	0.27	0.55[2]	0.21	0.56[2]
11/29/94	0.50[1]	0.58[2]	0.16	0.26	0.49[1]	0.55[1]	0.52[2]	0.61[2]	0.29	0.44
01/22/95	0.52[1]	0.29	0.41	0.09	0.52[1]	0.27	0.53[1]	0.26	0.46	0.11
04/13/95	0.40	0.39	0.48[1]	0.36	0.50[1]	0.40	0.41	0.45	0.51[1]	0.44
04/28/95	0.29	0.29	0.29	0.26	0.15	0.21	0.30	0.43	0.29	0.34
10/06/95	0.13	0.27	0.06	0.28	0.14	0.18	0.25	0.40	0.25	0.40
10/15/95	0.45[1]	0.47[1]	0.51[1]	0.43	0.51[1]	0.28	0.52[1]	0.30	0.63[2]	0.26
spring/fall mean	0.44	0.42	0.30	0.29	0.44	0.42	0.57[2]	0.57[2]	0.50[1]	0.52[1]
03/08/95	0.46	0.68[3]	0.54[2]	0.80[3]	0.59[2]	0.84[3]	0.60[2]	0.76[3]	0.64[2]	0.87[3]
03/14/95	0.54[2]	0.47[1]	0.52[2]	0.67[2]	0.49[1]	0.65[2]	0.50[1]	0.53[2]	0.55[2]	0.73[3]
spring high flow mean	0.54[2]	0.66[2]	0.67[3]	0.84[3]	0.68[3]	0.86[3]	0.67[3]	0.80[3]	0.76[3]	0.95[3]
all sample mean	0.59[2]	0.68[3]	0.74[3]	0.77[3]	0.74[3]	0.83[3]	0.71[3]	0.85[3]	0.86[3]	0.94[3]

* r^2's for individual low flow samples not shown.

[1] significant at $p<0.10$ [2] significant at $p<0.05$ [3] significant at $p<0.01$

Land Use Variable Set 1: X1=% agricultural land, X2=% developed land, X3=% wetland

Land Use Variable Set 2: X1=% row cropland, X2=% developed land, X3=% wetland

Land Use Variable Set 3: X1=% row cropland, X2=% pasture and hay, X3=% wetland

Land Use Variable Set 4: X1=% agricultural land, X2=presence/absence of cows-in-stream, X3=% wetland

Land Use Variable Set 5: X1=% row cropland, X2=presence/absence of cows-in-stream, X3=% wetland

Another limitation of the approach is that the pattern of r^2 values in Table 5 can not be evaluated in a strict statistical manner with corresponding inferences about the role of agricultural land, cropland, wetlands, etc. in determining phosphorus flux. First, these models employ only three out of the many independent variables possible. Other sets of possible regression models could have been tested. Other combinations of two, three or more independent variables were calculated, and are not reported here. Some of these models also result in high r^2 values. Which is more "correct?" Second, the relationship that each regression might suggest can be very different, or even contradictory. For example, the high r^2 values observed in November of 1994 suggests that TP export may be due to the extent of agricultural and developed land (land cover variable set 1), but not the amount of row cropland and developed land (predominantly residential, land cover variable set 2) in the studied watersheds.

While the individual regression results in Tables 4 and 5 are perhaps dangerous to interpret quantitatively, the overall approach seems to have utility when the results are considered as a whole. Thus it is the pattern and consistency of r^2 values that provide the evidence of meaningful relationships. The highest r^2 values in Table 5 are found during the spring runoff period using either TP concentration or TP export as the dependent variable. This suggests that under these high flow conditions, runoff from cropland (or agricultural land in general) determines the amount of TP in the streams. The landscape elements apparently contributing P to the stream water seem to consistently include some aspect of agricultural activity, in particular more intensive activities such as row cultivation and unrestricted access of the stream corridor to cows. A closer examination of the regression coefficients in Table 6 reveals a surprising degree of consistency when the coefficients are highly significant ($p < 0.01$). Regardless of the date of sampling or the use of means, the use of concentration or flux, or the variable set combination selected, the ratio of the coefficient for row cropland to wetland only varies from 1.2 to 3.6. This is perhaps 2 to 5 times higher than the ratio for all agricultural land. Together these two observations suggest that row cropland is the dominant agricultural land use contributing to P flux to surface waters, and wetlands are modest sinks for P. If we attempt to squeeze the maximum amount of information out of the data, because the unit of the dependent phosphorus variable is taken as g ha^{-1} day^{-1}, we can estimate that each hectare of row cropland yields 140 to 400 grams of P each day under the observed conditions of high spring flow (Q on March 8 and 14 were somewhat similar at the time of sampling, thus allowing this Q-independent estimate of daily P flux).

During most of the year, the observations in Table 5 suggest that land cover is not an important determinant of TP in stream water. This does not support our initial conception and suggests that at these times of year either 1) other aspects of the landscape control TP flux, or 2) perhaps in-stream phosphorus dynamics dominate TP flux.

3.2. Phosphorus Flux Is Attenuated By Wetlands

Our results (Tables 5 and 6) suggest that wetlands, as a general land cover type, can be an important sink for P emission from land surfaces. In addition, as a group, their unit area efficiency in P removal is not as great as the emission capacity of agricultural land, as suggested by comparison of regression coefficients. However, this may reflect the variability

Table 6. Coefficients in regression models of P concentration (mg l^{-1}) and export (g ha^{-1} day^{-1}) vs. land use variable sets. Concentration mean values are volume-weighted. Models are of the form: Y=b0+b1(X1)+b2(X2)+b3(X3)+e, where the X variables in each land use are described below.

		Land Use Variable Set 1				Land Use Variable Set 2				Land Use Variable Set 3			
Date	Variable	b1	b2	b3	-b1/b3	b1	b2	b3	-b1/b3	b1	b2	b3	-b1/b3
03/08/95	TP concentration	-0.05	-0.58	-8.4**	ns	10.7	-2.1	-9.3**	ns	15.6**	-2.0	-9.5***	1.6
03/14/95	TP concentration	7.7**	7.9**	2.1	ns	12.4**	1.7	-4.7*	2.6	11.1*	-1.1	-6.9**	1.7
Spring high	mean concentr.	0.00046	-0.005	-0.0075	ns	0.010*	-0.002	-0.008***	1.2	0.014**	-0.001	-0.0084***	1.6
Annual	mean concentr.	0.0015	-0.003	-0.0048**	ns	0.010***	-0.001	-0.005***	1.9	0.011***	0.0004	-0.0054***	2.0
03/08/95	TP flux	0.71***	-0.13	-1.1**	0.65	4.0***	0.46	-1.4***	2.8	3.06***	0.35*	-1.4***	2.2
03/14/95	TP flux	0.63*	0.57*	0.16	ns	1.4***	0.10	-0.40**	3.6	1.4***	-0.9	-0.50***	2.8
Spring high	mean flux	0.47***	-0.04	-0.86***	0.55	2.7***	0.34	-1.00***	2.7	2.1***	0.19	-0.98***	2.2
Annual	mean flux	0.063***	-0.07	-0.12***	0.51	0.33***	0.02	-0.14***	2.3	0.26***	0.03	-0.14***	1.9

Land Use Variable Set 1: X1=% agricultural land, X2=% developed land, X3=% wetland
Land Use Variable Set 2: X1=% row cropland, X2=% developed land, X3=% wetland
Land Use Variable Set 3: X1=% row cropland, X2=% pasture and hay, X3=% wetland
Significant at * 0.10, ** 0.05, and *** 0.01 levels.

in P uptake capacity of specific wetland types that make up the total. Many individual studies of wetlands have demonstrated that wetlands can act as important filters of water pollution, including phosphorus pollution (Johengen and Larock, 1993; Tilton and Kadlec, 1979; Dorioz and Ferhi, 1994). The diversity of wetlands in our landscapes is very great, and these individual studies may not be justifiably extrapolated to wetlands in general. Some wetlands may be ineffective in attenuating P and some might actually release P under certain conditions (Peverly, 1982). A survey we conducted of inlet and outlet SRP from 10% sample of randomly sampled wetlands in the LaPlatte River Basin indicated that net flux was positive for about 30% of sampled wetlands, and negative for 70 % of wetlands on one day during spring snowmelt (Goldsmith 1994, University of studies of wetlands have demonstrated that wetlands can act as important filters of water pollution, including phosphorus pollution (Johengen and Larock, 1993; Tilton and Kadlec, 1979; Dorioz and Ferhi, 1994). The diversity of wetlands in our landscapes is very great, and these individual studies may not be justifiably extrapolated to wetlands in general. Some wetlands may be ineffective in attenuating P and some might actually release P under certain conditions (Peverly, 1982). A survey we conducted of inlet and outlet SRP from 10% sample of randomly sampled wetlands in the LaPlatte River Basin indicated that net flux was positive for about 30% of sampled wetlands, and negative for 70 % of wetlands on one day during spring snowmelt (Goldsmith, 1994, University of Vermont, unpublished master's thesis). A similar observation was made for a selection of 20 wetlands of various morphological types in Lac Leman. No clear significant differences in phosphorus concentration between the inlets and outlets were observed (Parmeland, 1995).

To attempt to better understand wetland type and P dynamics, we developed a set of nine wetland variables, each capturing a slightly different aspect of the concept of wetland (Windhausen et al., 2003). Each of these variables was then successively included with the total agricultural land variable in regression models for each date sampled. The resulting r^2's continue to show that significant relationships between any expression of wetland and P flux only occur at high spring flows. On 8 March 1995, with the highest flows of the year, some differentiation among wetland types was observed (Table 7). The strongest relationship was observed for hydric soils, followed by an expression including all wetlands, lakes and ponds. Other r^2's were significant, but suggested a weaker relationship. Accepting, for the moment, a causal link to this relationship, we might infer that because the strongest relationships were observed for the most general wet surface classifications (hydric soils being the most general and including the greatest areal extent of land surface), perhaps wet places and topographic depressions, in general, serve to reduce P export. In this context, our efforts to detect which type of wetland may be the most efficient were not successful, perhaps due to the lack of an appropriate functional classification.

As observed in study 1, phosphorus flux, under some conditions, seems to be determined by land cover. This observation supports the common notion of export coefficients (Omernik, 1976; Beaulac and Reckhow, 1982). However, in almost every significant regression model relating TP flux to land cover, there are peripheral watersheds or outliers suggesting that land cover alone does not determine TP flux.

3.3 Phosphorus Flux Is Related To Agricultural Practices

Usually these points are taken as random error (sampling error, parameter variance, etc.) that gets in the way of the multiple regression analyses. However, also embedded in this residual term are the processes that determine TP flux that are not land cover related. For example, in the LaPlatte River watershed, one subwatershed consistently deviated from all regression models with unusually high P fluxes. Closer field observation of this watershed revealed one farm with consistently poor management practices, including poor management of stream banks and animal concentration areas in the surface drainage network. Foster et al. (1990) estimate 700 kg ha^{-1} yr^{-1} of sediment flux due to cattle disturbing stream banks in grasslands. In the LaPlatte River watershed, we also developed another measure of agricultural practice, a "cows in stream" variable. A high r^2 value for land use variable set 4 suggests that the presence of cows in the streams is an important determinant of P export (TP, see Table 6, set 4 or 5). Clearly, agricultural practice can be an important determinant of P fluxes. Unfortunately in our Vermont work, this was the only data on agricultural practices we had. This was not the case for our other studies in the Chablais in France.

Table 7. R^2's for regression models using agriculture and different wetland expressions as independent variables to predict phosphorus flux (kg ha^{-1} yr^{-1}) from 8 watersheds in the Lake Champlain Basin (1991 P flux, 1973 land use).

Explanatory variables	R^2
Agricultural land use used singly (ha)	0.39**
Wetland expressions used in conjunction with agricultural land	
wetlands, lakes, and ponds	0.70***
forested and scrub-shrub wetlands	0.57*
emergent and open water wetlands	0.40
mixed emergent and scrub-shrub wetlands	0.57*
non-agricultural wetlands, lakes and ponds	0.69***
lakes, temporarily flooded wetlands and seasonally flooded wetlands	0.62**
lakes, ponds, and wetlands isolated or along first order streams	0.61**
wetlands along second or third order streams	0.40
poorly and very poorly drained soils excluding agriculture	0.77***

* significant at $p<0.10$ ** significant at $p<0.05$ *** significant at $p<0.01$

In the Chablais study, using the same variables in the Vermont study (cropland, pasture, wetland), 58% of the variation in TP concentration is explained (r^2 in Table 8). However, additional residual variance is explained by using detail available through low altitude air photography and/or field surveys to classify cultivated cropland and pasture. For TSS, good relationships were found with variables expressing high altitude conditions and steep slopes (alpage or mountain pastures). For TP concentration additional variables also improve the fit (Table 8).

Because there were 17 available descriptions of land cover and cultivation practices, the number of possible combinations of dependent and independent variables was enormous. To simplify the analyses, step-wise multiple regression was used to select possible variable combinations, with a limitation of three independent variables. In addition, some models were chosen to evaluate initial notions of what might be determining P flux. The results of many regressions, some of which are shown in Table 8, suggests that a great many land cover and agricultural practice variables have a relationship to phosphorus concentrations in the drainage waters. While no clear pattern emerges from the regressions from the various dates, for the average TP concentration in the waters, the percentage of land in cultivated crops, regardless of crop, seems to be important. This crop-related P emission variable was often in combination with one other variable, $grain^3$, which is the area of exposed soil after wheat or cereal crop harvest with no cultivation or herbicide application. This set of practices results in good post-harvest field coverage due to regrowth of cereal crops and the new growth of weeds. $Grain^3$ has a negative coefficient suggesting that it reduces P emission. Together with the area in cultivated crops, the model explains almost 90% of the variation in TP concentration.

Thus, the amount of phosphorus in runoff is generally related to cultivated areas, but is more directly controlled by the condition of the soil and its susceptibility to superficial runoff, which is mediated by agricultural practices. Certain practices on cultivated fields (e.g., preparing a seed bed by cultivating the soil until it achieves a small, level, particle-size distribution) facilitate "crusting" of the soil (Vansteelant et al., 1997). This crusting was observed for the studied watersheds and, along with one other independent variable ($grain^3$) also explains almost 90% of the variation in P concentration. Alone, this variables explains over half of the variation in average P concentration (r^2=0.57), which is similar to the result for cultivated cropland (r^2=0.49). Different cultivation practices on cropland can lead to different soil conditions, and thus potentially this measure of crusted soil could lead to a differentiation of cropland by cultivation practices. However, in our study area, the amount of cultivated cropland and the amount of soil showing crusting correlated quite well (r = 0.98), thus preventing inferences about which explained patterns of P in the landscape better.

Additional information on the potential role of agricultural practices in controlling P fluxes requires a different approach than used above. All of the 17 independent variables can not be included in a regression model because in our case we had only 13 degrees of freedom. Thus some grouping of variables or other method of including the information on the independent variables is needed. One way we used to combine the land practice data was to calculate a simple additive index, Rsum, of the estimated risk of runoff from each type of field (14 agricultural field types). Individual components of risk were based on field observation of runoff due to the different combinations of cultural practices and included information on frequency of surface runoff events on each field type during a one year survey (Vansteelant et al., 1997). This variable explained just under half of the variation in P samples (Table 8). Unfortunately, in this attempt to consolidate many independant variables , Rsum correlated quite closely with both the amount of cultivated cropland (r = 0.86) and the amount of crusted soil (r = 0.92). This correlation makes it difficult to infer which one of these aspects of land practice is more related to P emission.

Table 8. Regression results using a step-wise and an intuitive approach for TP and TSS concentrations for watersheds in the Chablais during a high flow period in winter 1995. (-) before a variable indicates a negative relationship.

TP					TSS			
Observation	r^2	independent variables			r^2	independent variables		
by simple	land	cover						
Average	0.12	agriculture	forest		0.16	agriculture	forest	
Average	0.51**	cropland***	pasture		0.14	cropland	pasture	
Average	0.58**	cropland***	pasture	(-)wetland	0.14	cropland	pasture	(-)wetland
by	stepwise							
19 Jan 95	0.73***	roads***	(-)grain[3]***	pasture**	0.81***	alpage**	(-)grain[3]***	roads***
23 Jan 95	0.51*	pasture***	grain[1]	soil P conc*	0.45**	forest*	soil P conc.	
30 Jan 95	0.92***	(-)grain[3]***	fallow***	connected cropland*	0.69***	cereal**	(-)grain[3]***	soil P conc.
6 Feb 95	0.75***	grain corn***	plowed	roads***	0.22	grain corn	silage corn	grain[1]
16 Feb 95	0.83***	connected cropland***	(-)grain[3]***	cereal**	0.42	pasture	grain[1]	(-)grain[3]
24 Feb 95	0.74***	(-)grain[2]***	silage corn**	wetland**	0.53*	(-)grain[3]**	grain[2]	soil P conc.
6 Mar 95	0.83***	grain corn***	wetland***	road***	0.50*	forest	alpage*	(-)grain[3]
Average	0.90***	connected cropland***	(-)grain[3]***	forest	0.89***	alpage***	cereal***	(-)grain[3]***
by	intuition							
Average	0.89***	cropland***	(-)grain[3]***		0.54**	cropland	alpage***	
Average	0.87***	crusted soil***	(-)grain[3]***		0.58***	crusted soil*	alpage***	
Average	0.45***	Rsum***			0.0	Rsum		

* $p<0.10$ ** $p<0.05$ *** $p<0.01$

alpage – high elevation grasslands on mountain slopes and meadows; grain[1] - grain crop (wheat, cereals) with surface plowing after harvest; grain[2] - grain crop with no plowing and herbicide spraying after harvest; grain[3] - grain crop - soil covered with vegetation during the study period; fallow – would be covered by grass and weeds during study period; connected cropland - cropland area connected to the hydrologic network, often by drainage ditches; Rsum is derived from indices of risk of P in runoff, see text for details; alpage is high altitude semi-natural pasture used in summer (not actively used during the study period).

Our landscape observations appear to provide evidence that agricultural practice, as opposed to general land cover, may provide more information about sources of P emission in the landscape. We infer that the P emission gradient: forest - grassland - cropland is not a discrete series of surface types, but may be better represented as a series of overlapping distributions where land practice determines where in the distribution a particular parcel of land may be. For example a cultivated field with good year-round vegetative cover and a mulched surface can export lower amounts of P than an overgrazed pasture. Our approach, using widely dispersed surface water samples, allows a general evaluation of this comparison of practice vs. land cover.

3.4. Is Landscape Position An Important Modifier Of Function?

Information describing the relationship of elements within the landscape can provide important structural information describing processes affecting the transfer of phosphorus at the scale of the landscape. Perhaps most important among the possible structural descriptions is the location of the active hydrologic network and the relationship of individual sources and sinks of phosphorus to this network.

Observations in study 2 (Chablais) included information on the relationship of cultivated fields to the network of ditches and streams draining the landscape. These observations were made based on the idea that phosphorus moving over the surface of a field during a rain event only becomes a eutrophication threat when it enters the surface water network (Jordan et al., 1998). For each watershed, the percentage of fields adjacent to actively flowing ditches or streams (connection) was evaluated at the time of water sampling. In most cases, this surface water network consisted of man-made drainage ditches of various styles and constructions. If more than 50% of the fields were connected, we considered a critical threshold to be reached, and assumed connection was not a limiting factor (connection = 100%). This evaluation of connection to the hydrologic network was multiplied by the area of cultivated cropland to estimate the area of connected, cropland surface for each watershed. The correlation of TP concentration as a function of cropland alone yields an r^2 of 0.49, while use of the connected, cultivated cropland improves the r^2 to 0.73. This clear improvement suggests that the manner and degree to which farmers connect landscape elements to the hydrologic network plays a role in determining P flux to surface waters.

A more geographically extensive study (study 3) using eight watersheds ranging in size from 3,058 to 21,005 ha permitted another test of the idea that position in the landscape has important consequences for phosphorus export to surface waters. Taking agricultural and forest land cover as a phosphorus source area, and one expression of wetland presence as a sink variable, regressions on the dependent variable of annual phosphorus flux revealed a pattern suggesting that riparian wetlands near 1st- through 4th-order streams related best to the observed flux of P from the watersheds (r^2=0.88, significant at $p<0.05$, Table 9).

Table 9. Adjusted R values and associated probability of greater F from linear regression. These regressions related watershed characteristics to measured phosphorus loads from each watershed. Agricultural and forested area comprised the initial two independent variables. Wetland variables were added individually to the regression.

MODEL	Adjusted R^2	P>F
Agricultural and forested area alone	0.63	
Agricultural and forested areas with wetland variables:		
-Wetland quantity		
wetland area (Ha)	0.57	0.64
wetland (N)	0.70	0.23
total perimeter (Km)	0.62	0.67
-Wetland type		
deciduous forest (Ha)	0.65	0.32
coniferous forest (Ha)	0.58	0.59
mixed forest (Ha)	0.83	0.06
scrub-scrub/emergent(Ha)	0.57	0.61
Land use in buffer zones around wetlands:		
forest (ha)	0.59	0.51
agriculture (ha)	0.59	0.52
Wetland and streams:		
wetlands near first-order streams	0.76	0.13
wetland near first and second-order streams	0.79	0.10
wetlands near first, second, third-order steams	0.85	0.05
wetlands near first, second, third ,fourth-order steams	0.88	0.03
wetlands near first, second, third, fourth, fifth-order steams	0.78	0.10
total area of riparian wetlands	0.84	0.05

Here again, we used the strength of the regression (r^2 and significance level) to infer that wetlands in a particular spatial position are more important for P attenuation than wetlands located in other positions. While the numerical difference between using the 1st-through 4th-order riparian wetlands (r^2 =0.88) and all riparian wetlands (r^2=0.84) is unimportant and thus probably not justifiable as a basis for differentiating the two, the pattern of difference moving toward just lower-order streams or other measures of wetland presence suggests that riparian wetlands have some special status with regard to diffuse phosphorus pollution. Note this conclusion differs from that obtained from a more geographically restricted analysis of one of the watersheds (LaPlatte watershed reported above).

Finally, we inferred from the coefficients of the regression model (Pflux = 0.86 Agriculture + 0.64 Forest - 30 riparian wetland) that the sink power of riparian wetland areas (30) was many times greater than the source export of P from comparable-sized agricultural (0.86) or forested land (0.64). Depending on which wetland variable was used, the coefficient reflecting the strength of the P sink varied from –0.1 for mixed forest wetlands to –47 for wetlands proximal to 1st and 2nd order streams. This suggests that quantitative reliance on the coefficient as a measure of the per unit area annual flux of P is not justified without further corroboration. However, the consistency of the coefficient as a negative term that was considerably larger than the agricultural coefficient more strongly suggests that wetlands, and particularly riparian wetlands, are an important landscape element attenuating diffuse phosphorus pollution.

Our approach seems to allow a reasonable inference that landscape position is important for both sources and sinks of diffuse phosphorus. In both cases that we investigated, the important aspect of position is the relationship to the surface water network.

4. THE UTILITY OF THIS LANDSCAPE APPROACH.

The observations and analyses referred to above lead us to a variety of inferences about spatial and temporal patterns of diffuse P in the landscape. These inferences lead to 1) targeting land cover and agricultural practices most responsible for diffuse P pollution, 2) assessing at what times and under what conditions the risk is highest, and 3) refining our concept of diffuse P dynamics in the landscape to including structural components such as position with respect to the hydrologic network. Before concluding these inferences are worthy of being used as a framework for developing management plans, the rigor of these inferences must be considered.

4.1. Statistical Limitations

A tool should be used within the constraints of its assumptions. There are several such assumptions that are important in our use of multiple regression. Regression is a parametric procedure based on assumptions of normality of input variables and error terms. Thus the derived parameters of r^2 and the significance level of the model and its coefficients are very much dependent on the normality of the collected data. We relied heavily on these parameters in making inferences from the regression analyses presented in this paper. The normality of input variables is both easily calculated and often violated. The normality of the error term in the regression model is easily observed by examination of the model residuals. These residuals can be tested (e.g., Shapiro-Wilk W, JMP 3.2, SAS Institute). The problem arises if normality is not apparent and not easily compensated for by transformations of the data (e.g., log, arc sin, etc.). Models that did not meet this requirement could not be quantitatively compared. The homoscedasticity (constancy of variance) of the data is also important for performing hypothesis tests on the regression coefficients (Helsel and Hirsh, 1992). In our work several models required transformations of the data (log_{10}) in order to approximate this requirement. However, once the data are transformed, comparisons between models with transformed and untransformed data are not possible. Much of what we inferred depended on a comparison of r^2's and significance levels in large matrices of regression results. In some cases, necessary transformations were not made to permit construction of the larger comparative tables.

Often variables are not independent and thus care must be taken to use variables only in independent sets. For example cultivated cropland and connected cultivated cropland include some of the same land area and thus should not be used concurrently. Multicollinearity of independent variables presents a different kind of problem in regression, causing instability in the coefficients. Multicollinearity was assessed using the variance inflation factor (VIF, Neter et al., 1989), and values greater than 10 resulted in the exclusion of problem variables. The number of independent variables usable in a multiple regression model is dependent on the degrees of freedom (number of watersheds in our case). In most of our analyses we had about 15 observations and therefore limited the number of independent variables to three because of limitations due to the relatively few degrees of freedom available. While there is no firm rule in this case, we felt using more than three variables lead to unjustifiably high r^2's and do not help with data interpretation.

As noted above, our sets of independent and dependent observations on land and phosphorus permitted many combinations (up to many hundreds) of variables to be included in multiple regression models. The significance levels reported hold true for individual models, but just like for multiple comparisons in ANOVA , the joint significance level for all the possible combinations of independent variables can be very much lower. We did not compensate quantitatively for the effect of multiple trials/analyses on reported significance levels, especially since the number of trials was dynamic. Thus significance levels in our analyses only suggest the strength of relationship (the probability of achieving the same result from random data). Again, strict quantitative comparison of either the r^2 or the significance parameters is probably not warranted and generates a false sense of quantitative rigor.

4.2. The Validity Of Inferences From Regression

Deriving cause and effect relationships from significant correlations or regressions is inappropriate. Nevertheless, social scientists, epidemiologists, ecologists and others and having been using this approach to test ideas and generate hypotheses about the nature and function of their respective subjects of study for many years. Does this approach require additional justification for its use as a research tool in landscape ecology?

Common to all the inferences made above is the idea that a higher r^2 value infers a stronger relationship. While this statement may be statistically correct (the above statistical caveats taken into consideration), this can not be taken to mean there is a parallel, stronger causal relationship. For instance, in both the Vermont and France studies we observed stronger relationships of P concentration or flux with cropland than with general agricultural land. While this makes intuitive sense and is supported by a great deal of research (Dillon and Kirchner, 1975; Omernick, 1976; Belamie, 1986; CIPEL, 1988; Sharpley et al., 1993), we can not rigorously defend that this means cropland exports more P than general agricultural land. Concluding that cropland exports more P than agricultural land employs inductive logic not normally considered to be appropriate at the later stages of the scientific method. Induction stimulates the generation of hypotheses that are then tested using standard, rigorous, scientific methods with proofs and significance levels. This tension between the classical scientific method and conclusions based on inductive reasoning and circumstantial evidence is evident in the literature of diffuse pollution (e.g., Whigham et al., 1988; Beaulac and Reckhow, 1982; Omernik, 1976), but is age-old and not limited to landscape-level investigations. Our purpose in bringing it up here is to provide some context for our approach, which makes causal inferences from correlated information.

As stated in the beginning of this paper, our approach is based on our conceptual model of the system we seek to understand. The dynamics of phosphorus in the landscape is conceived of as a highly interactive, diverse system. It is this conception that leads us to start our research by sampling extensively and to make inferences from statistical relationships rather than seek to define general truths from intensive study of a few elements in the landscape. A more classical approach might take the following steps: 1) study a landscape element, e.g., and farm field under silage corn production on a particular soil-slope situation, 2) develop a model based on observational and/or experimental studies to define the flux of P from the field under a variety of climatic/seasonal conditions, 3) use this model to deduce the phosphorus flux from other fields exposed to some other set of cultural and environmental conditions (for example, see Haygarth and Jarvis, 1997). If these other fields exist within the parameter boundaries of the initial intensive study, then the extrapolation may work and the deduction is within legitimate and accepted scientific practice. On the other hand, if the other field lies outside this boundary (e.g., with respect to slope, soil type, climate, vegetation, cultural practices, etc.), then the extrapolation is perhaps flawed and certainly not part of accepted scientific practice. This field-scale approach also overlooks other elements in the landscape system such as the buffer and attenuation elements, and in-stream processing.

In the milieu of applied science, intuitive decisions based on inferences from whatever information is available are perhaps the norm rather than the exception. The application of landscape ecology to current environmental problems highlights this use of scientific knowledge because it is a theory-poor young science with great responsibilities for guiding critically needed practices and policies for land management. Because we see little prospect of conducting the requisite number of intensive studies on the diversity of landscape elements, we suggest that taking an inferential approach using the "flawed" logic of "correlation suggests cause" is a useful complement to the more classical approaches taken by quantitative science. In fact, we suggest that this evidence obtained via regression analysis is a necessary complement to experimental and process-based investigations that can not be justifiably extrapolated to the broad diversity of circumstances found in our complex landscapes. At this time, with respect to understanding diffuse phosphorus pollution, both approaches provide only evidence rather than proof.

6. CONCLUSIONS

Diffuse phosphorus pollution is a landscape-level phenomenon requiring a spatial-temporal approach to clarifying questions of sources, causes, timing, and sinks. P flux is unevenly distributed in time and space as a result of its origins from a complex interaction of human activity and the biophysical landscape. While our initial regressions support the idea that land cover controls P flux, more detailed land cover data suggest that agricultural practices within the same land cover plays an important part in determining P flux. At the field scale, clearly the management of soil surface structure and/or vegetative cover over the soil during times of high runoff risk is essential. Thus management of diffuse phosphorus pollution needs to consider another aspect of P dynamics in the landscape that is determined by socioeconomic factors (technical knowledge of the farmer, tools available, traditions, number of landowners, community structure, etc.).

Our initial approach using synoptic sampling of surface waters across the landscape supports the idea that cultivated land is the dominant source of diffuse P in our landscapes. Our observations also provide evidence that landscape elements can have a P sink function. While individual wetlands may be sources or sinks for P, natural wetlands as a group, collectively act to reduce the load of phosphorus to surface waters during critical times of year. Managed grassland can also provide similar functions (Haygrath and Jarvis, 1997). More detailed information about activities (as opposed to land cover information) also improves our understanding of diffuse phosphorus. Practices in cultivated fields, especially those that lead to soil "crusting"/compaction or to better connection of parcels to the hydrologic network, are an important determinant of P fluxes. Finally, our observations provide evidence that the position of source and sink elements affects the total amount of P coming from the landscape. In particular, the hydrologic network is a critical structural element when considering position. Keeping agricultural practices with the greatest risk of generating runoff at some distance from artificial or natural drainages reduces P flux to surface waters by lowering the probability of hydrologic connection. Again, the presence and arrangement of both agricultural and nonagricultural lands depends on social factors in the local and regional community. These and other observations lead us to suggest that phosphorus dynamics in the landscape can be envisioned as taking place in an interacting social and biophysical system, and it is this entire system that needs to be considered in the continuing effort to manage diffuse phosphorus pollution.

Thus we suggest that diffuse pollution can be envisioned as a function of:

1) *landscape area* - the area of corn culture, grassland, residences, etc.; usually described as land cover and observable from airplanes or satellites;
2) *attenuators* - the effective elements in a landscape can be an areal, linear or count units; for example, buffer strips, riparian areas, and wetlands reduce P flux from landscapes;
3) *activities/practices* - the landscape is still measured as area, but divided into more components as a function of type of activities taking place on the land; for example the area of corn under no-tillage, under conventional tillage, etc.; this level of detail is usually not observable from satellites or airplanes;
4) *position* - all areas of a land cover type are not equal contributors, spatial position mediates pollution; for example if a corn field is many kilometers away from the active stream network, the diffuse pollution does not reach a potential flux point.

These conclusions are based on inferences drawn from observing the relationship between phosphorus flux and descriptions of the landscape. Because the need for management of diffuse pollution is current and great, we suggest that management decisions based on this interim level of understanding are warranted. Understanding based on more rigorous experimental approaches or intensive observations of better controlled systems will eventually provide a more confident basis for management, but for the immediate future we suggest that developing more detailed descriptions of the landscape (environmental and social surveys, remote-sensing, and GIS) coupled with broad-scale water sampling can continue to provide much needed information on diffuse phosphorus dynamics and the risk of eutrophication of surface waters.

7. ACKNOWLEDGEMENTS

The research reported here has been supported by a variety of sources including McIntire-Stennis funding to the School of Natural Resources, the Lake Champlain Basin Program, and the INRA-Cemagref Programme AQUAE.

8. REFERENCES

Allen, T.F.H., and Starr, T.B., 1982, Hierarchy: Perspectives for ecological complexity. University of Chicago Press: Chicago.

American Public Health Association, 1985, Standard methods for the examination of water and wastewater, 17th ed. pp. 472-495. American Public Health Association: Washington, DC.

Anderson, J.R., Hardy, E.E., Roach, J.T., and Witmer, R.E., 1976, A Land Use and Land Cover Classification System for use with Remote Sensor Data. U.S. Geological Survey Professional Paper 964. U.S. Government Printing Office, Washington, D.C.

Appleton, J., 1993, Unpublished documentation. Chittenden County Regional Planning Commission, Essex Junction, Vermont.

Auzet, A.V., Boiffin, J., Papy, F., Maucorps, J, and Ouvry, J.F., 1990, An approach to the assessment of erosion forms and risk on agricultural land in the northen Paris Basin, France In: *Soils erosion and agricultural land.* John Wiley: Chichester. pp. 384-400.

Barroin, G., 1980, Eutrophisation, pollution nutritionnelle et restauration des lacs. Pollution des eaux continetales. Gauthier et Villars: Paris. Pp. 75-96.

Belamie, R., 1986, Les bassins versants ruraux et agricoles, outils pour la connaissance et le controle de la pollution par l'azote et le phosphore. Journées bassins versants. Paris, Ministere de l'Environnement, Mission eau- nitrates. 23 p.

Beaulac, M.N., and Reckhow, K.H., 1982, An examination of land use-nutrient export relationships. *Water Resources Bulletin* **18**:1013-1024.

Bovay, E., 1980, Ecologie et production agricole. Schweiz. Landw. Forschung *-La Recherche Agron. en Suisse* **19**: 1-48.

Boiffin, J., Papy, F, and Eimberk, M., 1988, Influence des systemes de culture sur le risques d'erosion par ruissellement concentré. Analyse des conditions de delanchement de l'erosion, *Agronomie* **8**(8): 663-673.

Bostrom, B., Persson G., and Broberg, B. .1988, Bioavailability of different forms of phosphorus in fresh water *Hydrobiologia* **170**, 133-155

Cassell, A., Dorioz, J.M., Kort, R.L., Hoffmann, J.P., Meals, D.W., Kistschel, D., and Braun, D.C., 1998, Watershed modeling : dynamic of phosphorus storage, cycling, transport, export. *J. Env. Qual.* Vol.**27**, 2, 293-298

CIPEL, 1984, Le Léman, synthèse des travaux de la Commission Internationale pour la Protection des Eaux du Leman contre la pollution, 1957-1982. Lausanne, Switzerland. 450 p.

CIPEL, 1988, Rapport de l'étude des pollutions diffuses dans le bassin lémanique. Commission Internationale pour la Protection des Eaux du Leman, Lausanne, Switzerland, 118 p.

Cowardin, L.M., Carter, V., Golet, F.C., and LaRoe, E.T., 1979, Classification of wetlands and deepwater habitats of the United States, U.S. Fish Wildl. Serv. FWS/OBS **79/31**. Washington, D.C: U.S. Government Printing Office

Detenbeck, N.E., Johnston, C.A., and Niemi, G.J., 1993, Wetland effects on lake water quality in the Minneapolis/St. Paul metropolitan area. *Landscape Ecology*, **8**(1): 39-61

Dillon, P.J., and Kirchner, W.B., 1975, The effects of geology and land use on the export of phosphorus from watersheds. *Wat.Res.* **9**: 135–148.

Dorioz, J.M., Cassell, A., Orand, A., and Eisenman, K., 1998, Phosphorus storage, transport and export dynamics in the Foron river watershed. *Hydrol. Processes* **12**: 285-309.

Dorioz, J.M., and Ferhi, A., 1994, Pollution diffuse et gestion du milieu agricole: transferts comparés de phosphore et d'azote dans un petit bassin agricole. *Wat. Res.* **28**(2): 395-410.

Dorioz, J.M., Orand, A., Pilleboue, E., Blanc, P., Colon, M., and Bosse, J.P., 1991, Prélèvement et échantillonnage dans les petits bassins-versants ruraux. *Rev. Fr. Sci. Eau* **4**: 211-238.

Dorioz, J.M., and Trevisan, D., 2001, Transferts de phosphore des bassins versants agricoles vers les eaux de surface: l'expérience du bassin lémanique (France) et sa portée générale. *Agrosol,* **12**(2): 85-97.

Federal Interagency Committee for Wetland Delineation. 1989, Federal Manual for Identifying and Delineating Jurisdictional Wetlands. U.S. Army Corps of Engineers, U.S. Environmental Protection Agency, U.S. Fish and Wildlife Service, and U.S.D.A. Soil Conservation Service, Washington, D.C. Cooperative technical publication.

Foster, I.D.L., Grew, R., and Dearing, J., 1990, Magnitude and frequency of sediment transport in agricultural catchments: A paired lake-catchment study in Midland England. pp., 153-171 In: Boardman, J., Foster, I.D.L., and Dearing, *J.A. Soil erosion on agricultural land.* John Wiley & Sons Ltd, London.

Gaynor, J.D., and Findlay, W.I., 1995, Soil and phosphorus loss from conservation and conventional tillage in corn production. *J. Env. Qual.* **24**:734-741.

Gilliam, J.W., 1994, Riparian wetlands and water quality. *J. Environ. Qual.* **23**: 896-900.

Goldsmith, L.A., 1994, A Landscape level evaluation of wetland structure and function in a large, multi-use basin. Unpublished Masters Thesis. University of Vermont, Burlington, Vermont

Haygarth, P.M., and Jarvis, S.C., 1997, Soil derived phosphorus in surface runoff from grazed grassland lysimeters. *Wat. Res.* **31**(1): 140-148.

Hamid, S., Dray, M., Ferhi, A., Dorioz, J.M., Normand, M., and Fontes, J.-C., 1989, Etude des transferts d'eau à l'intérieur d'une formation morainique dans le bassin du Léman - Transferts d'eau dans la zone non saturée. *J. Hydrol*, **109**: 369-385.

Heatwole, C.D., Campbell, K.L., and Bottcher, A.B., 1988, Modified CREAMS nutrient model for coastal plain flatwoods. *Transactions of the ASAE* **31**: 154-160.

Hecky, R.E., and Kilham, P., 1988, Nutrient limitation of phytoplankton in freshwater and marine environments: a review of recent evidence on the effects of enrichment. *Limnology and Oceanography* **33**: 796-822.

Helsel, D.R., and Hirsh, R.M., 1992, Statistical methods in water resources. Elsevier Science, New York.
Johengen, T.H., and Larock, P.A., 1993, Quantifying nutrient removal processes within a constructed wetland designed to treat urban runoff. *Ecol Engineering* **2**(4): 347-366.
Johnson, L.B., Richards, C., Host, G., and Arthur, J.W., 1997, Landscape influences on water chemistry in Midwestern stream ecosystems. Freshwater *Biology* **37**: 193-208.
Johnson, A.H., Bouldin, D.R., Goyette, E.A., and Hedges, A.H., 1976, Phosphorus loss by stream transport from a rural watershed: quantities, processes and sources. *J. Environ. Qual.* **5**: 148-157.
Johnston, C.A., Detenbeck, N.E., and Niemi, G.J., 1990, The cumulative effect of wetlands on stream water quality and quantity. A landscape approach. *Biogeochemistry* **10**:105-141.
Johnston, C.A., 1991, Sediment and nutrient retention by freshwater wetlands: Effects on surface water quality. CRC Critical Reviews in *Environmental Control* **21**: 491-565.
Jones, K.B., Neale, A. C., Nash, M.S., Van Remortel, R.D., Wickham, J.D., Riitters, K.H., and O'Neill, R.V., 2001, Predicting nutrient and sediment loadings to stream from landscape metrics: A multiple watershed study from the United States Mid-Atlantic Region. *Landscape Ecology* **16**: 301-312.
Jordan-Meille, L., Dorioz, J.M., and Wang, D., 1998, Analysis of the export of diffuse phosphorus from a small rural watershed. *Agronomie* **18**: 5-26.
Kennedy, E.J., 1984, Discharge ratings at gaging stations. In: Techniques of Water Resources Investigations of the United States Geological Survey. Book 3. *Applications of Hydraulics*. pp., 1-59. U.S. Government Printing Office, Washington, D.C.
Kling, G. W., Kipphut, G. W., Miller, M. M., and O'Brien, W.J., 2000, Integration of lakes and streams in a landscape perspective: The importance of material processing on spatial patterns and temporal coherence. *Freshwater Biology* **43**: 477-497.
Koro, N., Bernard, C., Laverdière, M.R., 1995, Contrôle du ruissellement, de l'érosion et des pertes de phosphore par les résidus de culture, sous pluie simulée. *Étude Gestion Sols* **2**:173-181.
Lake Champlain Management Conference, 1996, Opportunities for action - An evolving plan for the future of the Lake Champlain Basin. Lake Champlain Basin Program, Grand Isle, VT.
Leite, L. A., 1990, Réflexions sur l'utilité des modèles mathématiques dans la gestion de la pollution diffuse d'origine agricole. Thèse, Ecole Nationale des Ponts et Chaussées, Paris 6, 342 p.
Lenzi, M.A., and Di Luzio, M., 1997, Surface runoff, soil erosion and water quality modeling in the Alpone watershed using AGNPS integrated with a geographic information system. *European J. Agronomy* **6**: 1-14.
Logan, T.J., Glova, T. O., and Yacksich, S.M., 1979, Phosphate characteristics and bioavailability of suspended sediments from streams draining into lake Erie. *J. Great Lake Res.* **5**(2): 112-123.
McDowell, R., Sharpley, A.N., and Folmar, G., 2001, Phosphorus export from an agricultural watershed: Linking surface and transport mechanisms. *Journal of Environmental Quality*, **30**: 1587-1595.
McIsaac, G.F., Mitchell, J.K., and Hirschi, M.C., 1995, Dissolved phosphorus concentrations in runoff from simulated rainfall on corn and soybean tillage systems. *J. Soil Water Cons.*, **50**:383-387.
Meals, D.W., and Budd, L.F., 1998, Lake Champlain Basin nonpoint source phosphorus assessment. *J. Am. Water Res. Assoc.* **34**(2): 251-265.
Neter, J., Wasserman, W., and Kutner, M.H., 1989, Applied linear regression models. R.D. Irwin, Inc.: Homewood, Illinois. 667 pp.
Novotny, V. and Olem, H., 1994, Water quality: prevention, identification and management of diffuse pollution. van Nostrand-Reinhold: New-York.
Omernik, J. M., 1976, The influence of land use on stream nutrient levels. *EPA-600/3-76-014*. U.S. Environmental Protection Agency, Corvallis, Oregon, 105 pp.
Parmeland, E., 1995, Zones tampons et pollution diffuse dans le bassin versant du Léman. -DEA "Gestion des espaces Montagnards" Univ. Chambery.
Peverly, J.H., 1982, Stream transport of nutrients through a wetland. *J. Environ. Qual.* **11**:38-43.
Pilleboue, E and Dorioz, J.M., 1986, Mass-balance and transfer mechanisms of phosphorus in a rural watershed of lac Leman, France. In: Sediment and water interactions. *Springer-Verlag*: New York. pp. 91-102.
Pucket, L.J., 1995, Identifying the major sources of nutrient water pollution. *Environ. Sci. Tech.* **29**: 408-414.
Rantz, S.E., 1982, Measurement and Computation of Streamflow: Volume 1. Measurement of Stage and Discharge. U.S. Geological Survey Water-Supply Paper 2175. U. S. Government Printing Office, Washington, D.C.
Romkens, M.J.M., Nelson, D.W., and Mannering, J.V., 1973, Nitrogen and phosphorus composition of surface runoff as affected by tillage method. *J. Environ. Qual.* **2**(2): 292-295.
Ryden, J.C., Syers, J.K., and Harris, R.F., 1973, Phosphorus in runoff and streams. *Adv. Agro.* **25**:1-45.
Sharpley, A.N., Chapra, S.C., Wedepohl, R., Sims, J.T., Daniel, T.C., and Reddy, K.R., 1994, Managing agricultural phosphorus for protection of surface waters: issues and option. *J. Environ. Qual.* **23**: 437-451.

Sharpley, A.N., Daniel, T.C., and Edwards, D.R., 1993, Phosphorus movement in the landscape *J. Prod. Agric.* 492-500.
Sharpley, A.N., and Halvorson, A.D., 1994, The management of soil phosphorus availability and its impact on surface water quality. *In: Soil processes and water quality.* Lewis Publishers: Boca Raton, FL. R. Lal and B.A. Stewart (eds.) pp. 7-90.
Sims, J. T., Simard, R.R., and Joern, B.C., 1998, Phosphorus loss in agricultural drainage: historical perpective and current research. *J. Environ. Qual.* **27**: 277-293.
Smeltzer, E., and Quinn, S., 1996, A phosphorus budget, model, and load reduction strategy for Lake Champlain. *J. of Lake and Reservoir Management* **12**(3): 381-393.
Soranno, P.A., Hubler, S.L., Carpenter, S.R., and Lathrop, R.C., 1996, Phosphorus loading to surface waters: a simple model to account for spatial pattern of land use. *Ecol. Applic.* **6**(3): 865-878.
Trevisan, D., personal communication, unpublished report with Vansteelant, J-Y., Perron, L., Dorioz, J-M., and Roybin, D., 1995, Activite Agricole et Qualite des Eaux Superficielles dans le Bassin Lémanique Francais. Compte rendu final a Groupement d'Intéret Scientifique Alpes du Nord, Chambery. 90 p.
Tilton, D.L., and R.H. Kadlec, 1979, The utilization of a fresh-water wetland for nutrient removal from secondary treated waste water effluent. *J. Environ.Qual.* **8**: 328-334.
Tim, U.S., Mostaghimi, S., and Shanholtz, V.O., 1992, Identification of critical nonpoint pollution source areas using geographic information systems and water quality modeling. *Water Res. Bull.* 28(5): 877-887.
Tufford, L.D., McKellar, H.N., and Hussey, J.R., 1998, In-stream nonpoint source nutrient prediction with land use proximity and seasonality. *J. Eviron. Qual.***27**: 100-111.
Turner, M.G., Gardner, R.H., and O'Neill, R.V., 1995, Ecological dynamics at broad scales *BioScience suppl.* pp. s29-s35.
Urban, D.L., O'Neill, R.V., and Shughart, H.H. Jr., 1987, Landscape ecology. *BioScience* **37**: 119-127.
U.S.D.A., Natural Resource Conservation Service, 1995, Soil Survey Geographic (SSURGO) Data Base Data Use Information. Miscellaneous Publication Number 1527. U.S. Department of Agriculture.
U.S. Environmental Protection Agency, 1983, Methods for Chemical Analysis of Water and Wastes. EPA-600/4-79-020. U.S. Environmental Protection Agency, Cincinnati, Ohio.
U.S. Environmental Protection Agency, 1995, National water quality inventory -- 1994 report to Congress. EPA 841-R-95-005. U.S. Environmental Protection Agency, U.S. Government Printing Office: Washington, D.C. 497 pp.
Uusi-Kamppa, J., Braskerud, B., Jansson, H., Syversen, N., and Uusitalo R., 2000. Buffer zones and constructed wetlands as filters for agricultural phosphours. *Journal of Environmental Quality* **29**:151-158.
van Riemsdijk, W.H., Lexmond, T.M., Enfield, C.G., and van der Zee, S., 1987, Phosphorus and heavy metals: accumulation and consequences. In: Animal manure on grassland and fodder crops. H.G. van der Meer (ed.) pp.18-34.
Vansteelant J.-Y., Trevisan, D., Perron, L., Dorioz, J.-M., and Roybin, D., 1997, Conditions d'apparition du ruissellement dans les cultures annuelles de la région lémanique. Relation avec le fonctionnement des exploitations agricoles. *Agronomie* **17**: 65-82.
Verhoff, F.H., Melfi, A.D., and Yakish, S.M., 1982, An analysis of total phosphorus transport in river systems. *Hydrobiologia* **91/92**: 241-252.
Vermont Center for Geographic Information, Inc., 1997, Final Report: Development of Land Cover/Land Use Geographic Information System Data Layer for the Lake Champlain Basin and Vermont. Burlington, Vermont.
Vermont Department of Environmental Conservation & New York Department of Environmental Conservation., 1997, A phosphorus budget, model, and load reduction strategy for Lake Champlain: Lake Champlain Diagnostic-Feasibility Study. Waterbury, VT and Albany, NY.
Vollenveider, R.A., 1968, Scientific fundamentals of the eutrophication of lakes and flowing water with a particular reference to phosphorus and nitrogen as factor of eutrophication. Tech report, OCDE: Paris. 98 pp.
Wang, D., Levine, S.N., Meals, Jr., D.W., Hoffmann, J.P., Drake, J.C., and Cassell, E.A., 1999, Importance of instream nutrient storage to P export from a rural eutrophic river in Vermont, USA. In: Lake Champlain and Its Basin – A Summary of Six Years of Research. T. Manley (ed.).
Wendt, R.C.m and Corey, R.B., 1980, Phosphorus variations in surface runoff from agricultural lands as a function of land use. *J. Environ. Qual.* **9**(1): 130-136.
Weller, C.M., Watzin, M.C., and Wang, D., 1996, Role of wetlands in reducing phosphorus loading to surface water in eight watersheds in the Lake Champlain Basin. *Environ. Manage.* **20**(5):731-739.
Whigham, D.F., Chitterling, C., and Palmer, B., 1988, Impacts of freshwater wetlands on water quality: a landscape perspective. *Environ. Manage,* **12**(5): 663-671.

Windhausen, L, Braun, D., and Wang, D. 2003, A landscape scale evaluation of phosphorus retention in wetlands of the LaPlatte River basin, Vermont, USA. [this volume]

Young, R.A., Onstad, C.A., Bosch, D.D., and Anderson, W.P., 1989, AGNPS, Agricultural non-point-source pollution model for evaluating agricultural watersheds. *J. Soil Water Conservation* **44**: 168-173.

STATUS AND TRENDS OF PHOSPHORUS IN LAKE CHAMPLAIN AND ITS TRIBUTARIES, 1990-2000

Laura Medalie and Eric Smeltzer[1]

1. ABSTRACT

Reducing phosphorus loading to Lake Champlain in order to reverse eutrophication has been a priority of Vermont, New York, and Quebec since 1990. In-lake phosphorus concentration criteria, lake-segment watershed phosphorus load targets, and a Lake Champlain Phosphorus Total Maximum Daily Load (TMDL) have established reduction goals. Eleven years of tributary and lake phosphorus sampling and streamflow gaging have provided a set of data that can be used to determine preliminary status and trends of progress towards these goals.

Phosphorus levels consistently were above in-lake criteria values during 1990-2000 in the Main Lake, South Lake A, Missisquoi Bay, St. Albans Bay, and the Northeast Arm. Significant upward linear phosphorus trends were present during 1990-2000 in Malletts Bay, the Northeast Arm, and St. Albans Bay. A significant downward trend was present in the South Lake B segment. Substantial declines in Vermont point-source loads occurred after 1991. In 2000, the Vermont point-source loads were below the TMDL wasteload allocations in all lake segments except South Lake B and Otter Creek. Substantial declines in New York point-source loads also have occurred since 1991 in most areas, but current loads remain above the TMDL wasteload allocation for the South Lake B, South Lake A, Port Henry, and Main Lake segments. Estimated phosphorus loads have remained above the tributary target loads in all cases, with the exception of the LaPlatte River. The tributary loads estimated in recent years generally were higher than the 1991 base year loads. Eight of the 18 tributaries tested showed significant trends in flow-adjusted concentrations after reconciling results from the mixed and the time-series approaches. The LaPlatte River and Otter Creek, both in Vermont, showed downward trends. These two tributaries also showed the greatest magnitude of change. The Bouquet, Great Chazy,

[1]Medalie, Laura, U.S. Geological Survey, P.O. Box 628, Montpelier, VT 05601, Phone: 802-828-4512. E-mail: lmedalie@usgs.gov. Smeltzer, Eric, Vermont Department of Environmental Conservation, 103 South Main Street, Bldg 10 N, Waterbury, VT 05671-0408, Phone: 802-241-3770. Email: ericsm@dec.anr.state.vt.us.

Little Chazy, Salmon and Saranac Rivers, and Putnam Creek, all in New York, showed upward trends. Additional work relating trends to changes occurring in the tributary subbasins is needed to better understand the nature (upward or downward) of the trends and to help determine whether progress is being made in achieving the Lake Champlain Basin Program phosphorus reduction goals.

2. INTRODUCTION

Eutrophication is a major water-quality issue in Lake Champlain, and reduction of phosphorus has been identified as one of the top management priorities for the lake and its 8,234-mi^2 drainage area (Lake Champlain Management Conference, 1996). Management of eutrophication in Lake Champlain is guided by an agreement adopted jointly by the States of Vermont and New York and the Province of Quebec (Lake Champlain Management Conference, 1996; Vermont Department of Environmental Conservation [DEC] and New York State DEC, 1997; Smeltzer, 1999). The agreement consists of separate phosphorus-concentration criteria for each of the 13 lake segments shown in figure 1 and phosphorus loading targets for each lake-segment watershed. The in-lake criteria and lake-segment boundaries were adopted in the Vermont Water Quality Standards and endorsed by New York, Quebec, and Vermont in a 1993 water-quality agreement (Lake Champlain Phosphorus Management Task Force, 1993). The loading targets were established using a mass-balance modeling analysis and are consistent with attaining the in-lake criteria (Vermont DEC and New York State DEC, 1997). Subsequent to these efforts, a Lake Champlain Phosphorus Total Maximum Daily Load (TMDL) document was completed which defined specific point and nonpoint-source phosphorus-load allocations for each lake-segment watershed in the two States, as required by the Environmental Protection Agency (Vermont DEC and New York State DEC, 2002).

Efforts since 1990 to reduce phosphorus loading to Lake Champlain have included upgrades of wastewater-treatment facilities and implementation of agricultural nonpoint-source best-management practices (Lake Champlain Basin Program, 2000). The Lake Champlain Basin Program has made a substantial investment in long-term water-quality monitoring in order to measure progress in achieving the phosphorus-reduction goals over a 20-year time frame. Phosphorus monitoring since 1990 includes water samples from Lake Champlain, from point sources in the basin, and from 18 major tributaries in the basin (table 1 and fig. 2) covering about 5,700 mi^2 of drainage area (compared to 6,700 mi^2 of drainage area at the tributary outlets). After more than 10 years of monitoring, the data were examined to determine whether phosphorus is being significantly reduced in Lake Champlain—and in particular to determine whether phosphorus-loading targets for lake-segment watersheds and point-source and nonpoint-source phosphorus-load allocations for subbasins are being attained.

Numerous techniques are available for estimating loads and testing for trends. Estimated seasonal or annual loads are generally based on a model that utilizes continuous streamflow data and discontinuous concentration data. Estimated monthly, seasonal, or annual constituent loads can be analyzed for trends based on any statistical method for which the required model assumptions are met. However, streamflow is part of the load calculation, and any observed trends may be reflecting a trend in streamflow rather than

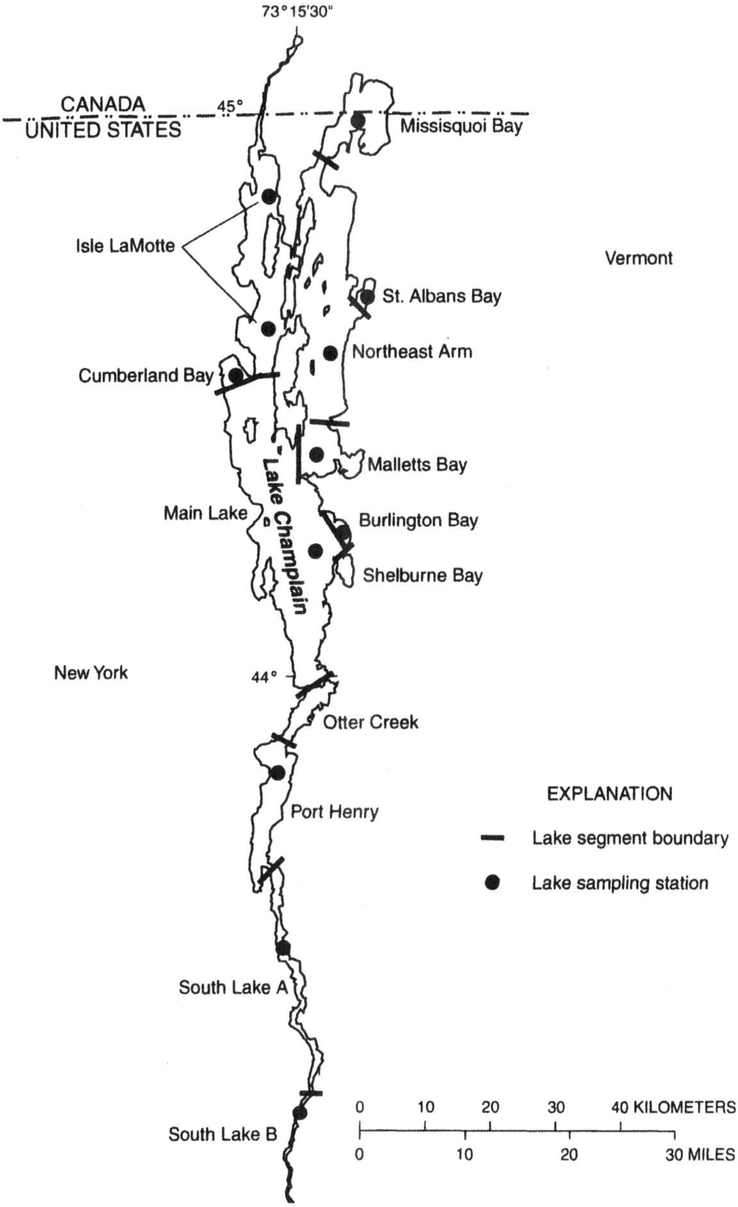

Figure 1. Lake segments and lake sampling stations in the Lake Champlain Basin. The Northeast Arm and Isle LaMotte segments are separated by islands.

Table 1. Station numbers, drainage areas, and number of phosphorus samples for Lake Champlain tributary water-quality sampling stations

[USGS, U.S. Geological Survey; No., number; mi^2, square miles]

Tributary name	USGS or MENV[a] Station No.	Drainage area at gaging station (mi^2)	Drainage area at tributary outlet (mi^2)	No. of phosphorus samples			
				1990	1991	Total 1992-2000	Median per year 1992-2000
Ausable	04275500	446	513	35	27	101	10
Bouquet	04276500	270	272	35	26	95	9
Great Chazy	04271500	243	299	35	22	92	9
Lamoille	04292500	686	723	65	27	109	11
LaPlatte	04282795	45	53	69	24	118	13
Lewis	04282780	77	81	68	26	69	11
Little Ausable	04273800	68	73	35	26	43	10
Little Chazy	04271815	50	53	36	25	97	9
Little Otter	04282650	57	73	69	26	109	12
Mettawee/Barge Canal	04280450	167	424	60	26	61	13
Missisquoi	04294000	850	865	64	30	103	12
Otter	04282500	628	944	72	31	115	12
Pike	030420	156	258	61	24	88	10
Poultney	04280000	187	263	59	26	107	11
Putnam	04276842	52	62	22	13	107	13
Salmon	04273700	63	68	36	25	98	10
Saranac	04273500	608	614	36	27	104	10
Winooski	04290500	1,044	1,063	71	30	99	12
Total		5,697	6,701	928	461	1,715	

[a] The Pike River station is operated by the Quebec Ministry of the Environment (MENV).

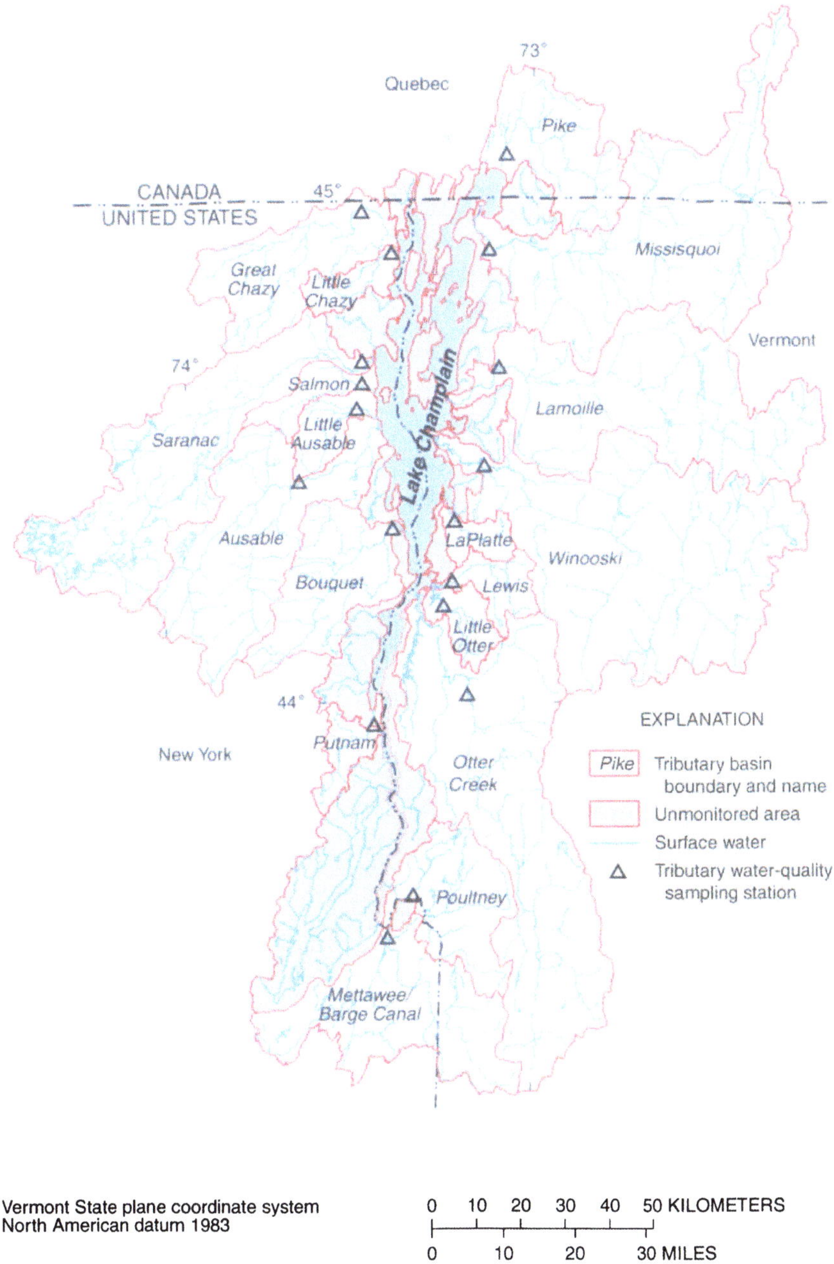

Figure 2. Lake Champlain Basin showing tributary boundaries and water-quality sampling stations.

increases in phosphorus from new sources or a degradation or improvement of management practices in the basin. Flow adjustment to constituent concentrations in order to correct for the compounding effects of streamflow can be accomplished in various ways. One common technique is to perform a simple parametric linear regression of concentration relative to streamflow. Residuals from the regression are flow adjusted because they represent the part of concentration that is not explained by streamflow (Helsel and Hirsch, 1992).

This paper presents results of various applications of statistical techniques to examine the status of in-lake concentrations, point-source loads, and tributary loads compared to TMDL allocations and the change over time, or trends, of tributary phosphorus loads and concentrations. Use of the term *phosphorus* in this paper always refers to total phosphorus.

3. DATA SOURCES

Phosphorus-monitoring data from Lake Champlain and its tributaries were obtained for the 1990-2000 period from the Lake Champlain Long-Term Water Quality and Biological Monitoring Program (LTMP) (Vermont DEC and New York State DEC, 2001), supplemented with comparable data from Vermont DEC and NY State DEC (1997). The sampling station network for the LTMP includes 12 lake stations and 18 tributary water-quality sampling stations shown in figures 1 and 2. All lake and tributary samples were analyzed for total phosphorus by persulfate digestion at the Vermont DEC Laboratory using U.S. Environmental Protection Agency (1993) method 365.1. Summary streamflow and phosphorus concentration statistics are presented in table 2 for each station. None of the data were below the method-detection level.

The lake stations were sampled using a Kemmerer- or Van Dorn-style sampler. Discrete-depth samples were combined to form vertical water-column composites of the upper mixed layer, determined from thermal profiles on the day of sampling. Lake stations were sampled approximately every 2-3 weeks during late April to early November, which resulted in 6 to 18 samples per year from each site during 1990-2000. In most cases, the lake sampling stations were located centrally within each of the phosphorus-management lake segments shown in figure 1. Because samples were not collected after 1991 in the Otter Creek or Shelburne Bay lake segments, no results are presented in this report for these segments. Two monitoring stations were located within the Isle LaMotte segment; the annual means from these two stations were averaged for analysis in this report.

Table 2. Summary streamflow and phosphorus-concentration statistics for Lake Champlain tributaries, 1990-2000

[cfs, cubic feet per second; mg/L, milligrams per liter]

Tributary name	Period of record	Streamflow: Mean annual flow, 1990-2000[a], cfs	Streamflow: Percentiles for period of record, cfs			Phosphorus concentration, mg/L: Range	Phosphorus concentration, mg/L: Percentiles		
			10th	50th	90th		25th	50th	75th
Ausable	1910-1968 1990-ongoing	800	165	350	1,510	0.007-0.712	0.014	0.021	0.052
Bouquet	1923-1968 1990-ongoing	366	60	153	700	.004-.508	.012	.026	.108
Great Chazy	1928-1968 1990-ongoing	367	45	140	635	.007-.496	.024	.041	.068
Lamoille	1929-ongoing	1,368	298	712	2,800	.009-.484	.016	.023	.045
LaPlatte	1990- ongoing	46.8	3.8	19	111	.016-1.11	.092	.155	.264
Lewis	1990- ongoing	105	17	60	231	.008-.772	.025	.047	.098
Little Ausable	1991- ongoing	63.8	9.3	33	124	.012-.712	.032	.054	.088
Little Chazy	1990-ongoing	62.9	4.8	29	160	.014-3.49	.037	.063	.098
Little Otter	1990-ongoing	62.1	5.1	25	156	.023-.440	.065	.100	.160
Mettawee/ Barge Canal	1990-ongoing	265	35	165	594	.014-.720	.043	.064	.102
Missisquoi	1990-ongoing	1,668	230	838	4,180	.015-.660	.028	.048	.093
Otter	1903-1907 1910-1920 1990-ongoing	1,078	260	636	2,330	.023-.498	.050	.077	.116
Pike						.016-.795	.058	.099	.178
Poultney	1928-ongoing	274	28	136	614	.012-.343	.030	.052	.120
Putnam	1990-ongoing	76.2	6.4	36	187	.004-.359	.009	.015	.027
Salmon	1959-1968 1990-ongoing	72.1	13	37	127	.006-.286	.014	.021	.049
Saranac	1903-1930 1943-ongoing	983	325	640	1,640	.010-.104	.017	022	.028
Winooski	1929-ongoing	1,957	358	1,000	4,000	.009-.756	.028	.048	.096

[a] Water year.

The tributary water-quality sampling stations were located as near to the river mouths as possible without encountering lake backwater. All tributary sampling stations were located on rivers with a U.S. Geological Survey or Quebec Ministry of the Environment streamflow-gaging station in continuous operation since at least 1991. Tributary samples were obtained manually from bridges using depth and velocity-integrating sampling devices (Edwards and Glysson, 1988). An effort was made to obtain the largest proportion of samples possible during high-flow conditions in order to improve the precision of tributary annual mass-loading estimates (Vermont DEC and New York State DEC, 1997). For example, on about two-thirds of the sampled days for the Winooski River, mean-daily streamflow was among the highest 25 percent of all mean-daily streamflow values for the 1990-2000 period. The number of samples collected from each of the tributary stations ranged from 13 to 72 per year during the more intensive monitoring period of 1990-91 (table 1), and from 4 to 20 per year during 1992-2000.

Wastewater-effluent monitoring data for treatment facilities within the Lake Champlain Basin were compiled from 60 plants in Vermont and 29 plants in New York (F. Dunlap, New York State DEC, pers. commun., 2001) during 1991-2000. This compilation included all municipal and industrial wastewater discharges containing appreciable amounts of phosphorus. Average daily flows and effluent phosphorus concentration results, generally reported on a monthly basis, were obtained from plant-operations reports and other sampling programs in each state. Data for 1991 were obtained from a previous study (Vermont DEC and New York State DEC, 1997). A gap in the compiled data record is present from 1992 to 1994. In the occasional case where no effluent phosphorus monitoring was conducted, concentrations were estimated from the available data from other years at the facility, or by using other assumptions. At least some effluent data were available for all but nine of the smaller facilities in Vermont.

In order to compare measured tributary phosphorus loads with management target loads, it was necessary to derive individual tributary loading targets from the lake-segment watershed target loads specified in the basin plan and the TMDL documents because some lake-segment watersheds include more than one monitored tributary. Many lake-segment watersheds also include unmonitored areas and wastewater that is discharged directly to the lake. Tributary loading targets derived for this purpose, shown in table 3, consist of the phosphorus allocation assigned to upstream point-source discharges plus a portion of the lake-segment nonpoint-source load allocation in the TMDL equal to the portion of the segment's measured 1991 base year nonpoint-source load contributed by that tributary.

Table 3. Derivation of tributary phosphorus-loading targets

[Column headings (A)-(E) are used to designate values utilized to calculate tributary target loads as specified in footnote d. For example, the tributary target load for Missisquoi Bay is 3.4 mt/yr + 93.0 mt/yr (75.1 mt/yr/151.9 mt/yr) or 49.4 mt/yr]; mt/yr, metric tons per year; VT, Vermont; QC, Quebec; NY, New York; DEC, Department of Conservation]

Lake segment watershed	State/ Province	(A) Lake segment 1991 nonpoint load[a] (mt/yr)	(B) Lake segment nonpoint allocation[b] (mt/yr)	Tributary	(C) Tributary upstream point allocation[b] (mt/yr)	(D) Tributary 1991 nonpoint load[a] (mt/yr)	(E) Tributary target load[d] (mt/yr)
Missisquoi Bay	VT/QC	151.9	93.0	Missisquoi	3.4[c]	75.1	49.4
				Pike	0.0[c]	44.4	27.2
Malletts Bay	VT	29.8	25.3	Lamoille	3.2	26.6	25.8
Shelburne Bay	VT	11.1	10.0	LaPlatte	0.3	7.6	7.1
Otter Creek	VT	58.9	44.1	Otter	11.4	47.4	46.8
				Lewis	0.0	5.2	3.9
				Little Otter	0.0	5.4	4.0
Main Lake	VT	60.3	51.3	Winooski	22.7	59.6	73.4
South Lake B	VT/NY	49.1	41.2	Poultney	1.3	14.1	13.2
				Mettawee/Barge Canal	1.6	33.7	29.9
South Lake A	NY	3.5	3.3	Putnam	0.0	1.3	1.2
Main Lake	NY	31.8	29.5	Bouquet	0.0	13.5	12.6
				Ausable	3.2	11.2	13.6
				Little Ausable	0.6	3.8	4.1
				Salmon	0.0	1.7	1.6
Cumberland Bay	NY	8.8	8.1	Saranac	6.0	7.7	13.1
Isle LaMotte	NY	20.9	18.9	Great Chazy	0.7	16.6	15.7
				Little Chazy	0.1	3.2	3.0

[a] Vermont DEC and New York State DEC, 1997.
[b] Vermont DEC and New York State DEC, 2002.
[c] Includes Vermont point sources only.
[d] $E = C + B * (D/A)$.

4. STATISTICAL METHODS

Various statistical methods were utilized in this study to assess the status and trends of in-lake phosphorus at lake sampling stations, the status of point-source loads to use as a basis of comparison to TMDL allocations, the status of tributary loads to use as a basis of comparison to TMDL allocations, and trends in tributary flow-adjusted concentrations.

4.1. Status and Trends of In-Lake Phosphorus at Lake Sampling Stations

Lake phosphorus-sample results were averaged by year for each station, and the presence of trends over time was tested using simple linear regression of the annual mean phosphorus concentrations and year (Snedecor and Cochran, 1967).

4.2. Status of Point-Source Loads

Point-source annual phosphorus loads were estimated by multiplying the annual average effluent flow by the annual average phosphorus concentration at each facility and then summing these loads for each lake segment. These annual loads simply were graphed over time and visually compared to the lake-segment TMDL wasteload allocation.

4.3. Status of Tributary Loads

Phosphorus loads from each tributary were estimated for calendar year 1991 by a previous loading study (Vermont DEC and New York State DEC, 1997). Phosphorus-concentration results during 1990-92 were applied to flows measured during 1991 to produce a hydrologic base-year loading estimate. During the subsequent monitoring period of 1993-2000, mean loading rates were calculated for 2-year intervals based on water years (ending on September 30 of the indicated year) in order to improve precision by increasing the sample size.

Mean annual tributary phosphorus loads were estimated using U.S. Army Corps of Engineers FLUX software (Walker, 1987, 1996; Vermont DEC and New York State DEC, 1997). For tributaries and 2-year-interval periods with significant ($p<0.10$) concentration versus streamflow regression relationships, loading estimates were based on log-log regressions between the phosphorus concentration and the average daily flow on the day of sampling (Walker, 1996, Method 6). Regression equations were stratified by streamflow to eliminate residual dependence, in cases where such dependence was demonstrated by visual examination of residuals versus flow plots. Where significant concentration versus flow relations were not established, phosphorus load was calculated based on the flow-weighted mean concentration times the mean flow for the averaging period (Walker, 1996, Method 2). An error-analysis procedure in the FLUX program was used to estimate 95-percent confidence intervals for the loading estimates.

Point-source loads from upstream wastewater discharges were subtracted from the total tributary loads that were estimated using the FLUX software to estimate the nonpoint-source component of the tributary loads.

The loading estimates produced by the FLUX program were adjusted proportionally for the additional drainage area downstream from each streamflow-gaging station so that the estimates would apply at the mouths of each tributary. Gaged tributary subbasin areas range from 39 to 99 percent of the total drainage area but in most cases represent over 80 percent of the area at the tributary mouth (table 1). Daily streamflows recorded at the gaging station on the New Haven River at Brooksville, VT, were added to the daily streamflows for the Otter Creek gage at Middlebury, VT, in order to supplement the hydrologic coverage within the Otter Creek subbasin.

4.4. Trends in Tributary Flow-Adjusted Concentrations

Two different approaches were used to thoroughly and appropriately investigate trends and to strengthen those statements of trend.

4.5. Mixed Approach

The mixed approach uses the parametric ESTIMATOR (Cohn and others, 1989; Gilroy and others, 1990; Cohn and others, 1992) program to adjust water-quality data for streamflow and season and then applies nonparametric methods to those results to test for trends. Helsel (1987) discusses some advantages of using nonparametric methods for trends testing. Adaptations specifically related to working with water-quality data further increase the benefits of using nonparametric methods. For example, the Mann-Kendall trend has been modified to address autocorrelation of data (Hamed and Rao, 1998) and serially correlated data (the correlation or dependence of residuals over time) with missing and censored values (Hirsch and Slack, 1984). The seasonal Kendall test for trend (Hirsch and others, 1982) provides modification for seasonal variations to the Mann-Kendall trend test (Zipper and others, 1992; Liebermann and others, 1989). The Kendall-Theil test (Langland and others, 2000) provides a further increase of power in the slope estimate of trend.

ESTIMATOR is a 7-parameter log-linear regression model that uses mean-daily streamflow to calculate loads. It is one of several models that the USGS has developed to incorporate the minimum-variance unbiased estimator (MVUE) (Bradu and Mundlak, 1970) to correct for potential retransformation errors when logarithms of variables are used.

The ESTIMATOR regression model is:

$$ln[C] = \beta_o + \beta_1\, ln\, [Q / \overline{Q}] + \beta_2\{ln\, [Q / \overline{Q}]\}^2 + \beta_3\, sin[2\pi T] \;+\; \beta_4 cos[2\pi T] \;+\; \varepsilon \qquad (1)$$

where

ln is the natural logarithm function;

C is estimated concentration, in milligrams per liter;

Q is measured streamflow, in cubic feet per second;

T is time, measured in decimal years;

$\overline{Q}$ is a centering variable for streamflow that is close to the mean;

β_x are parameters estimated by ordinary least squares regression;

β_o is a constant;

β_1 *and* β_2 describe the relationship between concentration and flow;

β_3 *and* β_4 describe seasonal variation in concentration data; and

ε is model error.

The residuals of the ESTIMATOR adjusted concentrations are the input data set for the Kendall-Theil monotonic linear trends test that is used to determine changes in the Lake Champlain phosphorus data during 1990-2000. For the Kendall-Theil test, the change in residual divided by the change in time is computed for all time/residual data pairs. Because the test for significance is based on adding the number of positive and negative slopes, and comparing the overall median slope to a Kendall's tau for significance of trend, the test validity does not require residuals to be normal and the test is resistant to outliers. These conditions are possible advantages of this approach over other approaches.

4.6. Time-Series Approach

The U. S. Geological Survey recently developed a parametric time-series approach (Vecchia, 2000) to trends testing for water-quality data. The time-series method uses a periodic autoregressive moving-average model designed specifically to address serial correlation and cross-correlation commonly seen with streamflow and concentration data. Strong serial correlation may result in a violation of the assumption that residuals are independent. If this violation occurs, there is a risk that the null hypothesis (that there is no trend) is rejected too easily (Helsel and Hirsch, 1992). A favorable attribute of the time-series model is the capacity to handle nonmonotonic trends and irregular sampling intervals.

The time-series trend model tests whether the mean of the log-transformed data remains constant over time. Output from the model provides an estimated coefficient and standard error for each specified step or linear trend interval, from which one can calculate the trend magnitude and p-value. Residual plots were examined to ensure that residuals were approximately normal with zero mean and constant variance. The general form of the discharge and concentration data modeled by the time-series approach (Vecchia, 2000) is:

$$X(t) = C_X + A_X(t) + S_X(t) + U(t) \tag{2}$$

where

X is the base-10 logarithm of discharge;

t is the time index, in decimal years;

C_X is a constant;

$A_X(t)$ is a time series that represents annual variability in X;

$S_X(t)$ is a time series that represents seasonal variability in X; and

$U(t)$ is a time series of deviations of $X(t)$ from $C_X + A_X(t) + S_X(t)$

and

$$Y(t) = C_Y + A_Y(t) + S_Y(t) + T_r(t) + W(t) \quad (3)$$

where

Y is the base-10 logarithm of concentration;

C_Y is a constant;

$A_Y(t)$ is a time series that represents annual variability in Y;

$S_Y(t)$ is a time series that represents seasonal variability in Y;

$Tr(t)$ is a trend in Y, and

$W(t)$ is a time series of deviations of $Y(t)$ from $C_Y + A_Y(t) + S_Y(t) + Tr(t)$.

Vecchia (2000) specifies that at least 60 concentration measurements during 15 years are needed in order to use the time-series model. However, 15 years of record is a conservative number, and the Lake Champlain data set comprised of at least 10 observations per year for 11 years should be sufficient for model validity, especially if streamflow is not tremendously variable during the period tested with the time-series approach compared to the period of record (Vecchia, U.S. Geological Survey, written commun., 2001). Streamflow variability was tested using statistics for USGS index station 04287000, Dog River at Northfield Falls, VT. Because of this station's average-size basin and stable rating (Slack and others, 1994), it was deemed to be representative of stations throughout Vermont. At the Dog River, mean annual flows for 1990 through 2000, the period used in this trends analysis, were 136 cubic feet per second (cfs) compared to 123 cfs for the rest of the period of record, 1935 through 1989. Mean annual flows for these two time periods are not significantly different according to results from the Wilcoxon sign-rank statistical test at the 95-percent confidence level. The other requirement of the time-series model, that no more than 10 percent of the measurements can be below the detection limit, was also met by the Lake Champlain tributary data. One advantage of the time-series model is that it is designed to be more powerful for detecting trends than a regression-based model such as the mixed approach. A second advantage is that it can accommodate user-specified multiple linear trends and step trends; however, the user is advised to define multiple model intervals based on factors unrelated to the data, such as a change in sampling location (Vecchia, 2000).

5. RESULTS

5.1. Status and Trends of In-Lake Phosphorus at Lake Sampling Stations

Mean annual phosphorus concentrations at the lake-sampling stations as compared to the in-lake water-quality criteria are shown in figure 3. Phosphorus levels consistently were above the criteria values during 1990-2000 in the Main Lake, South Lake A, Missisquoi Bay, St. Albans Bay, and the Northeast Arm lake segments (fig. 3). All the other lake segments had concentrations above the criteria values during some years.

A linear regression analysis of mean annual lake phosphorus concentration relative to year indicates that significant ($p<0.05$) upward linear trends were present during 1990-2000 in Malletts Bay, the Northeast Arm, and St. Albans Bay. A significant downward trend was present in the South Lake B segment. No other segments showed significant linear trends, although the plot for Cumberland Bay suggests a possible step reduction around 1996.

5.2. Status of Point-Source Loads

Phosphorus loads from point-source discharges to lake-segment watersheds in Vermont and New York showed overall declines from 1991 to 2000 in many subbasins (fig. 4). The sum of the loads from all facilities in each lake-segment watershed is compared in figure 4 with the phosphorus wasteload allocation in the Lake Champlain Phosphorus TMDL (Vermont DEC and New York State DEC, 2002).

Substantial declines in Vermont point-source loads occurred after 1991 (fig. 4a) as a result of phosphorus removal upgrades at many facilities. The Vermont point-source loads as of 2000 were below the TMDL wasteload allocations in all lake segments except South Lake B and Otter Creek. Substantial declines in New York point-source loads also have occurred since 1991 in most areas (fig. 4b), but 2000 loads remain above the TMDL wasteload allocation in the South Lake B, South Lake A, Port Henry, and Main Lake segments.

Overall, Vermont wastewater-treatment facilities reported reductions in their phosphorus discharges from 126 metric tons per year (mt/yr) in 1991 to 44 mt/yr in 2000 (65-percent reduction), while New York facilities reported a reduction from 58 mt/yr in 1991 to 39 mt/yr in 2000 (33-percent reduction).

5.3. Status of Tributary Loads

Estimated mean annual phosphorus load from each of the 18 monitored Lake Champlain tributaries for the 1991 hydrologic base year (Vermont DEC and New York State DEC, 1997) and the estimated 2-year means from 1993-2000 are compared to loading targets derived from the Lake Champlain Phosphorus TMDL watershed allocations and mean streamflow in figure 5. Estimated phosphorus loads have remained well above the tributary target loads in all cases, with the exception of the LaPlatte River. The tributary loads estimated in recent years generally were higher than the 1991 base year loads. Comparison of phosphorus load with mean streamflow for each time interval

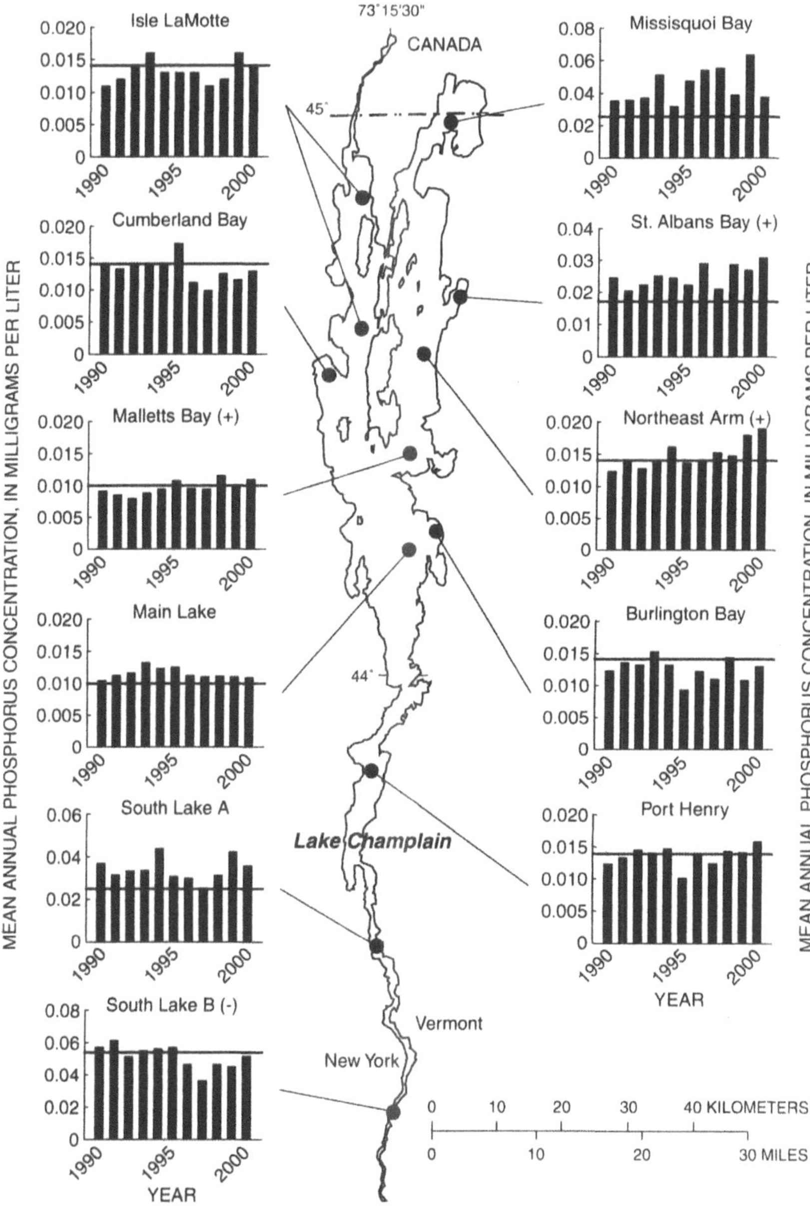

Figure 3. Mean annual phosphorus concentration at 12 lake-monitoring stations during 1990-2000 compared with the in-lake water-quality criteria (horizontal lines). A plus sign (+) indicates a significant ($p<0.05$) upward linear trend. A minus sign (-) indicates a significant downward linear trend.

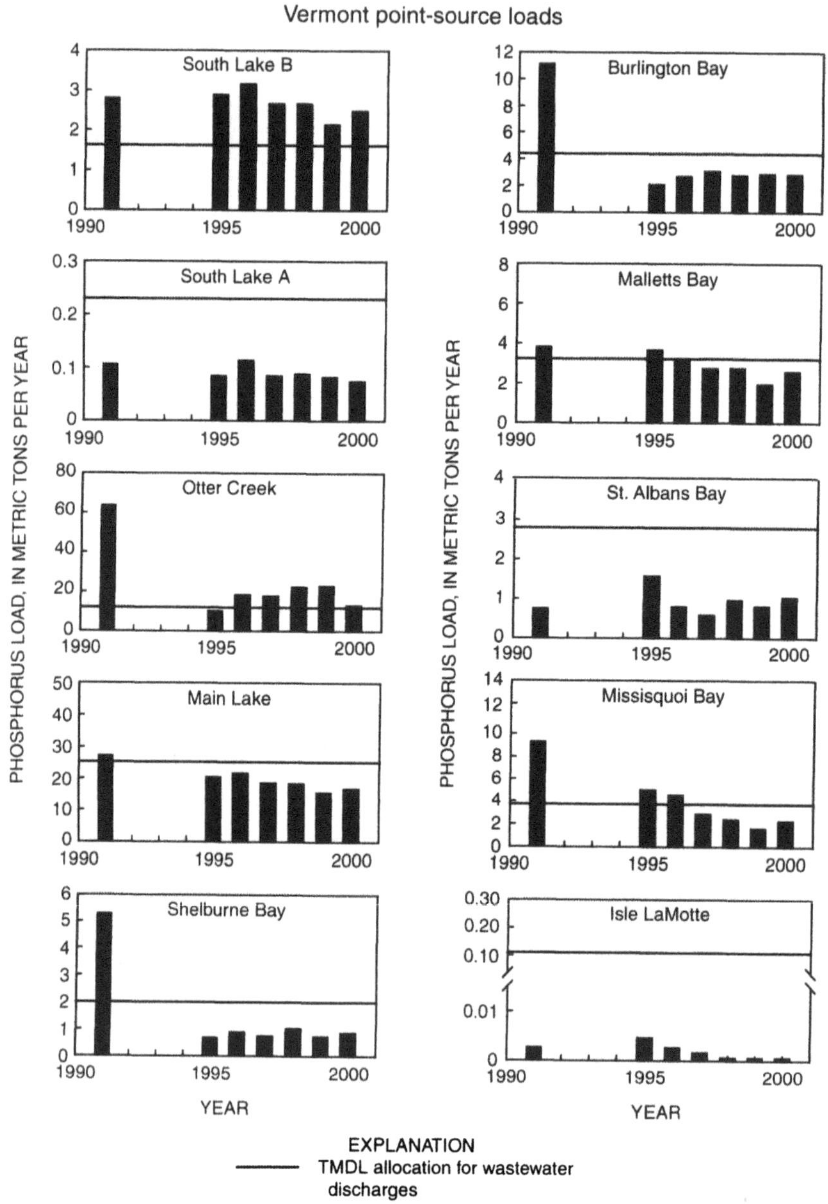

Figure 4a. Vermont point-source phosphorus loads to Lake Champlain segments during 1991 to 2000, in metric tons per year, compared with the TMDL allocation for wastewater discharges (Vermont DEC and New York State DEC, 2002).

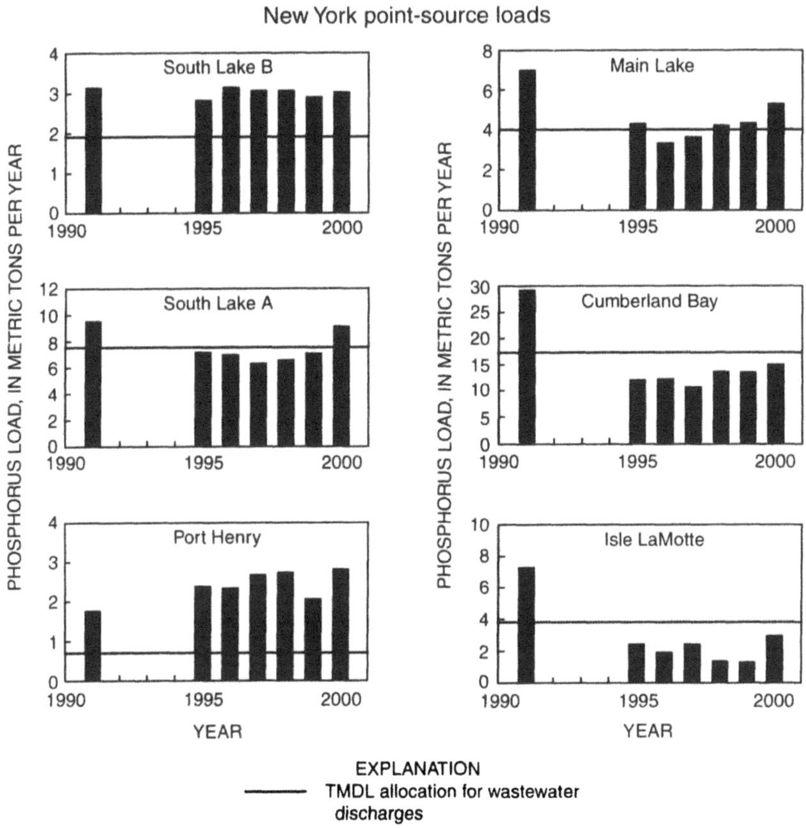

Figure 4b. New York point-source phosphorus loads to Lake Champlain segments during 1991 to 2000, in metric tons per year, compared with the TMDL wasteload allocation (Vermont DEC and New York State DEC, 2002).

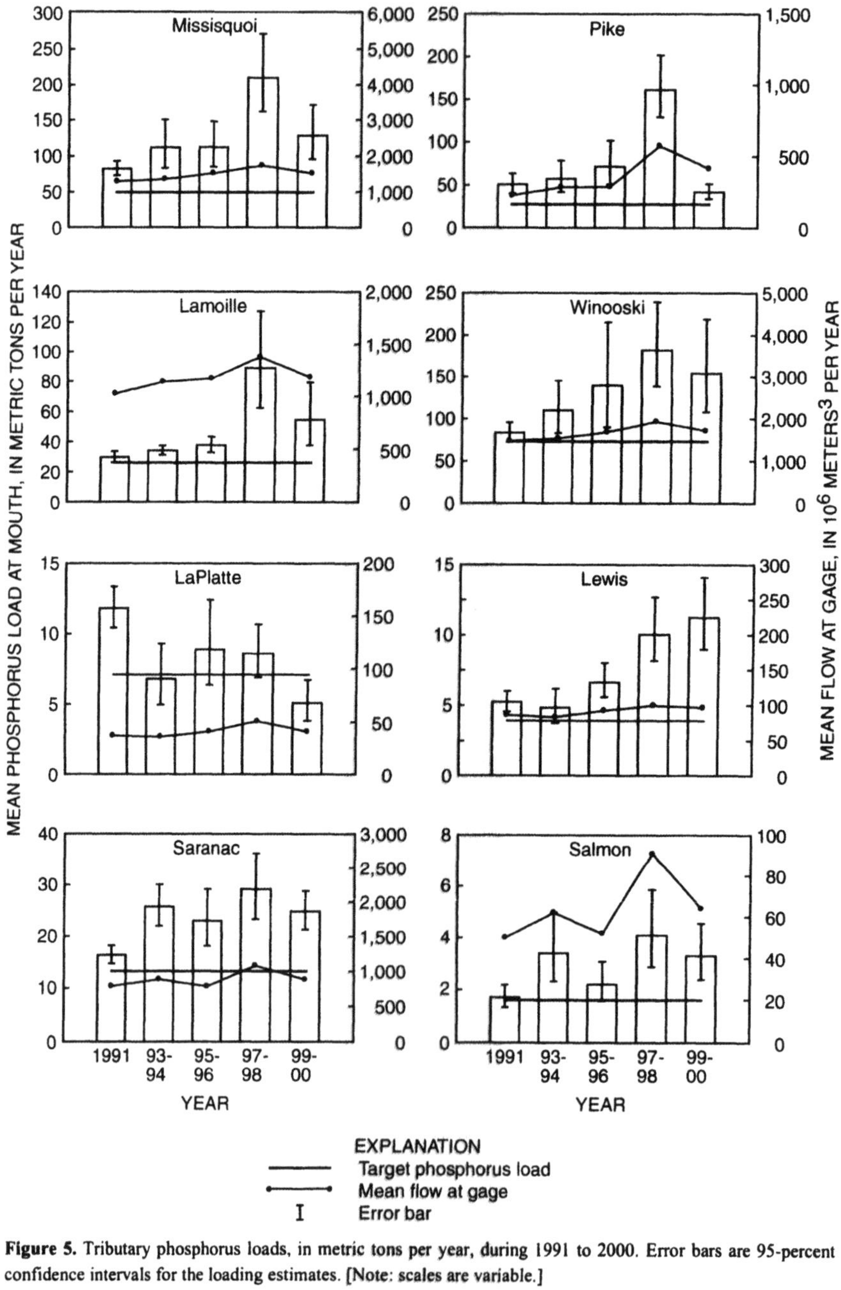

Figure 5. Tributary phosphorus loads, in metric tons per year, during 1991 to 2000. Error bars are 95-percent confidence intervals for the loading estimates. [Note: scales are variable.]

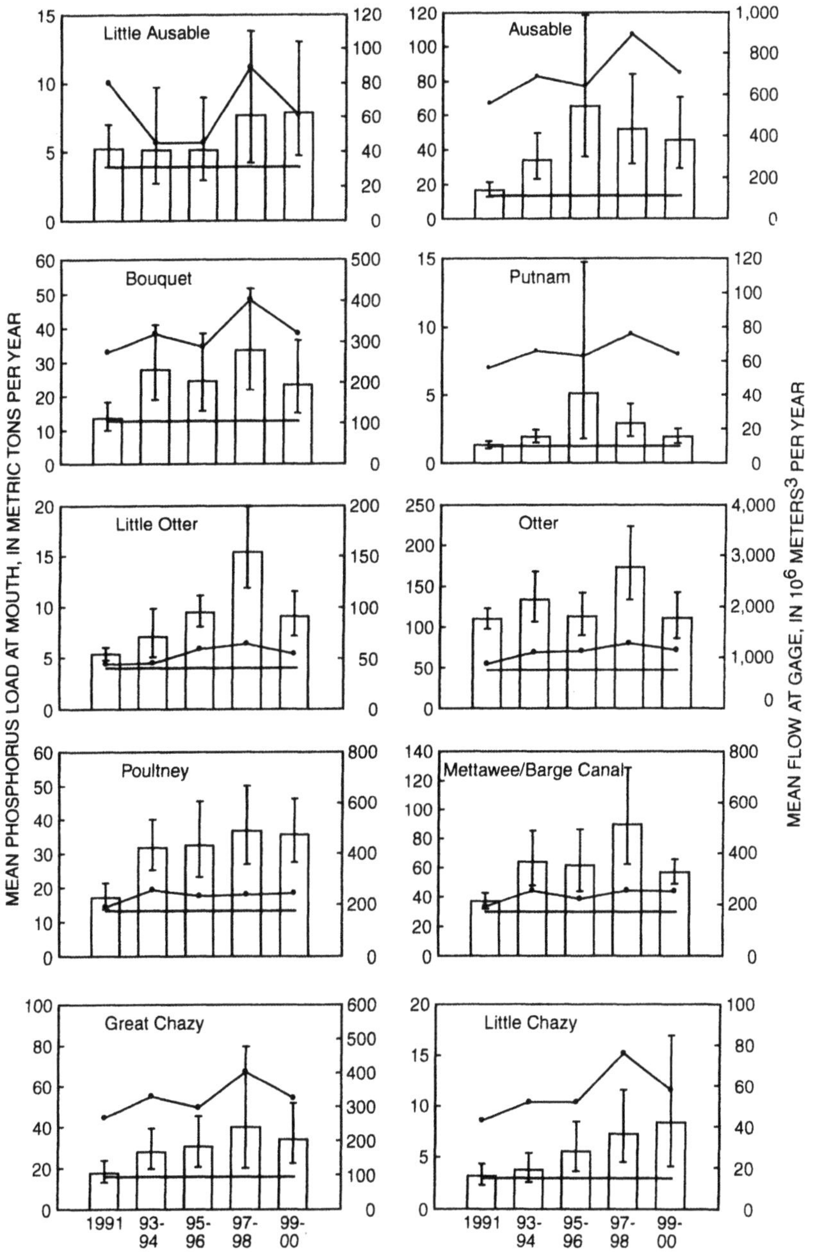

Figure 5. Tributary phosphorus loads, in metric tons per year, during 1991 to 2000. Error bars are 95-percent confidence intervals for the loading estimates [Note: scales are variable.]--Continued.

indicates that higher streamflows may be at least part of the explanation for the higher loading rates observed in recent years. For example, the time interval 1997-98 tends to have the highest streamflows as well as the highest loading rates.

Nonpoint sources are the largest contributor of phosphorus loading to Lake Champlain. Nonpoint sources dominate phosphorus loads in every tributary, representing 80-100 percent of the total load in the monitored tributaries (fig. 6). When direct lake discharges are taken into account, and assuming that nonpoint-source loads from unmonitored areas are proportional to the tributary nonpoint-source loads, it is estimated that nonpoint sources currently contribute about 93 percent of the phosphorus load to Lake Champlain.

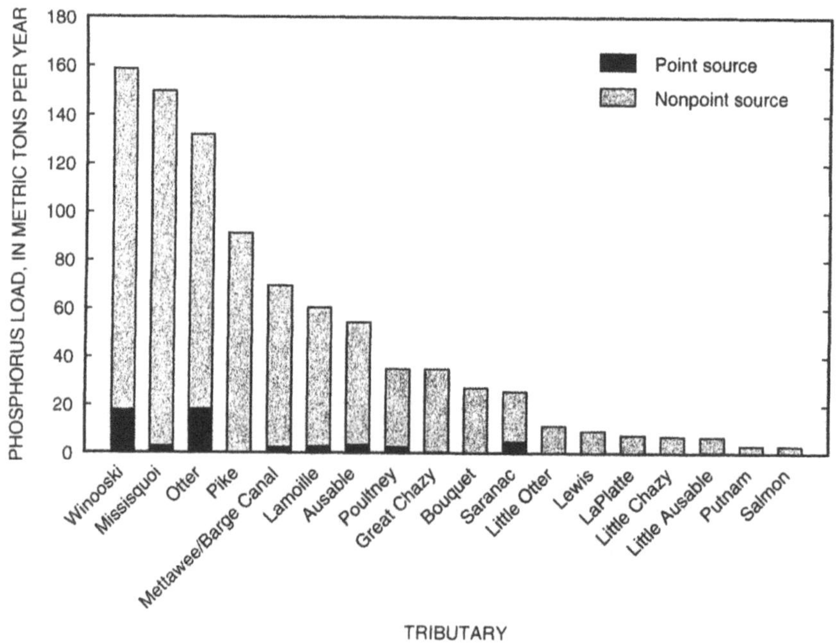

Figure 6. Average point- and nonpoint-source phosphorus loads, in metric tons per year, in monitored Lake Champlain tributaries during 1995-2000. No point-source data were available for the Pike River.

5.4. Trends in Tributary Flow-Adjusted Concentrations

Trends that were significant at the 95-percent confidence level from either the mixed approach or the time-series approach for trends in tributary phosphorus concentrations are shown in table 4 and graphically in figure 7. According to the mixed approach, six tributaries had significant ($p<0.05$) upward trends: the Ausable, Bouquet, Great Chazy, Little Otter, Putnam, and Saranac Rivers. The LaPlatte River had a significant downward trend. Results from the time-series approach showed that the Bouquet, Great Chazy, Little Chazy, Putnam, Salmon, and Saranac Rivers had significant upward trends. The LaPlatte River and Otter Creek had significant downward trends. Results from the two approaches are consistent for 13 of the 18 tributaries (table 4). Contradictory results between the two approaches were found for the Ausable River, Little Chazy River, Little Otter Creek, Otter Creek, and Salmon River.

Table 4. Trends in phosphorus concentrations for Lake Champlain tributaries, 1990 through 2000, based on the Kendall-Theil and time-series tests

[<, less than; shaded cells indicate significant trend at p-value less than 0.05]

Tributary name	Station No.	Mixed approach: Kendall-Theil test		Time-series test		
		Tau	p-value	Trend coefficient	p-value	Trend per year (percent)
Ausable	04275500	0.165	0.002	0.006	0.181	
Bouquet	04276500	0.138	0.010	0.013	0.013	+3.0
Great Chazy	04271500	0.152	0.006	0.015	0.014	+3.5
Lamoille	04292500	0.092	0.053	0.005	0.157	
LaPlatte	04282795	-0.325	<0.001	-0.056	<0.001	-12.1
Lewis	04282780	-0.010	0.839	-0.004	0.140	
Little Ausable	04273800	0.086	0.195	-0.012	0.167	
Little Chazy	04281815	0.101	0.061	0.016	<0.001	+3.7
Little Otter	04282650	0.103	0.029	0.003	0.213	
Mettawee/Barge Canal	04280450	0.037	0.442	-0.004	0.166	
Missisquoi	04294000	0.057	0.231	0.008	0.063	
Otter	04282500	-0.022	0.145	-0.025	<0.001	-5.6
Pike	030420	-0.049	0.339	0.001	0.430	
Poultney	04280000	0.089	0.066	0.009	0.051	
				000.012		
Putnam	04276842	0.158	0.005	0.012	<0.001	+2.8
Salmon	04273700	0.101	0.058	0.022	<0.001	+5.2
Saranac	04273500	0.200	<0.001	0.005	0.034	+1.2
Winooski	04290500	0.007	0.881	-0.004	0.235	

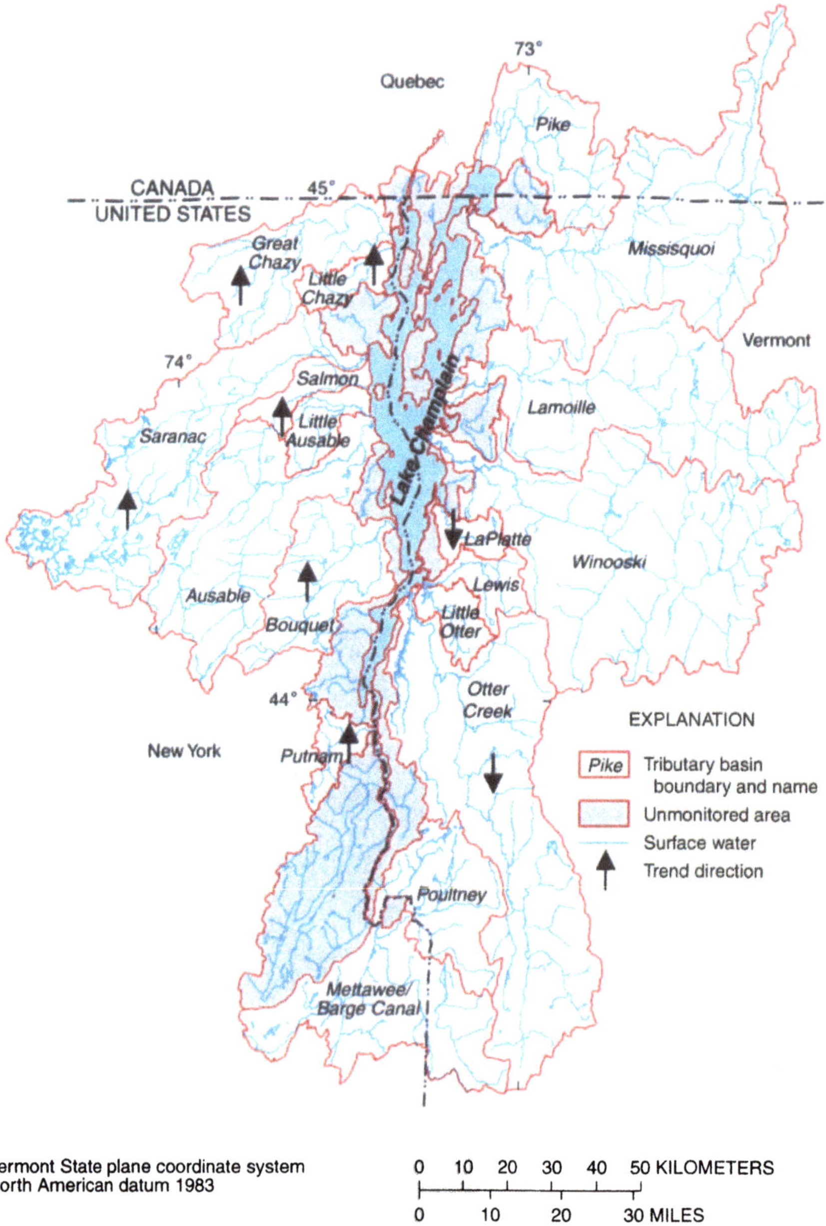

Figure 7. Lake Champlain Basin showing trends in total-phosphorus flow-adjusted concentrations. Lack of arrow implies no significant trend.

6. DISCUSSION

For the Little Chazy and the Salmon Rivers, the time-series test showed a strong upward trend (both p-values are <0.001) and the mixed test shows nearly significant trends (p-values of 0.061 and 0.058, respectively). The strength of the time-series test results for these two tributaries, the near-significance of the mixed-model test results, and visual inspection of plots (figs. 8b and 8e) of residuals from the mixed model, all indicate that an upward trend for both these sites is real. Since the time-series approach was designed to be more powerful for detecting trends than the mixed-model approach, it is not unusual for the time-series approach to detect a trend when no trend is detected using the mixed approach.

The Ausable River and the Little Otter Creek showed no trend in phosphorus concentration from the time-series test and a significant upward trend from the mixed-model test. Close visual inspection of the mixed-model residuals (figs. 8a and 8c) reveals a possibility for explaining the difference. At both sites, data plots (figs. 9a and 9b) show a very strong rise in residuals for a short period of time during the first part of the period of record, when data collection was most frequent. When the Kendall-Theil test is rerun using data from both tributaries without the first points of rapid rise (data after 10/3/1990 for the Ausable River and after 7/5/1990 for the Little Otter Creek), results strongly support the null hypothesis of no trend ($p=0.35$ and 0.20, respectively). Furthermore, the nonparametric and parametric plots for the entire period of record are convincingly free of trends. These data, taken together, indicate that for the 1990-2000 period, no significant trends occurred in the Ausable River and the Little Otter Creek.

Otter Creek also showed conflicting results for the two trends tests. An explanation for the different results may be illustrated in the mixed-model data plot (fig. 8d), which shows a break in the data between 1992 and 1993, possibly related to improvements at 2 wastewater-treatment facilities in this watershed. The Rutland City wastewater-treatment facility was upgraded in July 1993 and phosphorus effluent from a large industry contributing wastewater to the Middlebury facility was reduced in July 1994. The Kendall-Theil part of the mixed model balances out frequent but increasing data points from 1990 to 1992 and less frequent but decreasing data points from 1993 to 2000 and no overall trend is seen. Because the time-series test assigns three values per year based on a predictive model, there is no extraneous weight added in the trend test for the first 3 years of record.

The time-series test also has the capability to split up the record into discrete periods and to test for separate step and linear trends. Based on the record of treatment-plant upgrades in the Otter Creek subbasin, the null hypothesis is tested that there is a step trend during 1990-92 and that there are separate linear trends from 1990 to 1992 and 1993 to 2000. A significant upward linear trend during 1990-1992 and a significant downward linear trend during 1993-2000, are seen in figure 10.

The magnitudes of both this downward trend at Otter Creek, and the downward trend at the LaPlatte River, are larger than those for any of the other measured trends, whether upward or downward. The downward trend at the LaPlatte River was associated with the phosphorus removal upgrade of the Hinesburg Wastewater Treatment Plant in 1993 (Vermont DEC and NY State DEC, 1998).

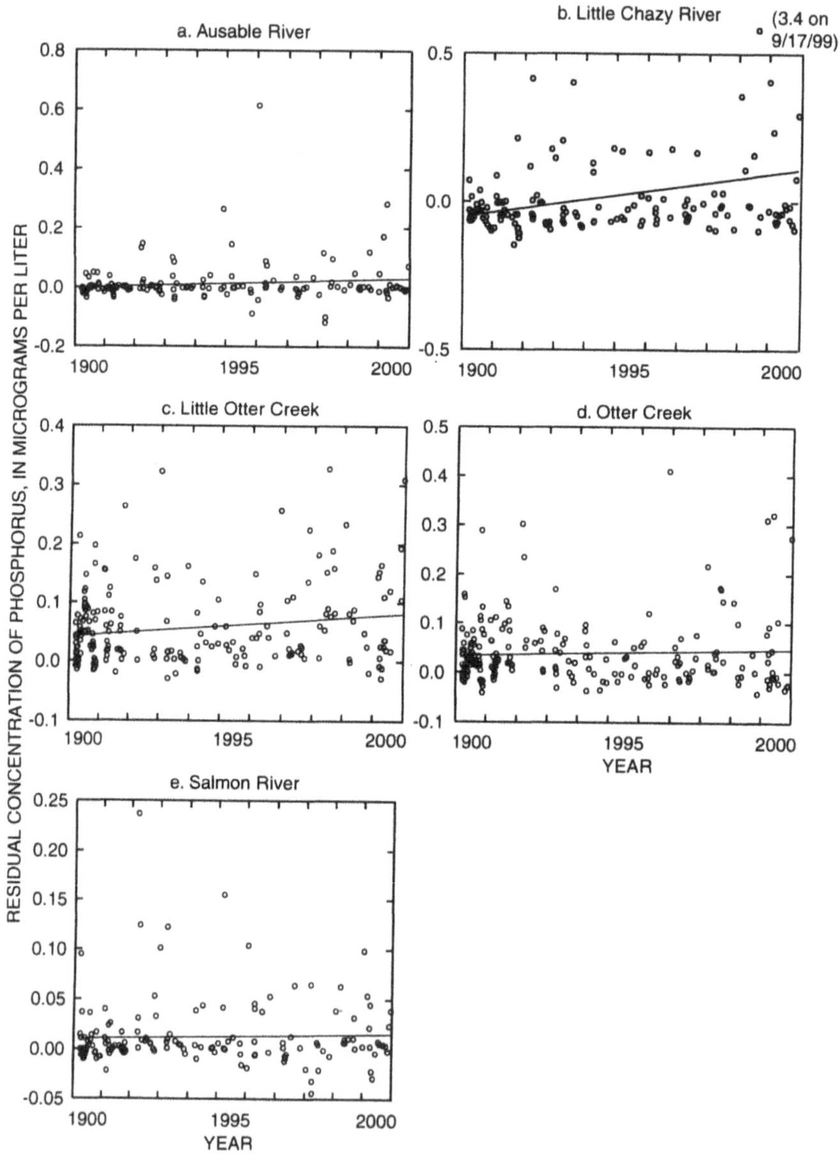

Figure 8. Mixed-model residuals (flow-adjusted phosphorus concentrations) for the (a) Ausable River, (b) Little Chazy River, (c) Little Otter Creek, (d) Otter Creek, and (e) Salmon River from 1990 to 2000.

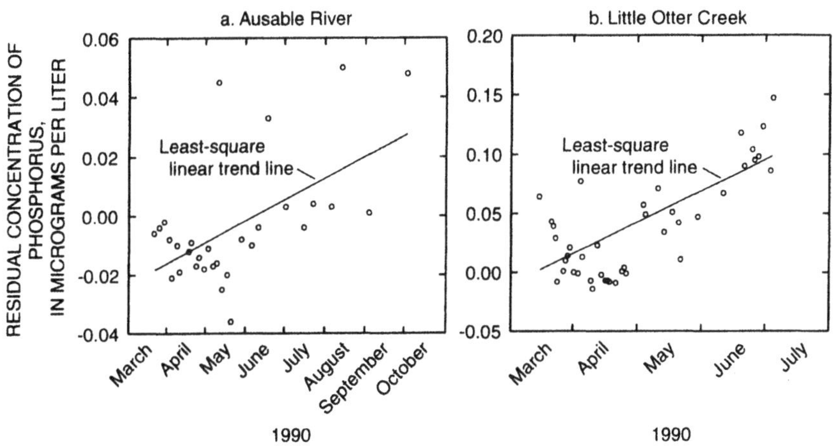

Figure 9. Mixed-model residuals (flow-adjusted phosphorus concentrations) for the early period of record for the (a) Ausable River and the (b) Little Otter Creek.

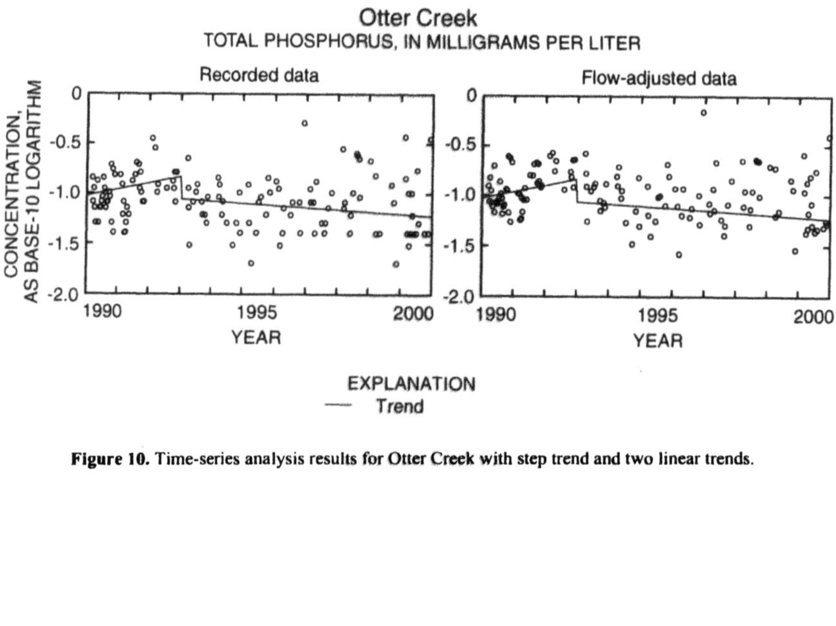

Figure 10. Time-series analysis results for Otter Creek with step trend and two linear trends.

7. CONCLUSIONS

Substantial management efforts took place during the 1990-2000 monitoring period to reduce phosphorus loading to Lake Champlain from point and nonpoint sources (Lake Champlain Basin Program, 2000). The wastewater-effluent monitoring data (fig. 4) document that major progress has been made in reducing point-source loads to the lake. Continued monitoring of all wastewater phosphorus discharges in Vermont, New York, and Quebec will be necessary to demonstrate full compliance with the wasteload allocations defined in the Lake Champlain Phosphorus TMDL (Vermont DEC and New York State DEC, 2002).

Most long-term trends are showing increasing phosphorus concentrations during the 1990-2000 monitoring period rather than the decreasing concentrations that had been expected. Because point-source loads now represent less than 10 percent of the total phosphorus load to Lake Champlain, likely explanations for the apparent lack of progress in meeting water-quality goals include insufficient implementation of nonpoint-source management practices or offsetting effects of land conversion to higher phosphorus-yielding uses in rapidly developing areas of the basin (Lake Champlain Basin Program, 2000). A confounding factor is that there could be lags in time between when land-based best management practices are implemented in stream watersheds and when effects in streams are seen in the Lake. Data in this report show that in general, there is no connection between time-series patterns in tributary loads (fig. 5) and in lake concentrations (fig. 3). The only possible exception might be in Missisquoi Bay where both the Missisquoi and the Pike Rivers show large increases in phosphorus loads in 1997-1998 and the Missisquoi Bay lake segment shows a moderate increase in 1999.

Phosphorus-reduction programs in the Lake Champlain Basin have not yet succeeded in attaining the phosphorus-reduction goals for the lake. Continued and expanded management efforts appear to be necessary, especially for nonpoint sources, as detailed in the Lake Champlain Phosphorus TMDL (Vermont DEC and New York State DEC, 2002).

8. ACKNOWLEDGMENTS

The Lake Champlain Long-Term Water Quality and Biological Monitoring Program is supported by grants from the Lake Champlain Basin Program to the States of Vermont and New York. Many staff of the Vermont DEC and the New York State DEC have contributed to this program by collecting samples, conducting laboratory analyses, and sharing data. We especially thank the following: Robert Bonham, Fred Dunlap, Catherine Eliopoulos, Karen Hyde, Neil Kamman, Daniel McAvinney, Joe Racette, Peter Stangel, Michaela Stickney, and John Townsend.

9. REFERENCES

Bradu, D., and Yair, M., 1970, Estimation in lognormal linear models, *J. American Statistical Assoc.*, **65**:329, pp. 198–211.

Cohn, T. A., Caulder, D. L., Gilroy, E. J., Zynjuk, L. D., and Summers, R. M., 1992, The validity of a simple statistical model for estimating fluvial constituent loads—An empirical study involving nutrient loads entering Chesapeake Bay, *Water Resources Research*, **28**:9, pp. 2353–2363.

Cohn, T. A., DeLong, L. L., Gilroy, E. J., Hirsch, R. M., and Wells, D. K., 1989, Estimating constituent loads, *Water Resources Research*, **25**:5, pp. 937–942.

Edwards, T. K., and Glysson, G. D., 1988, Field methods for measurement of fluvial sediment, *U. S. Geological Survey Open-File Report 86–531*.

Gilroy, E. J., Hirsch, R. M., and Cohn, T. A., 1990, Mean square error of regression-based constituent transport estimates, *Water Resources Research*, **26**:9, pp. 2069–2077.

Hamed, K. H., and Rao, A. R., 1998, A modified Mann-Kendall trend test for autocorrelated data, *J. Hydrology*, **204**, pp. 182–196.

Helsel, D. R., 1987, Advantages of nonparametric procedures for analysis of water quality data, *Hydrological Sci. J.*, **32**:2, pp. 179–190.

Helsel, D. R., and Hirsch, R. M., 1992, Statistical methods in water resources, Elsevier, New York.

Hirsch, R. M., and Slack, J. R., 1984, A nonparametric trend test for seasonal data with serial dependence, *Water Resources Research*, **20**:6, pp. 727–732.

Hirsch, R. M., Slack, J. R., and Smith, R. A., 1982, Techniques of trend analysis for monthly water quality analysis, *Water Resources Research*, **18**:1, pp. 107–121.

Lake Champlain Basin Program, 2000, Preliminary evaluation of progress toward Lake Champlain Basin Program phosphorus reduction goals: Prepared for Lake Champlain Steering Committee, Grand Isle, Vermont.

Lake Champlain Management Conference, 1996, Opportunities for action—An evolving plan for the future of the Lake Champlain Basin, Lake Champlain Basin Program.

Lake Champlain Phosphorus Management Task Force, 1993, Report prepared for Lake Champlain Steering Committee, New York State Department of Environmental Conservation, Adirondack Park Agency, Quebec Ministry of the Environment, and Vermont Agency of Natural Resources.

Langland, M. J., Blomquist, J. D., Sprague, L. A., and Edwards, R. E., 2000, Trends and status of flow, nutrients, and sediments for selected nontidal sites in the Chesapeake Bay watershed, 1985–98, *U. S. Geological Survey Open-File Report 99–451*.

Liebermann, T. D., Mueller, D. K., Kircher, J. E., and Choquette, A. F., 1989, Characteristics and trends of streamflow and dissolved solids in the Upper Colorado River Basin, Arizona, Colorado, New Mexico, Utah, and Wyoming, *U. S. Geological Survey Water-Supply Paper 2358*.

Slack, J. R., Lumb, A. M., and Landwehr, J. M., 1994, Hydro-climatic data network (HCDN)—Streamflow data set, 1874–1988, *U.S. Geological Survey Water-Resources Investigations Report 93–4076*.

Smeltzer, E., 1999, Phosphorus management in Lake Champlain, in: *Lake Champlain in Transition—From research toward restoration*, T. O. Manley and P. L. Manley, eds., American Geophysical Union, Washington, D. C., Water Science Application 1, pp. 435–451.

Snedecor, G. W., and Cochran, W. G., 1967, *Statistical Methods*, Sixth edition, Iowa State University Press, Ames, Iowa.

U. S. Environmental Protection Agency, 1993, Methods for the determination of inorganic substances in environmental samples, EPA/600/R-93/100, Washington, D. C.

Vecchia, A. V., 2000, Water-quality trend analysis and sampling design for the Souris River, Saskatchewan, North Dakota, and Manitoba, *U. S. Geological Survey Water-Resources Investigations Report 00–4019*.

Vermont Department of Environmental Conservation and New York State Department of Environmental Conservation, 1997, A phosphorus budget, model, and load reduction strategy for Lake Champlain (Lake Champlain diagnostic-feasibility study, final report).

Vermont Department of Environmental Conservation and New York State Department of Environmental Conservation, 1998, Long-term water quality and biological monitoring project for Lake Champlain—Cumulative data report for project years 1992–1996, Lake Champlain Basin Program Technical Report 26, Grand Isle, Vermont.

Vermont Department of Environmental Conservation and New York State Department of Environmental Conservation, 2001, Long-term water quality and biological monitoring project for Lake Champlain—Cumulative data report for project years 1992–2000, Prepared for Lake Champlain Basin Program, Waterbury, Vermont, and Ray Brook, New York.

Vermont Department of Environmental Conservation and New York State Department of Environmental Conservation, 2002, Lake Champlain Phosphorus TMDL, September 25, 2002, Waterbury, Vermont, and Albany, New York.

Walker, W. W., 1987, Empirical methods for predicting eutrophication in impoundments, Report 4—Applications manual, Tech. Rep. E-81-9, Prepared for U. S. Army Corps of Engineers Waterways Experiment Station, Vicksburg.

_____1996, Simplified procedures for eutrophication assessment and prediction: User manual. Instruction report W-96-2. (Updated April 1999). U. S. Army Corps of Engineers Waterways Experiment Station, Vicksburg.

Zipper, C. E., Holtzman, G. I., Rheem, S., and Evanylo, G. K., 1992, Trends in southwestern Virginia, 1970–1989: 1. Seasonal Kendall analysis, VPI-VWRRC-BULL 173, Virginia Polytechnic Institute and State University, Blacksburg.

A LANDSCAPE SCALE EVALUATION OF PHOSPHORUS RETENTION IN WETLANDS OF THE LAPLATTE RIVER BASIN, VERMONT, USA

Lisa J. Windhausen, David C. Braun, and Deane Wang*

ABSTRACT

We used a landscape scale approach to examine phosphorus retention in wetlands of the LaPlatte River basin (13,723 ha), Vermont. Total phosphorus (TP) export from 15 study catchments (149-1,396 ha) was measured on 18 dates, representing a range in seasons and hydrologic conditions. Multiple regression models were developed to relate TP export to 14 possible explanatory variables based on land cover/use, quantified using a geographic information system. Most wetland variables had significant ($p < 0.10$) negative relationships with TP export on at least 1 date. These relationships were strongest on 2 spring snowmelt events, when 31% of the annual TP export from the LaPlatte River basin occurred. Overall, the percentage of nonagricultural poorly and very poorly drained soils was the best representation of phosphorus sinks in the study catchments. Identifying lands with poorly drained soils and no known sources of phosphorus may be a more functional and simpler method of delineating P sinks in the landscape than identifying wetlands using jurisdictional definitions.

1. INTRODUCTION

Phosphorus (P) cycling in wetlands has received much attention in recent years because P is the nutrient most commonly associated with cultural eutrophication, and because of the general notion that wetlands can improve water quality. The ability of wetlands to reduce P transport from source areas to receiving waters has been demonstrated; wetlands have been used to reduce P transport from point sources and wastewater treatment systems (e.g., Boyt et al., 1977; Fetter et al., 1978; Tilton and Kadlec, 1979; Begg et al., 2001), and, more recently, from nonpoint sources of P (e.g.,

* School of Natural Resources, University of Vermont, Burlington, Vermont 05405, USA. Corresponding author: Deane Wang.

Lake Champlain: Partnership and Research in the New Millennium
Edited by T. Manley *et al.*, Kluwer Academic/Plenum Publishers, 2004

Mitsch and Reeder, 1992; Chambers et al., 1993, Weller et al. 1996; Uusi-Kamppa et al. 2000; Kao and Wu, 2001). Nutrient budget studies of individual wetlands suggest that a wetland's ability to retain P varies with season, hydrology, and wetland type (e.g., Klopatek, 1975; Lee et al.; 1975; Peverly, 1982; Gehrels and Mulamoottil, 1989; Mitsch and Gosselink, 1993; Kadlec and Reddy, 2001; Sanchez-Carrillo and Alvarez-Cobelas, 2001). In addition, prolonged or exceedingly high inputs of P to wetlands may cause P binding sites to become saturated, reducing the wetland's ability to retain P (Howard-Williams, 1985; Drizo et al., 2002). Because of the variability in P retention in wetlands (e.g., Bridgham et al., 2001), extrapolation of results from studies of specific wetlands to wide geographic regions may be inappropriate.

Wetland functions are commonly studied intensively at the scale of an individual wetland. Public policies and strategies for P management and wetland conservation, however, are usually developed on a regional basis. Thus, managers require tools to evaluate wetland function at this broader scale. Methods have been developed to evaluate wetland functions based on visual characteristics and limited field measurements (e.g., Adamus, 1983; Adamus et al., 1987; Amman and Stone, 1991). Efforts are also underway to develop regional classification systems based on wetland functions and values (e.g., Brinson, 1993). Existing functional evaluation methods relating nutrient retention and wetland characteristics, however, are often based on limited scientific evidence. Techniques are needed for collecting empirical data on P retention in wetlands at a landscape scale to refine regional functional evaluation methods (Weller et al., 1996).

The majority of catchment P studies has focused on identifying areas of high P export. Regression analysis is often used to establish relationships between catchment P export and the area of land use and land cover types (e.g., Omernik, 1976; Devito et al., 2000). Predictive equations have been developed and applied regionally. This information is used to prioritize areas for nonpoint source P management (Tim et al. 1992; Wickham et al. 2000). However, reliable coefficients for wetlands or other areas of P storage are not generally available, despite their potential influence on P export from catchments (Meade, 1982). Literature values for predicting P retention by the diversity of wetland types in a catchment are generally unavailable.

Multiple linear regression (MLR) analysis and a geographic information system (GIS) has been used to explore the collective function of wetlands at a landscape scale (Eckhardt and Moore, 1990; Johnston et al., 1990); generally finding some significant relationship between wetlands and nutrients in surface waters. An investigation of wetlands in the Lake Champlain basin, Vermont suggests that riparian wetlands are more important than other wetland types for reducing catchment P export (Weller et al., 1996). Because of the variability in wetland function across wetland types and physiographic regions, landscape scale evaluations of the collective function of wetlands within relatively homogeneous regions are needed to inform local planners when making regulatory decisions and prioritizing wetlands for conservation. Our study examined P retention in wetlands in an agricultural basin in northwestern Vermont to better understand the influence of different wetland types on P export during a range of seasons and hydrologic conditions.

2. METHODS

2.1 Study Area

Our study was conducted in the LaPlatte River basin, located 16 km south of Burlington, Vermont (Figure 1). The LaPlatte River originates in the foothills of the Green Mountains and flows northwesterly 24 km before emptying into Shelburne Bay, Lake Champlain. The 13,723-ha basin encompasses two physiographic regions, the Champlain Lowland and the foothills of the Green Mountains. The Champlain Lowland comprises approximately 80% of the basin and is dominated by limestone, dolomite, and shale bedrock overlain by post-glacial lacustrine deposits of sands, silts, and clays. Bedrock in the foothills of the Green Mountains, the eastern 20% of the basin, is primarily schist, gneiss, and phyllite. Glacial till occurs in smaller portions of the basin within both physiographic regions (U.S.D.A. Soil Conservation Service, 1989). Basin topography is dominated by the flat to gently sloping Champlain Lowland; elevation ranges from approximately 30 m above sea level at Shelburne Bay to 500 m in the foothills. The climate is cool and humid, with a mean annual precipitation of 84 cm, and a mean annual snowfall of 198 cm (National Oceanographic and Atmospheric Administration, 1995).

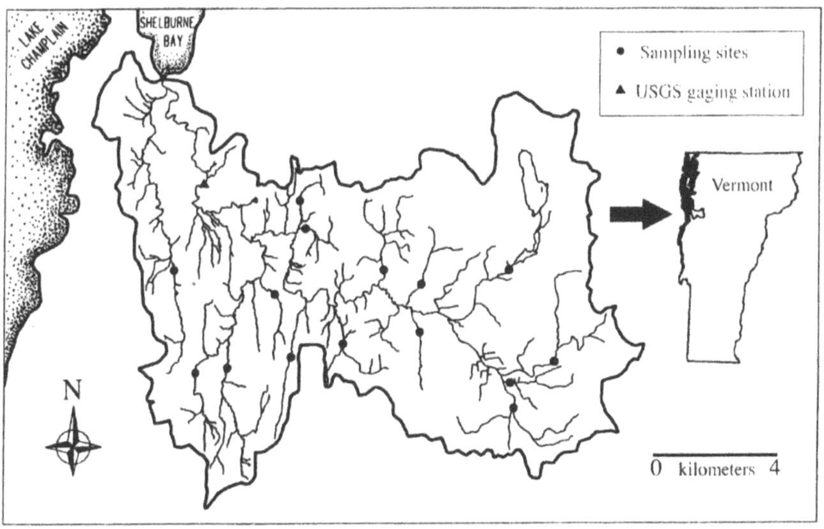

Figure 1. The stream network and sampling locations in the LaPlatte River Basin, Vermont.

Agriculture and forest characterize most of the LaPlatte River basin. Agricultural land predominates the Champlain Lowland and consists primarily of pasture and hay fields associated with the basin's 40 dairy farms (Meals, 1993). Hemlock-northern hardwood forests occupy most of the foothills of the Green Mountains and scattered patches throughout the Champlain Lowland. Urban land, primarily residential and recreational, comprises approximately 16% of the basin. Wetlands occupy approximately 7% of the basin and range in size from less than 1 to greater than 70 hectares. Wetlands are primarily associated with the riparian zone of the stream network; a small number occur in isolated depressions and on slopes within groundwater discharge zones.

We selected 15 relatively large catchments (149-1396 ha) within the LaPlatte River basin to evaluate the influence of wetlands on P export at the landscape scale. For measuring P export from each catchment, we selected individual stream reaches based on accessibility and the feasibility of measuring discharge (Figure 1). These stream reaches defined the outlet for each study catchment. All stream sampling sites were located upstream of town centers and known point source discharges of P, permitting the estimation of nonpoint source P export.

2.2 Spatial Analysis

The area of general land use and land cover types within each catchment was quantified from GIS coverages of catchment boundaries, land use, and wetlands using ARC/INFO (version 7.0.3, Environmental Systems Research Institute, Redlands, CA). Catchment boundaries were delineated on U.S.G.S. 7.5 minute topographic maps and verified in the field, particularly in areas of low relief. A GIS coverage was created by digitizing the catchment boundaries using ARC/INFO. This catchment boundary coverage was used to extract land use and land cover data for each study catchment. We used this data to develop explanatory variables for use in multiple regression analyses.

The base land use coverage selected for this study was developed from 1988 1:5000 panchromatic orthophotographs and updated with field surveys in 1990 as part of other studies. The overall accuracy of the linework was reported as +/- 25 m (Appleton, 1993). The land use coverage followed a classification system that was adapted for use in Vermont (Vermont Geographic Information System, 1992) from the Anderson land use and land cover system (Anderson et al., 1976).

We conducted further field surveys and orthophotograph analyses to provide a more accurate representation of the 1994-5 land use in the basin. Because the 1990 land use coverage provided no distinction among types of agricultural lands, the boundaries of all row cropland in the study area were mapped onto 1:5000 orthophotographs during summer field surveys in 1995, and digitized into the base coverage. In addition, a considerable amount of land classified as agricultural in the coverage was no longer in production; therefore, these areas were reclassified as transitional lands.

The documentation provided with the base land use coverage (Appleton, 1993) and preliminary field surveys in 1994 suggested that the coverage underestimated the area of wetland. A more recent coverage of wetlands in the entire LaPlatte River basin, prepared by Goldsmith (1994), was used to improve the base land use coverage. Goldsmith's wetland coverage was based on interpretation of National High Altitude Photograph (NHAP) color infrared imagery, enlarged to a scale of 1:20,000 from an original

Table 1. Description and codes for explanatory variables used in regression models. Explanatory variables are not completely independent and capture overlapping concepts, expecially with respect to wetland characteristics.

Explanatory Variables	Code
Land use and land cover	
Agriculture	AG
Row crops	CROP
Agricultural land excluding crops	NONCROP
Forest	FOREST
Single family residential and recreational	RESREC
Wetland	
Wetlands, lakes, and ponds	WET
Forested and scrub-shrub wetlands	WOODY
Emergent and open water wetlands	NONWOODY
Mixed emergent and scrub-shrub wetlands	PEM/SS
Nonagricultural wetlands, lakes, and ponds	NAWET
Lakes, temporarily flooded wetlands and seasonally flooded wetlands	RIVERINE
Lakes, ponds, and wetlands isolated or along first order streams	HEADWATER
Wetlands along second or third order streams	HIORDER
Poorly and very poorly drained soils excluding agriculture	HYDRIC

1:58,000. Wetland boundaries were rectified using 1988 1:5000 orthophotographs. The incorportation of Goldsmith's wetland coverage into the base land use coverage increased the represented wetland area in the LaPlatte River basin from 367 ha to 1023 ha. This wetland coverage was overlayed onto the base coverage. Areas within study catchments coded as wetland on the base coverage and not coded as wetland on Goldsmith's coverage were checked in the field; all were determined to be wetland areas and were retained in the final coverage. The WET explanatory (or independent) variable (Table 1) includes these revisions in wetland area represented in the land use coverage.

We categorized wetlands in various ways to better understand the role of different wetland types on catchment P export. Thus these predictor variable are not independent of each other, but attempt to capture different aspects of wetland characteristics. A commonly used classification of wetlands is by vegetation type (e.g, Cowardin et al., 1979). Therefore, we separated wetlands into three vegetation types, Woody, Nonwoody, and palustrine-emergent marsh/shrub-scrub PEM/SS (Table 1) and examined relationships between these wetland types and TP export.

Many wetlands in the LaPlatte River basin are used to graze livestock. Because livestock can contribute to high P export from catchments, wetlands used for pasture may act as P sources in the landscape, rather than P sinks. Therefore, we excluded the area of wetlands impacted by livestock from the total wetland area (WET), resulting in a new wetland variable, NAWET (Table 1).

P retention in wetlands may be related to periodic flooding regimes, depressions with distinct inflows and constricted outflows, diffuse channels, and dense vegetation. We identified wetlands with these characteristics from field surveys and used the area of these wetlands as a separate wetland variable. Because most of these wetlands were associated with stream channels, the variable was coded as RIVERINE (Table 1). Wetlands exluded from this category include wetlands located on slopes and wetlands with permanent standing water.

The overall influence of wetlands on catchment P export may be related to the spatial distribution of wetlands along the stream network. We distinguished among wetlands by their proximity to different stream orders. Wetlands without a connection to the perennial stream network or associated with first order streams were coded as HEADWATER (Table 1), and wetlands associated with second and third order streams were coded as HIORDER (Table 1).

Soil drainage data were used as an alternative to aerial photo-interpretation for inventorying wetlands in the study area. Most poorly and very poorly drained soils meet the hydric soils criteria in the Federal Manual for Identifying and Delineating Wetlands (Federal Interagency Committee for Wetland Delineation, 1989). The area of these drainage classes in each catchment was determined from an existing soils coverage of the LaPlatte River basin (U.S.D.A. Natural Resource Conservation Service, 1995). As with the NAWET variable, the area of agricultural land was excluded to eliminate agricultural sources of P. The resulting variable represented the area of nonagricultural poorly and very poorly drained soils, and was coded as HYDRIC (Table 1).

2.3 Water sampling and Chemical Analysis

Grab samples were collected at each sampling site (Figure 1) during 14 runoff events and 4 baseflow conditions from October 16, 1994 to October 15, 1995 (Figure 2). Sampling began the morning following major rain or snowmelt events and was completed within 10 hours. The tributaries were sampled in the approximate order of their position in the stream network: headwater sites were sampled first, followed by sites closer to the main stem of the LaPlatte River. The exact sampling order was varied on each date to reduce sampling bias. Water samples were packed in ice in the field and refrigerated at 4°C until analyzed.

Samples were analyzed for concentrations of total P (TP) and soluble reactive P (SRP), a P fraction equal to orthophosphate (PO_4) plus a variable amount of labile organic P. TP was analyzed colorimetrically on a spectrophotometer using the molybdenum blue method following digestion in a sulfuric acid-persulfate solution (U.S. Environmental Protection Agency, 1983). SRP was measured on samples filtered using a 0.45 μm membrane filter via the molybdenum blue method within 48 hours of sample collection. The lower limit of detection for TP and SRP concentrations was 0.005 mg l^{-1}.

2.4 Streamflow and Phosphorus Export

We developed discharge rating curves using paired measurements of stage and flow for each sampling site over a range of hydrologic conditions. Stage was measured at each site from fixed points above or within the channel on each sampling date. Flow was measured on selected dates using a velocity-integrating flow meter according to the conventional mid-section method (Rantz, 1982). Where stream depths were too low to use the current meter (i.e., less than 4 cm), stream velocity was measured by timing floating objects over a known distance. At certain sites, low flow measurements were obtained by recording the time it took to fill a 4 l bucket below a culvert.

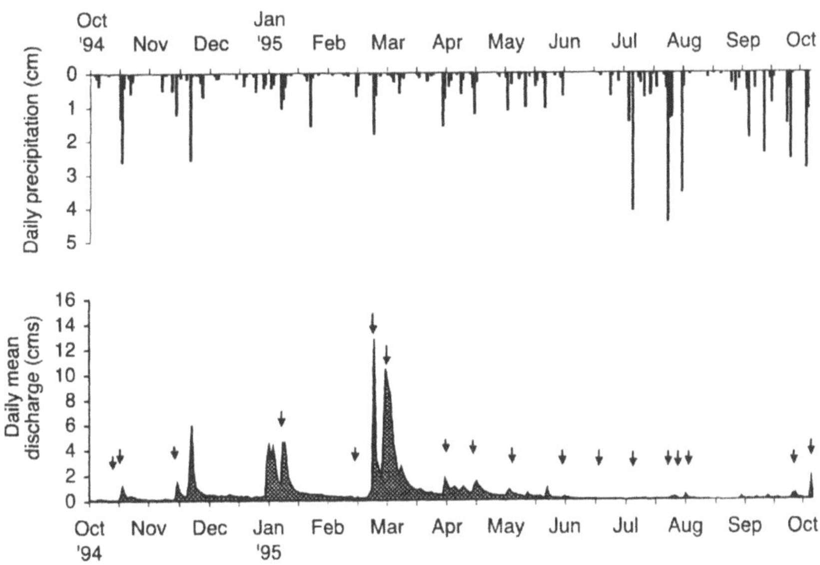

Figure 2. Daily precipitation in South Burlington, Vermont (10 km from the study area) and mean daily discharge of the LaPlatte River during the study period, October 16, 1994 to October 15, 1995. Arrows indicate stream sampling dates.

Discharge rating curves were developed for each sampling site according to Kennedy (1984). Exponential functions best described the flow vs. stage relationships at all sampling sites (range in r^2 = 0.86-0.99). Flow calculated from the rating functions using measured stage on each sampling date was multiplied by concentrations of TP and SRP to determine instantaneous exports from each study catchment. Where ice prevented accurate stage measurement or where extreme flow conditions altered the flow vs. stage relationship established in the rating curves, measured flow was used in the export calculation. TP and SRP exports were divided by the catchment area associated with each sampling site to normalize for catchment size in the statistical analyses.

Annual P export at the U.S.G.S. gaging station on the LaPlatte River main stem was estimated to compare the relative P loading during the events sampled. Total P concentration was determined for samples collected at the gaging station on each sampling date. P concentration data were also obtained from the Vermont Agency of Natural Resources (Smeltzer, 1996) for additional dates in the study year. The concentration data from both sources were used with corresponding mean daily discharge data provided by U.S.G.S. (Hammond et al., 1996) to estimate annual P export. The mean daily discharge on one date, March 8, 1996, was estimated by a hydrograph smoothing technique because of complications resulting from ice jamming at the gaging station. Annual P export was calculated according to a flow-stratified Beale Ratio Estimator method (BRE; Richards and Holloway, 1987; Preston et al., 1989). Daily P export on each sampling date was divided by the annual P export to determine the proportion of the annual P export on each date.

2.5 Data Analysis

We applied both simple and multiple linear regression analyses to explore relationships between P export and wetlands at the landscape scale. The dependent variable for all regressions was total P export (g ha^{-1} d^{-1}) and the explanatory variables were the 14 land use and land cover types, expressed as percentages of each catchment (Table 1). The number of explanatory variables in each regression model was limited to three to maintain adequate degrees of freedom. Relationships were evaluated by interpreting coefficients of determination (R_a^2), adjusted for the number of explanatory variables in the model, and p-values.

We used regression analysis in an exploratory manner to reduce the total number of explanatory variables to a reasonable subset for developing the final regression models. As an initial screening measure for selecting explanatory variables, we compared results of 2-variable models using TP export as the dependent variable and each land use and land cover type as the explanatory variable. For comparing wetland types, we developed separate models using 2 explanatory variables, the total agricultural variable and each wetland variable. Variables that explained a significant ($p < 0.10$) portion of TP export variance were selected for the final analyses. Multiple regression models using all combinations of the selected subset of explanatory variables were then evaluated to identify the combination that maximized the explained variance in TP export on each sample date.

Multicollinearity among explanatory variables can cause difficulties when interpreting regression coefficients and results of hypothesis tests. We avoided including variables expressing the same landscape characteristics in the same multiple regression models to reduce the effects of multicollinearity. In addition, we tested all combinations of explanatory variables for multicollinearity using the variance inflation factor (VIF, Neter et al., 1989). We excluded all combinations of explanatory variables having VIF values greater than 10 from the results presented.

We performed additional diagnostic tests on each regression model to determine if the assumptions necessary for conducting hypothesis tests and making inferences related to the slope coefficients were satisfied. We performed tests of normality (Shapiro-Wilks, $p < 0.05$) on the residuals and examined scatter plots of the residuals vs. predicted values. We applied a $\log_{10}$ transformation on the dependent variable, TP export, in some models to satisfy the constant variance assumption, thereby increasing our ability to interpret the models. The final regression models presented have both normally-distributed and homoscedastic residuals (Table 4). Given the exploratory nature of this work, no control for experiment-wise error was attempts.

3. RESULTS AND DISCUSSION

3.1 Land Use and Land Cover Data

Summary statistics calculated for each study catchment showed that most catchments were dominated by either agriculture or forest (Table 2). Forest land ranged from 3-77% of the study catchments, total agricultural land ranged from 5-75%, and residential and recreational lands ranged from 6-27%.

Wetlands represented smaller portions of the catchments than agriculture and forest (Table 2). Wetlands and surface waters (WET) ranged from 1-39% of the study catchments. Forested and scrub/shrub wetlands (WOODY) occupied less than 1% of most catchments, while NONWOODY and palustrine, emergent marsh/shrub-scrub (PEM/SS) wetlands were more evenly distributed. Approximately half of the catchments had a higher proportion of wetlands along second and third order streams (HIORDER) than along first order streams or in isolation from the stream network (HEADWATER). Most wetlands in the study catchments were associated with the stream network and had seasonal or temporary saturation water regimes (RIVERINE, 1-39%). The area of nonagricultural poorly and very poorly drained soils (HYDRIC) ranged from 4-29% of the catchments. Portions of this area included lands classified as residential, recreational, and forest lands in the base coverage.

3.2 Phosphorus Data

Stream P concentrations showed high spatial and temporal variability throughout the LaPlatte River basin during the study period. Total P (TP) concentrations ranged from 0.005 to 1.420 mg l^{-1} (Figure 3), while mean values at each site ranged from 0.020 to 0.550 mg l^{-1}. Most of the 223 observations of TP concentrations measured during the study period exceeded the 0.010 mg l^{-1} nutrient critera for rivers and stream (U.S. Environmental Protection Agency, 2001).

In general, high TP and SRP concentrations, and high SRP:TP ratios, were measured in streams draining agricultural catchments (e.g., Catchments 7, 8, and 9, Figure 3). Low SRP:TP ratios can be associated with agricultural catchments (Prairie and Kalff, 1986) due to the high particulate P often introduced where erosion potential of exposed soil particles is high. High SRP:TP ratios, however, have also been observed in heavily fertilized catchments. Exceptionally high TP concentrations were measured in a stream draining Catchment 15 (Figure 3a), which contained only 27% agricultural land. This stream received runoff from a ditch draining a concentrated livestock area. This livestock area acted as a point source of P because of the high concentrations (Figure 3a) produced from a relatively small area. Because this site was an easily identified outlier in preliminary regression models and repeatedly exerted a disproportionate influence on the calculation of regression equations, it was excluded from the results presented.

Table 2. Land use and land cover percentages in 15 catchments of the LaPlatte River Basin. Catchment size is in hectares. Variable codes are described in Table 1.

ID	ha	AG	CROP	NON CROP	FOREST	RES REC	WET	WOODY	NON WOODY	PEM /SS	NA WET	RIV ERINE	HEAD WATER	HI ORDER	HYDRIC
1	1396	11.2	0.1	11.1	48.6	22.9	15.1	4.0	1.5	1.2	14.8	6.3	14.2	0.9	26.6
2	790	4.9	0.0	4.9	76.7	16.5	1.0	0.7	0.0	0.2	0.7	1.0	0.8	0.3	7.2
3	714	33.7	4.3	29.4	56.7	7.4	1.9	1.3	0.6	0.0	1.4	1.8	1.6	0.3	7.7
4	323	4.2	0.2	54.1	27.1	7.1	4.5	0.6	3.9	3.8	0.9	4.4	1.2	3.4	8.0
5	925	44.0	2.2	41.8	32.5	14.8	6.6	3.1	3.4	0.1	4.6	4.6	2.7	4.0	10.0
6	171	43.9	0.0	43.9	19.2	26.7	10.2	0.2	8.7	2.5	3.8	9.5	6.8	3.4	9.6
7	174	68.7	16.2	52.5	10.0	9.2	12.0	6.5	4.0	1.4	8.4	10.2	10.5	1.4	10.2
8	190	55.1	3.1	52.0	23.5	16.3	3.7	0.0	2.3	1.4	2.2	3.6	0.9	2.8	7.1
9	647	74.5	9.5	65.0	10.7	7.1	5.6	0.1	1.9	3.6	4.7	5.5	0.5	5.1	4.3
10	363	41.4	8.1	33.3	24.9	11.1	19.8	0.0	0.1	19.7	19.7	19.7	0.1	19.7	16.0
11	426	39.4	2.6	36.8	20.9	11.7	13.6	0.2	3.6	9.8	13.3	13.4	3.5	10.0	23.1
12	515	24.5	2.3	22.2	61.8	9.5	2.2	0.1	2.1	0.0	0.0	2.0	0.2	2.0	4.7
13	149	48.0	5.0	43.0	3.2	9.5	39.3	23.6	0.4	15.2	39.2	38.9	39.3	0.0	29.4
14	478	31.9	3.7	28.2	43.9	6.2	14.9	3.9	10.4	0.6	14.9	1.9	14.9	0.0	17.3
15	317	27.0	5.6	21.3	46.5	20.9	2.3	0.3	2.0	0.0	0.1	2.0	0.8	1.5	10.2

The temporal variability observed in instream P concentrations was partially related to the magnitude of streamflow. The highest TP concentrations and the highest TP exports (Figure 3) occurred on the highest flow event sampled, March 8, 1995 (Figure 2). Estimates of the total annual streamflow and P export at the U.S.G.S. gaging station (Figure 1) showed that approximately 10% of the total annual discharge and 31% of the total annual TP export (2405 kg) occurred on the two snowmelt dates sampled, March 8, 1995 and March 14, 1995. A similarly disproportionate export of P during high flow events was previously observed in the LaPlatte River by Meals (1985) and in other river systems (e.g., Johnson et al., 1976; Ellis et al., 1978; Monke et al., 1981; Lowrance et al., 1984; Paschal and Sherwood, 1987; Pionke et al., 1988; Longabucco and Rafferty, 1989).

Relationships between TP concentration and streamflow at each tributary were examined further using linear regression analysis. Regression models were developed using TP concentration (mg l^{-1}) as a dependent variable and measured streamflow (m^3 s^{-1}) as an explanatory variable. Only six models had significant positive linear relationships ($p < 0.10$) and each relationship was dominated by either one or two outlying data points. Log_{10} transformation of TP concentrations and discharge did little to improve the relationships.

The weak relationships observed between TP concentration and streamflow in some catchments are the result of variability in TP concentrations measured during spring and fall medium flow events and summer thunderstorm events (Figure 3a). For example, two spring rain events, April 13, 1995 and April 28, 1995, produced the two lowest mean TP concentrations for all sites (Figure 3a) despite the moderately high streamflow response (Figure 2). In addition, the intermittent flow following summer thunderstorms often produced exceptionally high TP concentrations during low discharge events (Figure 3b).

The high TP concentrations observed during many low streamflow events may be the result of a shift in flowpath dominance, from overland to subsurface flow. High concentrations of nutrients observed in stormflow following drought conditions by Mulholland et al. (1990) was attributed to leaching of vegetation in riparian zones, leaching of accumulated organic matter in drier portions of the channel, and hydrologic flushing of nutrients concentrated in soil water within the near-stream zone. Conversely, the low TP concentrations measured during the late spring rain events in our study may be the result of previous flushing of P in soil water during the snowmelt season. Thus, streamflow may be a poor predictor of P concentration when site- and season-specific conditions have a dominant effect on processes of instream P concentration.

3.3 Phosphorus Sources and Sinks in the Landscape

Results of the preliminary regression analyses examining relationships between TP export and agricultural, forest, and residential variables show that selecting a single variable for representing P source areas for the majority of dates sampled was not possible. Residential and forest variables were not significantly ($p < 0.10$) related to TP export on any date. Agricultural variables had significant positive relationships with TP export on many dates ($p < 0.10$), suggesting that these areas were sources of P. Different variables, however, had their strongest relationship with TP export on different dates. Therefore, all three agricultural variables, AG, CROP, and NONCROP, were selected for multiple regression analyses.

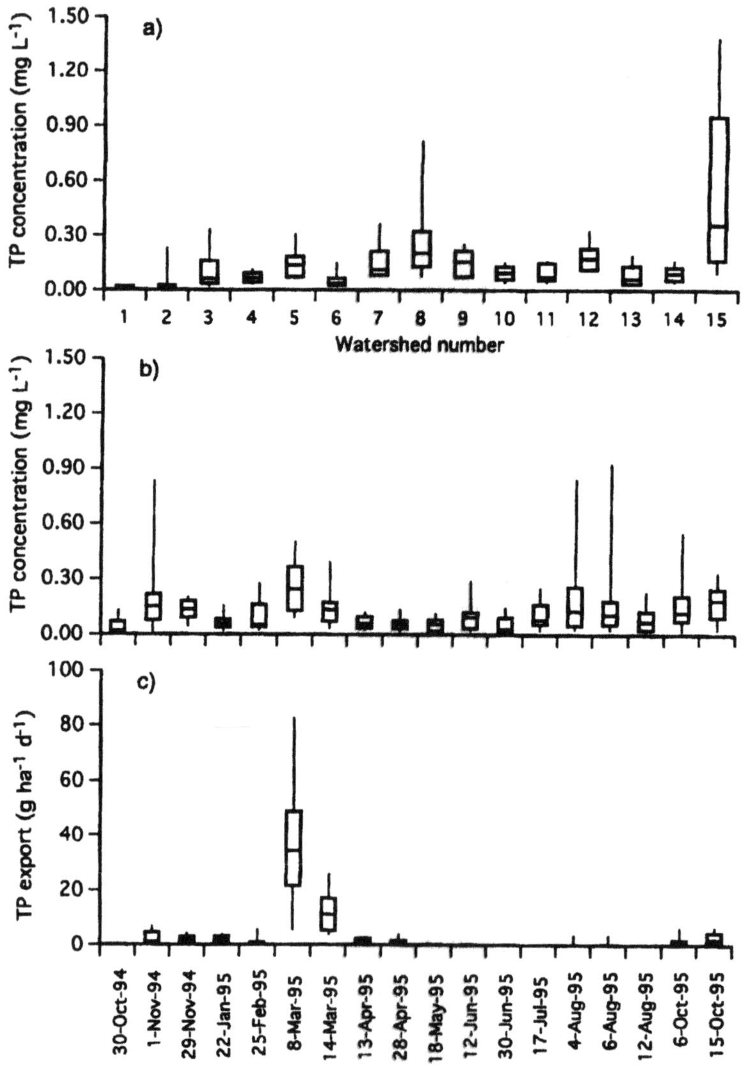

Figure 3. Spatial and temporal variability in total phosphorus (TP) data: a) TP concentrations at each catchment outlet for all dates sampled; b) TP concentrations on each smapling date for all catchments; c) TP export on each sampling date for all catchment outlets. Box diagram shows median value (center), upper and lower quartile (top and bottom of box), and high and low values (top and bottom of line).

Results of preliminary multiple regressions using TP export, wetland and agricultural variables suggest that one wetland variable, HYDRIC, was a better representation of P sinks in the catchments than other wetland variables during a majority of the dates sampled. The variable, HYDRIC, had significant, negative relationships with TP export on 7 of the 18 dates sampled ($p < 0.10$, Table 3). These relationships were usually stronger than relationships between TP export and other wetland variables (Table 3). To examine the effect of changing the agricultural variable on the strengths of the relationships, three wetland variables, WET, NAWET, and HYDRIC, were selected for further analyses.

For our final analyses, we compared multiple regression models using all combinations of the three selected agricultural variables (AG, CROP, NONCROP) and the three selected wetland variables (WET, NAWET, HYDRIC) to maximize the explained variance in TP export on each sample date. Our results of the regression models with the highest, significant R_a^2 for each date ($p < 0.10$, Table 4) show significant relationships between TP export and agricultural and wetland categories on 8 sample dates, with R_a^2 ranging from 0.24 to 0.83. As with the preliminary regressions, models using HYDRIC as an explanatory variable have higher R_a^2 values and lower p-values than models using WET or NAWET. The consistency of this result adds further support to the idea that areas of nonagricultural poorly and very poorly drained soils were the best representation of P sinks in the study catchments of all wetland variables explored.

The weaker relationships between TP export and the area of wetlands identified from aerial photo interpretation (WET) as compared to the area of wetlands delineated from soil drainage class (HYDRIC) may be explained by the difficulties in identifying wetlands from remotely sensed data. The color infrared imagery used to identify wetlands in the study catchments was taken in early spring prior to deciduous leaf production, facilitating the identification of surface waters and enhancing the ability to distinguish among vegetation types. Evergreen forested wetlands and temporarily flooded wetlands, however, are particularly difficult to identify (Federal Interagency Committee for Wetland Delineation, 1989). In addition, many poorly and very poorly drained areas cleared of characteristic wetland vegetation are not technically wetlands, according to some jurisdictional definitions of wetlands (e.g., Federal Interagency Committee for Wetland Delineation, 1989). Many of these areas were classified as residential and recreational lands in our land use coverage. Our results show that including forested, residential and recreational areas with poorly and very poorly drained soils in a wetland variable (i.e., HYDRIC) strengthened the negative relationship with TP export, suggesting that these areas function as P sinks in the study catchments.

Table 3. Adjusted r^2 values for regression models of TP export vs. agricultural and wetland variables. Where there are no values the r^2 is less than 0.10. Each model contains the total agriculture (AG) variable and one wetland variable. The single variable model with AG predicting TP is included for comparison. Variable codes are described in Table 1.

DATE	AG	WET	WOODY	NON WOODY	PEM /SS	NA WET	RIV ERINE	HEAD WATER	HI ORDER	HYDRIC
11/01/94	0.07	0.24*	0.18		0.15	0.27*	0.18	0.15		0.30
11/29/94	0.32**	0.40	0.30	0.21	0.53**	0.44*	0.40	0.29	0.38	0.43
01/22/95	0.20	0.13	0.22		0.13	0.14	0.12	0.23	0.29	0.13
02/25/95					0.12					
03/08/95	0.34**	0.65***	0.49*	0.29	0.49*	0.64***	0.55**	0.54**	0.29	0.73**
03/14/95		0.18*			0.17*	0.17*				0.23*
04/13/95		0.15			0.22*	0.25*			0.15	0.25*
04/28/95				0.12						0.14*
05/18/95										
06/12/95										
08/04/95										
08/06/95										
08/12/95		0.11					0.16			
10/06/95										0.17*
10/15/95		0.13			0.14	014				0.21*

* p<0.10; ** p<0.05; *** p<0.01

Table 4. Coefficients in regression models of TP export vs. agricultural and hydric soil variables. Models are in the form: TP ($g\ ha^{-1}\ d^{-1}$) = $a + bX_1 + cX_2 + dX_3$, where X_i are the explanatory variables; both two and three variable models are shown. In some cases the dependent variable (TP) was transformed to $\log_{10} + 2$). r_a^2 is adjusted r-square.

						Coefficients (b, c, or d) in the model			
Date	n	trans	r^2	r_a^2	(a)	HYDRIC	AG	CROP	NONCROP
Spring									
03/08/95	14	No	0.87	0.83**	40.3**	-1.87**		2.30**	0.24
03/14/95	14	Yes	0.68	0.62**	3.24**	-0.03**		0.03#	
04/13/95	14	Yes	0.58	0.50**	2.11**	-0.03*			0.001
Fall									
11/01/94	14	Yes	0.72	0.64**	1.62**	-0.04*		-0.03	0.03**
11/29/94	14	No	0.52	0.43*	1.24**	-0.06	0.04*		
10/06/95	14	Yes	0.56	0.42*	1.74**	-0.05*		-0.06	0.02#
10/15/95	14	Yes	0.53	0.44*	2.01**	-0.04*			0.02#
Winter									
01/22/95	13	Yes	0.32	0.19	1.84**	-0.001			0.01#
02/25/95	14	Yes	0.35	0.24#	1.17*	-0.02	0.02#		

\# p<0.10; * p<0.05; ** p<0.01

Similar landscape scale studies using regression analysis found significant relationships between nutrient concentration or export and catchment variables, including wetland types. Some of these have suggested that wetland position in the landscape influences the ability of a wetland to improve water quality (e.g., Whigham et al., 1988; Johnston et al., 1990; Detenbeck et al., 1993; Weller et al., 1996). Our results show that models using wetland variables expressing position in the landscape (HEADWATER, HIORDER) and measures of wetland vegetation type (WOODY, NONWOODY, PEM/SS) resulted in poor model fits. One exception to this pattern is the PEM/SS classification, which was negatively correlated to P export on four dates; the relationships, however, were relatively weak. For our study, distinguishing among wetland types for the purpose of classifying wetlands according to their ability to retain P was not possible. Measures of total wetland area (WET, NAWET, HYDRIC) were the most successful wetland variables for identifying P sinks in the landscape (Table 3). Therefore, our results suggest that all wetlands in the landscape, particularly defined by soil drainage class, are important for buffering the impact of P export from agricultural land to surface waters.

3.4 Seasonal patterns of phosphorus storage

A distinct seasonal pattern was observed in the ability of catchment variables to predict P export. R_a^2 values of the multiple regression models ranged from 0.24 to 0.83, with the highest values and the lowest p-values occurring in models representing spring snowmelt events (Table 4). This seasonal pattern may be explained by seasonal fluctuations in runoff producing areas and instream processes related to flow and biological activity.

Seasonal variability in runoff patterns can determine the contribution of various areas in a catchment to P export. During the highest streamflows in the early spring snowmelt period, much of the land surface area is saturated and may contribute overland flow and its associated P load. Many studies have suggested that this area expands and contracts throughout the year in relation to temperature, and precipitation intensity, duration, and frequency (e.g., Dunne and Black, 1970; Dunne et al., 1975; Bernier, 1985; O'Loughlin, 1986; Gburek, 1990; Eshleman et al., 1993; Montgomery and Dietrich, 1995, Kadlec and Reddy, 2001). Storm runoff during snowmelt conditions can increase the opportunity for land use practices and land cover types throughout a catchment to influence streamflow chemistry because of the relatively large runoff producing zones. Our relationships between TP export and land use and wetland variables that represented most of the area of the catchments were strongest during these saturated conditions (Table 4), suggesting that runoff producing zones were relatively large.

High flow events during snowmelt conditions may also create high opportunities for P retention in wetlands. Mitsch et al. (1979) found that 23% of the annual particulate P deposited in an alluvial swamp in southern Illinois occurred during a 5-day spring flood event, while Yarbro et al. (1984) found an increase in P retention in the Creeping Swamp in North Carolina as the area of inundation increased. Our study found strongest negative relationships between TP export and HYDRIC during the two spring snowmelt events sampled. Under snowmelt conditions, these hydric soils areas should produce saturation overland flow, according to the saturation overland flow models. It is likely, however, that erosion of P-laden sediments was restricted to steep saturated slopes and stream channels, and that areas with non-agricultural hydric soils acted more as sediment deposition areas, than as sediment source areas even during the highest flow events sampled.

Our relationships between TP export and agricultural and wetland variables were weakest during the summer and winter events. Precipitation during most summer events was greater than spring and fall events, however, streamflows were considerably lower. Nutrient budgets constructed for individual wetlands have suggested that wetlands are sinks for P predominantly during the lower flow growing season, and are sources for P during higher flow events in the fall and spring because of sediment scouring (Klopatek, 1975; Lee et al., 1975; Gehrels and Mulamoottil, 1989). On the whole, our results examining P retention in wetlands at a landscape scale did not follow this pattern. During the summer thunderstorm events pulses of P were most likely exported on relatively short time scales from areas near or within the stream channel, preventing the opportunity for P export from or retention in areas further from the channel. Simple generalizations of source and sink dynamics at the landscape scale under summer conditions may be precluded by site and season-specific conditions.

4. CONCLUSIONS

A landscape scale approach was used to identify catchment attributes that represent phosphorus sources and sinks in the landscape and to evaluate the collective function of these sources and sinks during a range in seasons and hydrologic conditions. Our results suggest that the best representation of P sinks in the landscape was the area of nonagricultural poorly and very poorly drained soils. This area had significant negative relationships with TP export in 6 of the 15 multiple regression models.

Soil drainage class was a more useful criterion for delineating P sinks in the study catchments than wetlands identified through color infrared photo interpretation. Wetlands identified from remotely sensed data may not adequately represent P sink areas because of the difficulties identifying evergreen forested and temporarily flooded wetlands, the inclusion of sloping wetlands, and the exclusion of low-lying areas devoid of wetland vegetation. Managers should consider that seasonally saturated areas lacking wetland vegetation may be important areas for P retention. While soil drainage class was useful for delineating the area of P sinks in this study, more research is needed to evaluate this relationship in other regions and, to develop functional evaluation methods with regional applicability that address other wetland functions and classification systems.

Results further suggest that wetlands function as P sinks during the highest P export season. The strongest negative relationships between wetlands and P export occurred during two spring snowmelt dates, when an estimated 31% of the annual TP export from the LaPlatte River occurred. Because runoff producing zones can be relatively large during the snowmelt season, land use in the entire catchment has a greater opportunity to influence streamflow chemistry than during lower flow seasons. Relationships between P export and both source (agricultural) and sink (wetland) variables were not significant on any summer date. Other processes, such as biological storage and release mechanisms operating throughout the catchment, or instream and near-stream P cycling may explain P export variation during the growing season.

Catchment modeling of P export used to identify sources and sinks of P and to predict reductions in P export following implementation of nutrient control measures should incorporate P sink variables to more accurately describe the relationship between P export and land use. Future landscape scale evaluations of P sources and sinks could also assist catchment managers by documenting more intensive temporal P export and storage patterns in small catchments within a season or during an event and by examining the influence of spatial patterns of P export and storage areas on P transport through the landscape.

5. REFERENCES

Adamus, P.R., 1983, *A Method for Wetland Functional Assessment. Volume I. FHWA Assessment Method.* Report Number FHWA-IP-82-24, U.S. Department of Transportation, Washington, D.C.

Adamus, P.R., ARA, Inc, Clairain, Jr., E.J., Smith, R.D., and Young, R.E., 1987, *Wetland Evaluation Technique (WET) Volume II: Methodology.* U.S. Army Engineer Waterways Experiment Station, Vicksburg, Mississippi.

Amman, A.P., and Stone, A.L., 1991, *Method for the Comparative Evaluation of Nontidal Wetlands in New Hampshire,* New Hampshire Department of Environmental Services, Concord, New Hampshire.

Anderson, J.R., Hardy, E.E., Roach, J.T., and Witmer, R.E., 1976, *A Land Use and Land Cover Classification System for use with Remote Sensor Data,* U.S. Geological Survey Professional Paper 964. U.S. Government Printing Office, Washington, D.C.

Appleton, J., 1993, Unpublished documentation, Chittenden County Regional Planning Commission, Essex Junction, Vermont.

Begg, J. S., Lavigne, R. L., Veneman, P. L. M., 2001, Reed beds: Constructed wetlands for municipal wastewater treatment plant sludge dewatering, *Water Sci. Technol.* **44**: 393-398.

Bernier, P.Y., 1985, Variable source areas and storm-flow generation: An update of the concept and a simulation effort, *J. Hydrol.* **79**:195-213.

Boyt, F.L., Bayley, S.E. and Zoltek, Jr., J., 1977, Removal of nutrients from treated municipal wastewater by wetland vegetation, *J. Water Poll. Control Fed.* **349**:789-799.

Bridgham, S. D., Johnston, C. A., Schubauer-Berigan, J. P., and Weishampel, P., 2001, Phosphorus sorption dynamics in soils and coupling with surface and pore water in riverine wetlands, *Soil Sci. Soc. Amer. J.* **65**(2): 577-588.

Brinson, M.M., 1993, *A Hydrogeomorphic Classification for Wetlands,* Technical Report Number WRP-DE-4. U.S. Army Engineer Waterways Experiment Station, Vicksburg, Mississippi.

Chambers, J.M., Wrigley, T.J., and McComb, A.J., 1993, The potential use of wetlands to reduce phosphorus export from agricultural catchments, *Fertil. Res.* **36**:157-164.

Detenbeck, N.E., Johnston, C.A., and Niemi, G.J., 1993, Wetland effects on lake water quality in the Minneapolis/St. Paul metropolitan area, *Landscape. Ecol.* **8**:39-61.

Devito, K. J., Creed, I. F., Rothwell, R. L., Prepas, E. E., 2000, Landscape controls on phosphorus loading to boreal lakes: Implications for the potential impacts of forest harvesting, *Can. J. Fish. Aquatic Sci.* **57**(10): 1977-1984.

Drizo, A., Comeau, Y., Forget, C., and Chapuis, R. P., 2002, Phosphorus saturation potential: A parameter for estimating the longevity of constructed wetland systems, *Environ. Sci. Tech.***36**(21): 4642-4648.

Dunne, T., and Black, R.D., 1970, Partial area contribution to storm runoff in a small New England watershed, *Water Resour. Res.* **6**(5):1296-1311.

Dunne, T., Moore, T.R., and Taylor, C.H., 1975, Recognition and prediction of runoff-producing zones in humid regions, *Hydrol. Sci.* **3**:305-327.

Eckardt, B.W., and Moore, T.R., 1990, Controls on dissolved organic carbon concentrations in streams, Southern Quebec, *Can. J. of Fish. Aquat. Sci.* **47**:1537-1544.

Ellis, B.G., Erickson, A.E., and Wolcott, A.R., 1978, *Nitrate and Phosphorus Runoff Losses from Small Watersheds in Great Lakes Basin,* EPA-600/3-78-0128, U.S. Environmental Protection Agency, Washington, D.C.

Eshleman, K.N., Pollard, J.S., and Kuebler O'Brien, A., 1993, Determination of contributing areas for saturation overland flow from chemical hydrograph separations, *Water Resour. Res.* **29**(10):3577-3587.

Federal Interagency Committee for Wetland Delineation., 1989, *Federal Manual for Identifying and Delineating Jurisdictional Wetlands,* U.S. Army Corps of Engineers, U.S. Environmental Protection Agency, U.S. Fish and Wildlife Service, and U.S.D.A. Soil Conservation Service, Washington, D.C. Cooperative technical publication.

Fetter, Jr., C.W., Sloey, W.E., and Spangler, F.L., 1978, Use of a natural marsh for wastewater polishing, *J. Water Poll. Control Fed.* **50**:290-307.

Gburek, W.J., 1990, Initial contributing area of a small watershed, *J. Hydrol.* **118**:387-403.

Gehrels, J., and Mulamoottil, G., 1989, The transformation and export of phosphorus from wetlands, *Hydrol. Proc.* **3**:365-370.

Goldsmith, L.A., 1994, *A Landscape Level Evaluation of Wetland Structure and Function in a Large, Multi-use Basin,* Master's Thesis, University of Vermont, Burlington, Vermont.

Hammond, R.E., Coakley, M.F., Kierstad, C., and Kiah, R.G., 1996, *Water Resources Data for New Hampshire and Vermont, Water Year 1995,* U.S. Geological Survey, Pembroke, New Hampshire.

Howard-Williams, C., 1985, Cycling and retention of nitrogen and phosphorus in wetlands: A theoretical and applied perspective, *Freshwater Biol.* **15**:391-431.

Johnson, A.H., Bouldin, D.R., Goyette, E.A., and Hedges, A.M., 1976, Phosphorus loss by stream transport from a rural watershed: Quantities, processes, and sources, *J. Environ. Qual.* **5**(2):148-157.

Johnston, C.A., Detenbeck, N.E., and Niemi, G.J., 1990, The cumulative effect of wetlands on stream water quality and quantity. A landscape approach, *Biogeochemistry* **10**:105-141.

Kadlec, R. H.;, Reddy, K. R., 2001, Temperature effects in treatment wetlands. *Water Environ. Res.* **73**(5): 543-557.

Kao, C. M., and Wu, M. J., 2001, Control of non-point source pollution by a natural wetland, *Water Sci. Technol.*, **43**(5): 169-174.

Kennedy, E.J., 1984, Discharge ratings at gaging stations, in: *Techniques of Water Resources Investigations of the United States Geological Survey. Book 3. Applications of Hydraulics*, U.S. Government Printing Office, Washington, D.C., pp. 1-59.

Klopatek, J.M., 1975, The role of emergent macrophytes in mineral cycling in a freshwater marsh, in: *Mineral Cycling in Southeastern Ecosystems*, Howell, F.G., Gentry, J.B., and Smith, M.H., eds., ERDA Symposium Series CONF-740513, Springfield, Virginia, pp. 357-393.

Lee, G.F., Bentley, E., and Admundson, R., 1975, Effect of marshes on water quality, in: *Coupling of Land and Water Systems*, Hasler, A.D., ed., Springer, New York, pp. 105-127.

Longabucco, P., and Rafferty, M.R., 1989, Delivery of nonpoint source phosphorus from cultivated mucklands to Lake Ontario, *J. Environ. Qual.* **18**(2):157-163.

Lowrance, R.R., Todd, R.L., and Asmussen, L.E., 1984, Nutrient cycling in an agricultural watershed: II. Streamflow and artificial drainage, *J. Environ. Qual.* **13**(1):27-32.

Meade, R.H., 1982, Sources, sinks, and storage of river sediment in the Atlantic drainage of the United States, *J. Geol.* **90**(3):235-252.

Meals, D.W., 198, Monitoring changes in agricultural runoff quality in the LaPlatte River watershed, Vermont, In: *Perspectives on Nonpoint Source Pollution*, EPA-440/5-85-001, U.S. Environmental Protection Agency, Washington, D.C., pp. 185-190.

Meals, D.W., 1993, Assessing nonpoint phosphorus control in the LaPlatte River watershed, *Lake Reservoir Manage.* 7:197-207.

Mitsch, W.J., Dorge, C.L., and Wiemhoff, J.R., 1979, Ecosystem dynamics and a phosphorus budget of an alluvial cypress swamp in southern Illinois, *Ecology* **60**:1116-1124.

Mitsch, W.J., and Gosselink, J.G., 199, *Wetlands*, 2nd Edition, Van Nostrand Reinhold, New York.

Mitsch, W.J., and Reeder, B.C., 1992, The role of wetlands in the control of nutrients with a case study of western Lake Erie, in: *Ecological Engineering: An Introduction to Ecotechnology*. Mitsch, W.J. and Jorgensen, S.E., eds., John Wiley & Sons, New York, pp. 129-158.

Monke, E.J., Nelson, D.W., Beasley, D.B., and Bottcher, A.B., 1981, Sediment and nutrient movement from the Black Creek watershed, *Trans. Am. Soc. Agric. Eng.* **24**(2):391-395.

Montgomery, D.R., and Dietrich, W.E., 1995, Hydrologic processes in a low-gradient source area, *Water Resour. Res.* **13**(1):1-10.

Moustafa, M. Z., 2000, Do Wetlands Behave Like Shallow Lakes in Terms of Phosphorus Dynamics? *J. Amer. Water Res. Assoc.* **36**: 43-54.

Mulholland, P.J., Wilson, G.V., and Jardine, P.M., 1990, Hydrogeochemical response of a forested watershed to storms: Effects of preferential flow along shallow and deep pathways, *Water Resour. Res.* **26**(12):3021-3036.

National Oceanographic and Atmospheric Administration., 1995, *Local Climatological Data. Annual Summary with Comparative Data. Burlington, Vermont*, National Climatic Data Center, Asheville, North Carolina.

Neter, J., Wasserman, W., and Kutner, M.H., 1989, *Applied Linear Regression Models*, R.D. Irwin, Inc., Homewood, Illinois.

O'Loughlin, E.M., 1986, Prediction of surface saturation zones in natural catchments by topographic analysis, *Water Resour. Res.* **22**(5):794-804.

Omernik, J. M., 1976, *The Influence of Land Use on Stream Nutrient Levels*, EPA-600/3-76-014, U.S. Environmental Protection Agency, Corvallis, Oregon.

Paschal, Jr., J.E., and Sherwood, D.A., 1987, *Relation of Sediment and Nutrient Loads to Watershed Characteristics and Land Use in the Otisco Lake Basin, Onondaga County, New York.* Water-Resources Investigations Report 86-4026, U.S. Geological Survey, Ithaca, New York.

Peverly, J.H., 1982, Stream transport of nutrients through a wetland, *J. Environ. Qual.* **11**:38-43.

Pionke, H.B., Hoover, J.R., Schnabel, R.R., Gburek, W.J., Urban, J.B., and Rogowski, A.S., 1988, Chemical-hydrologic interactions in the near-stream zone, *Water Resour. Res.* **24**(7):1101-1110.

Prairie, Y.T., and Kalff, J., 1986. Effect of catchment size on phosphorus export. Water Resour. Bull. 22(3):465-470.

Preston, S.D., Bierman, Jr., V.J., and Silliman, S.E., 1989, An evaluation of methods for the estimation of tributary mass loads, *Water Resour. Res.* **25**(6):1379-1389.

Rantz, S.E., 1982, *Measurement and Computation of Streamflow: Volume 1. Measurement of Stage and Discharge*, U.S. Geological Survey Water-Supply Paper 2175, U. S. Government Printing Office, Washington, D.C.

Richards, R.P., and Holloway, J., 1987, Monte Carlo studies of sampling strategies for estimating tributary loads, *Water Resour. Res.* **23**(10):1939-1948.

Sanchez-Carrillo, S., and Alvarez-Cobelas, M., 2001, Nutrient dynamics and eutrophication patterns in semi-arid wetland: The effects of fluctuating hydrology. *Water, Air, Soil Pollut.***131**: 97-118.

Smeltzer, E., 1996. Unpublished data. Vermont Agency of Natural Resources. Waterbury, Vermont.

Tilton, D.L., and Kadlec, R.H., 1979, The utilization of a fresh-water wetland for nutrient removal from secondarily treated waste water effluent, *J.Environ. Qual.* **8**:328-334.

Tim, U. S., Saied, M., and Shanholtz, V.O., 1992, Identification of critical nonpoint pollution source areas using geographic information systems and water quality modeling, *J. Amer. Water Res. Assoc.* **28**: 877-887

U.S.D.A. Natural Resource Conservation Service., 1995, *Soil Survey Geographic (SSURGO) Data Base Data Use Information.* Miscellaneous Publication Number 1527, U.S. Department of Agriculture.

U.S.D.A. Soil Conservation Service., 1989, *Soil Survey of Chittenden County, Vermont*, U.S. Department of Agriculture, Vermont Agricultural Experiment Station, and The Vermont Department of Forests and Parks. U.S. Government Printing Office, Washington, D.C.

U.S. Environmental Protection Agency., 1983, *Methods for Chemical Analysis of Water and Wastes.* EPA-600/4-79-020, U.S. Environmental Protection Agency, Cincinnati, Ohio.

U.S. Environmental Protection Agency, 2001, Ambient Water Quality Recommendations - Information Supporting the Development of State and Tribal Nutrient Criteria - Rivers and Streams in Nutrient Ecoregion VIII. EPA 822-B-01-015, December 2001.

Uttormark, P.D., Chapin, J.D., and Green, K.M., 1974, *Estimating Nutrient Loadings to Lakes from Non-point Sources*, EPA-660/3-74-020, U.S. Environmental Protection Agency, Washington, D.C.

Uusi-Kamppa , J., B. Braskerud, H, Jansson, N. Syversen, and R. Uusitalo, 2000, Buffer zones and constructed wetlands as filters for agricultural phosphorus, *J. Environ. Qual.* **29**:151-158.

Vermont Geographic Information System., 1992, *Policies, Standards, Guidelines, & Procedures Handbook*, Vermont Center for Geographic Information, Inc. (VCGI), Burlington, Vermont.

Weller, C.M., Watzin, M.C., and Wang, D., 1996, Role of wetlands in reducing phosphorus loading to surface water in eight watersheds in the Lake Champlain Basin, *Environ. Manage.* **20**(5):731-739.

Whigham, D.F., Chitterling, C., and Palmer, B., 1988, Impacts of freshwater wetlands on water quality: A landscape perspective, *Environ. Manage.* **12**:663-671.

Wickham, J. D., Riitters, K. H., O'Neill, R. V., Reckhow, K. H., Wade, T. G., and Jones, B.J., 2000, Land Cover As a Framework for Assessing Risk of Water Pollution, *J. Amer. Water Res. Assoc.* **36**: 1417-1422.

Yarbro, L.A., Kuenzler, E.J., Mulholland, P.J., and Sniffen, R.P., 1984, Effects of stream channelization on exports of nitrogen and phosphorus from North Carolina Coastal Plain watersheds, *Environ. Manage.* **8**:151-160.

THE OCCURRENCE OF CYANOBACTERIAL TOXINS IN LAKE CHAMPLAIN

Gregory L. Boyer, Mary C. Watzin, Angela D. Shambaugh, Michael F. Satchwell, Barry H. Rosen, and Timothy Mihuc*

ABSTRACT

Cyanobacteria are common in the plankton of most lakes, including Lake Champlain. Certain species of cyanobacteria can produce toxins that are harmful to people and their pets. In 2000 and 2001, we examined the occurrence of potential toxin-producing species of cyanobacteria in Lake Champlain. Samples were collected from locations throughout the lake and at the intake locations for five water treatment plants. Drinking water is a potential route of exposure for the thousands of people who draw their drinking water from the lake. When toxin-producing species of cyanobacteria were found, samples were also collected for toxin analysis. Microcystins were measured using a combination of different assays including an enzyme-linked immunoassay or ELISA, inhibition of the protein phosphatase 1A, and by high performance liquid chromatography (HPLC). Anatoxin-a was determined by HPLC and confirmed by HPLC coupled with mass spectroscopy. The PSP toxins (saxitoxin, neosaxitoxin, and gonyautoxins 1-4) were measured by HPLC with fluorescent detection.

At least six species of potential toxin producing cyanobacteria were commonly found in the Lake Champlain, including *Aphanizomenon flos-aquae*, *Anabaena flos-aquae* and *Microcystis aeruginosa*. Microcystins were the predominate toxin and were observed at low levels in nearly half of the samples collected. Except for several samples collected in

* Gregory L. Boyer and Michael F. Satchwell, Department of Chemistry, State University of New York, College of Environmental Science and Forestry, Syracuse, NY, 13210. Mary C. Watzin and Angela Shambaugh, University of Vermont, Rubenstein Ecosystem Science Laboratory, School of Natural Resources Burlington, VT, 05401. Barry H. Rosen, USDA-NRCS, Watershed Science Institute, Burlington VT, 05405 (current address: US Fish and Wildlife Service, 1339 20th St., Vero Beach, FL 32960). Timothy Mihuc, Lake Champlain Research Institute, Plattsburgh State University, Plattsburgh, NY 12901.

the north lake, microcystin concentrations were quite low, well below the 1 μg L^{-1} World Health Organization's guideline for the protection of human health. Concentrations of anatoxin-a were likewise generally low, but when this toxin did occur, it sometimes reached levels that could potentially exceed threshold values affecting human health. Low levels of toxins were also detected in both raw and finished water samples collected from the water treatment plants. Highest concentrations of toxin were found in the treatment plants that have shallow intake pipes. Treatment plants with deeper water intakes showed lower concentrations of cyanobacterial toxins. The general occurrence of toxins suggests that additional research is needed to fully assess the risks of exposure for people.

1. INTRODUCTION

Cyanobacteria, or blue-green algae, are common in nature and found in nearly all aquatic environments. In recent years, outbreaks of toxic blue-green algae have drawn increasing attention in both the scientific and general press. Cyanobacteria blooms themselves are not unusual, and are often associated with taste and odor problems in drinking waters (AWWA, 1989). In addition, blooms of certain species can produce extremely toxic compounds that may result in animal or human fatalities. A bloom of neurotoxic *Anabaena circinalis* in the Darling River of Australia killed an estimated 10,000 livestock and required emergency drinking water be supplied to several towns and villages (Falconer, 2001). In recent years, toxic outbreaks of cyanobacteria have also been reported closer to home in Lake Erie (Brittain et al., 2000), Lake Ontario (Howell et al., 2002) and Lake Champlain (Rosen et al., 2001; Yang et al., 2001; Watzin et al., 2002). While human intoxications are rare, an outbreak of a hepatotoxic microcystin-producing species in the water supply for a hemodialysis plant in Caruaru, Brazil, resulted in at least 75 human fatalities, despite extensive treatment of the incoming water (Jochimsen et al., 1998; Pouria et al., 1998). This outbreak prompted the World Health Organization to develop and adopt a provisional human health guideline of 1.0 μg microcystin-LR per liter in drinking water (WHO, 1998). Similar values have been developed and adopted by Canada and Australia (Falconer, 2001). Guidelines to protect human health for the neurotoxins produced by cyanobacteria are under development (M. Burch, personal communication).

Cyanobacterial toxins fall into several diverse structural categories. The microcystin peptides (Figure 1) are extremely potent hepatotoxins and their production has been associated both with acute human toxicity (Carmichael, 1998; Kuiper-Goodman et al., 1999) and an increased incidence of hepatic cancer (Fujiki et al., 1996). More than 70 different cyclic peptide structures having the microcystin or nodularin-ring system have been reported and they vary widely in their biological activity. Microcystins are produced by a number of different species (Table 1), though the primary source is likely to be from *Microcystis*, *Anabaena*, or *Oscillatoria* species.

Separate from the hepatotoxic cyclic peptides are the neurotoxic alkaloids. These consist of anatoxin-a, anatoxin-a (S), and saxitoxin or related analogs called the Paralytic Shellfish Poisoning (PSP) toxins. The most important of these compounds is probably anatoxin-a (Figure 1). Anatoxin-a was originally isolated from *Anabaena flos-aquae* (Devlin et al., 1977), but is also reported in *A. planktonica*, *Oscillatoria* species, and *Cylindrosperum* species. Toxic cyanobacteria may also contain the neurotoxins saxitoxin (STX) and neosaxitoxin (neoSTX). These toxins are identical to those produced by some "red-tide" marine dinoflagellates and accumulate in shellfish that feed on toxic

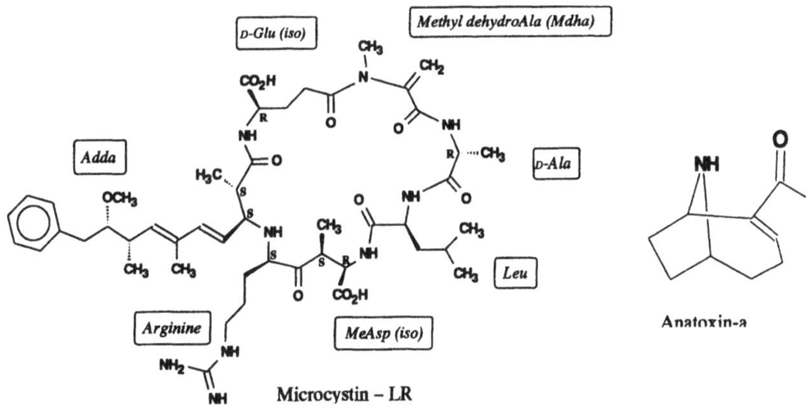

Figure 1: The structure of the hepatotoxin microcystin-LR and the neurotoxin anatoxin-a. Both toxins were found in Lake Champlain. While anatoxin-a has only a single chemical congener (homo-anatoxin-a), there are more than 70 known microcystins that vary widely in their biological activity.

dinoflagellates. Human consumption of these now-toxic shellfish results in a clinical condition known as paralytic shellfish poisoning or PSP. Freshwater or cyanobacterial PSP toxins where first reported from an *Aphanizomenon flos-aquae* species isolated from Kezar Lake in New Hampshire (Sawyer et al., 1968), but have also been found in *Anabaena circinalis, Lyngbya wollei,* and a Brazilian isolate of *Cylindrospermopsis raciborskii. Cylindrospermopsis raciborskii* from Hungary, Australia and many of the Florida reservoirs instead produces another toxin, the hepatotoxic alkaloid cylindrospermopsin (Chorus and Bartram, 1999).

Table 1. Cyanobacterial toxins and their sources. Those species identified in Lake Champlain phytoplankton net tows are indicated in bold.

Toxin and Type	Cyanobacterial Source [a]
Microcystins (peptide hepatotoxin)	***Microcystis aeruginosa,*** *M. viridis, M. botrys,* ***Anabaena flos-aquae,*** *A. lemmermannii,* ***A. circinalis*** *Anabaenopsis millerii, Oscillatoria limosa, Nostoc* sp., and others.
Anatoxin-a (alkaloid neurotoxin)	***Anabaena flos-aquae, A. planctonica, Oscillatoria spp.,*** *Planktothrix sp.,* ***Microcystis sp., Aphanizomenon flos-aquae,*** and others.
PSP toxins (alkaloid neurotoxin)	***Anabaena circinalis, Aphanizomenon flos-aquae*** [b], *Cylindrospermopsis raciborskii, Lyngbya wollei.*
Anatoxin-a(s) (organophosphate neurotoxin)	***Anabaena flos-aquae,*** *A. lemmermannii.*
Cylindrospermopsin (alkaloid hepatotoxin)	*Cylindrospermopsis raciborskii, Aphanizomenon ovalisporum, Raphidiopsis curvata, Umezakia natans.*

[a]from Chorus and Bartram, 1999.
[b]The taxonomy of this PSP-producing species is currently under revision (Li et al., 2000)

This large variety of toxins and potentially toxic species illustrates the need for careful monitoring and analysis of any water body that may serve as potable water for human consumption. In 1999, the death of at least three dogs in the Point Au Roche region of Lake Champlain was attributed to consumption of water containing toxic blue-green algae (Yang et al., 2001). This event emphasized the need to know both the abundance of potentially toxic species, as well as the production of cyanobacterial toxins in Lake Champlain. A collaborative study was established using the resources of the University of Vermont, State University of New York (SUNY) at Plattsburgh and SUNY College of Environmental Science and Forestry (ESF) to address this need. This paper summarizes the initial results of that collaboration.

2. METHODS

3.1. Sample Collection

Multiple collection strategies were used by the different institutions to collect samples for toxin analysis and species abundance. In 2000 and 2001, 1-liter grab samples were collected from 16 sites along the shores of Lake Champlain in mid August. These samples were filtered through 47mm 934 AH glass fiber filters in the field, frozen on dry ice, and returned to the lab for toxin analysis. From June through October in 2000 and 2001, both 63 μm plankton net vertical tow samples and water samples were collected from five stations in the Burlington Bay region and from the two largest water treatment plants on the lake, the City of Burlington Water Treatment Plant and the Champlain Water District Water Treatment Plant. In 2000, water samples were collected from three additional water treatment plants, the Swanton Water Department, North Hero Water System, and the Tri-town Water System plant in Addison County, Vermont. In 2000, one liter of untreated raw water and one liter of finished drinking water were collected biweekly at each water plant from late July through September. Raw water was the water taken into the water treatment facility for processing. Finished water was water distributed to consumers after processing, which varied by treatment facility. In 2001, 5 to 15-liter samples of both raw and finished water were collected at 2-4 week intervals from August through October. Water was filtered through a C-18 solid phase extraction cartridge for concentration of toxins.

Capitalizing on the fieldwork already in place for monitoring water quality in Lake Champlain, the Vermont and New York Departments of Environmental Conservation in 2000 also collected 63 μm phytoplankton net vertical tow samples approximately every 2-3 weeks from most of the offshore stations as indicated in Figure 2. Occasional samples were collected from some of these sites in 2001. One-liter grab samples were also collected at some sites. The plankton net samples were examined under the light microscope to determine if cyanobacterial species capable of producing toxins were present. Presence of these cyanobacteria initiated analysis of toxins as described below.

In 2001, 20-liter grab samples were also collected from several lake stations using a high-speed peristaltic pump and immediately filtered using a 150 mm filter holder equipped with a 934AH glass fiber filter. Toxin distribution with depth was determined at two sites (Treadwell Bay and Ausable Shoals) at near noon by collecting 20-L grab samples at 0, 1, 3, 5 m, and approximately 0.5 meter above the bottom (10 or 15m respectively) using a submersible pump. Separate samples for chlorophyll analysis were

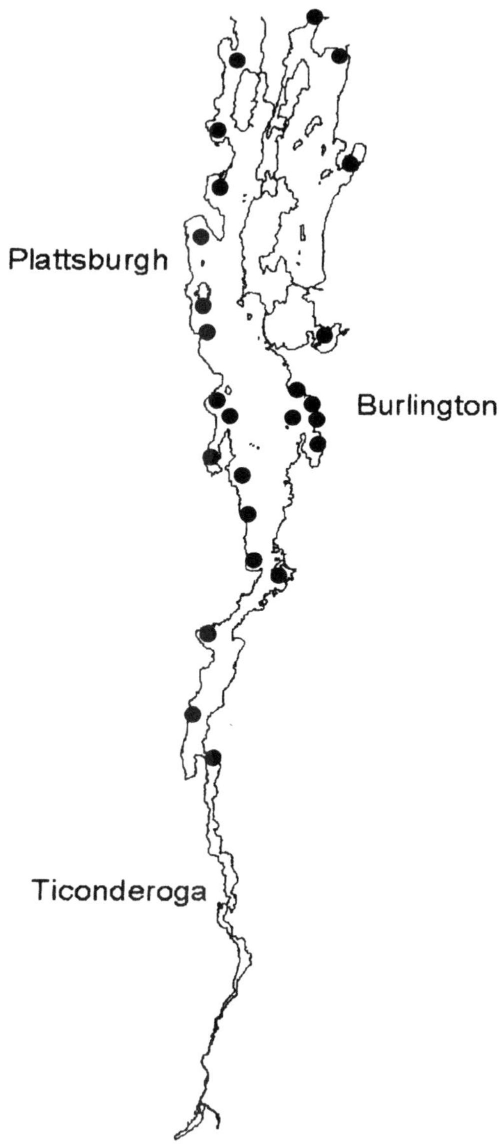

Figure 2. Location of the combined sampling sites for 2000 and 2001.

filtered on to 47 mm glass fiber filters and immediately frozen on dry ice. Upon return to the lab, chlorophyll a was determined using the spectrophotometric procedure of Parson et al. (1984). *In vitro* phycocyanin concentrations were determined in the field using a Turner Designs 10-AU fluorometer.

3.2. Toxin Analysis and Identification

Filters for toxin analysis were extracted by sonication with 10 ml of 50% methanol either with (ESF) or without (UVM) 1 % acetic acid, clarified by centrifugation, and the extract used for analysis of the different toxins. Microcystins were measured using a combination of different assays including an enzyme-linked immunoassay or ELISA (EnviroLogix Inc, Portland ME), inhibition of the protein phosphatase 1A (PPIA: Carmichael and An, 1999), and by high performance liquid chromatography (HPLC) coupled with either UV or mass selective (MS) detection (Harada, 1996). Anatoxin-a was determined by HPLC after derivatization with 7-fluoro-4-nitro-2,1,3- benzoxadiazole (NBD-F) (James and Sherlock, 1996) and confirmed by HPLC – MS of the NBD-derivative. The PSP toxins (saxitoxin, neosaxitoxin, and gonyautoxins 1-4) were measured by HPLC with fluorescent detection after either post-column chemical oxidation (PCRS: Oshima, 1995) or using the electrochemical oxidation system (ECOS: Boyer and Goddard, 1999) to form fluorescent derivatives.

3.3. Phytoplankton Identification

Phytoplankton net samples were preserved in either 1% Lugol's solution or 0.5 % glutaraldehyde in the field and returned to the laboratory for examination. Initial samples were evaluated qualitatively, simply noting the taxa present. Once blue-green algae were identified in the samples and observed routinely, evaluation became quantitative. Individual alga in the samples were identified and enumerated, and cell densities calculated for each taxon.

For both qualitative and quantitative evaluations, an aliquot of a well-mixed sample was placed into a phytoplankton counting chamber (Phycotech Inc, volume 0.06 mL) and allowed to settle for 5 minutes. Chamber bottoms were evaluated using an Olympus CK-2 or IX70 inverted microscope at 100x. For qualitative samples, the entire chamber was scanned and algal taxa present were recorded. In addition, example fields from the samples were photographed to provide a permanent record of the key species. For quantitative samples, individual algal cells and colonies were identified and counted. Counting continued until 100 individuals of the most abundant taxa had been identified, or 10 fields had been examined from each of 5 aliquots. The mean number of cells per natural unit was determined by counting cells in up to 10 colonies of each taxa observed. The number of natural units counted in a sample was multiplied by the mean number of cells per natural unit to estimate cell densities. All cell densities are expressed per unit unconcentrated lake water.

3. RESULTS

4.1. Identification of Potentially Toxigenic Species of Cyanobacteria

Lake Champlain has a highly diverse phytoplankton flora; an earlier survey (1991) identified over 215 taxa in the lake (Shambaugh et al., 1999). At least six species of cyanobacteria, identified as known toxin producers from the literature (Table 1), were commonly found in the Lake Champlain during the summers of 2000 and 2001. *Aphanizomenon flos-aquae* was moderately abundant in the northern part of the lake.

Anabaena flos-aquae formed a brief bloom during late July 2000 near the *RV Melosira* boat slip in Burlington Harbor and this species was a likely source of toxicity for the dog deaths that occurred on the Main Lake during that time. Similarly, *Anabaena circinalis*, *Anabaena planktonica* and *Oscillatoria* spp. were observed in several of the samples. All are reported to produce a number of different toxins (Chorus and Bartram, 1999). *Microcystis aeruginosa* was found in many of the net tows and is a potential source for the microcystin detected in many of the samples. Two other species, *Gloeotrichia echinulata* and *Coelosphaerium naegelianum*, also formed locally abundant blooms. These species have been reported to produce unknown compounds that cause irritation or gastrointestinal distress (AWWA, 1989). Reports of these symptoms were not monitored as part of this study.

4.2. Identification of Cyanobacteria Toxins Found in Lake Champlain

Since both toxic and nontoxic isolates are known for all the toxin-producing species, the identification of a potentially toxic species in a water sample does not necessarily imply that a toxin will be produced. Therefore, samples were also collected for toxin analysis using a combination of different assays. The use of multiple assays is important, especially for such toxins as the microcystins where a number of toxic and non-toxic forms are known to exist (Ward and Codd, 1999).

The qualitative results of the microcystin analyses for samples collected in 2000 and 2001 are summarized in Table 2. Microcystins were the predominant toxin and were observed at low levels in nearly half of the samples collected. The major difference between year 2000 and year 2001 was a difference in sampling strategy (20-L grab vs. 1-L grab) that led to an increased sensitivity of analysis when expressed per liter lake water. Net-tow samples showed a similar high frequency of toxicity (>75%) in both years. Despite the high number of samples that contained measurable amounts of microcystins, their overall concentrations were generally quite low and well below the WHO advisory limit of 1 $\mu g\ L^{-1}$. The highest concentration observed in the 20-L grab samples was 0.025 μg microcystin-LR equivalents L^{-1} in Treadwell Bay in mid August 2001 (see Figure 5 discussed below). This corresponded to approximately 3 ng microcystin-LR per μg chlorophyll. In contrast, highly enriched net tows collected from the Missisquoi Bay region in mid-September 2001 during a cyanobacteria bloom showed extremely high concentrations of microcystins on a per extract volume basis (60-150 $\mu g\ L^{-1}$). When corrected for enrichment, these values were approximately 5 ng microcystin-LR per μg chlorophyll. Although it is not possible to determine which species or individual colonies were producing the toxins, the similarity in toxin content per unit chlorophyll between the bloom and non-bloom events makes it likely that the high concentrations of microcystins observed in Missisquoi Bay were the result of the large cyanobacterial biomass that occurred during the bloom, and not due to the occurrence of a more highly toxic variant of *Microcystis*.

Concentrations of anatoxin-a were generally low, but when this toxin did occur, it sometimes reached levels that could potentially exceed the threshold values affecting human health. In some cases, the levels of anatoxin-a far exceeded those found for microcystin. For example, a bloom in Burlington Harbor in August 2000 contained over 740 ng anatoxin-a μg^{-1} chl *a* and 3 ng microcystin-LR μg^{-1} chl *a*. Similar high concentrations of anatoxin-a were observed in samples sent to ESF for analysis from a camp owner whose dogs had died in August 2000 after playing in lake water at approximately the same time as a "blue-green algal bloom." Subsequent analysis of those samples showed lake water containing values of anatoxin-a >1.5 μg L^{-1} (Yang et al., 2001). While the WHO has not adopted a human health guideline for anatoxin-a in drinking water, the Australian National Plan is currently recommending a provisional guideline value of 1-3 μg L^{-1} for cyanobacterial neurotoxins (M. Burch, personal communication) in drinking water. Values of anatoxin-a measured in Lake Champlain approached or exceeded this threshold.

The PSP toxins saxitoxin and neosaxitoxin (Table 2) and the organophosphate toxin anatoxin-a (S) were not observed in any of the Lake Champlain samples tested (Zuniga, et al., 2002). These toxins are much less common than microcystins and anatoxin-a and appear to be less of a threat for Lake Champlain waters. To date, the samples have not been tested for the hepatotoxin, cylindrospermopsin; however, efforts in this direction are currently underway.

Table 2. The occurrence of cyanobacterial toxins in Lake Champlain.

Toxin and Assay	Total number of samples	Toxic (total) 1-L grabs	Toxic (total) 20-L grabs	Toxic (total) Net tows[a]	% Toxic (All samples)
2000					
Microcystins					
ELISA[1]	42	0 (16)	-	20 (26)	48%
PPIA[1]	46	0 (20)	-	19 (26)	41%
Anatoxin-a					
HPLC-FD[2]	47	2 (20)	-	3 (27)	11%
PSP-toxins					
HPLC-FD[3]	46	0 (20)	-	0 (26)	0%
2001					
Microcystins					
ELISA[4]	41	-	5 (19)	19 (22)	59%
PPIA[4]	41	-	8 (19)	20 (22)	68%
Anatoxin-a					
HPLC-FD[5]	41	-	11 (19)	5 (22)	39%
PSP-toxins					
HPLC-FD[3]	22	-	0 (0)	0 (22)	0%

[a]Toxicity is expressed in per liter concentrated extract, not lake volume.
In 2001, our detection limits were lower than those in 2000 because larger sample volumes were collected. These detection limits are given in the footnotes below.
[1] Toxicity defined as greater than 0.15 μg microcystin-LR equivalents L^{-1}
[2] Toxicity defined as greater than 0.1 μg anatoxin-a L^{-1}
[3] Toxicity defined as greater than 0.06 μg saxitoxin equivalents L^{-1}
[4] Toxicity defined as greater than 0.01 μg microcystin-LR equivalents L^{-1}
[5] Toxicity defined as greater than 0.001 μg anatoxin-a L^{-1}.

4.3. Distribution of Cyanobacterial Toxins in Lake Champlain

Initial studies in 2000 of the occurrence of cyanobacteria in the areas of the lake that served as source water for drinking water treatment plants showed highest concentrations in late summer (Figure 3). The concentration of microcystin in these waters was loosely related to the concentration of chlorophyll; however, there were several notable

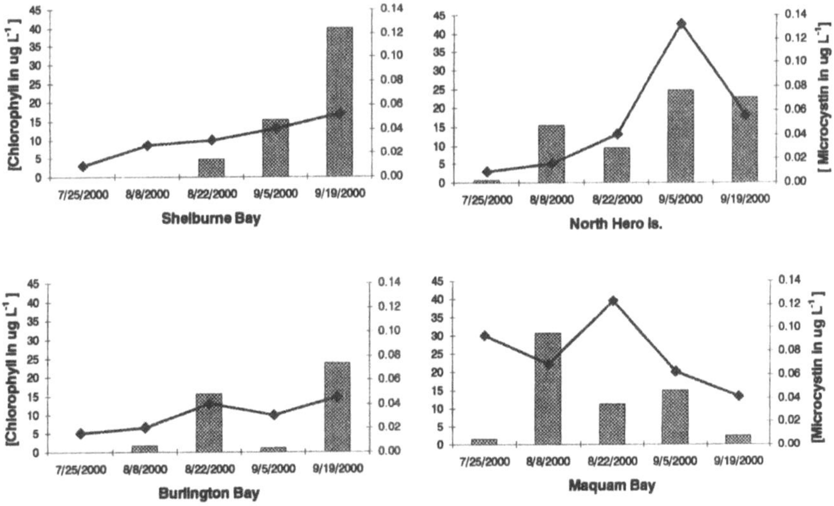

Figure 3. Change in chlorophyll and microcystin content over time in 2000 in the source waters that service four water treatment plants. Chlorophyll content (lines) is given in μg L^{-1} on the left axis. Microcystin content (bars), as determined by the protein phosphatase inhibition assay using concentrated phytoplankton net-tow samples, is given using the scale on the right axis. Microcystin concentrations were normalized to per liter concentrations from the net tows as described in Rosen et al. (2001). Shelburne Bay is the source water for the Champlain Water District, Burlington Bay for the City of Burlington, and Maquam Bay for the Swanton Town treatment plant.

exceptions to this generalization. At the North Hero, Swanton (Maquam Bay) and City of Burlington (Burlington Bay) treatment plant sites, the highest levels of microcystin did not correspond with the highest chlorophyll concentrations. Although the chlorophyll concentrations were similar (~15-18 μg L^{-1}) at all four sites on September 19, 2000, the microcystins concentrations varied almost 10-fold between the different sample sites.

Few cyanobacterial blooms were observed in Burlington Bay during 2001, therefore; only limited sampling was conducted at the water treatment facilities in August and September when cyanobacterial densities were highest. Microcystin was not detected in the surface waters above the water treatment facilities on August 27 (Figure 4). On August 1, August 4 and September 25, ambient concentrations of microcystin in the plankton in the surface waters were below 1 μg L^{-1}. Anatoxin-a was detected in the plankton above both intakes on September 4 (0.002 – 0.005 μg L^{-1}). Cyanobacteria represented between 50 and 100% of the cells observed on these dates.

Microcystin was detectable in the raw water intakes at both facilities on all the dates sampled using at least one analytical technique (0.001 - 0.157 μg L^{-1}, Table 3). Microcystins were not measurable or detected in very low concentrations in the finished water at both facilities on all dates.

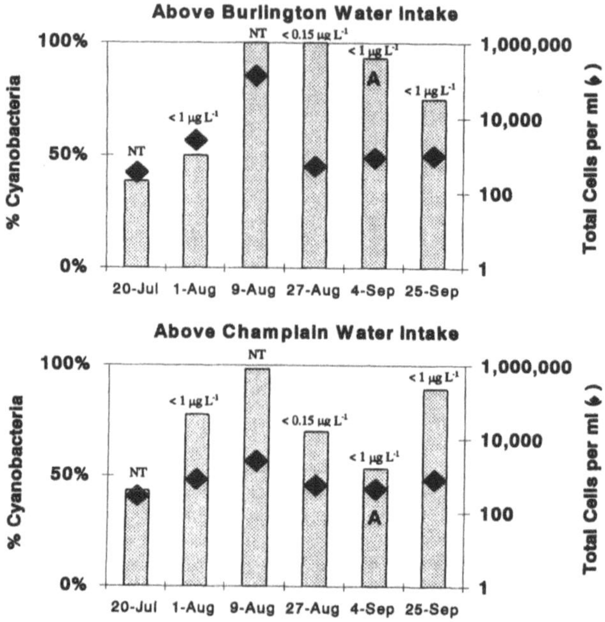

Figure 4. The total cell counts, relative cyanobacterial abundance and toxin results in surface waters above the Burlington Bay and Champlain water facility intake structures during 2001. Bars indicate the percent of total cells represented by cyanobacteria. Diamonds indicate total phytoplankton cell densities. The numbers above the bars represent the microcystin concentration as not tested (NT), <0.15 µg L^{-1} (i.e. below detection using ELISA) or < 1 µg L^{-1} (i.e. below the WHO's advisory limit). The "A" in bars for September 4th indicates that anatoxin-a was detected in these samples.

Table 3. Toxin concentrations in µg L^{-1} for water collected from Burlington Bay water treatment facilities in 2001. Raw water samples represent untreated water collected from the intake pipe. Finished water represents treated water leaving the facility for distribution.

Collection Date	Sample Location	Microcystin Concentrations PPIA	ELISA	Anatoxin-a HPLC-FD
8/01/01	Burlington Water – Finished	0.001	0.000	0.000
8/01/01	Burlington Water – Raw	0.157	0.078	0.000
8/01/01	Champlain Water – Finished	0.001	0.000	0.000
8/01/01	Champlain Water – Raw	0.094	0.076	0.000
9/05/01	Burlington Water – Finished	0.003	0.002	0.001
9/05/01	Burlington Water – Raw	0.126	0.007	0.000
9/05/01	Champlain Water – Finished	0.008	0.004	0.003
9/05/01	Champlain Water – Raw	0.001	0.001	0.000
10/09/01	Champlain Water – Finished	0.001	0.001	0.000
10/09/01	Champlain Water – Raw	0.102	-[a]	0.000
10/15/01	Burlington Water – Finished	0.005	0.003	0.000
10/15/01	Burlington Water – Raw	0.008	0.006	0.000

[a] sample not analyzed

In October, after isothermal conditions had developed in Burlington Bay, raw water concentrations at the Burlington facility decreased compared to earlier in the season. Raw water concentrations at the Champlain facility remained low, as seen in August and September. Although not detected in the raw water, anatoxin-a was detected at extremely low concentrations in finished water at both facilities on September 4th (0.001– 0.003 μg L^{-1}), at the same time that anatoxin-a was detected in the surface plankton above the intake structures.

The spatial distribution of cyanobacteria toxins in Lake Champlain was examined over a two-day period (August 14th -15th) in 2001 (Figure 5). For the sites sampled, microcystin concentrations showed a generally northern trend with a 3-fold higher concentration observed in the northern end of the lake. The highest concentrations

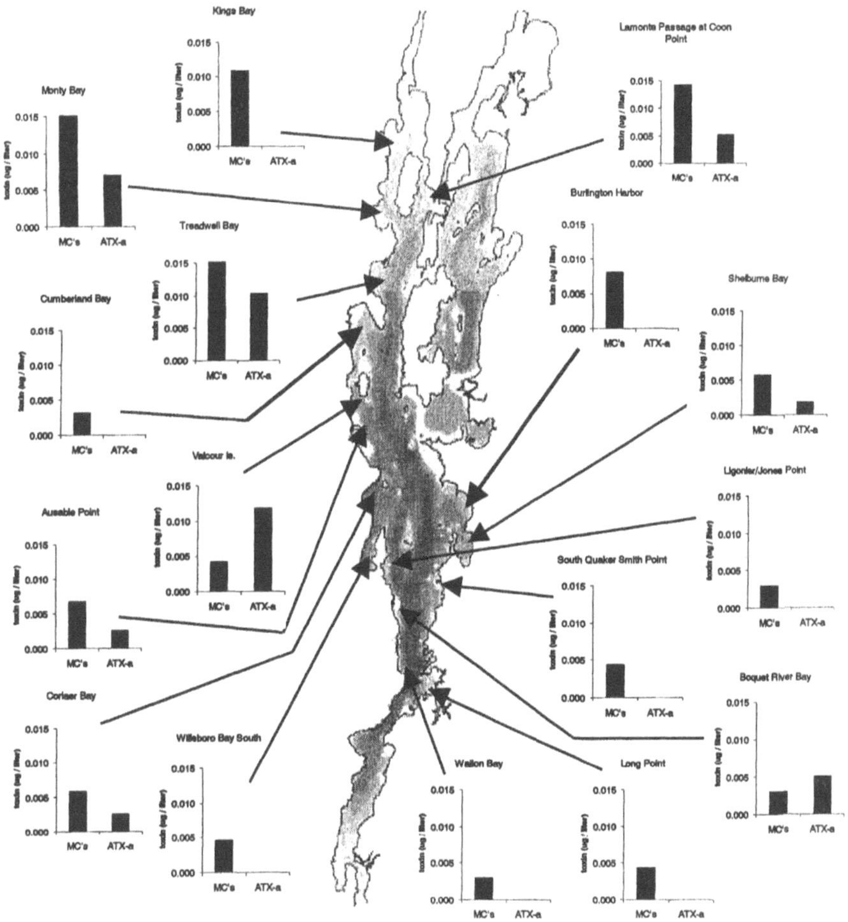

Figure 5. The spatial distribution of cyanobacteria toxins in Lake Champlain on August 14 and 15, 2001. Microcystins and anatoxin-a were measured using the PPIA and HPLC-FD respectively in 20-L grab samples.

observed during this time period were at Monty Bay, Treadwell Bay, and LaMotte Passage at Coon Point. For logistical reasons, samples were not taken in the northeast arm and Missisquoi Bay, however, we would expect, based on the high concentrations of microcystins observed in plankton samples (see below) taken on September 18th and 19th in that region, that high concentrations of microcystins probably would have been observed in these locations.

Anatoxin-a concentrations were also uniformly low in grab samples taken during this same time period, well below the Australian advisory level of 1-3 μg L^{-1} (M. Burch, personal communication). In contrast to that observed with the microcystin concentrations, anatoxin-a did not show a north-south trend in its distribution; rather the locations of higher anatoxin-a concentrations were scattered around the lake

For two sites, Treadwell Bay and Ausable Shoals, the distribution of cyanobacteria toxins with depth was also examined as part of the lake wide survey (Figure 6). Both Treadwell Bay and Ausable Shoals showed clear indication of stratification with the thermocline below 5 m. The distribution of the cyanobacterial pigment phycocyanin was very different between the two sites. Treadwell Bay showed a clear surface maximum, which corresponded to a *Gloeotrichia* bloom that was occurring at that time. Ausable Shoals showed a much lower and flatter distribution in both its phycocyanin and microcystin concentrations. For both sites, the anatoxin-a concentrations were extremely low and confined to the top 3 m of the water column.

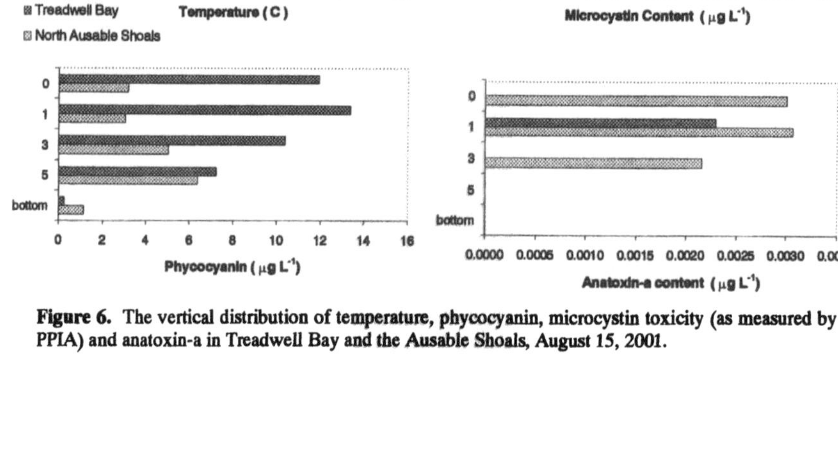

Figure 6. The vertical distribution of temperature, phycocyanin, microcystin toxicity (as measured by PPIA) and anatoxin-a in Treadwell Bay and the Ausable Shoals, August 15, 2001.

4.4. Occurrence of Cyanobacterial Toxins in Missisquoi Bay

In late August and September 2001, an extended cyanobacterial bloom occurred in Missisquoi Bay, along the Canadian shoreline near Phillipsburg. This bloom produced high concentrations of microcystin, exceeding 1 μg L^{-1} (S. Blais, personal communication). One phytoplankton net sample was collected by the Vermont DEC from the US side of Missisquoi Bay during this time, near the Vermont 78 boat launch. It showed high concentrations of microcystins (~6 ng microcystin μg chl a^{-1}). The HPLC-UV trace for this sample is shown in Figure 7. The sample contained a number of microcystins that were tentatively identified on the basis of their molecular weights and retention times. In contrast to what has previously been reported for Lake Erie (Brittain et al., 2000), the toxin in the sample collected from the Vermont 78 boat launch was not predominantly microcystin-LR. Instead, the sample included a more polar microcystin

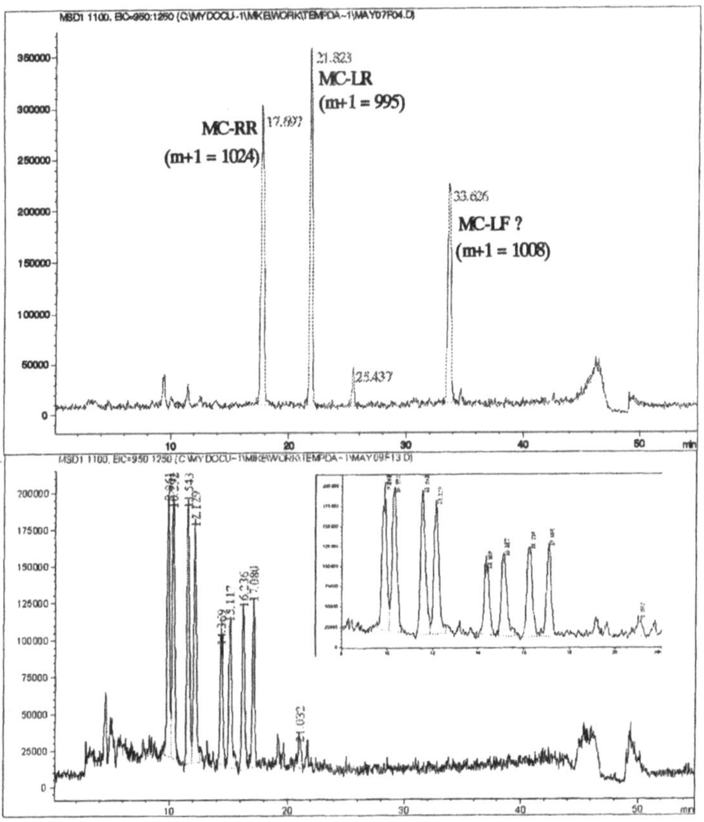

Figure 7. Extracted ion channel chromatograms of a mixed microcystin standard (A) and the field sample from Missisquoi Bay Vermont 78 boat launch (B), obtained using electrospray HPLC-MS. The molecular weights (m + 1) of the major ion in each standard are given in (A). The insert in (B) is an expanded portion of the 9-22 minute region.

variant. Toxins that eluted in the 14 to 17 minute range included the DhA variant of microcystin-LR and microcystin-YR along with the DhA variant of microcystin-FR. Further confirmation of the structure of these microcystins is in progress.

4. DISCUSSION

5.1. The Occurrence of Potential Toxin-Producing Cyanobacteria

Cyanobacteria were a regular component of the plankton of Lake Champlain in both 2000 and 2001, especially in late summer and fall, before turnover. Visible cyanobacterial surface blooms were not common in the open waters of the lake, but did occur in the shallow bays and boat slips around the lake, especially in northern Lake Champlain and Missisquoi Bay. Both potential toxin-producing and non-toxin producing taxa were observed in the plankton. The most common potential toxin-producers include several *Anabaena spp.*, *Aphanizomenon flos-aquae*, and *Microcystis aeruginosa*. These patterns are consistent with historical data from Lake Champlain (Myer and Gruendling, 1979; Brown et al., 1991; 1992; Shambaugh et al., 1999) and with the ecology of these organisms. Cyanobacteria prefer warmer water temperatures and tend to form surface blooms or scums only in protected areas, when winds are calm (Dokulil and Teubner, 2000; Paerl et al., 2001).

5.2. Toxin Production in Lake Champlain and Potential Human and Animal Exposure to Toxins

Both anatoxin-a and microcystin(s) were detected throughout Lake Champlain, but in most instances, concentrations were low and well below the current guidelines available for the protection of human health. The exception to this was several samples from the Northeast Arm, Missisquoi Bay, and a boat slip in Burlington Harbor. The environmental conditions that prevented the accumulation of cyanobacteria at the surface in many areas of the lake probably also prevented toxin concentrations from reaching levels of concern. It is possible that under different environmental conditions, greater bloom formation and higher toxin production could occur.

Low levels of toxins were also detected in both raw and finished water samples collected from the water treatment plants in 2000 and 2001. Microscopic analysis confirmed the presence of cyanobacteria in the raw water at these plants. In 2000, higher concentrations of toxin were generally found in the Swanton and North Hero treatment plants, both of which have intake pipes located in relatively shallow water. Because the intake structures for the Champlain Water District and the City of Burlington are located in more than 15 m of water, and cyanobacteria tend to accumulate in the upper water column, the likelihood that large quantities of cyanobacteria or their toxins would enter these facilities seems low. However, large amounts of water are processed by such facilities. It is possible to artificially concentrate cyanobacteria in the pre-filtering process, resulting in detectable toxin concentrations (Chorus and Bartram, 1999; Lahti et al., 2001). If cyanobacterial densities remain high during fall turnover, it is also possible that algae and toxins would be transported down to the intake pipe.

Several high profile cases of human poisoning by cyanobacterial toxins in drinking water have occurred in recent years (Kuiper-Goodman et al., 1999). In 1979, the hospitalization of 140 children on Palm Island off Queensland Australia was required as the result of a massive outbreak of gastroenteritis traced to the drinking water supplied from a reservoir containing *Cylindrospermopsis raciborskii*. More recently (1996), a cyanobacterial bloom occurred in the water supply for a hemodialysis center in Caruaru, Brazil. Despite extensive pretreatment (sand, carbon, resin and microfiltration), more than 117 of 136 patients (86%) experienced symptoms and at least 75 patients (55%) died from the incident. Measurable concentrations of microcystins, produced during the cyanobacterial bloom, were found in the blood serum of all exposed patients and in the liver tissues of the fatalities (Jochinsen et al., 1998; Pouria et al., 1998). In the United States, adverse human affects have included severe gastroenteritis reportedly from contaminated drinking water associated with a bloom of *Schizothrix calcicola* (Lippy and Erb, 1976). Cyanobacterial toxins have also been associated with the poisoning of birds and livestock (Chorus and Bartram, 1999). During 1999 and 2000, large die-offs of waterfowl occurred in Lake Erie and Lake Huron, the result of type E or C avian botulism (Domaske and Obert, 2001). While the connection with toxic cyanobacteria is tenuous, some investigators have suggested that microcystin-producing cyanobacteria may sensitize the birds to avian botulism (Murphy et al., 2000).

The connection between an anatoxin-a producing bloom in Lake Champlain and dog fatalities is more direct. In 1999 and again in 2000, ingestion of a toxic cyanobacteria resulted in the deaths of several dogs. In addition to drinking the water, the dogs probably ingested significant amounts of cyanobacteria by cleaning their fur after contact with the bloom (Ian Falconer, personal communication). Regardless of the mode of intake, the dogs died within hours with convulsions and respiratory paralysis, symptoms typical of anatoxin-a poisoning. Samples of the lake water and algal debris on the shore in the vicinity confirmed the presence of elevated levels of anatoxin-a ($> 1\ \mu g\ L^{-1}$) near this site (Yang et al., 2001). Although people are likely to avoid entering water with a visible surface algal scum, pets will not avoid this water, and on hot days, seem the most likely candidates for poisoning in Lake Champlain.

5. SUMMARY

Cyanobacteria were a regular component of the plankton of Lake Champlain in both 2000 and 2001. Both hepatotoxins and neurotoxins were found throughout the lake when cyanobacteria were abundant. Microcystins were the predominate toxin and were observed at low levels in nearly half of the samples collected. In general, microcystin concentrations were well below the $1\ \mu g\ L^{-1}$ WHO guideline for the protection of human health. The exception to this was several samples from the Northeast Arm, Missisquoi Bay, and a boat slip in Burlington Bay. Concentrations of anatoxin-a were likewise generally low, but when this toxin did occur, it sometimes reached levels that could potentially exceed threshold values affecting human health.

Low levels of toxins were also detected in both raw and finished water samples collected from the water treatment plants in 2000 and 2001. Highest concentrations of toxin were found in the treatment plants that have shallow intake pipes. Treatment plants with deeper water intakes showed lower concentrations of cyanobacterial toxins. These observations suggest that placement of intakes pipes is an important parameter affecting human exposure to cyanobacterial toxins through drinking water. The general occurrence of toxins suggests that additional research is needed to fully assess the risks of exposure for people.

6. ACKNOWLEDGEMENTS

We would like to acknowledge the help of Elizabeth Patchett (ESF), Xingye Yang (ESF), Emily Brines (UVM), Felicity Smith (UVM), Lisa Ferber (UVM), Trevor Carpenter (PSU) and Jamie Lescinski (PSU) for help in collecting and analyzing the samples for toxicity and species abundance. We would also like to recognize the valuable contributions of Captain Jeff Jones (PSU) and the crew of the *RV Monitor*, and Captain Dick Furbush (UVM) and the crew of the *RV Melosira* for all their valuable assistance. Some of these samples were also collected by field personnel from New York DEC and Vermont DEC and their help was also greatly appreciated. Financial support was provided by New York Sea Grant (to GLB), the Lake Champlain Basin Program (to MCW and BHR), the Centers for Disease Control (to MCW and BHR), and Green Mountain Power Corporation (to MCW).

7. REFERENCES

AWWA. 1989. Taste & Odor in Drinking Water Supplies. American Water Works Association,

Boyer, G. L., and G. D. Goddard. 1999. High Performance Liquid Chromatography (HPLC) coupled with post-column electrochemical oxidation (ECOS) for the detection of PSP toxins. Natural Toxins. 7: 353-359.

Brittain, S. M., J. Wang, L. Babcock-Jackson, W. W. Carmichael, K. L. Rinehart, and D. A. Culver. 2000. Isolation and characterization of microcystins, cyclic heptapeptide hepatotoxins from a Lake Erie Strain of *Microcystis aeruginosa*. J. Great Lakes Res. 26: 241-249.

Brown, E. A. Duchovnay, A. McIntosh, A. Shambaugh, and A. Williams. 1991. 1991 Lake Champlain Biomonitoring Report. Vermont Water Resources and Lake Studies Center. School of Natural Resources, University of Vermont.

Brown, E. A. Duchovnay, A. McIntosh, A. Shambaugh, and A. Williams. 1992. 1992 Lake Champlain Biomonitoring Report. Vermont Water Resources and Lake Studies Center. School of Natural Resources, University of Vermont.

Carmichael, W. W. 1998. Microcystin concentrations in human livers: Estimation of human lethal dose-lessons from Caruaru, Brazil. Abstracts, 4th International Conference Toxic Cyanobacteria, September, 27. Beaufort, NC.

Carmichael, W. W. and J. An. 1999. Using an enzyme linked immunosorbent assay (ELISA) and a protein phosphatase inhibition assay (PP1A) for the detection of microcystins and nodularins. Nat. Toxins. 7: 377-385.

Chorus, I., and J. Bartram. 1999. Toxic Cyanobacteria in Water: A guide to their public health consequences, monitoring and management. E&FN Spon, New York, NY, 416 pp.

Devlin, J. P., O. E. Edwards, P. R. Gorham, N. R. Hunter, R. K. Pike, and B. Stavric. 1977. Anatoxin-a, a toxic alkaloid from *Anabaena flos-aquae* NRC-44h. Can J. Chem. 55: 1367-1371.

Dokulil, M. T., and K. Teubner. 2000. Cyanobacterial dominance in lakes. Hydrobiologia. 438:1-12.

Domaske, H.M. and E.C. Obert. 2001. Avian botulism in Lake Erie. Great Lakes Research Review. 5: 1-5.

Falconer, I. R. 2001. Toxic cyanobacterial bloom problems in Australian waters: risks and impacts on human health. Phycologia. 40:228-233.

Fujiki, H., E. Sueoka, and M. Suganuma. 1996. Carcinogenesis of microcystins. In: "Toxic Microcystis" M. F. Watanabe, K. Harada, W. W. Carmichael, and H. Fujiki, Eds., CRC Press, Boca Raton, FL, pp. 203-232.

Harada, K. 1996. Chemistry and detection of microcystins. In: "Toxic Microcystis" M. F. Watanabe, K. Harida, W. W. Carmichael, and H. Fujiki, Eds., CRC Press, Boca Raton, FL, pp. 103-148.

Howell, T., L. Heintsch, and J. Winter. 2002. Abundance and distribution of cyanobacteria in Hamilton Harbour and adjoining areas of Lake Ontario on September 6 and 20, 2001. Abstract presented to

Environmental Monitoring and Reporting Branch, Ontario Ministry of Environment and Energy. Spring 2002

James, K. J., and I. R. Sherlock. 1996. Determination of the cyanobacterial neurotoxin, anatoxin-a, by derivatization using 7-fluoro-4-nitro-2,1,3- benzoxadiazole (NBD-F) and HPLC analysis with fluorimetric detection. Biomed. Chromatogr. 10: 46-47.

Jochimsen, E. M., W.W. Carmichael, J. S. An, D.M. Cardo, S.T. Cookson, C.E. Holmes, M.B.D. Antunes, D.A. Demelo, T.M. Lyra, V.S.T. Barreto, S. M. F. O. Azevedo, and W.R. Jarvis. 1998. Liver failure and death after exposure to microcystins at a hemodialysis center in Brazil. New Eng. J. Med. 338:873-878.

Kuiper-Goodman, T., I. Falconer, and J. Fitzgerald. 1999. Human Health Aspects. In: "Toxic Cyanobacteria in Water." I. Chorus, and J. Bartram, Eds., World Health Organization, London, pp. 113-153.

Lahti, K., J. Rapala, A.L. Kivimaki, J. Kukkonen, M. Niemela and K. Sivonen. 2001. Occurrence of microcystins in raw water sources and treated drinking water of Finnish waterworks. Water Sci. Tech. 43: 225-228.

Li, R., W. W. Carmichael, Y. Liu, and M. M. Watanabe. 2000. Taxonomic re-evaluation of *Aphanizomenon flos-aquae* NH-5 based on morphology and 16s rRNA gene sequences.

Lippy, E.C. and J. Erb, 1976. Gastrointestinal illness at Sewikley PA. J. Am. Water Works Assoc. 68: 606-610.

Murphy, T.P, A. Lawson, C. Nalewajko, H. Mirkin, L. Ross, K Oguma, and T. McIntyre. 2000. Algal toxins – initiators of avian botulism? Environ. Tox. 15: 558-567.

Myer, G. E., and G. K. Gruendling. 1979. Limnology of Lake Champlain. Lake Champlain Basin Study, New England River Basins Commission, Burlington, VT.

Oshima, Y. 1995. Postcolumn derivatization liquid chromatographic method for paralytic shellfish toxins. J. AOAC Int. 78: 528-532.

Paerl, H. W., R. S. Fulton, P. H. Moisander, and J. Dyble. 2001. Harmful Freshwater Algal Blooms, with an emphasis on cyanobacteria. The Scientific World. 1:76-113.

Parsons, T. R., Y. Maita, and C. M. Lalli. 1984. A Manual of Chemical and Biological Methods for Seawater Analysis. Pergamon Press, NY, pp 101-104.

Pouria, S., A. de Andrade, J. Barbosa, R. L. Cavalcanti, V. S. T. Barreto, C. J. Ward, W. Preiser, G. K. Poon, G. H. Neild, and G. A. Codd. 1998. Fatal microcystin intoxication in haemodialysis unit in Caruaru, Brazil. Lancet. 352:21-26.

Rosen, B. H., A. Shambaugh, M. Watzin, G. Boyer, F. Smith, L. Ferber, C. Eliopoulos, and P. Stangel. 2001. Evaluation of Potential Blue-green Algal Toxins in Lake Champlain. Technical report prepared for Lake Champlain Basin Program, July 2001.

Sawyer, P. J., J. H. Gentile, and J. J. Sasner, Jr. 1968. Demonstration of a toxin from *Aphanizomenon flos-aquae*. Can. J. Microbiol. 14: 1199-1204.

Shambaugh, A., A. Duchovnay, and A. McIntosh. 1999. A Survey of Lake Champlain Plankton. In: "Lake Champlain in Transition: From Research toward Restoration" T. O. Manley, and P. L. Manley, Eds., Amer. Geophys. Union, Wash. DC., pp. 323-340.

Watzin, M.C., A.L. Shambaugh, E.K. Brines, and G.L. Boyer. 2002. Monitoring and evaluation of Cyanobacteria in Burlington Bay, Lake Champlain: Summer 2001. Technical report prepared for Lake Champlain Basin Program, July 2002.

Ward, C. J., and G. A. Codd. 1999. Comparative toxicity of four microcystins of different hydrophobicities to the protozoan, *Tetrahymena pyriformis*. J. Appl. Microbiol. 86:874-882.

World Health Organization. 1998. Guidelines for Drinking Water Quality. 2nd ed. Addendum to Volume 2, Health Criteria and Other Supporting Information. World Health Organization, Geneva.

Yang, X., M. F. Satchwell, and G. L. Boyer . 2001. The identification of anatoxin-a from a toxic blue-green algae bloom in Lake Champlain, USA. Abstracts, Fifth international Conference on Toxic Cyanobacteria, Noosa Lakes, Queensland, AU. July 15, 2001.

Zuniga, D., M. F. Satchwell, and G. L. Boyer. 2002. Development of an acetylcholinesterase inhibition assay for the detection of anatoxin-a(s) in New York waters. Abstracts, SUNY-ESF Graduate and Undergraduate Research Symposium. Syracuse, NY April 28 2002.

WHERE HAVE ALL THE ROTIFERS GONE? ZOOPLANKTON COMMUNITY PATTERNS IN LAKE CHAMPLAIN FROM 1992-2001

Karen J. Carling [1], Timothy B. Mihuc[1], Clifford Siegfried[2], Robert Bonham[3] and Fred Dunlap[3]

ABSTRACT

This study focused on zooplankton community patterns in five Lake Champlain sites from 1992-2001. Analysis of summer (July-August) zooplankton data demonstrates a decline in rotifer species abundance and richness at all study sites. Long-term patterns show a shift in zooplankton community composition from dominance by rotifers, to increased densities of cyclopoids in the late 1990's. Between 1994 and 1997 rotifer species richness had declined by 5-9 species at all study sites. Some rotifer taxa (*Synchaeta* spp., *Filinia* spp., and *Lecane* spp.) have not been observed at any study site since 1996. Rotifer species composition has shifted from taxa indicative of eutrophic conditions to those, characteristic of oligotrophic waters. Coincidental with the shift in community composition is the introduction and proliferation of zebra mussels into Lake Champlain in the mid 1990's. Possible zebra mussel mediated effects or other causes for the observed community shifts are discussed.

1. INTRODUCTION

Lake Champlain represents a unique large lake ecosystem as one of the largest freshwater bodies in the U.S. and as part of the Great Lakes ecosystem connected to the St. Lawrence River drainage. Part of the unique nature of the Lake Champlain ecosystem is the relatively recent invasion of the zebra mussel (*Dreissena polymorpha*) since 1993-some 5 to 10 years later than in the Great Lakes proper. Zebra mussels were abundant in

[1]Lake Champlain Research Institute, Plattsburgh State University, Plattsburgh, New York 12901
[2]New York State Museum, Cultural Education Center, Albany, New York 12230
[3]New York State Department of Environmental Conservation, Region 5, Route 86, Ray Brook, New York 12977

Lake Champlain: Partnership and Research in the New Millennium
Edited by T. Manley *et al.*, Kluwer Academic/Plenum Publishers, 2004

Lake Champlain by 1997at most lake sites (Eliopoulos and Stangel, 2001). In addition, Lake Champlain has yet to receive numerous other invasive species, including several zooplankton (*Bythotrephes cederstroemi, Cercopagis pengoi, Daphnia lumholtzi*) and invasive fishes (alewife- *Alosa psuedoharengus,* erasian ruffe- *Gymnocephalus cernuus,* round goby- *Neogobius melanostomus*). Lake Champlain also represents a very dynamic ecosystem with respect to hydrodynamics (Manley et al., 1999 and this volume).

While past research on biotic communities exists, most notably Meyer and Gruendling's "Limnology of Lake Champlain" (1979), there still exists a vast opportunity to explore and understand biotic communities, particularly zooplankton, in Lake Champlain. Previous investigations of Lake Champlain zooplankton found a community with characteristics of a deep, oligotrophic, temperate lake (Muenscher, 1929, Keen and Potash, 1978, Meyer and Gruendling, 1979, Shambaugh et al., 1999). The purpose of this study was to compare Lake Champlain zooplankton community composition over a 10- year period at five lake sites located along the 200 km north-south lake axis. This study utilized biological samples collected by the New York State Department of Environmental Conservation (NYSDEC) and New York State Museum as part of the Lake Champlain monitoring program from 1992-2001. The New York State Museum and the Lake Champlain Research Institute at Plattsburgh State University of New York performed zooplankton identification. Community level comparisons were made within and among taxa from three major zooplankton groups: Rotifera, Cladocera and Copepoda.

2. METHODS

2.1. Study Sites

The Lake Champlain Monitoring Program has measured water quality variables and collected biological samples for plankton throughout Lake Champlain since 1992. Our comparison of zooplankton communities focused on five study sites representing a gradient from north to south. Site 04 is a south lake site, site 19 is sampled in the main lake, site 34 is considered the Northeast arm; separated from the main lake by the "gut", a narrow shallow channel between North and South Hero islands, site 36 is sampled in the North lake, and site 50 is sampled in Mississquoi Bay. A brief description of each site appears in Table 1.

Table 1. Characteristics of the 5 Lake Champlain study sites.

Station #	Latitude (North)	Longitude (West)	Depth (meters)	Trophic Status*
04	43 57.10	73 24.47	10	Eutrophic
19	44 28.26	73 17.95	100	Mesotrophic
34	44 42.49	73 13.61	50	Mesotrophic
36	44 45.37	73 21.30	50	Mesotrophic
50	44 00.80	73 10.43	04	Eutrophic

*Source- Lake Champlain Basin Atlas (1999)

2.2. Sampling Procedures

Sampling procedures utilized in the biological monitoring conducted by NYSDEC are based on handbooks published by Vollenweider (1969) and Edmondson (1971) for sampling zooplankton. For this study, only mid-summer (July-August) samples were used in all cases except for 1995 where no summer data was available and a later sample was used (collected on Sept. 4, 1995). We compared summer community composition and abundance data for each of the five study sites from 1992-2001.

In the field, zooplankton samples were collected using vertical net tows through the entire water column by, NYSDEC (in particular R. Bonham who has been involved with the collected of field samples since the program began), and NY State Museum staff. Duplicate samples were collected using a 30cm diameter; 153um mesh net, pulled vertically from just above the sediments to the surface. Net retrieval rate was 1 meter per second. Samples were washed into bottles and carbonated water was added to nacrotize the zooplankton. After about five minutes, a formalin-rose Bengal solution was added.

Laboratory identification of all zooplankton taxa was performed (primarily by K. Carling for all samples after 1993 and C. Siegfried prior to 1993) (Siegfried et al, 1995) on 1ml sub samples taken with a Hensen-Stempel pipette and a 1ml Sedgwick-Rafter Cell used for counting. Additional 1ml sub samples were counted until at least 100 zooplankton individuals were counted for each sample. Identification was made to the lowest possible taxon using standard taxonomic keys (Balcer et al., 1984, Ruttner-Kolisko, 1974, and Thorp and Covich, 1991) with verification as needed by regional experts. Zooplankton abundance ($\# \cdot m^{-3}$) was calculated using the subsample density ($\# \cdot ml^{-1}$) to estimate the total density per 100ml sample and converting that data to density per cubic meter using the ratio of the total filtered net volume (net tow depth x surface area of net opening) to the sample volume. Data were graphed as total density ($\# \cdot m^{-3}$) for each site through the ten-year period. In addition Sorenson's coefficient of community similarity was determined for major taxonomic groups (Cladocera, Copepoda, Rotifera) as in Brower et al. (1990). Sorenson's similarity was compared for the years before (1992-1996) and after (1997-2001) *Dreissena polymorpha* colonization using Mann-Whitney U nonparametric *t*-tests (a = 0.05).

3. RESULTS

3.1. Species Richness Trends Over Time

Over 50 separate taxa of zooplankton were identified over the entire study period in Lake Champlain, 13 Copepoda, 17 Cladocera and 19 Rotifera taxa were recorded. Species richness varied throughout the study period with a decline in rotifer richness at all sites during the study period (Figure 1). Some rotifer species that were consistently part of the community were not recorded in Lake Champlain after 1996, such as *Synchaeta* spp., *Filinia* spp., and *Lecane* spp. (Figure 1a.). In the early 1990s rotifers represented 80% of the total zooplankton abundance. Beginning in 1994, and continuing through 2001, rotifer species richness has declined in all study sites. By 1997 rotifer richness in all sites was less than half of what it was in 1994 representing less than 40% and as low as 5% of the total zooplankton community abundance (Figure 1a; Species

Richness, 4; Species Abundance). In contrast copepod and cladoceran communities have been and continue to be highly variable populations over the 10-year period.

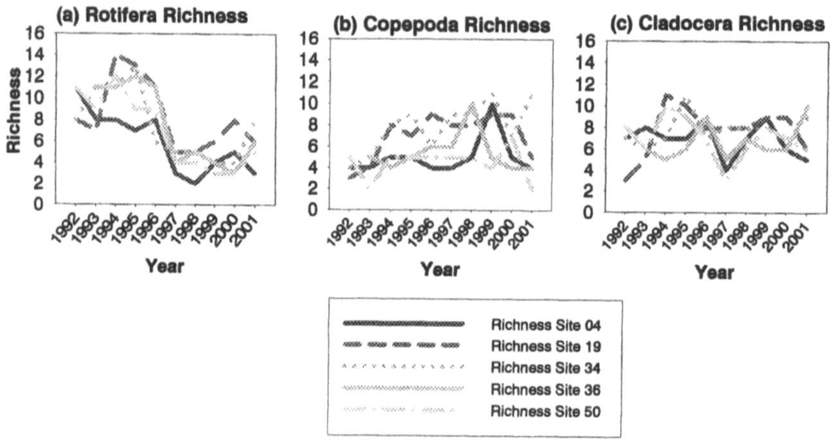

Figure 1. (a) Rotifera, (b) Copepoda, and (c) Cladocera species richness from 1992-2001.

3.2. Zooplankton Community Composition

Copepoda maintained a similar community structure with no significant ($P > 0.05$) change throughout the 10- year period at any study site (Table 2a, Figure 2). Cyclopoids were the most abundant group dominated by *Diacyclops thomasi* and *Mesocyclops edax*. *Acanthocyclops robustus* was also present in high abundance in some sites (e.g., sites 50 and 19, Figure 2). Calanoids were found but in smaller numbers than cyclopoids. *Leptodiaptomus* spp. were the most abundant calanoid with the largest numbers recorded in deep water sites (19 and 36). *Limnocalanus macrurus*, a large bodied calanoid copepod generally found in the hypolimnion of large, deep northern lakes (Balcer et al., 1984) were found in moderate densities (Figure 2).

Cladocera also demonstrated a similar community structure ($P > 0.05$) with no change throughout the 10- year period at any study site (Table 2b). The dominant Cladocera were filter feeders from the Bosminidae and Daphnidae families such as *Bosmina longirostris*, *Eubosmina coregoni*, *Daphnia retrocurva*, *Daphnia mendotae* and *Ceriodaphnia* spp. The limnetic, filter feeder *Diaphanosoma* spp. was also present in relatively high abundance (Figure 3).

Rotifer abundance and community composition, however, indicate major shifts in community composition from 1992-2001. Rotifers exhibited a consistent decrease in overall abundance and species richness over the 10-year study period (Figure 1a. Species Richness, Figure 4. Species Abundance). The rotifer community of Lake Champlain has shifted from dominance by *Polyarthra vulgaris* in the early 1990s to *Asplanchna* spp. and the colonial species *Conochilus unicornis* in the late 1990s. The spiny rotifers *Keratella cochleris* and *Kellicottia longispina* were present in high abundance throughout all study years. The rotifers *Lecane* spp, *Synchaeta* spp. and *Filinia* spp. were found only in samples before 1996 (Figure 4). Results of the MannWhitney U *t*-test ($P < 0.05$) show that the rotifer community of Lake Champlain has changed since the invasion of zebra mussels (Table 2c).

Table 2. Means based on Sorenson's community similarity

Table 2a. Copepod mean similarity

Site Number	Pre-zebra mussels	Post-zebra mussels
04	0.597	0.543
19	0.67	0.684
34	0.374	0.549
36	0.634	0.546
50	0.675	0.511

Table 2b. Cladoceran mean similarity

Site Number	Pre-zebra mussels	Post-zebra mussels
04	0.601	0.661
19	0.638	0.581
34	0.569	0.609
36	0.583	0.659
50	0.591	0.559

Table 2c. Rotifer mean similarity

Site Number	Pre-zebra mussels	Post-zebra mussels
04	0.435	0.273
19*	0.846	0.264
34*	0.619	0.297
36*	0.631	0.141
50*	0.643	0.189

*($P<0.05$)

3.3. Trends Among Sites

Site 19 represents the deepest lake site at a depth of 100m (Table 1). It consistently had the highest zooplankton density throughout the 10-year period. Total zooplankton density was lowest in the two shallow sites (04 and 50) for all three groups of plankton (Cladocera, Copepoda, Rotifera). Structural characteristics of the copepod community from shallow to deep sites shifted in dominance from *Mesocyclops edax* to *Diacyclops thomasi* respectively. The rotifer *Kellicottia longispina* was found in low densities in shallow sites (04 and 50) as compared to much greater densities in the deeper water sites (19, 34, and 36).

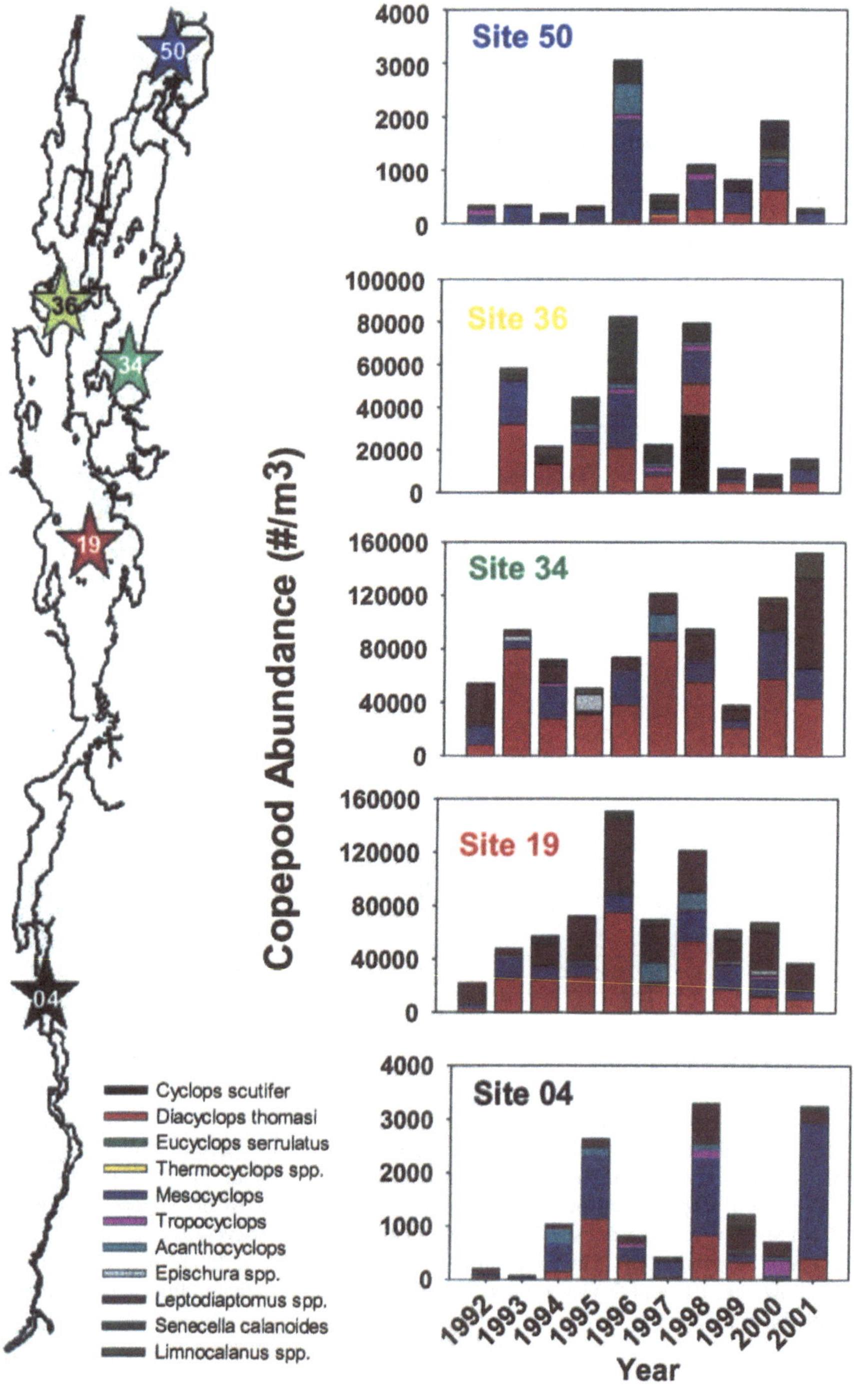

Figure 2. Copepod abundance through time (1992-2001) at each of the five study sites

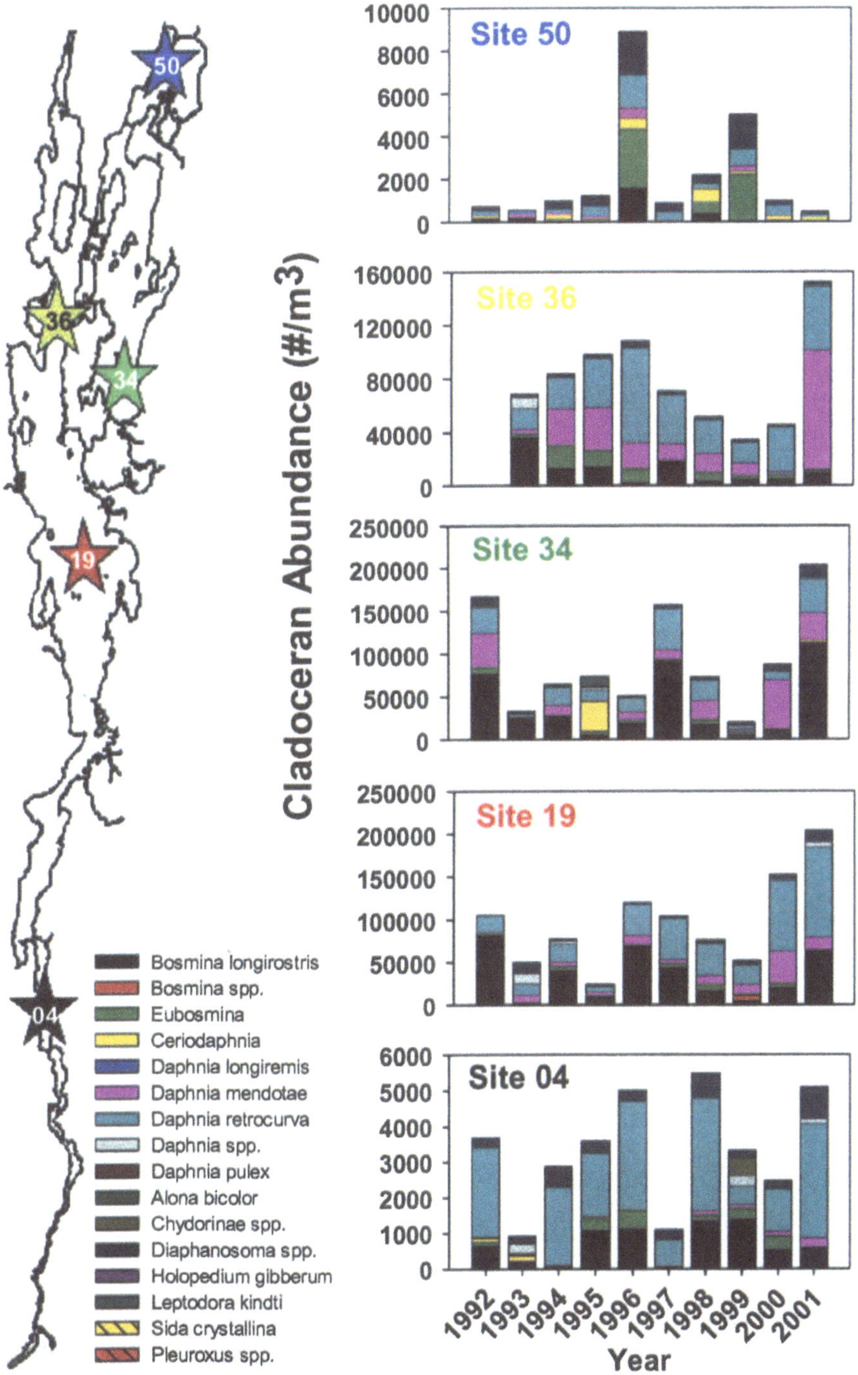

Figure 3. Cladoceran abundance through time (1992-2001) at all five study sites.

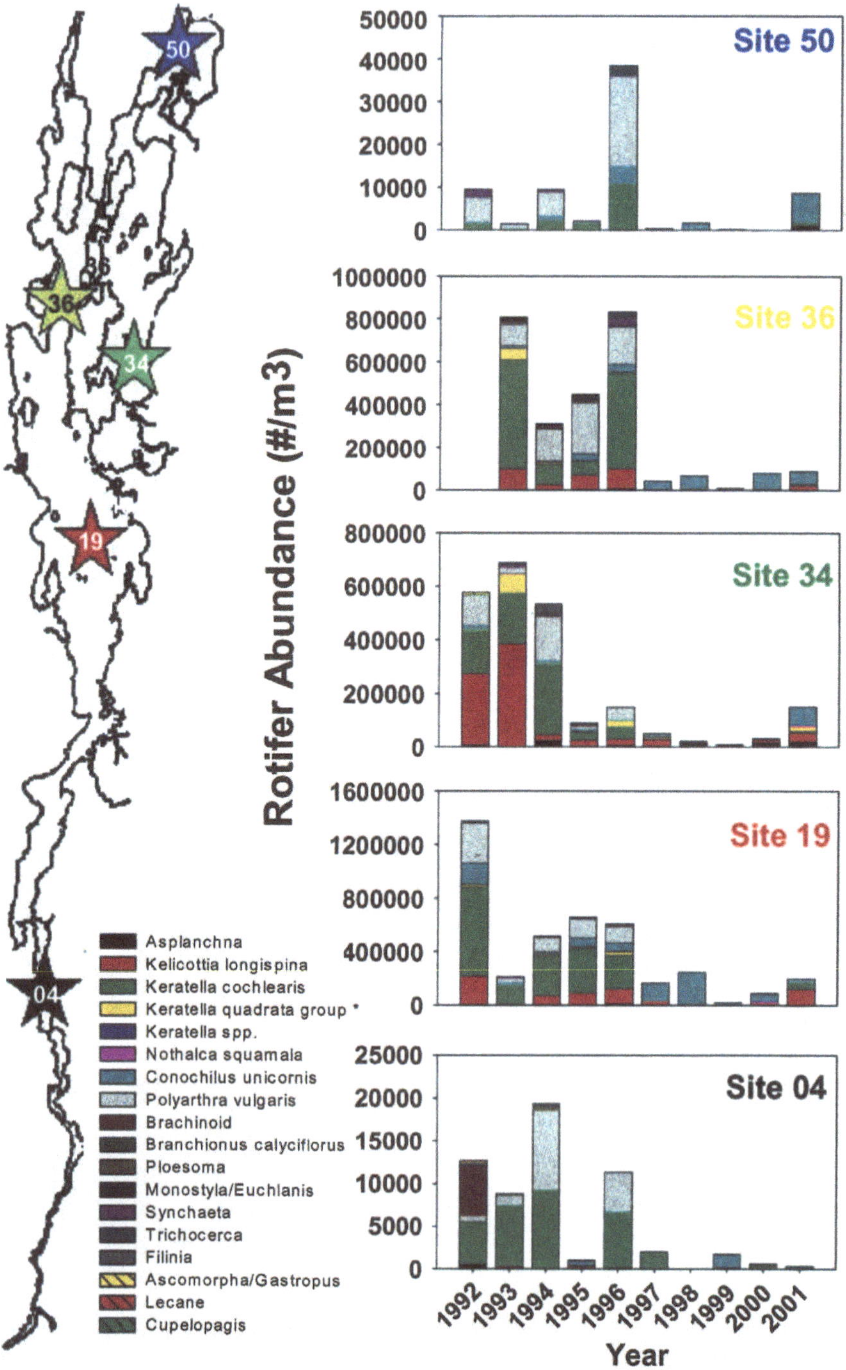

Figure 4. Rotifer abundance through time (1992-2001) at each of the five study sites.
*Ruttner-Kolisko (1974)

4. DISCUSSION

Lake Champlain zooplankton communities in the early 1990s are very similar to those found in previous studies for all three major taxonomic groups: Rotifera, Copepoda, and Cladocera (Muenscher, 1930, Meyer and Gruendling, 1979, and Shambaugh et al., 1999). Microzooplankton, specifically rotifers, were the most abundant zooplankton in Lake Champlain throughout the study period (1992-2001). The abundance of rotifers throughout this study is consistent with Meyer and Gruendling's (1979) historical comparison of zooplankton, which determined that qualitatively the crustacean zooplankton communities of Lake Champlain had not changed in 45-50 years; our findings add another 20 years to the qualitative consistency. Copepod and cladoceran communities remain similar to previous studies; however, the observed shift in rotifer community composition has ended a nearly 70-year trend.

In 1930, Muenscher reported that *Polyarthra vulgaris* was a widely distributed and dominant rotifer. Similar results were found by Keen and Potash (1978), and in our data until 1997. From 1997-2000, *Polyarthra vulgaris* was not identified in any sites of Lake Champlain, but in 2001 low numbers of *Polyarthra vulgaris* were recorded. From 1997-2001 *Conochilus unicornis* was the most dominant rotifer at all study sites. *Polyarthra vulgaris* is a eutrophic indicator in other lake systems (Makarewicz, 1993). Rotifers *Synchaeta* spp. and *Filinia* spp. which have not been identified in any samples since 1996 are also indicative of eutrophic conditions, while *Conochilus unicornis*, currently the most abundant rotifer, is characteristic of oligo-mesotrophic waters (Makarewicz, 1993).

Consistent dominance by rotifers in this and previous studies suggests that intense predation on larger bodied zooplankton (cladocerans and copepods) by planktivorous fish in Lake Champlain could be a major factor in structuring zooplankton communities such that "communities are dominated by small-bodied forms such as rotifers, bosminids, and cyclopoid copepods." (Mazumder et al., 1992). However, other systems with similar declines in rotifer populations after the invasion of zebra mussels have not related the changes to increased planktivory in fact, Lake Erie had a decline of planktivores after the invasion (MacIsaac et al., 1995).

4.1. Zebra mussel invasion

Since the invasion of zebra mussels (1st found in 1993) the zooplankton community composition of Lake Champlain has significantly ($P < 0.05$) changed. These shifts in the community can be directly attributed to a decline in rotifers (Table 2) while larger bodied, stronger swimming zooplankton such as copepods and cladocerans have remained similar in community structure (Table 2a and 2b). MacIsaac et al. (1995) reported similar results after the zebra mussel invasion of Lake Erie; small rotifer density declined by 75%, while abundance of larger zooplankton, mainly copepod and cladoceran was not changed ($P > 0.05$).

Zebra mussels are likely to have contributed to the shifts in the Lake Champlain rotifer community. MacIsaac et al. (1995) directly correlated the reduction of rotifers to suppression by zebra mussels; rotifers were cleared from suspension at the same rate as phytoplankton, while the clearance rate of larger bodied zooplankton was zero. Estimating an average filtration rate of 1L per day, (Bunt et al., 1993), a density of 100,000 adult zebra mussels colonizing 1m of the entire Lake Champlain shoreline zebra mussels could filter the entire volume of the lake, 25.8 km^3 (Meyer & Gruendling, 1979) once every 268.75 days. This is a conservative estimate given that adult zebra mussels have been recorded in densities of 1,000,000+/m^2 at the Burlington Boat House (Eliopoulos & Stangel, 1998) and zebra mussels often inhabit far beyond 1m of the lake shoreline. This supports the conclusion that, direct filtration of rotifers by zebra mussels is a plausible explanation for the decline of rotifers in Lake Champlain.

Pace et al. (1998) note that factors such as "flow and temperature were unrelated to the changes in zooplankton communities" and report that zebra mussels suppress zooplankton by direct consumption; zebra mussels select for particles between 5 and 45um (MacIsaac et al., 1995) a system dominated by small zooplankton, such as Lake Champlain, would suffer large declines when invaded.

Rotifer community shifts could also be related to patterns in phytoplankton composition in Lake Champlain. Caraco et al. (1997) suggest that phytoplankton biomass may have increased in the Hudson River after zebra mussel invasion due to the increase in water transparency. This may result in alterations to phytoplankton community composition such as shifts in species, which is related to shifts in zooplankton composition. Heavy grazing by bivalve mussels may favor inedible species such as cyanobacteria. Recent work on Lake Champlain phytoplankton suggests that Lake Champlain rotifers were less efficient at harvesting green algae versus diatoms, blue-green algae, and cryptophytes (Levine et al., 1999). The observed decline in rotifers, therefore, could reflect an increase in green algae in Lake Champlain. However, little evidence exists to support the idea that the Lake Champlain phytoplankton assemblage has shifted to green algae. In fact, diatoms still dominate the phytoplankton community in 2000 and 2001 (unpublished data) as they did in earlier studies (Muenscher, 1930, Meyer and Gruendling, 1979, Shambaugh et al., 1999).

The transparency of the water column is often affected by zebra mussel filtration; Secchi penetration data from Lake Champlain used to measure water transparency was highly variable throughout the study period. However, there was an increase in secchi depth at site 04 from 0.8m in 1993 to 2.4m by 1998 (R. Bonham, unpublished data). Due to the depths of main lake sites 19 and 36, zebra mussel filtering has likely had less impact on water transparency than at shallow sites 04 and 50. Wu and Culver (1991) found in a study of Lake Erie that Secchi transparency was only increased for a short period of time after adult zebra mussel colonization, while rotifer populations remained depleted.

Rotifers are inclined to respond more quickly to environmental and water quality changes than copepods and cladocerans (Makarewicz, 1993). The shifts in rotifer community from species characteristic of eutrophic waters in the early 1990s, to species characteristic of oligotrophic waters following 1996, indicate that zebra mussels may have altered the water quality of Lake Champlain, resulting in reduced planktonic productivity from 1997-2001.

Comparing the results of studies by MacIsaac et al. (1992, 1995) on Lake Erie and Carco et al. (1998) on the Hudson River to the zooplankton data from Lake Champlain indicate that zebra mussels through direct or indirect suppression have altered the rotifer community of Lake Champlain. The water quality and biological monitoring data collected by the Lake Champlain monitoring program will serve as an important tool in understanding zebra mussels mitigated affects along with other dynamics of the lake as it changes. Monitoring programs such as this are valuable in bringing about awareness of changes that may have occurred in lakes before and after invasions of exotic species, such as zebra mussels. Further analysis of the 10-year patterns observed herein will likely provide useful indicators, both physical and biological, for long-term trajectories in the integrity of the Lake Champlain ecosystem.

5. ACKNOWLEDGEMENTS

Funding for this study was provided by, the Lake Champlain Basin Program through the U.S. Environmental Protection Agency and the New York Department of Environmental Conservation. Many thanks to the employees of the Lake Champlain Research Institute for hours of excellent work in the field and in the lab: Ian Ater, Adam Bouchard, Trevor Carpenter, Virginia Damuth, Jeff Jones, Marti Kroll, Dustin Lewis, Shannon Margrey, Tracy McGregor, Stephanie Stone and David Welch. Special thanks to the reviewers and those who provided comments on the manuscript especially Muriel LaFave and Mark Malchoff.

6. REFERENCES

Bacler, M., Korda, N., and Dodson, S. A., 1984, *Guide to the Identification and Ecology of the Common Crustacean Species.* University of Wisconsin Press, pp. 74-101.

Bonham, R., 2002, Personal communication. New York State Department of Environmental Conservation, Ray Brook, New York.

Bower, J. E., Zar, J. H., and von Ende, C. N., 1990, *Field and Laboratory Methods for General Ecology* (3rd ed.), Wm. C. Brown Publishing, Dubuque, IA, p. 167.

Bunt, C. M., and MacIsaac, H. J., 1993, Pumping rates and projected filtering impacts of juvenile zebra mussels (*Dreissena polymorpha*) in western Lake Erie, *Can. J. Fish. Aquat. Sci.* **50**: 1017-1022.

Caraco, N. F., Cole, J. J., Raymond, P.A., Strayer, D.L., Pace, M.L., Findley, S.E.G., and Fischer, D.T., 1997, Zebra mussel invasion in a large turbid river: phytoplankton response to increased grazing, *Ecology.* **78**:588-602.

Edmondson, W. T., (ed.), 1971, *A Manual on Methods for the Assessment of Secondary Productivity in Fresh Waters*, Blackwell Sci. Publ., Oxford, p. 358.

Eliopoulos, C., and Stangel, P., 1994-2001. Lake Champlain Zebra Mussel Monitoring Program Reports, Vermont Dept. Environ. Conserv., Waterbury, VT.

Gannon, J. E., and Stemberger, R. S., 1978, Zooplankton (especially crustaceans and rotifers) as indicators of water quality, *Trans. Amer. Microsc. Soc.* **97**:16-35.

Keen, R. E., and Potash, M., 1978, Determination of the secondary production and population dynamics of the zooplankton of the Lake Champlain ecosystems. Report to the Vermont Water Resources Research Center and U.S. Department of Interior, Office of Water Resources and Technology, Project No. A-028-VT, p. 77

Levine, S. N., Shambaugh, A. D., Pomeroy, S. E., and Braner, M., 1997, Phosphorus, Nitrogen, and Silica as controls on phytoplankton biomass and species composition in Lake Champlain, *J. Great Lakes Res.* **23**:131-148.

Levine, S. N., Borchardt, M. A., Braner, M., and Shambaugh, A. D., 1999, The impact of zooplankton grazing on phytoplankton species composition and biomass in Lake Champlain. *J. Great Lakes Res.* **25**: 61-77.

MacIsaac, H., Sprules, G., Johannsson, O., and Leach, J. H., 1992, Filtering impacts of larval and sessile zebra mussels (*Dreissena polymorpha*) in western Lake Erie, *Oecologia.* **92**:30-39.

MacIsaac, H., Lonner, C. J., and Leach, J. H., 1995, Suppression of microzooplankton by zebra mussels: importance of mussel size, *Freshwater Biology.* **34**: 379-387.

Makarewicz, J. C., 1993, A lakewide comparison of zooplankton biomass and its species composition in Lake Erie, 1983-87, *J. Great Lakes Res.* **19**(2):257-290.

Manley, T., Hunkins, J. H., Saylor, J. H., Miller, G. S., and Manley, P. L., 1999, Aspects of summertime and wintertime hydrodynamics of Lake Champlain, in: *Lake Champlain in Transition: From Research*

Toward Restoration, T. O. Manley and P. L. Maley, ed., American Geophysical Union, Washington, DC, pp. 67-115.

Manley, T. O., and Maley P. L., 1999, Preface, in: Lake Champlain in Transition from Research Toward Restoration, T. O. Manley and P. L. Maley, ed., American Geophysical Union, Washington, DC, p. vii.

Manley, T.O. This volume. Hydrodynamics of the south main lake and south lake, Lake Champlain, in: *Lake Champlain in Transition*: , T. O. Manley, P. Manley, and T. B. Mihuc, ed., Kluwer Publishing, City.

Mazumder, A., Lean, D. R. S., and Taylor, W. D., 1992, Dominance of small filter feeding zooplankton in the Lake Ontario foodweb, *J. Great Lakes Res.* **18**(3): 456-466.

Muenscher,W. G., 1930, Plankton studies in the Lake Champlain watershed, in: *A Biological Survey of the Lake Champlain Watershed*, N.Y. State Conservation Dept., Supplement to the 19th Annual Report.

Myer, G. E., and Gruendling, G. K., 1979, *Limnology of Lake Champlain*, Lake Champlain Basin Study, New England River Basins Commission, Burlington, VT.

Pace, M., Findlay, S., Fischer D., 1998, Effects of an invasive bivalve on the zooplankton community of the Hudson River, *Freshwater Biology*. **39**:103-116.

Ruttner-Kolisko, A., 1974, *Plankton Rotifers Biology and Taxonomy* (G. Kolisko, Trans.), E. Schweizerbart'sche Verlangsbuchhandlung, Stuttgart, p. 76.

Shambaugh, A., Duchovnay, A., and McIntosh, A., 1999, A survey of Lake Champlain's plankton, in: *Lake Champlain in Transition: From Research Toward Restoration*, Water Science Application 1, American Geophysical Union, Washington, pp. 323-340.

Siegfried, C., et al., 1995, *Long-term Water Quality and Biological Monitoring Program for Lake Champlain*. Interm data report.for Lake Champlain management conference, Water Management Division, U.S. Environmental Protection Agency.

Thorp, J., and Covich, A., (ed.), 1991, *Ecology and Classification of North American Freshwater Invertebrates*, Academic Press, San Diego, California.

Vollenweider, R. A., (ed.), 1969, *A Manual on Methods for Measuring Primary Production in Aquatic Environments*, Blackwell Sci. Publ., Oxford, p. 211.

Wu L., and Culver, D., 1991, Zooplankton grazing and phytoplankton abundance: an assessment before and after invasion of Dreissena polymorph,. J. Great Lakes Res. 17(4): 425-436.

USE OF GIS AND REMOTE SENSING TO DEVELOP INDICATORS OF PHOSPHORUS NON-POINT SOURCE POLLUTION IN THE PIKE RIVER BASIN

Julie Deslandes*, Aubert Michaud, and Ferdinand Bonn

ABSTRACT

The Pike River is amongst the main watercourses discharging into Missisquoi Bay, a spur of Lake Champlain extending into Quebec. Its watershed drains some 630 km^2, of which 99 km^2 are located in Vermont. The objective of this study is to adapt agrienvironmental indicators to the landscape scale to provide an holistic perspective of phosphorus sources and transfer in the Pike River watershed. The development of agrienvironmental indicators of non-point P pollution was supported by remote sensing, spatial integration of data and the Soil and Water Assessment Tool (SWAT). These indicators were evaluated on subwatersheds of about 5 km^2. The validation of the indicators at the subwatershed scale was supported by water quality records from the monitoring of 18 tributaries and sections of the Pike River. The agricultural land use, topographic attributes, the runoff SCS curve number, the mass balance of P and the P soil saturation were significantly correlated with P concentrations measured under peak streamflow conditions, in response to spring and fall long-duration precipitations, snowmelt and saturated or frozen soil conditions. It was found that 73 % of the variation for bioavailable P sampled within the hydrological network is explained by the P mass balance and subwatershed agricultural land use.

1. INTRODUCTION

Intensification and specialization of agricultural systems in Quebec over the last three decades have increased pressure on soil and water resources quality. Management of nonpoint source contamination of surface waters has become a major issue in numerous rural basins.

Phosphorus dynamics in the agricultural, terrestrial and aquatic ecosystems of the Pike River Bay watershed have been researched at various scales of study, from

* Julie Deslandes, Research and Development Institute for the Agri-Environment (IRDA), 2700 Einstein street, Ste-Foy (Quebec) Canada, G1P 3W8

monitored runoff plots to experimental watersheds (Jamieson *et al.*, 2003; Michaud *et al.*, 2002). The objective of this study is to adapt agrienvironmental indices to the landscape scale to provide an holistic perspective of phosphorus sources and transfer to land use planners, extension personnel and the farming community of the Pike River watershed. The basin, a tributary of Missisquoi Bay, has been described as particularly vulnerable to agricultural nonpoint sources phosphorus exports (Budd and Meals, 1994; Hegman *et al.*, 1999).

The development of spatially explicit indices adapted to large watersheds presents important challenges in terms of the acquisition, management and spatial integration of data. The site level representation of the vulnerability of lands to phosphorus losses is built upon the concept of the P-index (Lemunyon and Gilbert, 1993). The P-index is a simple indexing method which consists of attributing a relative weight to different key factors that reflect the vulnerability to non-point source P losses (at the field-scale): erosion, surface runoff, soil richness, rates and management of organic and inorganic fertilizers. Landscape level indices, though developed at a coarse resolution, remain targeted on the delineation of source factors (summary of inputs, P saturation of soil) and P transport factors (erosion, surface runoff). Their evaluation and update are based on existing geo-referenced databases or imply reasonable acquisition costs and time to insure the easy transfer of the approach developed. These limitations meant that the derivation of these indices required the subdivision of the Pike River watershed into sub-watersheds of about 5 km^2.

2. METHODS

Study site

The Pike River watershed is a sub-watershed of the Richelieu River. The Pike River is among the main watercourses discharging into Missisquoi Bay, a spur of Lake Champlain extending into Quebec (Figure 1). The river has been identified as one of the principal contributors of P to the Missisquoi Bay (Hegman *et al.*, 1999). Its watershed drains some 630 km^2, of which 99 km^2 are located in Vermont. The watershed can be split into two portions of equal area. The headwaters of the watershed (from the Canada-USA border to Bedford) are situated in the Appalachian piedmont where elevation ranges from 20 to 710 m and slope varies from 2.8 to 38 degrees. This portion of the watershed does not lend itself to intensive agriculture with a land cover of woods, orchards and forage plants. The other portion, extending downstream from Bedford, comprises the majority of the watershed's agricultural activities as well as its industrial core. The majority of the population in the region, which has remained near 9000 since 1979 (Caumartin and Vincent, 1994), is in this portion of the watershed. The elevation of this part of the watershed, varies from 20 to 130 m with a mean slope of 0.6 degrees.

In the watershed as a whole, loam is the dominating soil texture (covering 32 % of the total area). Some 280 km^2 or 45 % of the watershed is devoted to agricultural land use. Of these cultivated lands, slightly more than half are planted to wide row-crops (made up mostly of corn [*Zea mays* L.] and to a lesser extent by forage crops).

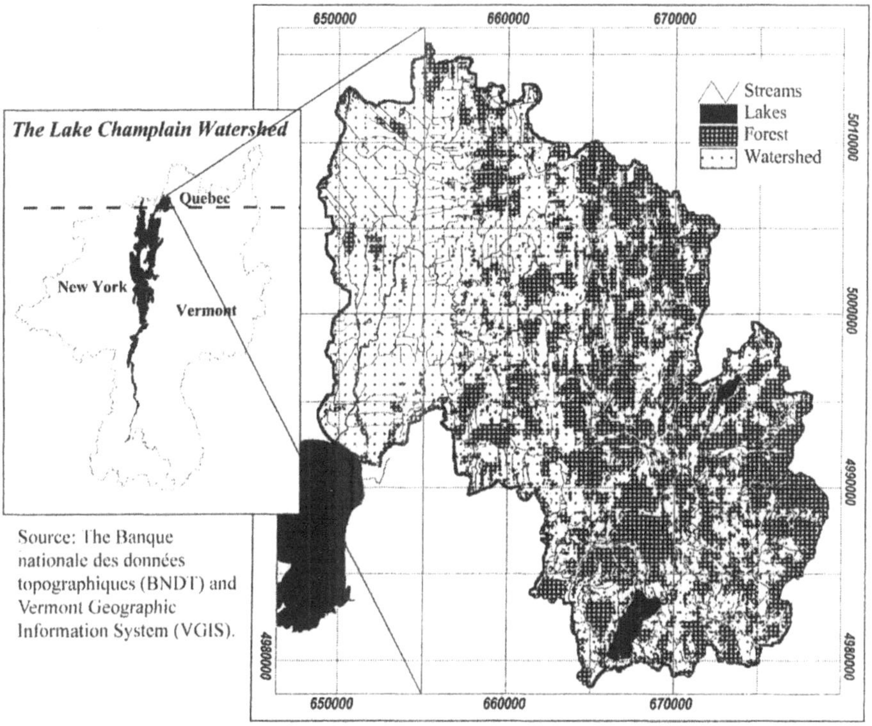

Figure 1. The Pike River Watershed of the Lake Champlain Basin

Study approach

Agrienvironmental Indicators development is based on three-steps approach, namely:

- evaluation of transport parameters from acquisition, spatial integration, and modeling of biophysical and land use data;
- estimation and localization of source variables (phosphorus mass balances and terrestrial stocking); and
- validation of the indicators through event sampling of stream water quality (TSS, TP, RSP, BioP).

Modeling of phosphorus transport Indicator

Considering the scale of study, a modeling approach was selected for spatial representation of transport parameters. Model selection was based on following criteria:

- ability to handle a large watershed, heterogeneous in land use, soil type and landforms;
- ability to operates without stream hydrometric gauging;
- ability to assess erosion and surface runoff on an annual basis, and model Quebec's winter conditions;
- user friendliness and transferability to other biophysical environments.

The Soil and Water Assessment Tool (SWAT) developed by Arnold *et al.* (1993) was selected. SWAT is a distributed model which operates on a continuous basis, with a daily time-step. It was developed to predict with reasonable accuracy the effect of management alternatives on the generation of sediment and nutrient loads in ungauged watersheds.

SWAT is an enhancement of the Soil and Water Integrated Model (SWRRB; Arnold *et al.*, 1990) which allowed the modeling of large heterogeneous rural watersheds made up of several hundred sub-watersheds. In the SWAT model the water budget subroutine controls all physical processes modeled within the watershed. Watershed hydrology simulation occurs in two phases:

- Land phase: the quantity of water, sediments, pesticides and nutrients reaching the main channel from surface flows is determined,
- Stream phase: the movement of these compounds through the hydrographic network to its outfall is characterized.

The SWAT model estimates erosion using the Modified Universal Soil Loss Equation (MUSLE; Williams, 1975) and runoff by the SCS (Soil Conservation Service) curve number method (SCS-USDA, 1972). Modeling of erosion and surface runoff processes is supported at the hydrological response units (HRU). Individual HRU's present unique combinations of spatially-integrated land use, soil and topography characteristics. HRU are further spatially integrated at designated sub-watershed scale for simulation. The current study only simulated the land phase in the SWAT model.

Data acquisition and entry

The input parameters required by SWAT to model the movement of water, sediments, and nutrients as well as plant growth are: hydrography, topography, soil properties, land use, cultural practices and meteorological record.

Hydrography. A vector map of the hydrographical network was obtained from the Quebec topographic database (BQDT) and the *Vermont Center for Geographic Information* (Linear surface water features from RF 1:20,000 orthophotos - Chittenden, Franklin and Grande Isle counties).

Topography. To obtain the most accurate digital elevation model (DEM), a variety of data sources for elevation were obtained, digitized and integrated into the GIS. Points from the geodesic network, the National and Quebec topographic databases, municipal drain maps from Agriculture, Fisheries and Food department of Quebec, as well as from the National Elevation Database (NED) of the United States Geological Service (USGS) were used. These data were integrated into the Arcview software package and the DEM estimated by the spherical kriging method for the whole study area (pixel size, 30 m x 30 m, see Figure 2a). The overall DEM precision ranges between 2 to 5 meters.

Soil properties. For the Quebec portion of the Pike River watershed, 1: 63,360 digital soils maps were integrated in the GIS and, the physicochemical inputs used by the SWAT model were documented. Individual soil series properties, including particle size, hydraulic conductivity, bulk density, percent organic content and percentage of rock fragments were also drawn from regional data of the National soil quality inventory (MAPAQ, 1990). Available water data were developed from the Ranges of AWC in relation to soil texture and organic matter table in the National Soil Survey Handbook (USDA, 2001). Soil erodibility factors from Bernard (1996) developed for typical watershed soil series were estimated using the Wischmeier (1971) monograph. All data were converted in English units (Foster *et al.*, 1981) for SWAT operation. Lastly, the albedo was set at 0.01 for all soil series.

For the American portion of the watershed, available soil series data were at 1:126,720 and 1:20,000 scales. Notwithstanding the difference in scale, numerical data at the 1:20,000 scale and associated soil characteristics were acquired from the State Soil Geographic database (STATSGO) and combined with the Quebec data in the GIS.

Land use. An image Landsat ETM+ from July 5, 1999 was used to establish land use patterns over the area of land within the watershed. Radiometric, atmospheric and geometric corrections were applied to the image. The classes adopted were: corn, soybean (*Glycine max* Merr.), cereal crops, pasture/prairie, wetlands, forest, and orchards. Maximum likelihood method was used in this classification (Figure 2b). The image classification rate for the image was 78 % (Cattai, 2002).

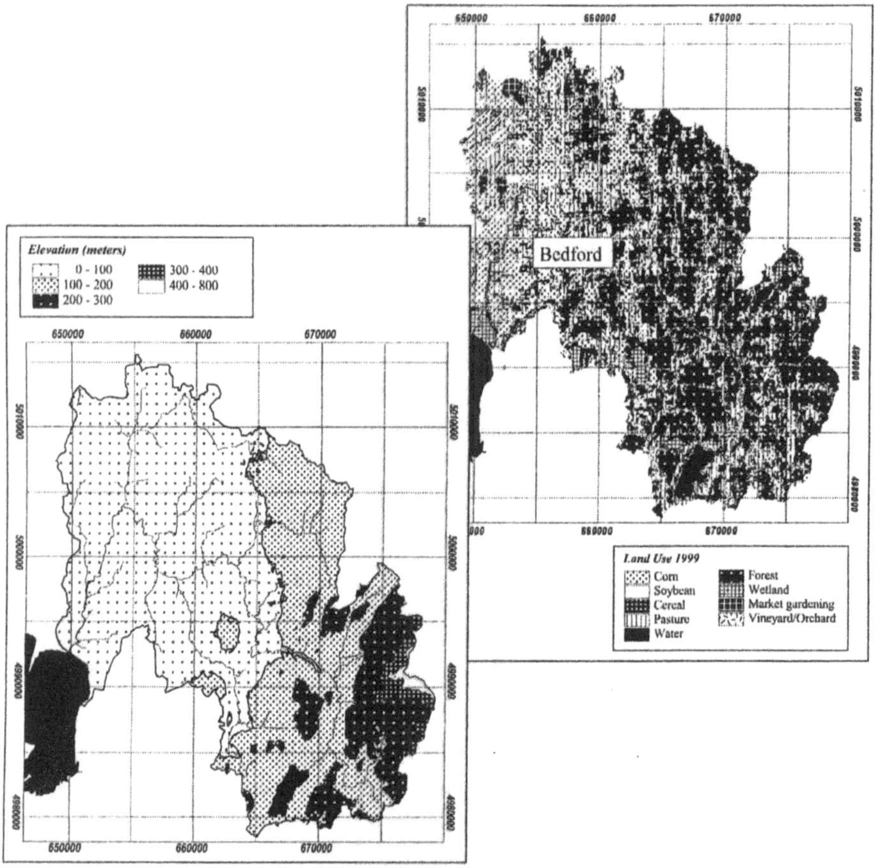

Figure 2. (a) Digital Elevation Model DEM ; (b) Land use of the Pike River Watershed

Meteorological data. SWAT requires daily precipitation and temperature data, as well as solar irradiation, wind velocity and relative humidity data. Daily precipitation and temperature data (1999-2001) were obtained from the Philipsburg station located at 45°02' N and 73°0 45' W. The weather station is located about 10 km southwest of the city of Bedford, roughly centred with respect to the Pike River watershed. Solar irradiation, wind speed, and relative humidity data from the Plattsburg, N.Y. station, located about 60 km SW from the centre of the watershed, provided the other climatic inputs used in SWAT.

Modeling of phosphorus source Indicators

Variables used to describe phosphorus sources were derived only for the Quebec portion of the watershed.

Phosphorus soil saturation

The percent phosphorus saturation of the soil is an expression of phosphorus availability, and the level of environmental risk associated with water pollution (Giroux and Tran, 1996). Recent studies have shown that the solubility and desorption of P from soil is inversely proportional to the soil's ability to bind P (Giroux and Tran, 1996; Pote *et al.*, 1996). This parameter is calculated as:

$$\textit{soil P binding capacity} = \frac{\textit{soil P content}}{2.24 \times \textit{soil Al content}} \quad (1)$$

where the P and Al contents (kg ha^{-1}) are measured from samples extracted with Mehlich-3 extractant.

Soil analysis databases from the Quebec Ministry of Agriculture, Fisheries and Food (MAPAQ) (1998-2000) and Agridirect (1995-1996), a private company, were used to document the soil P saturation on the Pike River watershed. In total, 1328 analyses of soil P saturation (P Mehlich-3/Al Mehlich-3) were spatially integrated at municipal scale.

Soil values from the two sets of samples 1995-1996 and 1998-2000, were not statistically different (ANOVA, $\alpha = 0.05$). Table 1 represents the median, mean and standard deviation of soil P saturation estimated for each municipality of the watershed. Because of the high standard deviations, the median values were used to compute

Table 1. Median, mean and standard deviation of soil P saturation (Mehlich-3 P/Al) measured for municipalities located within the Pike River watershed

Municipality	Soil P saturation %			
	n[1]	Median	Mean	Std. Dev.[2]
Bedford/ Bedford Township	83	8.8	11.9	21.0
Dunham	212	4.2	6.3	6.2
Frelighsburg	57	4.0	6.0	5.4
Paroisse de Notre-Dame-de-Stanbridge	88	7.4	10.5	9.0
Rainville	68	14.2	14.5	8.1
Saint-Alexandre	214	9.0	10.2	6.2
Saint-Amrand	134	6.2	7.0	5.7
Sainte-Brigide-d'Iberville	86	7.0	8.7	6.3
Sainte-Ignace-de-Stanbridge	131	11.8	14.1	10.8
Sainte-Sabine	55	6.8	8.3	6.2
Saint-Pierre-de-Véronne-à-Pike-River	103	7.3	8.5	5.6
Saint-Sébastien	45	9.5	10.6	6.7
Stanbridge East	27	7.7	8.3	5.6
Stanbridge Station	25	6.7	7.8	7.5

[1] *n* = number of samples, [2] Std. Dev. = standard deviation

indicators of terrestrial phosphorus at subwatershed scale Soil saturation indicators, at subwatershed scale, were derived by averaging the municipal-scale data, spatially intersecting the sub-basin, weighted for respective cropped area.

P Mass balance

The soil surface phosphorus mass balance estimates the difference between the nutrient contributions of organic and inorganic fertilizers, or other inputs that enter the soil and crop nutrient uptake (OCDE, 1997; Michaud and Beaudet, 1998). Farm census data (MAPAQ, 2001) allow a spatially explicit accounting of plant uptake as well as organic and inorganic inputs. Spatial data treatment and representation at the subwatershed scale respected the confidentiality of records and was carried out in three steps:

1) The spatial representation of plant P uptake of phosphorus (kg) was derived from the census data crop acreages through delineation of circular zones of influence. The overlap of these zones of influence with subwatershed boundaries was then established. Uptake (kg) was then reassigned to the subwatersheds according to the respective overlap.

2) A similar approach using circular zones to delineate for phosphorus inputs (kg) was employed in conjunction with the following equation:

$$\text{Zone of influence} = \frac{\text{on-farm inorganic inputs} + \text{on-farm organic inputs}}{\text{agricultural vs. watershed area weighted inputs}} \quad (2)$$

The maximum radius of influence was set at 5 km, which corresponds to the maximum distance traveled for surplus manure land spreading within the region.

To take into account the export of manure, farm census secondary locations (production units on different sites) were used. The spatial allocation of phosphorus inputs was then based on an even distribution of the primary centroid data to the corresponding secondary centroids, because production data for the latter were not available. The spatial representation of phosphorus inputs was finally based on overlapping zones of influence with subwatershed boundaries.

3) Thus, for each subwatershed, the overall P budget is calculated as:

$$\text{inputs} = \frac{[\text{subwatershed inputs (kg)}] - [\text{subwatershed crop uptake (kg)}]}{\text{cropped area of the subwatershed (ha)}} \quad (3)$$

Validation

The validation of the transport and source was supported by water quality records from the monitoring of 18 tributaries and sections of the Pike River. Stream water was sampled (Figure 3) ten times during typical peak flows during spring and fall long-duration precipitation event, snowmelt, and saturated or frozen soil conditions.

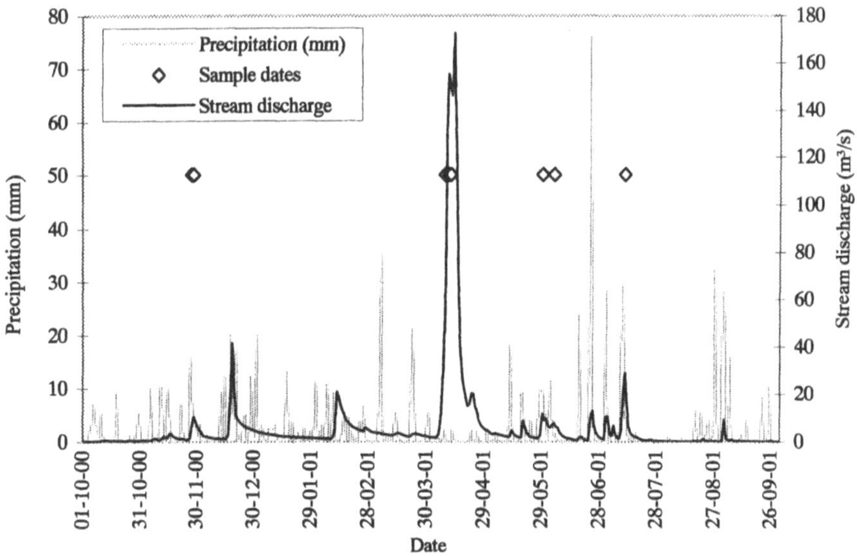

Figure 3. Sampling programmes 2001-2002 on the Pike River Watershed

The sampling station network was inspired by Caumartin and Vincent (1994), supplemented by a few gauging stations (Figure 4). It is assumed that the phosphorus concentration revealed by event sampling, under long and intense hydrological events, expresses the relative importance of stream loadings associated with each subwatersheds. This assumption is based on the flow-dependency of phosphorus concentration data documented for the study area (Michaud and *al*, 2002). Head-to-head comparisons of stream phosphorus concentration and subwatershed indicators provided the basis for validation.

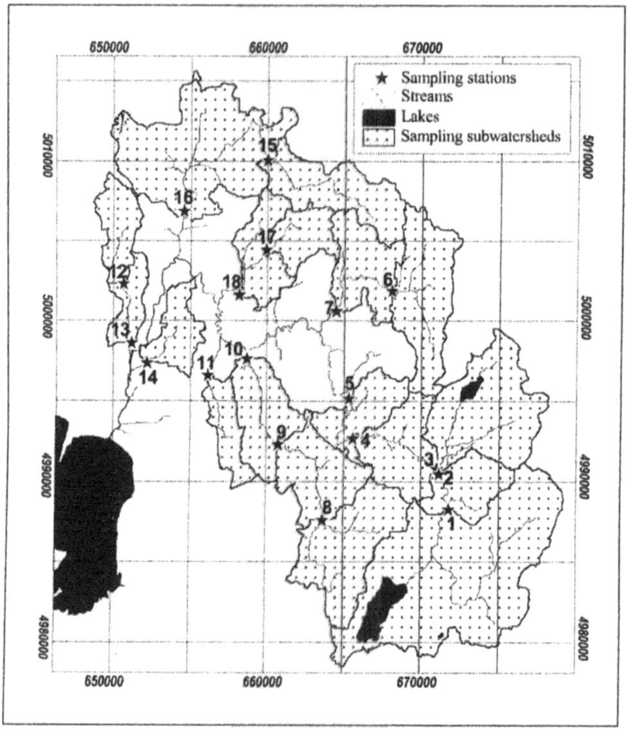

Figure 4. Sampling station location on the Pike River Watershed

Chemical analytical methods

Water samples were kept under 4°C until analysis in laboratory. Total suspended sediments (TSS) was determined through filtration method (<0.45 μm). Sample pH was measured with a combined electrode Accumet meter, model 950. Reactive soluble phosphorus (RSP), bioavailable phosphorus (Bio P) and total phosphorus (TP) concentrations were determined colorimetrically using the molybdenum-blue method of Murphy and Riley (1962), following respectively filtration (<0.45 μm), 0.1 N NaOH extraction (Sharpley *et al.*, 1991) and persulfate digestion. Soil P concentrations were determined by extraction of 1 g soil subsamples using the Mehlich-3 extractant (Mehlich, 1984).

Statistical methods

An analysis of variance relating indicators and water quality data was performed using the PROC MIXED procedure (Littell, 1996). The requirement for normality was met through a log-transformation of the phosphorus concentration data. The model-adjusted means of phosphorus concentrations (total, bioavailable, and orthophosphate) were then used as dependent variables in multiple linear regressions, using the various indicators values as independent variables. This regression analysis was done with the REG procedure in SAS (Littell,, 1996).

3. RESULTS AND DISCUSSION

Transport Indicator

SWAT simulation results at the subwatershed scale indicate that erosion is generally low over the entire land area (Figure 6a). Most of the subwatersheds (95 %) showed soil loss rates below the agronomic threshold allowed in Quebec (6 Mg yr^{-1}). Figure 5 illustrates the sensitivity of the SWAT erosion simulations for distinct HRU as a function of crop and slope.

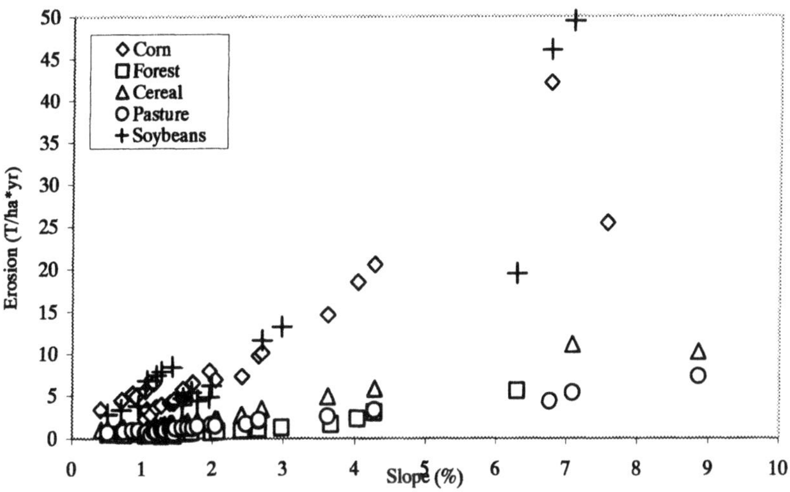

Figure 5. Bulk erosion evaluated by SWAT (MUSLE) expressed as a function of crop and slope for the HRU composed of Milton soil series

The SWAT model estimates the mean annual surface runoff over the entire watershed at 229 mm yr^{-1}. This estimate lies within the low range of the Canadian hydrological survey that indicates a value between 200 and 600 mm yr^{-1} for the area. For the current simulation year (1999), total measured precipitation (884.3 mm yr^{1}) was below the 40-year climatic normal of 1057.8 mm yr^{-1} set by Environment Canada for the Philipsburg station. Notwithstanding the fact that SWAT was developed to model non-calibrated watersheds with reasonable accuracy, the runoff volume for the reference year was more likely under predicted. The relative sensitivity of the HRU's and subwatersheds to erosion and surface runoff are however the parameters of interest in this study.

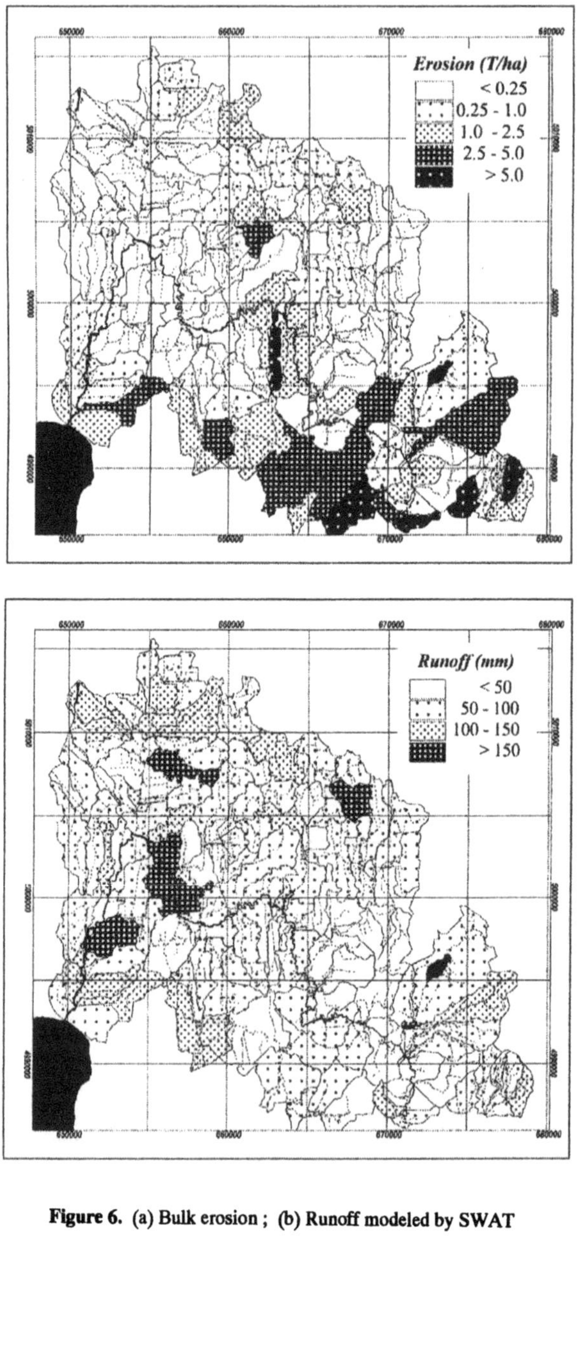

Figure 6. (a) Bulk erosion ; (b) Runoff modeled by SWAT

Source Indicators

Table 2 presents the overall summary of P Mass balance assessed for all the agricultural lands in the Pike River watershed (spatially distributed on the Figure 7b). Roughly 65 % of inputs originate from farm manure and, on a mass basis, correspond to the theoretical crop uptake.

Table 2. P_2O_5 inputs to the topsoils of the Pike River watershed

Inorganic P inputs (Mg)	Organic P inputs (Mg)	Total P inputs (Mg)	P uptake by crop (Mg)	Net P input (Mg)	Agricultural area (ha)	P Mass Balance (kg ha^{-1})
233	435	668	426	242	28 000	8.6

Following the spatial pattern in soil P saturation, a relatively small number of the subwatersheds of the Pike River show a tendency for levels to exceed the vulnerability threshold of 10 % (Figure 7a). A strong correlation between the soil P-richness index and the P inputs (r= 0.69, p=0.01) was noted, indicating a net P enrichment of soils particularly in the region downstream from Bedford.

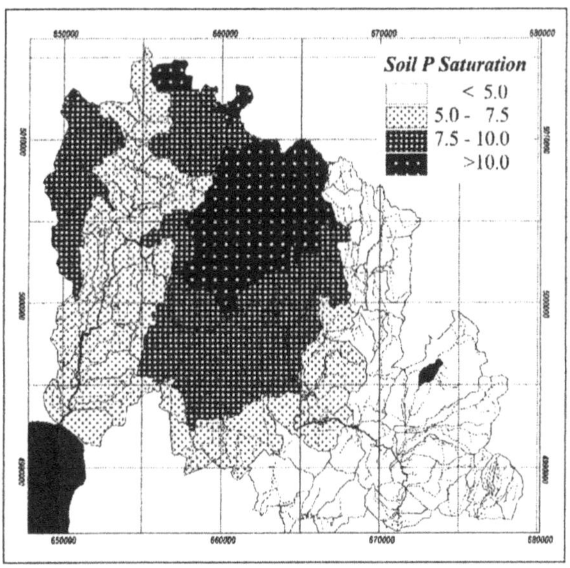

Figure 7a. % Soil P saturation spatialised in Arcview

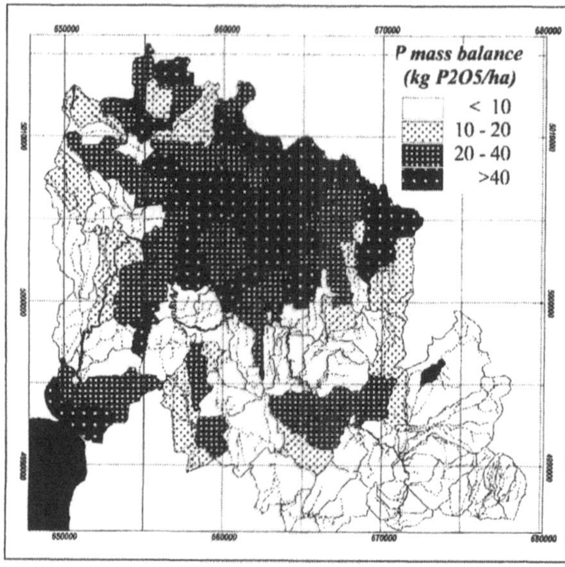

Figure 7b. P Mass balance (kg/ha) spatialised in Arcview

Water sampling

Figure 8 illustrate both spatial and temporal variability in Total stream P concentration monitored for the 18 sampling stations network. The high range in Total P concentrations, varying from 30 to 730 ug l^{-1}, as well as elevated standard deviations, reflect contrasting land uses and variability in hydrologic activity among subwatersheds. Globally, stream P concentration present a spatial pattern, where subwatersheds located upstream of Bedford (1 to 6) present relatively low P concentrations in reflection of a dominant forested land use. Contrasting with upper part of Pike River watershed, subwatersheds located downstream of Bedford (7 to 18) are dominantly under agricultural land use and present higher stream P concentrations. Elevated standard deviations in Total P concentrations for these subwatersheds indicate that water quality is particularly sensitive to variations in hydrologic conditions under agricultural land use.

Validation of the indicators

Correlation among indicators and stream P concentration

Correlations were performed (RSP, TP, Bio P) between the log mean P from event sampling of the 18 subwatersheds and the source and transport indicators aggregated for each subwatersheds (Table 3).

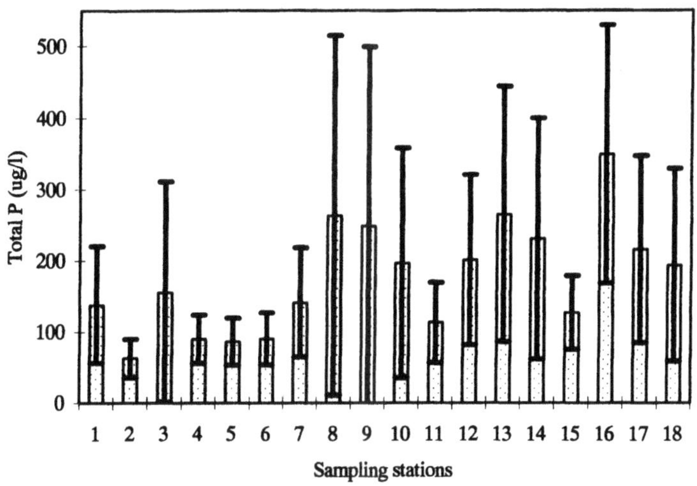

Figure 8. Total P concentrations at the 18 sampling stations and for the 10 sampling programs on the Pike River Watershed

Table 3. Correlation matrix between Log-transformed median P concentrations and source and transport indicators evaluated for 18 subwatersheds in the Pike River Basin

	Transport Indicators			Source Indicators		Landscape Attributes	
	Runoff	**Erosion**	**P Sat.**	**Balance**	**% Agri.**	**Elev.**	**Slope**
RSP	0.69 **	-0.55 **	0.65 ***	0.61 **	0.87 ***	-0.81 ***	-0.70 ***
Bio P	0.66 ***	-0.57 **	0.63 ***	0.55 **	0.86 ***	-0.77 ***	-0.67 ***
TP	0.56 **	-0.48 *	0.72 ***	0.44 *	0.78 ***	-0.85 ***	-0.69 ***

*, **, ***:Significatively correlated at 10 %, 5 %, 1 % (bilateral); NS: non significantly correlated at 10 %

Table 4. Correlation matrix between the source and transport indicators evaluated for the 18 subwatersheds in the Pike River Basin

	Erosion	P Sat.	Balance	% Agri.	Elev.	Slope
Runoff	NS	NS	NS	0.74 ***	NS	NS
Erosion		-0.49 *	NS	-0.46 *	0.55 **	0.79 ***
P Sat.			0.59 ***	0.53 **	-0.67 **	-0.70 ***
Balance				NS	NS	-0.46 *
% Agri.					-0.70 ***	-0.54 **
Elev.						0.88 ***

*, **, ***:Significatively correlated at 10 %, 5 %, 1 % (bilateral); NS: non significantly correlated at 10 %

Landscape (mean elevation and slope) and percentage agricultural land use patterns are all significantly linked to P concentrations variability within the Pike River watershed. These land use and landform gradients are also significantly correlated, which makes it difficult to contrast the effects of P sources and transport factors on stream P. The spatial overlap reflects the geophysical configuration of the watershed. The western, downstream part of watershed presents an intensive production profile and dominant agricultural land use. The relative extent and intensity of agricultural land decreases within higher, upstream parts of the watershed, where the land rating value for agriculture presents more limitations.

The runoff indicator (CN-SCS) reflects the percentage of agricultural land use gradient (r=0.74, p=0.01) and most probably the lower permeability of soils from the downstream, flat portion of watershed. The contribution of runoff as a transport indicator is particularly relevant for soluble and bioavailable P concentration patterns (r= 0.69, p=0.05 for RSP and r=0.66, p=0.01 for BioP).

The spatial distribution of the erosion indicator (MUSLE) reflects the landscape erosivity of the upstream part of watershed (r=-0.55, p=0.05 for RSP, r=-0.57, p=0.05 for Bio P and r=-0.48, p=0.10 for TP. A similar trend was observed within experimental Ruisseau-aux-Castors watershed (Michaud and *al,* 2002), a 11 km² basin located in the downstream part of Pike River basin.

Soil P saturation and mass balance used as P sources indicators are also significantly correlated (r=0.59, p=0.10) to the stream P concentrations (Table 3). The correlation likely reflects the enrichment of soils in response to the intensity of farming. Overlapping spatial gradients in landscape and agricultural land use implies that the soil P enrichment dynamic follows Landscape attributes, as suggested by significant correlations between soil P saturation and watershed mean slope (r=-0.70, p=0.05) and mean elevation (r=-0.67, p=0.01).

Linear regressions

Table 5 illustrates selected regressions equations indicate that 86 % of the variation for Bio P, 87 % of that for RSP and 79 % of that for TP from sampled streams is explained by the P balance indicator in conjunction with the variable for mean elevation. Regressions analysis have also suggested that: 66 % of the variation for total P, 75 % of that for orthophosphate and 73 % of the variation for bioavailable P sampled within the hydrological network is explained by the P mass balance (at soil surface) and subwatershed agricultural land use (%). The P sources indicators and elevation attribute thus contribute, in a complementary fashion, to explain the pattern of P concentration in the hydrologic network. As suggested by the correlations among independent indicators, subwatershed elevation translate both landform and land use spatially overlapping gradients, and likely reflects the near-soil surface hydrological activity within the basin. As observed in table 4, the transport (erosion and runoff) and the source factors don't correlate well, allowing a combination of both to be used to predict stream P concentration, but this combination of variables don't improve significantly the model. This may suggest that the aggregated, erosion and runoff data at the subwatershed scale tends to dilute the individual field risk and don't always really reflect what is going on the field. This is not to be neglected because as less as 10 % of a watershed can contribute to more than 90 % of its P export in surface runoff (Sharpley *et al.* 1994 ; Daniel *et al.,* 1994).

Table 5. Linear regression

Dependant variable (x)	Independent variable		Model					
	Variable 1 (a_1)	Variable 2 (a_2)	F value	Pr>F	R^2	b_0	b_1	b_2
Bio P	Mean elevation	P Balance	39.25	<0.001	0.86	1.931	-0.002	0.007
	P Balance	% Agricultural land use	17.56	0.002	0.73	1.591	0.007	0.003
RSP	Mean elevation	P Balance	43.39	<0.001	0.87	1.756	-0.003	0.008
	P Balance	% Agricultural land use	19.45	0.001	0.75	1.223	0.01	0.005
TP	Mean elevation	P Balance	24.31	<0.001	0.79	2.141	-0.063	0.004
	P Balance	% Agricultural land use	12.62	0.0009	0.66	1.926	0.007	0.002

Linear regression model :$x. = b_0 + b_1{*}a_1 + b_2{*}\, a_2$

Orthogonal Contrast analysis

A orthogonal contrast analysis was performed on the water quality data in order to separate out the effects of P sources and transport indicators (using PROC MIXED procedure from SAS, Littell, 1996). The analysis among subwatersheds 2-3-4-5 vs. 7-18 and 10 vs. 15 (Table 6b) contrast P sources within relatively comparable transport indicator values. Stream P concentrations within subwatersheds 7 and 18 are higher than upstream watersheds group 2-3-4-5, which is attributed to an higher mass P balance (from approximately 10 to 40 kg P/ha). A similar trend is observed for orthogonal contrast pair 10 and 15 within the downstream part of watershed. Conversely, the contrast between subwatersheds 6 and 13 highlights a significant runoff indicator effect on stream P , under similar P mass balance values. However, contrasting land uses and corresponding differences in runoff predictions among subwatersheds groups 12-14 and 11-16 did not express a significant difference in stream P data, likely reflecting a buffer effect of elevated P mass balances.

Table 6a. Orthogonal contrast analysis :Test of the Fixed Effects

Effect	DF	Num DF	Den F Value	Pr> F
Station	15	134	8.44	<0.001

Table 6b. Contrast analysis of station effect on model adjusted (ls) total P means.

Contrast	DF	Num DF	Den F Value	Pr> F
2-3-4-5 vs 7-18 [a]	1	134	17.92	<0.001
6 vs 13 [b]	1	134	18.28	<0.001
12-14 vs 11-16 [c]	1	134	2.12	0.2281
10 vs 15 [d]	1	134	2.72	0.0492

[a, d] Relatively same runoff but different P mass balances (see Figure 10)

[b] Relatively same P balances but different runoff (see Figure 10)

[c] Drastic difference in % agricultural land use (respectively means of 50 % and 94 %)

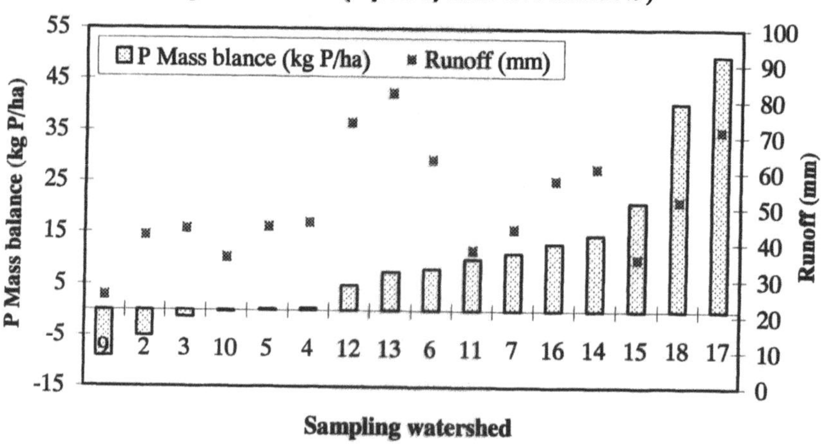

Figure 10. Runoff and P Mass balance of the sampling watershed on the Pike River watershed

4. CONCLUSION

The complexity of the transfer of nonpoint source contaminants within large watersheds requires that a compromise must be reached between detailed models, which require an inordinately large number of inputs, and simpler models with low interpretative value. This study demonstrates the feasibility of generating P source and transfer indicators from the integration of (1) generally available census and (2) geophysical data, coupled with (3) satellite imagery. Tools such as remote sensing and geographical information systems (GIS) were found to be indispensable in supporting spatial data integration and management. While the SWAT model seemed to under predict the surface runoff, it provided an efficient framework for discriminating, on a fine spatial basis, physical processes involved in nonpoint P transfer. Further development of a systematic approach to assess sources of nonpoint P pollution at the regional scale should include a systematic indexing method to establish critical risk classes and a validation of the optimum subwatershed scale for data integration.

The P source and transport indicators provided valuable and complementary explanations of spatial distribution of stream P data. Strategic planning and management of nonpoint sources of phosphorus can benefit from GIS technology that can be readily used by regional stakeholders to identify and prioritize the areas that are most at risk. Subwatershed-level spatial data provides an efficient framework for land planning related issues, while the HRU data are particularly relevant for farm scale extension work. The approach is transferable to most regions of rural Quebec, where similar input data are available.

REFERENCES

Arnold, J.G., P.M. Allen, and G.T. Bernhardt (1993) A Comprehensive Surface-Groundwater Flow Model, Journal of Hydrology 142: 47 -69.

Arnold, J.G., J.R. Williams, A.D. Nicks, and N.B. Sammons (1990) SWRRB A basin scale simulation model for soil and water resources management, Texas A&M Press. College Station, TX, 255 p.

Bernard, C. (1996) Estimation de l'érodabilité des principales séries de sol du Québec, à l'aide du nomographe de Wischmeier , Agrosol 9(2) : 6-12.

Budd, L. and D.W. Meals (1994) Lake Champlain Nonpoint Source Pollution Assessment. Technical Report No. 6, Lake Champlain Basin Program, Grand Isle, Vermont.

Cattaï, J. (2002) Amélioration des classifications de l'utilisation du territoire par des approches multi-temporelle, stratifiée et champ-par-champ d'images du satellite Landsat, mémoire de maîtrise non-publié, Département des lettres et sciences humaines, Université de Sherbrooke.

Caumartin, J. and R. Vincent (1994) Diagnostic environnemental de la rivière aux Brochets, Ministère de l'Environnement et de la Faune du Québec, 93 p.

Commission sur la gestion de l'eau (2002) Revoir de fond en comble la stratégie d'assainissement agricole, Gouvernement du Québec, Rapport No GQM00256, 30 p.

Jamieson, A., C.A. Madramootoo and P. Enright (2003) Phosphorus losses in surface and subsurface runoff from a snowmelt event on an agricultural field in Quebec, Canadian Biosystems Engineering 45:1.1-1.7.

Daniel, T.C., A. N. Sharpley, D. R. Edwards, R. Wedepohl, and J. L. Lemunyon (1994) Minimizing surface water eutrophication from agriculture by phosphorus management. Journal of Soil and Water Conservation Supplement 49: 30-38.

Foster, G.R., D.K. McCool, K.G. Renard and W.C Moldenhauer (1981) Conversion of the Universal Soil Loss Equation to SI Mertric Units, Journal of Soil and Water Conservation, 36:355-359.

Giroux, M, and T.S. Tran (1996) Critères agronomiques et environnementaux liés à la disponibilité, la solubilité et la saturation en phosphore des sols agricoles du Québec, Agrosol 2 : 51-57.

Hegman, W., D. Wang and C. Borer (1999) Estimation of Lake Champlain basin wide nonpoint source phosphorus export: Grand Isle, Vt., Lake Champlain Basin Program, Technical Report no. 31, 81 p.

Lemunyon, J.L., and R.G. Gilbert (1993) The Concept and Need for a Phosphorus Assessment Tool, Journal Proc. Agric. 6(4): 483-486.

Littell, R. C., Milliken, G. A., Stroup, W. and R. D. Wolfinger (1996) SAS System for Mixed Procedure, Cary, NC: SAS Institute Inc., pp 1-29

Mehlich, A. (1984) Mehlich-3 soil extractant: A modification to Mehlich-2 extractant. Commun. Soil Sci. Plant Anal. 15:1049-1416.

Ministère de l'Agriculture, des Pêcheries et de l'Alimentation du Québec (1990) Inventaire des problèmes de dégradation des sols agricoles du Québec, Région agricole 6 Richelieu Saint-Hyacinthe, Gouvernement du Québec, Bibliothèque nationale du Québec, 119 p.

Michaud, A.R. and P. Beaudet (1998) Analyse de la pression agroenvironnementale à l'échelle du territoire. Réseau Érosion de l'ORSTOM, Bulletin 18, Congrès conjoint AQSSS-ORSTOM : Le sol et l'eau : deux ressources à gérer en interrelations, p. 491-505.

Michaud, A.R., R. Lauzier and M.R. Laverdière (2002) Temporal and spatial variability in nonpoint source phosphorus: The Ruisseau-aux-Castors case study, Pike river basin, Quebec. Lake Champlain Research Consortium, Symposium 2002: Lake Champlain in the new millennium. Saint-Jean-sur-Richelieu, Québec.

Organisation de coopération et de développement économique (1997) Indicateurs environnementaux pour l'agriculture. Service des publications, OCDE, Paris, 72 p.

Murphy, J., and J.R. Riley (1962) A modified single solution method for the determination of phosphates in surface waters. Anal. Acta 27:31-36.

Pote, D.H., T.C. Daniel, A.N. Sharpley, P.A. Moore, Jr., D.R. Edwards, and D.J. Nichols (1996) Relating extractable soil phosphorus to phosphorus losses in runoff, Soil Science Society of America Journal 60:855-859.

Sharpley, A.N., W.W. Troeger and S.J. Smith (1991) The measurement of bioavailable phosphorus un agricultural runoff. J. Env. Qual. 20:235-238.

Sharpley, A.N., S.C. Chapra, R. Wedepohl, J.T. Sims, T.C. Daniel, K.R. Reddy (1994) Managing Agricultural Phosphorus for Protection of Surface Waters : Issues and Options, J. Env. Qual., 23, 437-451.

Williams, J.R. (1975) Sediment routing for agricultural watersheds, Water Resources Bull. 11(5) : 965-974.

Wischmeier W.H., Jonhson C.B. and Cross B.V. (1971) A soil erodibility nomograph for farmland and construction sites,. J. Soil and Water Conservation 26(5): 189-192.

U.S. Department of Agriculture, Natural Resources Conservation Service (2001) National Soil Survey Handbook, title 430-VI. [Online] Available :http://www.statlab.iastate.edu/soils/nssh/.

U.S. Department of Agriculture, Soil Conservation Service (1972) National Engineering Handbook, Hydrology Section 4, Chapters 4-10.

INFLUENCE OF SUBSTRATE COMPOSITION ON DISTRIBUTION OF EASTERN SAND DARTERS (*AMMOCRYPTA PELLUCIDA*) IN THE POULTNEY RIVER

Douglas E. Facey and Shannon M. O'Brien *

ABSTRACT

Eastern sand darters (*Ammocrypta pellucida*) are a threatened species in both New York and Vermont and have become rare in other parts of their range, in part due to their dependence on clean, sandy substrates in large rivers. Field observations suggest that eastern sand darters are more likely to be found in fine sand than in coarser sand or gravel. The goal of this study was to evaluate the habitat used by eastern sand darters in the Poultney River, and characterize more specifically the sizes of sediment particles that they seem to use most often. Between July 30 and August 7, 2001, we collected sediment and fishes, and measured water velocity and depth, in 99 different sample plots located along a one-mile stretch of the Poultney River on the New York-Vermont border. The areas we sampled were all less than one meter deep and had a velocity that was less than 1 m/sec. Our study indicated that eastern sand darters seem to be more common in areas with a substrate made up of a high percentage of particles between 0.12 and 0.54 mm in size. Fewer eastern sand darters were found in areas that had high percentages of particles greater than 1.0 mm. This suggests that eastern sand darters are very selective about the habitat in which they live and might, therefore, be affected by fluctuations or changes in the composition of the substrate within that habitat.

1. INTRODUCTION

Eastern sand darters generally occur in streams and rivers with fine sand substrate, although they have been reported from wave protected, sandy shoals of Lake Erie (Langlois 1954). Smith (1985) reported that rivers supporting eastern sand darters should

* Department of Biology, Saint Michael's College, Colchester, VT 05439

have "... currents slow enough to retain sand but fast enough to prevent deposition of fine silt." Historical information suggests that there has been a decrease in the abundance of this species throughout much of its range, perhaps due to habitat degradation (Trautman 1957, 1981). Several land use practices result in increased siltation of rivers and streams, making formerly sandy areas no longer suitable for eastern sand darters (Scott and Crossman 1973, Smith 1985, Daniels 1993, Holm and Mandrak 1996). There is very little historical information about the occurrence of eastern sand darters in Vermont, making it difficult to compare current populations to previous levels (Facey 1998). The sections of the few Vermont rivers known to support the species are characterized by having fine sandy substrate, although eastern sand darters have occasionally been captured in areas of sand and gravel.

Extensive observations and sampling in the Mettawee River, NY in 1982 and 1984 showed that eastern sand darters were found only in areas with substrates that were at least 90% sand (Daniels 1993). Daniels also noted that eastern sand darters are often found in fine sand substrates within 20 m downstream of a bend in the channel, on the depositional side of the river, within 5 m of shore, in water less than 0.5 m deep and with a mean water-column velocity of less than 20 cm/sec. Similar observations were made by Facey (1995, 1998) in several rivers in Vermont. Daniels (1993) also observed that 90% of all sand darters in an artificial stream selected sand (0.25-0.5 mm particle size) over gravel (1-2 cm) or rubble (8-15 cm). It appears, therefore, that the eastern sand darter is quite specific in its habitat needs, requiring sandy substrates in areas of relatively low to moderate velocity. However, personal observations by several researchers (D. Bouton, R. Daniels, D. Facey) over several years indicate that not all sandy substrates are equally likely to have eastern sand darters; the fish seem more likely to be caught in softer sand in shallow areas with very low current. To try to determine whether the specific consistency of the sandy substrate seemed to affect utilization by eastern sand darters, we analyzed sediment composition and the distribution and abundance of eastern sand darters along a one-mile stretch of the Poultney River.

2. METHODS

Between July 30 and August 7, 2001, we sampled sediment and fishes at 17 sites within a one-mile stretch of the Poultney River. This section of the river is immediately upstream from the mouth of the Hubbardton River, and lies on the border between New York and Vermont. The sites were selected to represent the range of available habitats less than one meter deep within this one-mile section of the river. We had to restrict sampling to sites shallower than one meter because seining could not be done effectively at deeper sites. At each site, we sampled between four and seven different 3 X 10 m plots that were close to one another but did not overlap. Plots were selected to represent the range of substrates available at the site, but substrate composition was quite consistent within a given plot. In some cases we had to sample smaller plots due to water depth or the presence of obstructions in the river. The number of plots was determined by the size of the area that was shallow enough to seine.

We measured the water depth at the center of each plot to provide an estimate of the mean depth, and we determined surface velocity by measuring the time needed for a partially submerged tennis ball to float the length of the plot. We also collected a 200 ml sample of substrate from the center of each plot; these samples were from the upper four to six cm of substrate and appeared to be representative of the surface substrate composition of the plot.

To collect fishes, we used a 2-ft high seine with 3/16'' mesh for the five sites (27 plots) furthest downstream, and then switched to a 4-ft high bag seine with ¼'' mesh wings and 3/16'' mesh collecting bag in the middle for the remaining plots. We seined each plot once, from downstream to upstream, recorded the number of individuals of each fish species caught, and measured the standard length (mm) of all eastern sand darters.

After the field collections were completed, each of the substrate samples was dried and sifted through a series of sieve screens to separate each into different substrate size components. The sieve screen sizes were #5 (4.1 mm), #10 (1.9 mm), #18 (1.0 mm), #35 (0.54 mm), #60 (0.23 mm), and #120 (0.12 mm). Each size component was weighed and the percent composition by weight was calculated for each substrate size category of each sample.

To determine whether or not eastern sand darters were randomly distributed among the plots, we used a chi-square goodness of fit test with the null hypothesis that the number of plots with a certain number of fish (0, 1, 2, etc.) would resemble a Poisson distribution (Dytham 1999). A significant chi-square result would indicate that the fish were not randomly distributed. To determine whether depth, water velocity, or substrate composition seemed correlated to sand darter abundance, we grouped the plots based on the number of eastern sand darters that we caught within each plot (0, 1, 2, 4, 6, or 7) and determined the average water depth, water velocity, and percent composition for each substrate size category for plots with that sand darter density.

3. RESULTS

We found a total of 42 eastern sand darters in 19 of 99 different plots. The plots were located in 11 of the 17 total sites along the one-mile stretch of the Poultney River. From the 19 total plots containing sand darters, nine plots contained only one sand darter, six plots contained two sand darters, two plots contained four sand darters, one plot contained six sand darters, and one plot had seven sand darters. All of the eastern sand darters were between 48 mm and 57 mm standard length (mean= 52.14 mm, SD= 2.47 mm), suggesting that they were three years old (see Facey 1998).

The chi-square analysis showed that the 42 eastern sand darters captured were not randomly distributed among the 99 plots sampled (P=0.00002, 1 df). Because the probability of capturing three or more fish in one plot was extremely low, we combined all plots with two or more fish into a single category in order to have an expected number of plots of at least five for each category and calculate the chi-square statistic. The Poisson distribution indicated that if the fish were randomly distributed, we would expect 64.5 plots with no fish, 27.7 plots with 1 fish, and 6.8 plots with 2 or more fish. Our results showed 80 plots with no fish, nine plots with one fish, and 10 plots with two or more fish. This resulted in a chi-square statistic of 17.85. Normally a chi-square test with three categories would have two degrees of freedom, but in this case we subtracted an additional degree of freedom because one of the categories was combined (Dytham 1999).

Table 1. Mean depth (m), mean velocity (m/s), and mean percent composition of substrate by weight (± SD) of plots with different densities of eastern sand darters (number per 30 m^2).

	Density of eastern sand darters (number per 30 m^2)					
	0	1	2	4	6	7
Number of Plots	80	9	6	2	1	1
Mean Depth (m)	0.36 (±0.14)	0.23 (±0.10)	0.24 (±0.08)	0.32 (±0.25)	0.28	0.42
Mean Velocity (m/s)	0.26 (±0.18)	0.27 (±0.18)	0.18 (±0.11)	0.29 (±0.10)	0.19	0.18
Percent substrate < 0.12 mm	3.18 (±3.17)	1.63 (±1.10)	2.39 (±1.48)	2.66 (±1.01)	1.34	3.01
Percent substrate 0.12 – 0.23 mm	10.26 (±10.76)	6.09 (±3.79)	7.79 (±5.34)	17.98 (±2.74)	16.30	9.39
Percent substrate 0.23 – 0.54 mm	26.89 (±15.78)	28.14 (±13.72)	29.59 (±22.29)	60.35 (±43.79)	43.79	22.80
Percent substrate 0.54 – 1.0 mm	14.29 (±12.26)	19.12 (±10.12)	14.12 (±8.43)	13.15 (±3.04)	27.32	23.78
Percent substrate 1.0 – 1.9 mm	10.48 (±6.54)	14.30 (±8.52)	18.05 (±12.63)	2.15 (±2.18)	8.48	16.61
Percent substrate 1.9 – 4.1 mm	12.45 (±7.85)	12.11 (±7.76)	17.28 (±11.81)	1.96 (±1.91)	2.65	11.72
Percent substrate > 4.1 mm	22.45 (±19.17)	18.62 (±19.62)	10.77 (±8.19)	1.72 (±1.56)	0.12	12.67

With so few plots having more than just one or two eastern sand darters, we could not statistically compare the water velocity, water depth, or substrate characteristics of the plots with different fish densities. There was no apparent pattern suggesting a correlation between water depth or water velocity and the density of eastern sand darters (Table 1). There were, however, some trends in the relationship between fish density and substrate composition. In general, the density of eastern sand darters was lowest in plots with the largest percent of substrate greater than 1.9 mm, and fish density increased as the percent of substrate greater than 1.9 mm decreases, except for the single plot with seven eastern sand darters (Table 1). Similarly, the three plots with four or six eastern sand darters had considerably higher percentages of substrate 0.12-0.23 mm and 0.23-0.54 mm than did the plots with two or fewer fish. Here again, the single plot with seven eastern sand darters went against the general trend by having levels of substrate in these size categories that were similar to the plots with two or fewer fish. Our data also show that the plots with only one or two eastern sand darters were quite similar in substrate characteristics to the plots with no eastern sand darters; except for the percent of substrate greater than 4.1 mm, which was clearly highest in the plots with no eastern sand darters.

Each site had more than one plot, and there was considerable variability in the number of eastern sand darters caught in different plots within a site, despite the close proximity of these plots to one another. To try to determine whether substrate differences might be related to eastern sand darters using some plots over adjacent and nearby plots, we compared the sediment composition of the individual plots with four or more sand darters with the sediment composition of the other plots at the same site. Three of the four plots with four or more eastern sand darters supported the overall trend seen in Table 1; the plots with the most eastern sand darters had higher percent composition of sediment particles from 0.23 to 0.54 mm and a very low percent composition of coarse sediment particles than nearby or adjacent plots with fewer or no eastern sand darters.

4. DISCUSSION

We sampled 99 plots at 17 different sites with a variety of different substrate compositions along a one-mile section of the Poultney River. Eastern sand darter density tended to be higher in areas with a high percentage of fine and medium sand (particles 0.12 to 0.54 mm) and low percentages of coarse substrate (particles over 1.9 mm), suggesting that eastern sand darters prefer areas with relatively fine sandy substrate (Table 1). However, the fact that we caught a total of only 42 eastern sand darters and that only 10 of the 99 plots sampled yielded more than one eastern sand darter made it impossible to perform suitable statistical analyses and draw strong conclusions.

Previous studies have indicated that the preferred habitats of eastern sand darters are areas with fine sandy substrate, slow velocities and shallow water (Daniels 1993, Facey 1998). In an artificial stream simulation, Daniels (1993) provided eastern sand darters with a choice of sand (0.25-0.5 mm), gravel (1-2 cm), or rubble (8-15 cm). He then varied the depth and velocity of the water in the holding tank to imitate actual river conditions for different times of year. Daniels reported that under all depth and velocity conditions between 84% and 90% of fish were found on the sandy substrate. Less than 3% of the observations found the eastern sand darters on the rubble. All plots with sand darters in our study had at least some proportion of coarser substrate (> 1.0 mm), but this simply reflects the fact that substrates are often heterogeneous in rivers.

Sediment composition is very dependent on the position of the area in the river because of the way water flow patterns determine deposition. Daniels (1993) reported that most eastern sand darters in the Mettawee River, NY were caught or observed within 20 m downstream of a bend in the channel, on the depositional side of the river, within 5 m of shore, in water less than 0.5 m deep and with a mean water-column velocity of less than 20 cm/sec. Facey (1995, 1998) noted similar patterns in habitat selection among eastern sand darters in the Winooski, Lamoille, and Missisquoi Rivers of Vermont. In our study, only five of the 19 plots yielding eastern sand darters fit the description of Daniels (1993), including two of the four plots with at least four eastern sand darters. However, this may have been in part due to our ability to sample mid-river sites due to extremely low water levels during the summer of 2001. Earlier surveys in both the Winooski and Missisquoi Rivers found some eastern sand darters at mid-river sandbars when water levels were low enough to seine these sites (Facey 1995). The low water level and velocity during the summer of 2001 may have also affected the location and relative abundance of substrates suitable for eastern sand darters.

Welsh and Perry (1998) studied darters in the Elk and Birch Rivers in West Virginia, and found eastern sand darters in pools in the Birch River. The microhabitats of the 10 eastern sand darters observed had a mean velocity of 0 cm/sec, the mean depth was 32.4 (±6.4) cm, and the substrate was composed of particles ranging from 0.06 to 2 mm. These results support our conclusions that eastern sand darters occur in areas with substrates of relatively small particle size, which typically corresponds to areas with low water velocity.

The range of the eastern sand darter extends from the St. Lawrence River and the Lake Champlain drainage southwesterly to southwestern Ontario, southeastern Michigan, and throughout much of the Ohio River basin to West Virginia and Kentucky (Page 1983; Page and Burr 1991), but populations have declined in many parts of that range. Among the concerns regarding the decrease in eastern sand darter populations has been the impact of habitat degradation due to siltation, and it has been suggested that eastern sand darters will not use muddy or heavily silted areas (Scott and Crossman 1973, Smith 1985, Daniels 1993, Holm and Mandrak 1996). We were not able to evaluate this, however, as there were no truly muddy or heavily silted areas in the section of the Poultney River that we studied. It is possible that recent improvements in water quality in some areas, particularly decreases in silt loads, have expanded suitable habitat for the eastern sand darter. Greeley (1930) reported no eastern sand darters found during a 1929 survey of the Mettawee River (NY) and Poultney River (NY-VT); field notes apparently indicated that both rivers had mud bottoms (Daniels 1993). Sandy stretches of both of these rivers have supported eastern sand darters since at least the 1980s (Daniels 1989, 1993; Bouton 1991).

The results of this study provide a greater degree of resolution to the observations of previous studies that eastern sand darters require sandy habitat. We have been able to more specifically identify the substrate particle sizes in areas with somewhat different densities of eastern sand darters. Unfortunately, the small number of fish caught made it impossible to make more definitive conclusions. This study will be followed by a similar study in the Winooski River (Vermont), to see if a similar pattern exists, and by a laboratory study in which eastern sand darters will be placed in aquaria with four different substrates available (0.23-0.54 mm, 0.54-1.0 mm, 1.0-1.9 mm, and 1.9-4.1 mm) to see if they show a significant preference.

5. ACKNOWLEDGEMENTS

This project was generously funded by a grant from The Nature Conservancy's Freshwater Initiative Small Grants Program, the Clare Booth Luce Scholarship Fund, and the Lake Champlain Research Consortium. The project was directed out of the Nature Conservancy's Southern Lake Champlain Valley Office, and Nature Conservancy staff Mary Droege and Paul Vidovich provided valuable assistance in the field. Special thanks to Joanne and David Calvi for their generous hospitality throughout the duration of the field research.

REFERENCES

Bouton, D. M. 1991. A summary of 1987-91 survey data collected on the eastern sand darter (Ammocrypta pellucida) in the Mettawee River, Poultney River, and Little Salmon – Salmon River systems. Unpublished report of the New York State Department of Environmental Conservation, Division of Fish and Wildlife. March 31, 1991.

Daniels, R. A. 1989. Significance of burying in Ammocrypta pellucida. *Copeia* 1989: 29-34.

Daniels, R. A. 1993. Habitat of the eastern sand darter, Ammocrypta pellucida. *J Freshwater Ecol* 8: 287-295.

Dytham, C. 1999. Choosing and Using Statistics: a Biologist's Guide. Blackwell Science, Malden, Massachusetts.

Facey, D. E. 1995. Status of the Eastern sand darter (Ammocrypta pellucida) in Vermont. Unpublished report to the Vermont Chapter of The Nature Conservancy. Montpelier, VT. August, 1995.

Facey, D. E. 1998. Status of the Eastern sand darter (Ammocrypta pellucida) in Vermont. *Can Field-Nat* 112: 596-601.

Greeley, J. R. 1930. Fishes of the Lake Champlain watershed. Pages 44-87 in E. Moore (ed.) *A biological survey of the Champlain watershed*. New York State Department of Conservation. Albany, New York.

Holm, E., and N. E. Mandrak. 1996. The status of the Eastern sand darter, Ammocrypta pellucida, in Canada. *Can Field-Nat* 110(3): 462-469.

Langlois, T. H. 1954. *The west end of Lake Erie and its ecology*. J. W. Edwards, Inc., Ann Arbor, Michigan.

Page, L. M. 1983. *Handbook of darters*. TFH Publishers, Inc. Neptune City, New Jersey.

Page, L. M., and B. M. Burr. 1991. *Freshwater Fishes*. Peterson Field Guide Series. Houghton Mifflin Company, Boston, Massachusetts.

Scott, W. B., and E. J. Crossman. 1973. *Freshwater fishes of Canada*, Bulletin of the Fisheries Research Board of Canada 184.

Smith, C. L. 1985. *The inland fishes of New York State*. New York State Department of Environmental Conservation, Albany, New York.

Trautman, M. B. 1957. *The fishes of Ohio*. Ohio State University Press, Columbus, Ohio.

Trautman, M. B. 1981. *The fishes of Ohio*. Ohio State University Press, Columbus, Ohio.

Welsh, S. A., and S. A. Perry. 1998. Habitat partitioning in a community of darters in the Elk River, West Virginia. *Env Biol Fishes* 51: 411-419.

LAKEBED POCKMARKS IN BURLINGTON BAY, LAKE CHAMPLAIN: I. HYDRODYNAMICS AND IMPLICATIONS OF ORIGIN

Patricia L. Manley[1], T. O. Manley[1], Mary C. Watzin[2], and Josh Gutierrez[1]

ABSTRACT

Using side-scan sonar and sub-bottom profiles, a 40 m diameter, ~4 m deep lakebed pockmark was investigated in Burlington Bay, Lake Champlain, Vermont. Five piston cores (2 inside, 1 on the rim and 2 outside of the pockmark) were collected and analyzed for magnetic susceptibility, physical properties and ^{210}Pb and ^{137}Cs concentrations. A yearlong subsurface mooring, equipped with an Acoustic Doppler Current Profiler (ADCP) and vertical temperature chain, was placed in the deepest section of the pockmark (27.7 m). A second temperature chain was placed along the pockmark floor and over its rim. At the base of the mooring, 35 mm stereo cameras and a Savonius rotor current meter (~ 0.3 m above the lake floor) were installed. Bottom photographs show episodic events of suspended sediment. Temperature data show the presence of anomalously cold temperature water near the rim of the pockmark during stratified periods and anomalously warmer water temperatures during isothermal winter type conditions. Coarser grain size and bulk density occur inside the pockmark with an associated decrease in water content in comparison to sediment outside the pockmark. Radionuclide results show no detectable net accumulation of sediment within the rim of the pockmark, slight accumulation in the deepest part of the pockmark, and only a few tenths of a mm/yr accumulation outside. This pockmark was most likely formed by groundwater seepage and is subsequently modified by continuing groundwater seepage and bottom-flowing currents. Taylor column and frictional boundary layer dynamics are believed to play a role in the circulation within this feature.

[1] Geology Department, Middlebury College, Middlebury, VT 05753
[2] Rubenstein Ecosystem Science Laboratory, School of Natural Resources, University of Vermont, Burlington, VT 05401

1. INTRODUCTION

Sea-bed pockmarks are found in a variety of settings including estuaries (e.g. Maine - Scalon and Knebel, 1989; Fader, 1991; Kelley et al., 1994; Barnhardt and Kelley, 1995; Barnhardt et al., 1997; Fleischer et al., 2001) continental margins and shallow seas (e.g. Judd and Hovland, 1992; Baltic Sea – Whiticar and Werner, 1981;Barents Sea – Solheim and Elverhøi, 1985; Borneo – Demyttenaere et al., 2000; California – Paull et al., 2002; North Sea – Harrington, 1985; Cole et al., 2000; Voring Plateau – Bouriak et al., 2000) and ocean settings (e.g. Scotia Sea– King and MacLean, 1970; Blake Outer Ridge – Flood, 1981; Paull and Dillion, 1981). Roughly circular in shape and few meters to several hundred of meters in diameter with up to 30 m in depth, pockmarks are usually the result of upward migration of a fluid through fine-grained sediments (Hovland and Judd, 1988). Similar sized features are being identified in freshwater environments (e.g. Berkson, 1973; Johnson, 1980; Flood and Johnson, 1984; Pickrill, 1993; Blasco et al., 1998; Grant and Morrison, 1998; Blasco and Keyes, 2000) as more swath mapping and investigation of lakes occurs.

Large fields of pockmarks were identified in several locations in Lake Champlain, specifically in Cumberland Bay (North, 1997) and Burlington Bay (Pedersen, 1992; Sayward, 1996; Gutierrez, 2001) (Figures 1 and 2). Encompassing up to 2 km^2 in area, these bottom sediment bedforms, if active, may be manifestations of a significant source of upwelling gas or groundwater into the lake that was not previously defined. In addition, the upwelling fluid may affect the benthic communities within and outside the pockmark.

Project GULP (Groundwater Upwelling in Large Pockmarks) was initiated in 1999 to investigate the hydrology and ecology associated with the largest pockmark (named the General) in Burlington Bay. The "General" was instrumented for a period of one year. The purpose of this investigation was to determine if the feature is active, and if so, what was the main constituent of the upward migrating fluid and what its impact is on the local bottom communities (see paper by Watzin et al., this volume).

2. BACKGROUND AND PREVIOUS WORK

2.1. Regional Setting And Hydrodynamics

Lying between New York, Vermont, and Quebec, Lake Champlain is the sixth largest freshwater lake in the United States (Figure 1). The lake is 170 km in length and ~20 km wide at its widest point. It extends from Whitehall, New York, in the south, to St. Jean, Quebec, in the north, where it flows into the Richelieu River, which connects to the St. Lawrence River and eventually the Atlantic Ocean.

Metamorphosed rocks of the Adirondack Mountains along with sandstone and carbonate rocks are located on the western margin of the lake. The eastern shore has exposed carbonates, sandstones and shales. The elongate shape of the basin as well as distinct change in bedrock lithology suggest that the basin may be fault bounded (Hunt et al., 1972). Though ice occupied the basin several times during the Pleistocene, no Pleistocene deposits older than Wisconsinan have been identified (Hunt et al., 1972). The present day Champlain basin was modified during the last glacial period (Chapman, 1942; Connally and Sirkin, 1973; Teller, 1987) and underwent several stages during deglaciation. Proglacial Lake Vermont (~13,000 ybp) filled the Champlain valley to an elevation of 183 m and drained southward at Coveville, NY (known as the Coveville stage). As the glacial retreat continued to the north, lake level lowered as an outlet near Fort Ann, NY was opened (Fort Ann stage). Continuing its northward retreat into Canada, isostatic lowering of the regional land surface, permitted a marine incursion through the St. Lawrence lowland into the basin forming the Champlain Sea (~10,200 ybp). Lasting only 2,000 years, the Champlain Sea became Lake Champlain as the water freshened with the exclusion of the marine influx due to the rising of the northern portion of the basin from isostatic rebound (e.g. Hillaire-Marcel, 1974; Corliss et al., 1982).

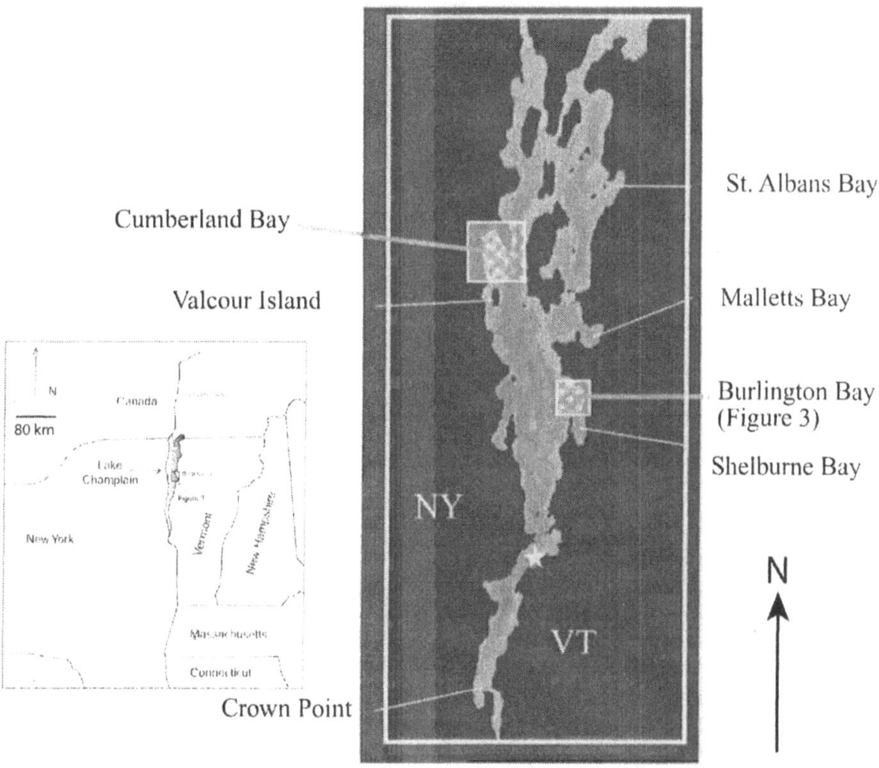

Figure 1. Map of Lake Champlain showing locations of large pockmarks in Cumberland and Burlington Bays. Star locates a smaller region containing large pockmarks near Diamond Island.

These three stages deposited distinct sedimentary units found in Lake Champlain today: Lake Vermont sediment, Champlain Sea sediment and Lake Champlain sediment. Lake Vermont sediments are non-fossiliferous with alternating layers of dark gray and dark brown clays devoid of organic material. Champlain Sea sediments contain marine foraminifera in a dark gray mud while Lake Champlain sediments are characterized by diatomaceous, poorly sorted muds and clays ranging in color from dark gray to grayish brown (Chase and Hunt, 1972).

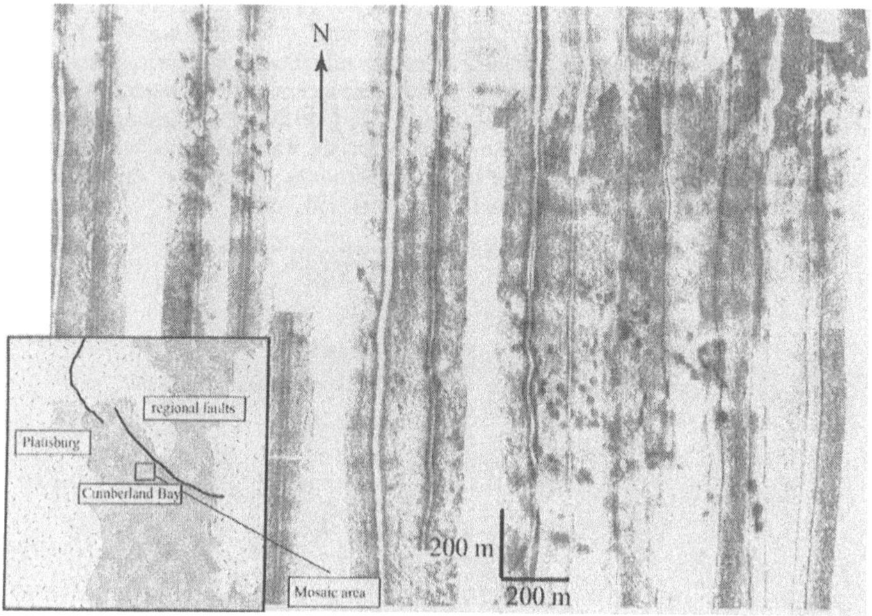

Figure 2. Side-scan mosaic of pockmark field in Cumberland Bay. Dark returns are reflective material and represent an increase in % sand-sized material in the sediment. Light returns (white) are more dominated by mud. The dark circular features are the pockmarks having an average diameter of ~20 m. Notice the alignment of these features with the structural trends of the area (insert).

The lake is underlain by numerous N-S oriented faults that are often crossed by east-striking high-angle faults (Hunt, 1972; 1977; 1981). In the Burlington region, a major thrust fault, known as the Champlain Thrust, is exposed at Lone Rock Point, VT. Along this fault Cambrian-aged Dunham Dolostone was emplaced over Ordovician-aged Iberville Shale during the Acadian Orogeny (Stanley, 1987). This fault runs N-S and is west of the study area (Figure 3).

During periods of stratification (early spring to late fall), the circulation within Lake Champlain is primarily governed by a uninodal internal seiche with a period of approximately 4.25 days (Myer and Gruendling, 1979; T. Manley et al., 1999). Bottom currents up to 50 cm/s are observed during times of severe atmospheric storms that can depress the metalimnion to depths over 60 m. These high-speed currents erode and deposit sediments along the lake bottom and form distinct sedimentary bedforms (P. Manley et al., 1999). Additionally, the internal seiche can flood deep, cold hypolimnic waters into shallow bays (Thompson, 1994; Sayward, 1996), thus, having an impact on sediment redistribution and biological communities.

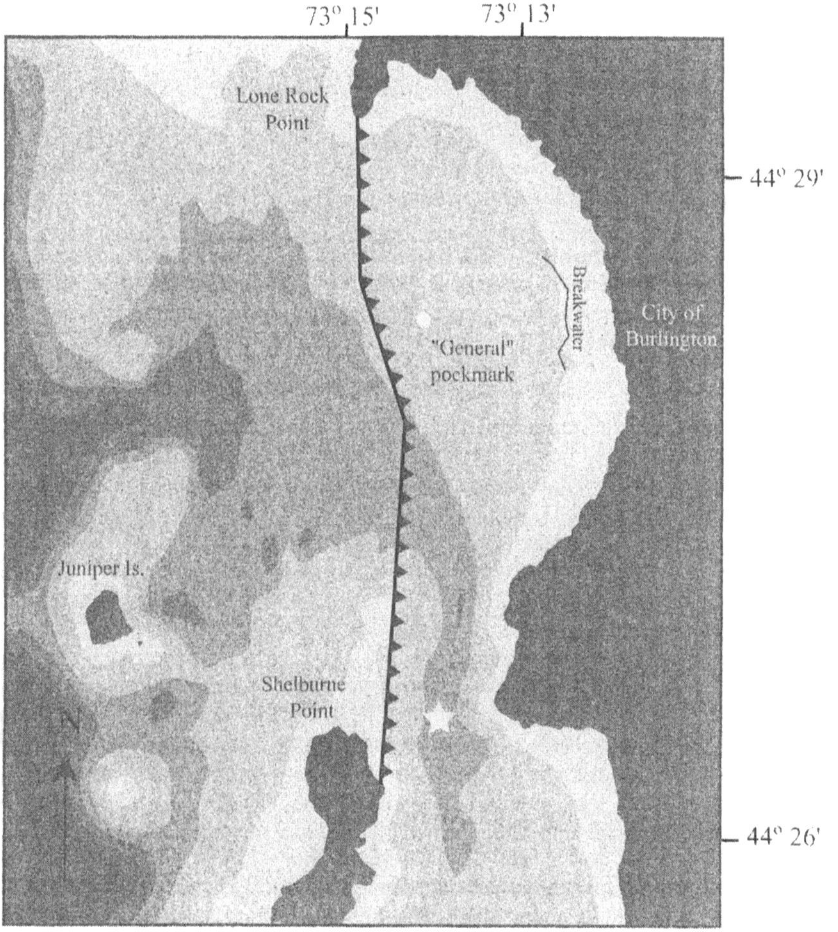

Figure 3. Location of the study area with respect to the Champlain Thrust. The mooring (circle) was located on the north rim of the pockmark known as the "General". The Shelburne Bay mooring is denoted by the star located east of Shelburne Point. Bathymetric contour shading interval is 10 m.

2.2. Pockmarks

Pockmark formation theories include nesting activities of fish (Johnson et al., 1984), disruption by a foreign body, differential rates of sediment compaction, change in sediment composition, human interference and fluid seepage (Hovland and Judd, 1988, references therein). Of these processes, continuous or sporadic pore fluid seepage through fine-grained sediments is the most widely recognized mode of formation of pockmarks. In this scenario, an impermeable layer in the sediment stops the migration of the pore fluid. Fluid pressure builds behind the impermeable layer, sometimes doming the overlying sediments. When the pressure overcomes the shear strength of the overlying sediments, the pore fluid erupts, suspending the overlying sediment into the water column where bottom currents remove the finest sediment particles (clays). The coarser (silt and sand) sediments settle within the vicinity of the formed pockmark (Hovland and Judd, 1988).

Hovland and Judd (1988) have synthesized our understanding of the known existence of pockmarks, their associated settings, and noted that the escaping fluid may be of thermogenic, biogenic, volcanic, hydrothermal, or groundwater in origin. Those pockmarks associated with upward migrating gas commonly display acoustic wipeout zones in seismic records as well as mid-water reflections (plumes) (Fleischer et al., 2001). Cemented seabed sediments often associated with gas migration can be analyzed for carbon isotopes to determine the source of the gas (thermogenic or biogenic). Researchers have also noted pockmark alignment with known basement faults (e.g., Platt, 1977; Foland et al., 1999).

2.2.1. Pockmarks in Lake Champlain

The Lake Champlain whole lake survey was initiated in 1996 with the goal of imaging the entire bottom of Lake Champlain with side-scan sonar. With nearly 90% of the survey done to date, pockmarks are found throughout Lake Champlain. In particular, large fields of these features are observed in Cumberland and Burlington Bays (Figure 1).

In Cumberland Bay there are two separate fields of pockmarks with spatial extents of 1 and 2 km^2, respectively. Having an average diameter of 20 m in both fields, these pockmarks are arranged in long linear chains (Figure 2). North (1997) determined the grain size as well as pore water chemistry differences between the sediments inside and outside the pockmark using 2 m piston cores. North's (1997) study showed slightly coarser grain sizes inside (~7ϕ) versus outside (~9ϕ) the features as well as isovalues for the elements Ca, K, Na and Mg in the pore waters with depth (suggesting a vertical mixing of interstitial waters near the center of the pockmarks). The linear alignment of the pockmarks with basement fault structures of the region suggests that fluid migration along fault planes might be responsible for their formation.

In Burlington Bay, pockmarks are located roughly 150 m east of the Champlain Thrust (Figure 3) in ~30 m water depth and are often arranged in linear chains (Figure 4a). The largest pockmark in Burlington Bay, the "General", has a diameter of nearly 40 m and a depth of approximately 4m. Nearly circular in shape, with an inside volume of ~3000 m^3, it is marked by an acoustically dark (reflective) outer ring and a light (non-reflective) center on the side-scan image (Figure 4a). The north and south rims are asymmetric, with 10 m relief in the north and 5 m in the south (Figure 4b). The pockmark has a slightly inclined floor with the deepest section occurring at the edge of the northern rim. Pedersen (1992) did grain size analyses (pipette and settling column method) from push cores inside, at the rim and outside the pockmark. He found that for surface sediments, finer grained sediments are located outside the pockmark (mean 5ϕ), the rim had a mean of 2.3ϕ, and the floor of the pockmark had the coarsest sediment with a mean of ~1.5ϕ. Mid-water reflections (plume) venting from the pockmark were observed during initial investigations by the first author during August 1991.

Sayward's (1996) analysis of pockmarks in Burlington Bay described 3 distinct groups: 1) < 5 m diameter pockmarks with a density of 328 per km^2; 2) 5 – 10 m diameter discrete pockmarks with a density of 30 per km^2; and 3) >10 m diameter with a frequency of 3 per km^2. Sayward (1996) determined that several other sedimentary bedforms such as lineations, furrows and sediment waves indicated the presence of strong bottom currents in Burlington Bay. Additionally, Sayward investigated the hydrodynamic system of Burlington Bay and documented that both wind and internal seiche driven currents affect the movement of water throughout the bay. She determined the mean circulation pattern in Burlington Bay was counterclockwise and contour following. The average speed was ~4 cm/s but ranged up to 30 cm/s. Large atmospheric events forced the internal seiche to flood the bay with deep hypolimnic water lasting up to 1.5 - 3.5 days.

3. METHODS

Side-scan sonar imagery (Klein 595 system) and standard 200 kHz PDR profiles were used to image the pockmark during the 1999 and 2000 field seasons (Figures 4 and 5).

A

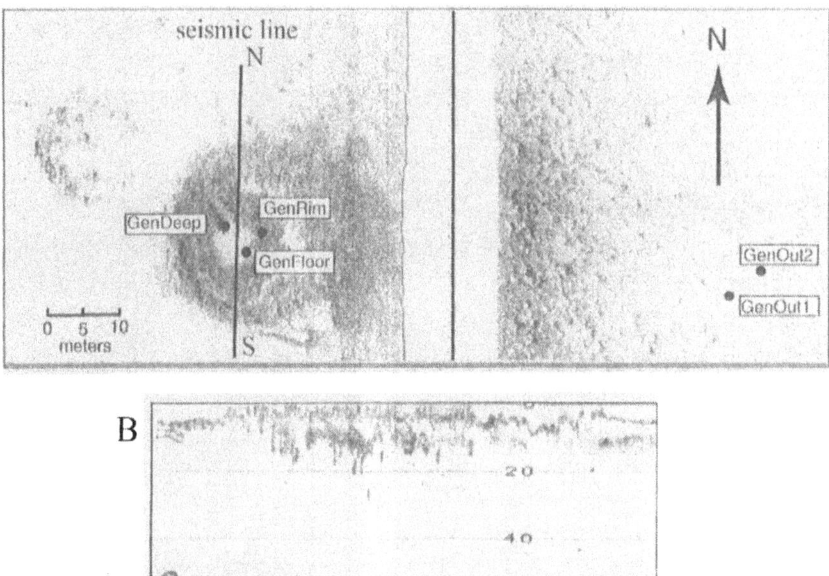

B

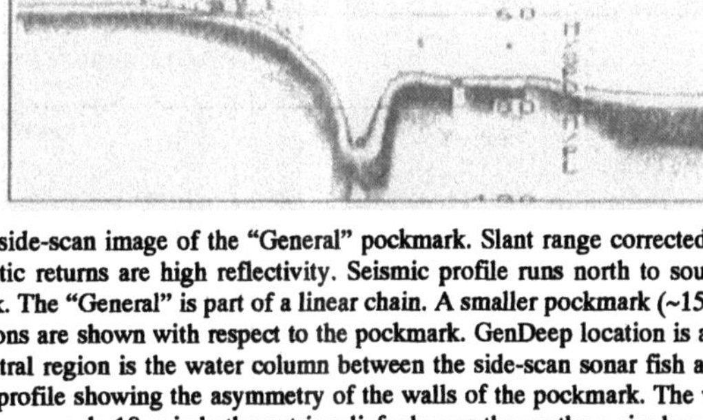

Figure 4. A) 100 kHz side-scan image of the "General" pockmark. Slant range corrected with a 75 m range on each side. Dark acoustic returns are high reflectivity. Seismic profile runs north to south through the deepest section of the pockmark. The "General" is part of a linear chain. A smaller pockmark (~15 m diameter) is located to the NW. Core locations are shown with respect to the pockmark. GenDeep location is also where the mooring was located. White central region is the water column between the side-scan sonar fish and the lake bottom. B) 200 kHz echo sounder profile showing the asymmetry of the walls of the pockmark. The vertical scale is in feet. The northern rim possesses nearly 10 m in bathymetric relief whereas the southern rim has ~5m.

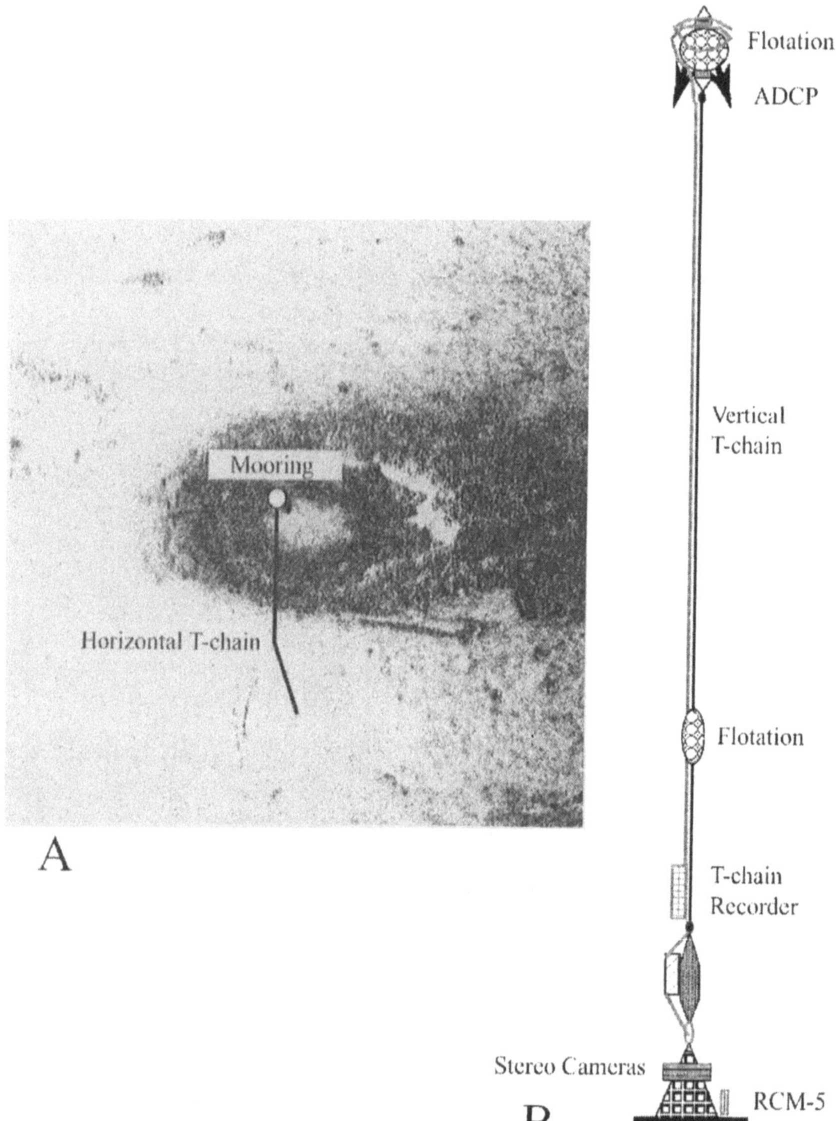

Figure 5. A) Location of sub-surface mooring and horizontal T-chain with respect to the pockmark viewed on side-scan imagery. B) Diagram of mooring showing equipment location.

A sub-surface mooring was deployed in the deepest portion at the northern edge of the rim and floor of the pockmark. The mooring gathered data from September 2, 1999 to July 25, 2000 and was equipped with vertical and horizontal Aanderaa temperature chains (T-chain), a downward looking RDI Acoustic Doppler Current Profiler (ADCP), Benthos Photosea 5000 stereo cameras and an Aanderaa RCM5 current meter at the base of the mooring (Figure 5b). Five 2 m piston cores (two inside, one on the rim and two outside) were collected and analyzed for magnetic susceptibility, physical properties and concentration of ^{210}Pb and ^{137}Cs radionuclides (Figure 4a). In addition to these data, SCUBA divers conducted video surveys as well as chemical and biological testing within the pockmark (Watzin et al., this volume).

ADCP data were collected every half-hour for the entire survey. The data provided backscatter intensity, current speeds for east/west, north/south and vertical components. The vertical and horizontal Aanderaa t-chains contained sensors every 4 m and recorded temperatures every two hours. The vertical temperature chain extended from 1 m above the base of the mooring to within three meters of the lake surface. The horizontal temperature chain (44 m) was laid along the lake bottom and extended from the base of the mooring to the south, up the rim and out of the pockmark (Figure 5a). Individual sensors were kept ~5 cm off the sediment by thin gauge wire hangers pushed into the sediment. Benthos stereo cameras took photographs of a 1 m x 1 m frame located at the base of the mooring every 12 hours until the main battery failed 45 days into the deployment. The RCM-5 current meter located at the base of the mooring recorded bottom water temperature, current speed every hour for the entire survey.

3.1. Piston Cores

Five piston cores were collected. Cores GenFloor and GenDeep were taken from the floor of the pockmark, with GenDeep located at the deepest part. Core GenRim was taken on the rim, or wall, of the pockmark and cores GenOut1 and GenOut2 were taken outside the pockmark (Figure 4a). All cores were analyzed at 5 cm intervals for physical properties (bulk density, water content, porosity and void ratio) and magnetic susceptibility was measured using a Barrington loop. Cores GenDeep, GenRim and GenOut2 were additionally sampled at a 1-cm interval over the top 10 centimeters for ^{210}Pb and ^{137}Cs presence.

^{210}Pb is a natural radionuclide with a half-life of 22.26 years and is part of the ^{234}U decay series. Supported ^{210}Pb is derived from the decay of ^{226}Ra in sediment particles. Unsupported (or excess) ^{210}Pb is added to the sediments from the decay of ^{222}Rn gas in the atmosphere. By measuring the total ^{210}Pb in the samples and subtracting the supported ^{210}Pb (estimated as the total ^{210}Pb in deep core sections) yields excess ^{210}Pb that decays exponentially, thus, allowing an assessment of the chronology of sediment accumulation in the core. In a core with a low accumulation rate, hiatus, physical mixing or bioturbation, the detection of excess ^{210}Pb can identify sediments that have accumulated approximately in the past century. ^{137}Cs is an anthropogenic radionuclide derived from global fallout from the atmospheric testing of nuclear weapons that began around 1954. Peak ^{137}Cs activity is associated with the fallout maximum of 1963 and 1964 (Bopp, R., Pers. Comm.). In cores with low sedimentation rates or non-continuous accumulation, ^{137}Cs profiles can be dominated by mixing and bioturbation but can still be useful. If present, ^{137}Cs identifies sediment with a significant component of particles that were deposited since the 1950s. Both ^{210}Pb and ^{137}Cs are not mobile in sediment and thus are excellent isotopes to determine recent sedimentation.

Two years after the completion of the field program, an EdgeTech FS-SB-216S Sub-bottom Profiler obtained sonar images of the pockmark and vicinity. The 216S system is a wideband frequency modulated profiler with a swept frequency of 2-16 kHz. A south to north seismic line started in Shelburne Bay and crossed the "General" pockmark in Burlington Bay.

4. OBSERVATIONS

4.1. Video and SCUBA Survey

Video and SCUBA surveys of the pockmark were obtained on July 25, 2000 during the mooring recovery and on August 16, 2000. These surveys confirmed that coarser-grained sediments occur within the pockmark in comparison to outside the pockmark. SCUBA divers determined that the central portion of the pockmark was firm to the touch, sandy in nature, and contained clinkers (remains of coal combustion from 1800's steamboats). Outside the pockmark, the diver could easily push his arm up to the elbow into the muddy sediment. There was no evidence of lithified (cemented) sediments within the pockmark or along the rim. Loose debris such as beer cans, branches and aquatic plants coalesced at the center of the pockmark, with little or no debris found around the rim.

Piston Cores

No piston cores (~2 m in length) showed evidence of gas expansion or small gas cracks within the sediment. Core GenFloor (202 cm in length) was taken from the floor of the pockmark, at a depth of approximately 26 m (Figure 4a). The top 7 cm were composed mainly of coarse sand with some clinkers and gravel. In general the core is olive-gray (5Y 4/1) silt with small sand layers or laminae throughout the entire core.

Core GenRim (207 cm in length) was taken from the rim of the pockmark at a depth of approximately 24.5 meters (Figure 4a). The top 10 cm of the core is composed of a mixture of clinkers, gravel, olive-gray (5YR 4/1) silty sand and the remainder is a dark greenish gray (5GY 4/1) silty clay containing no sand layers.

The top 4 cm of core GenDeep (205 cm in length), located at the deepest part of the pockmark (29 meters), is composed of coarse-grained greenish black (5GY 2/1) silty sand, clinkers, and some gravel. The remainder of the core is composed of alternating layers of dark greenish gray (5GY 4/1) to dark gray (5G 4/1) silty clay. Small pieces of organic matter are located at 17-35 cm and two pieces of wood, oriented vertically, occur at 42-45 cm and 61-62 cm.

Two cores located outside the pockmark were taken at ~20 m water depth. Core GenOut1 (210 cm in length) is characterized by greenish gray (5GY 4/1) clay throughout the core. The second core, GenOut2 (204 cm in length), contains gray (5GY 4/1) silt with small amounts (<5%) of sand, gravel and clinkers in the top 7 cm of the core. The remainder of the core consists of gray (5GY 4/1) silty clay with few laminations of darker gray units.

For comparison, Figure 6 contains the grain size analyzes of push cores analyzed by Pedersen (1992). The push cores were obtained by SCUBA diver and are located at the center of the pockmark, the rim and then outside the pockmark on a south to north transect. The push core from the center of the pockmark is consistently coarser than the surface sediment found outside the pockmark. The rim core has a surficial grain size that is comparable to the inside of the pockmark. For those samples >10 cm below the surface, the rim push core has grain sizes similar to the finer sediment located outside the pockmark.

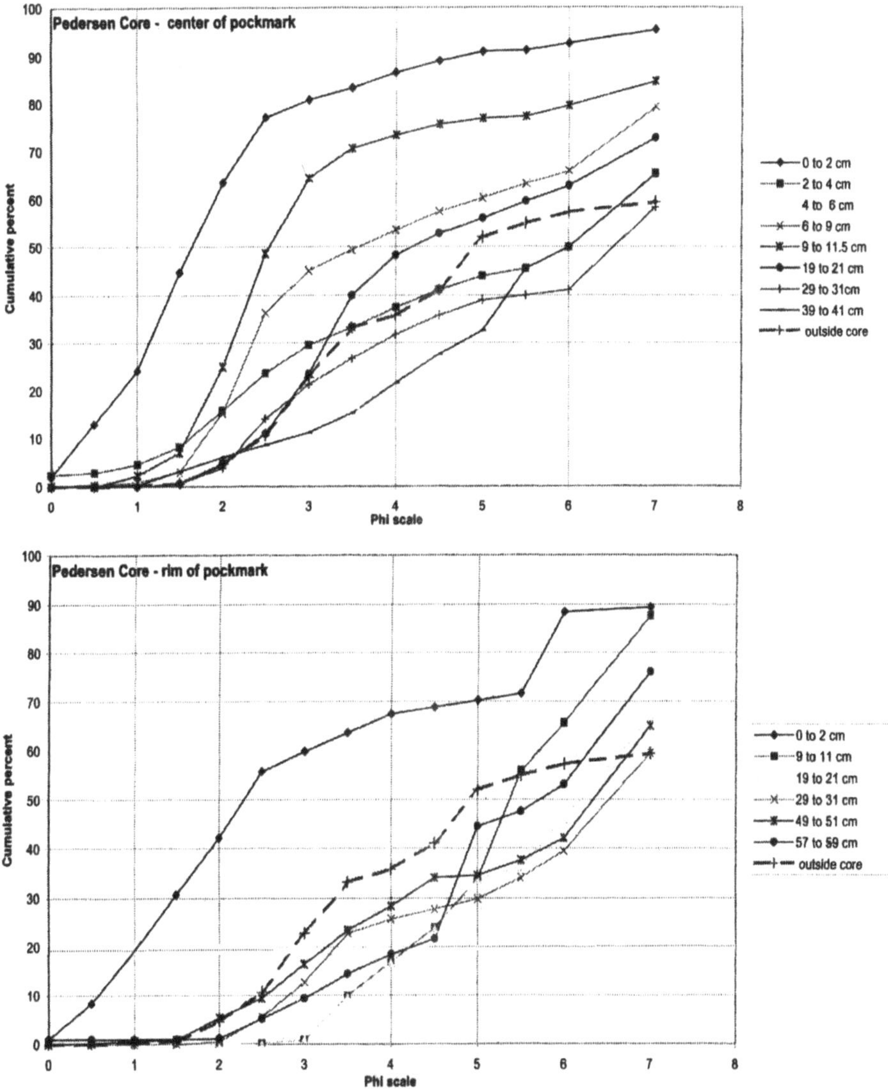

Figure 6. Cumulative percent plots of grain size for the push cores taken in 1992. The center of the pockmark is very coarse grained for the upper 20 cm of the core and then was silt dominated at depth. The grain size for the rim of the pockmark is coarsest at the surface and then becomes more silt/clay dominated at depth. The single surface sample obtained from the outside of the pockmark (outside core) had a grain size that is silt/clay dominated and was comparable to those samples found at the rim of the pockmark.

Plots of magnetic susceptibility and physical properties for GULP cores show that cores within the pockmark are consistently coarser, have higher magnetic susceptibility and lower water content than those outside the pockmark (Figure 7).

4.2. ^{210}Pb and ^{137}Cs dating

In the deepest portion of the pockmark, at the base of the north rim and the floor, there is barely detectable ^{137}Cs in the top 5 cm of GenDeep, and no detectable excess ^{210}Pb (Table 1). This implies minimal net sediment accumulation (< 0.1 mm/yr) with mixing dominating the profile. On the rim of the pockmark, no ^{137}Cs or excess ^{210}Pb was detected within the top 0 - 1 cm; therefore no modern accumulation. On the outside of the pockmark, to the east (GenOut2), there is a strong ^{137}Cs signal along with excess ^{210}Pb. The peak ^{137}Cs near the surface indicates a mixing-dominated profile with a net accumulation rate of a few tenths of a mm/yr (Table 1).

4.3. ADCP and Vertical T-Chain

Eleven months of data (September 1999 – July 2000) were organized into monthly color-filled contour plots of current speed (N-S, E-W and vertical), backscatter intensity and temperature. Of the eleven monthly plots, only September 1999 and February 2000 are provided in this paper (Figures 8 and 9, respectively).

Vertical temperature for September (Figure 8), June and July show well defined stratified conditions, during which the hypolimnion (cold, deep water) lies beneath the epilimnion (shallow, warm water). October through May show temperature transitional characteristics from stratified to unstratified conditions (Figure 9). Throughout periods of stratification, internal lake modes were frequently observed in the records as a somewhat periodic replacement of the warmer bay-water with colder, denser water from the deeper parts of the Main Lake located to the west. These oscillatory events possessed periods that ranged from 1.8 to 7.4 days, however the dominant range was from 5.5 – 7.4 days with a mean of 6.4 days. This observed average period of 6.4 days reflects neither the typically accepted Main Lake internal standing mode of 4.3 days or the atmospherically forced modes of 7.1 and 10.7 days (T. Manley et al., 1999). During weaker stratification periods of the lake (e.g., caused by cooler summers), T. Manley et al. (1999) noted that the change of density contrast (i.e., across the metalimnion) from 0.002 to 0.001 g/cm^3 would be sufficient to change the Main Lake internal mode from 4.3 days to 6.1 days.

Over the span of observations, there was an exceptional event that occurred from 16-18 September 1999. During this period, the ADCP and vertical temperature chain recorded a rapid rise of hypolimnic waters to the surface. Prior to this event, currents were to the NW with little if any vertical component. During the event on the 17th, southeast currents are accompanied with a noticeable downward velocity. Immediately following the event, strong currents flowing to the northwest were associated with a significant vertical current reaching a maximum upward velocity of ~7 cm/s. This upward flux is the largest ever documented to date for the 10 years of hydrodynamic measurements taken in Lake Champlain (T. Manley et al., 1999). While there were no other thermal events of this nature found in the year long data set, there were fluctuations in the depth of the metalimnion, but they were not as large, nor were they associated with large values of vertical velocity.

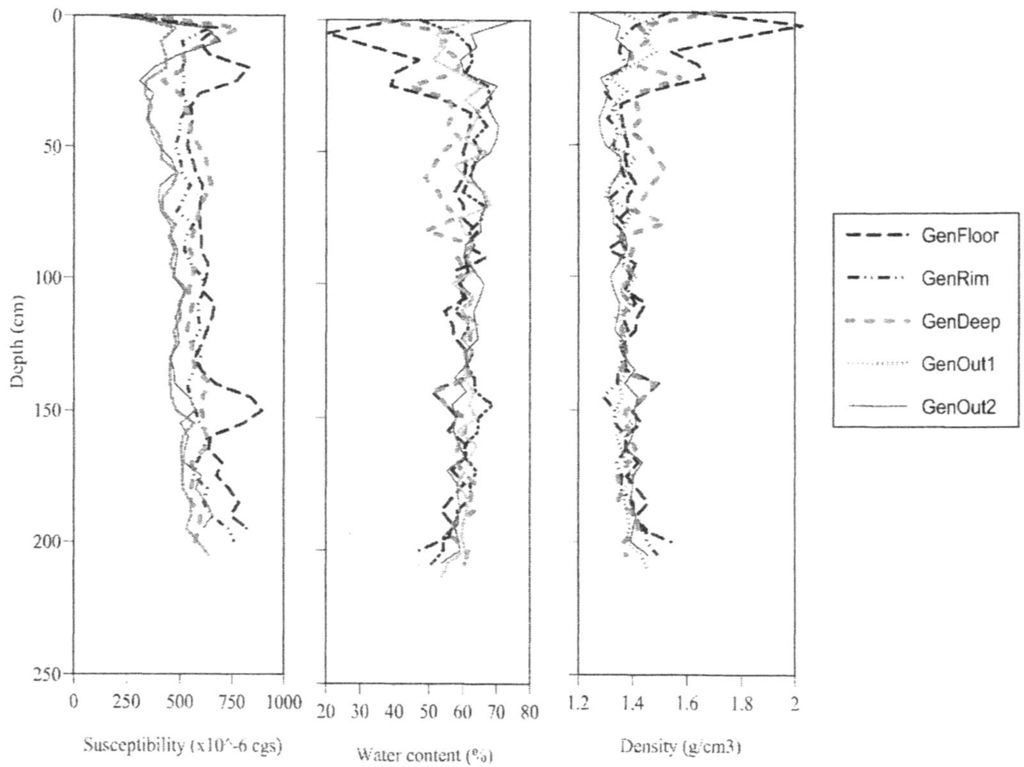

Figure 7. Comparison of magnetic susceptibility, water content and density for all piston cores. Note that cores located in the pockmark (GenFloor, GenDeep and GenRim) have higher magnetic susceptibility, density and lower water content than those outside (GenOut1 and GenOut2).

Table 1. Table showing levels of ^{137}Cs and ^{210}Pb in the top 10 cm of cores GenOut2, GenDeep and GenRim.

Depth (cm)	^{137}Cs (pCi/kg)	^{210}Pb(dpm/g)
	GenOut2	
0-1	2674	8.73
1-2	998	3.26
2-3	365	3.41
3-4	138	2.63
4-5	178	1.98
5-6	85	2.24
6-7	97	0.84
7-8	-8	1.91
8-9	-31	1.51
9-10	37	0.40
	GenRim	
0-1	2	-0.40
1-2	1	-0.43
2-3	-52	-0.83
3-4		
4-5	-2	0.23
5-6	-37	-0.12
6-7		
7-8		
8-9	44	0.32
9-10		
	GenDeep	
0-1	86	0.59
1-2	98	-0.14
2-3	42	-1.05
3-4	86	0.59
4-5		
5-6		
6-7		
7-8		
8-9	-76	0.27
9-10		

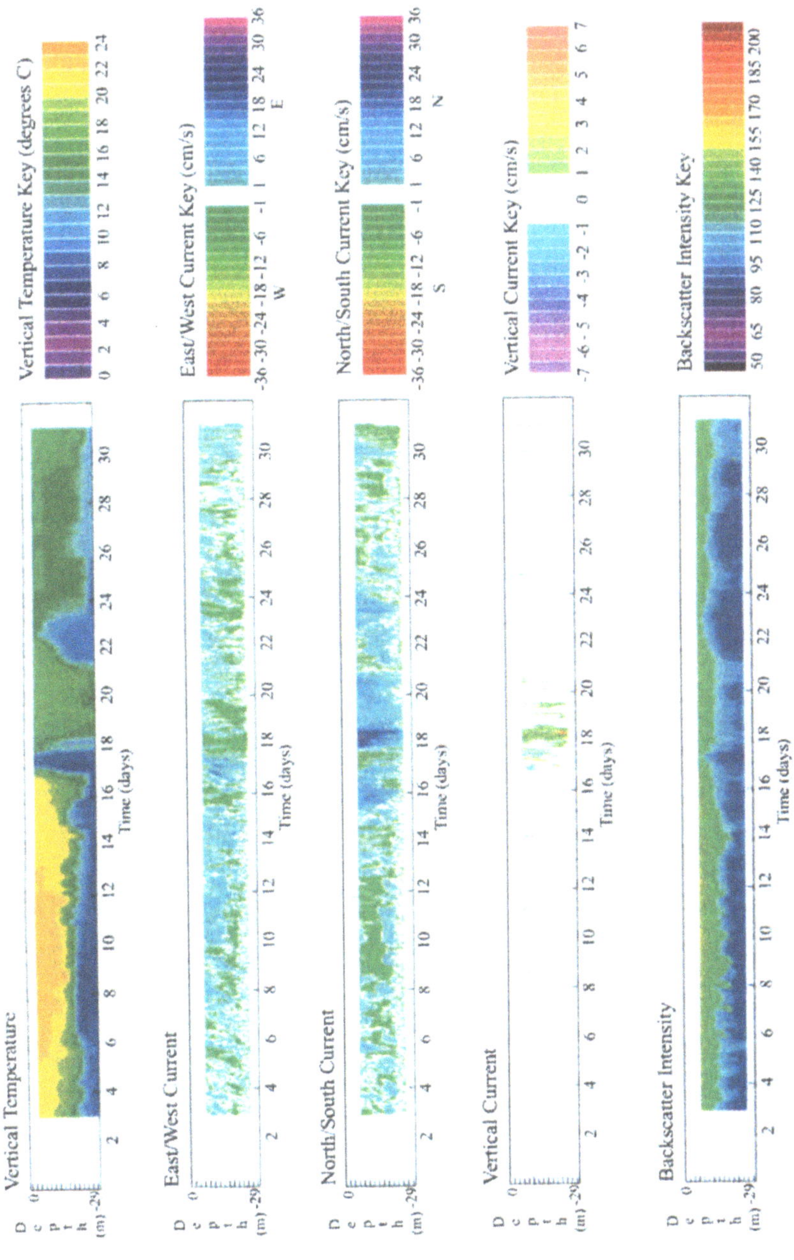

Figure 8. September 1999 ADCP and vertical temperature chain data. Note the thermal event occurring on September 17th and the high vertical velocity (~7 cm/s) on September 18th.

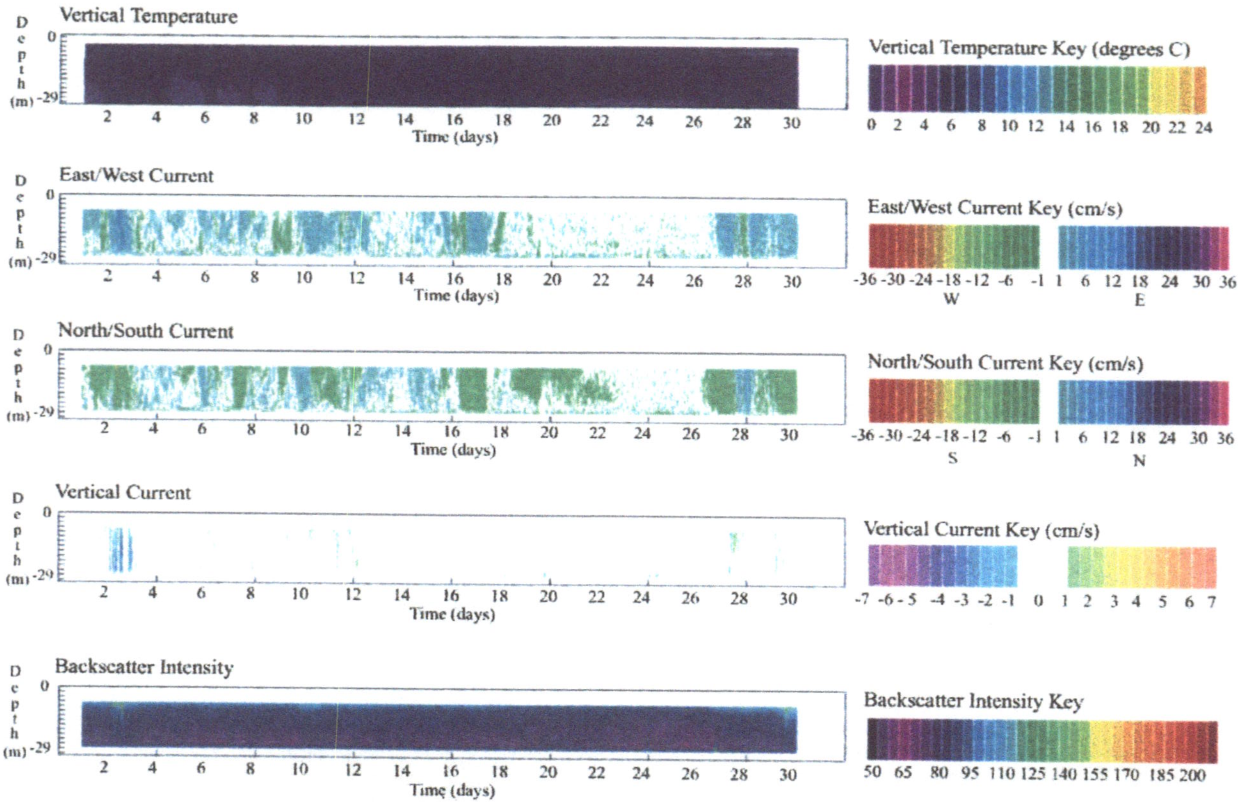

Figure 9. February 2000 ADCP and vertical temperature chain data. Note the nearly isothermal structure of the water column with the exception of the anomalously warmer water at the bottom.

4.4 Horizontal T-Chain

To gain some relational perspective between the vertical and horizontal components of temperature, data were combined, gridded, and color-filled contoured every two hours. Results outside a range of 5 m from any temperature sensor were removed thereby deleting any extrapolated or excessively interpolated results. Representative plots (Figure 10) have the x-axis as the horizontal distance from the mooring (deepest part of the pockmark with variations in bottom topography taken into account) and the y-axis as depth from the surface of the lake. The south rim of the pockmark is located from 12 m to 16 m along the horizontal temperature chain and distances greater than 16 m are outside the pockmark.

A typical stratified water column is shown in Figure 10a. Six days later (Figure 10b), the coldest water was located near the outer perimeter of the floor and rim of the pockmark. Stratification within the water column was later reestablished on the 1st of October (Figure 10c); the coldest water was again concentrated on of the floor and rim. While there was some residual cold water at the mooring site (deepest point) the coldest water was located along the outer perimeter of the floor and most of the rim. Same-depth temperature comparison between the second and fourth temperature sensors (vertical and horizontal t-chains, respectively with the first sensor being located at the mooring anchor) show a marked difference of 4.4 °C and emphasizes the non-uniformity of the water column over short distances of less than 20m. This trend of stratification and cooling pulses on the outer perimeter of the floor and mid to lower section of the rim continues throughout September and October. After the lake became isothermal, anomalous warm pulses were associated with the near-rim portion of the pockmark (Figures 10d and 10e). However, the number of occurrences and temperature differences dropped significantly. As stratification redeveloped in the following spring and summer, the pattern of large thermal differences repeated itself (e.g., Figure 10f).

Plots of both temperature and temperature differences for the second vertical (2V) and fourth horizontal (4H) temperature sensors (away from the mooring anchor) over a 37-day period (September 1 – October 11, 1999) are provided in Figure 11. These two sensors are at nearly the same depth but the horizontal sensor is located at the middle of the southern rim, approximately 20 m away (Figure 10d). Nearly identical temperature structure can be seen several times during the record (e.g., Julian days 246-251, 280-283) when there were gradual variations in the water column. These similar temperature traces shows that the rim sensor was not buried in the sediment. Only once during a large thermal excursion (internal seiche activity) did both sensors closely record the same temperature (event b in Figure 11). However, several thermal events (a, c, d and e) have marked dissimilarities. As temperature increased at the mooring (due to metalimnion depression), the rim sensor located 20 m away 1) warmed gradually (event c), 2) remained fairly constant (events a and e), or 3) cooled over time (event d and the later portion of event b). Similar event structures were seen during the summer of 2000 when internal seiche activity was present. After mid-October and prior to mid-May, the magnitude and frequency of the temperature differences between the two sensors decreased rapidly until they could no longer be considered due to limitations of instrument accuracy. The only significant event where warmer water was found at the rim (average of 0.15 °C warmer) was observed from February 22 to 25. After February 25th, both sensors showed similar temperatures until the onset of stratification in early May.

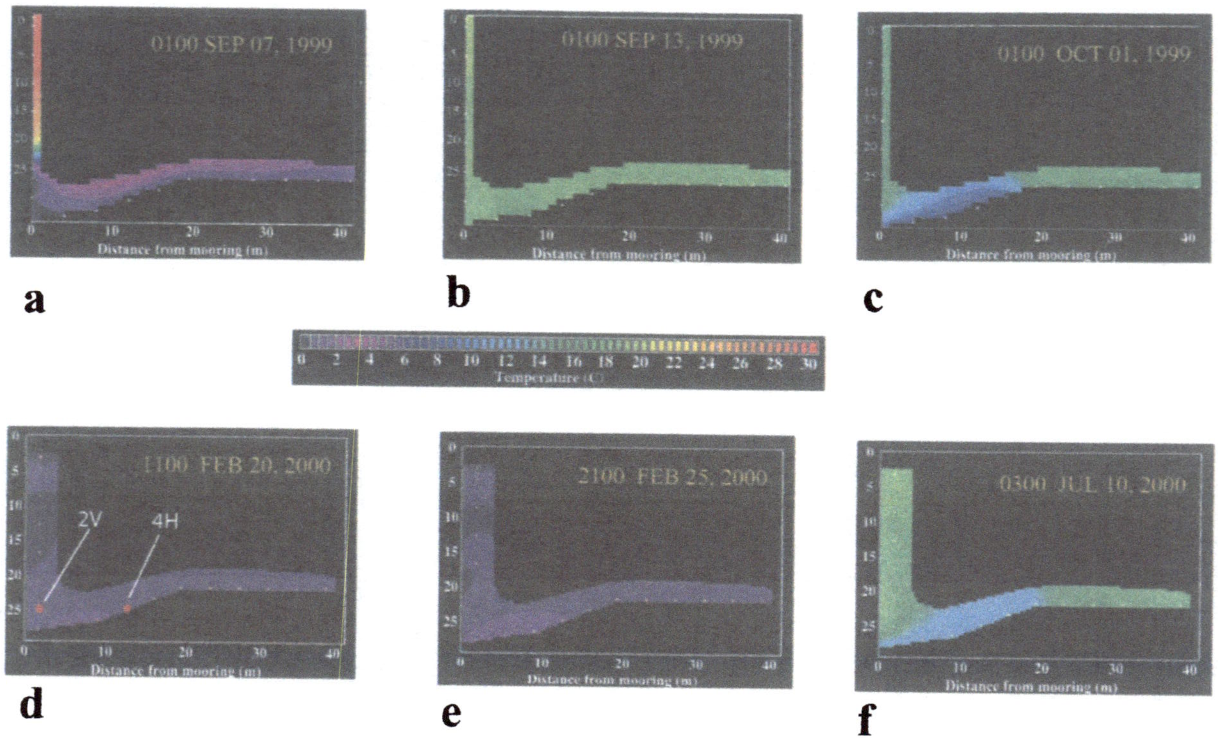

Figure 10. Plots of vertical and horizontal temperature data. Vertical temperature data is located at the mooring at the deepest point of the pockmark. The horizontal temperature data represents the distance from the mooring along the bottom of the pockmark and over the south rim. Distances greater than 16 m are outside the pockmark. Plot a) shows the stratified conditions of the water column, b) and c) anomalous cold water along the rim and not at the base of the mooring, d) and e) isothermal conditions during February but the rim had anomalously warmer water, f) stratified conditions returned the following year with cooler water along the rim.

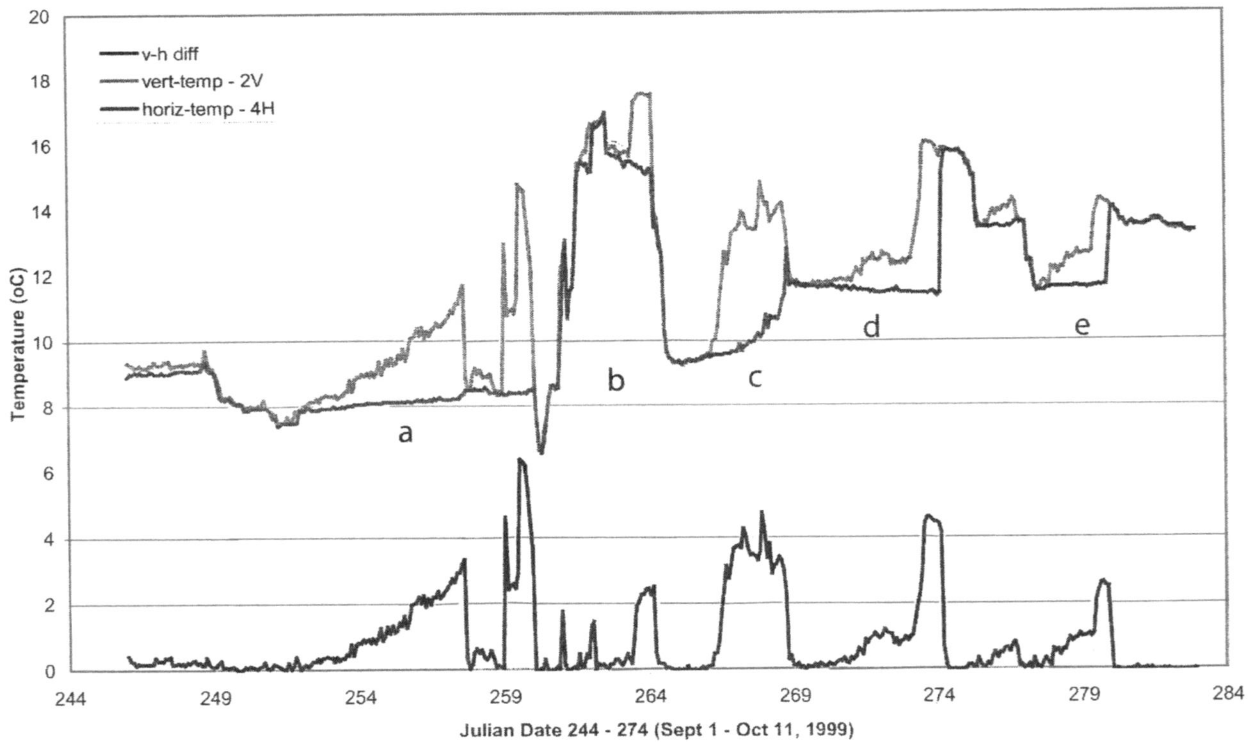

Figure 11. Plots of temperatures observed at the second vertical (2V) and fourth horizontal (4H) temperature sensors away from the mooring anchor and at approximately the same depth. Temperature differences between these two sensors are provided in the lower portion of the graph. Thermal events a-e are described more fully in the text.

4.5. Benthos Stereo Cameras

The 45 days of photos show varying degrees of sediment resuspension: from completely clear to a completely obscured water column. Some of these resuspension events are correlated with episodes of vertical current, such as the event of September 17, 1999 and the vertical currents on October 7th, 9th and 14th, whereas other resuspension events were not correlated with any vertical current.

Four photographs taken during the cold-water influx event on September 17, 1999 are shown in Figure 12. Figure 12a was taken on September 18th at 1300 hours and showed moderately high levels of sediment resuspension but the frame could still be seen. The photo on September 19th at 0100 hours (Figure 12b) showed slightly less resuspension than the previous photo, but by 1200 hours there was large-scale resuspension of sediment occurring (Figure 12c), and the frame is not visible at all. Finally, 12 hours later on September 20th at 0000 hours (Figure 12d), the frame was completely visible with no sediment resuspension. The large amount of vegetation caught by the camera frame was *Potamogeton* (Watzin, M., Pers. Comm.).

5. DISCUSSION

5.1. "General" pockmark origin

Pockmarks are generally found in regions of hydrocarbon generation and accumulation, and gases associated with these features are thermogenic in origin (Hovland and Judd, 1988). Other reports of pockmarks have demonstrated that the gas is biogenic in origin (e.g. Garcia-Gil et al., 2002) and in rare instances, pockmarks are inferred to form from pore-water escape (e.g. Whiticar, 2002).

Lake Champlain has accumulated its bottom sediments since the last glacial maximum (LGM) in a tectonically stable region. Basement consists of Precambrian metamorphosed crystalline rocks or Cambrian/Ordovician-age carbonates, sandstones, and shales with little hydrocarbon potential. Though possible, it seems unlikely that deep-seated thermogenic gases cause pockmarks in Lake Champlain.

Lake Champlain and Champlain Sea sediments average between 4 to 6% organic matter whereas Lake Vermont sediments are low in organic content usually less than a few percent (Chase and Hunt, 1972; Hunt, 1980). Thibodeau (1996) determined that a minimum of 9% organics in the sediment is needed for methane gas to be produced and found methane-enriched sediments only northeast of Valcour Island in Lake Champlain. Acosta (1984) states that 7% organic matter is sufficient for generating gas that can cause acoustic wipeout in seismic profiles. The EdgeTech seismic survey imaged up to ~20 m of sediment thickness in the region of the pockmark. The absence of acoustic blanking throughout the profiles demonstrated that no significant accumulation of biogenic gas was in this region (Figure 13). If gas were present, the coarse grained nature of the sediment inside the pockmark would allow the gas to migrate through. Thus there would be no build up of pressure needed to resuspend sediment.

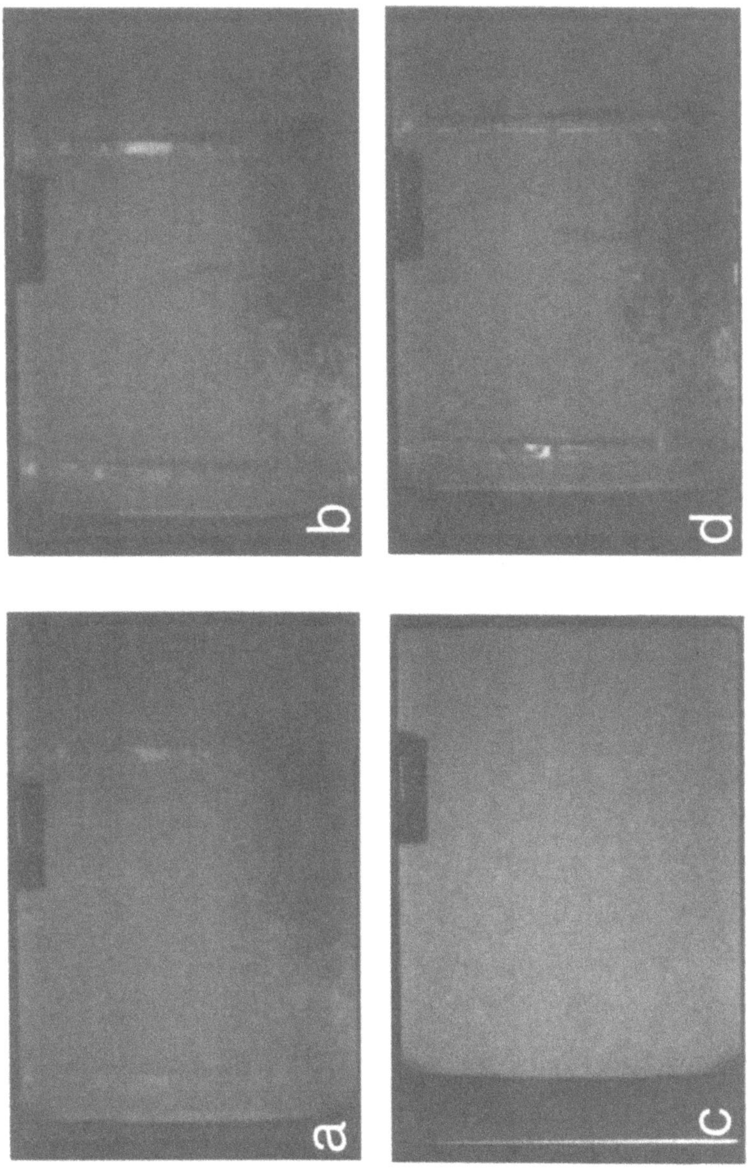

Figure 12. Bottom photographs taken during the 1999 field season. Camera reference frame is 100 cm square; a portion of which is shown in the photograph. Large and small crosses are fiducial marks used to correlate stereo pairs. Time sequence of frames are as follows: a) September 18 at 1200 hour; b) September 19 at 0000 hour; c) September 19 at 1200 hour; d) September 20 at 0000 hour.

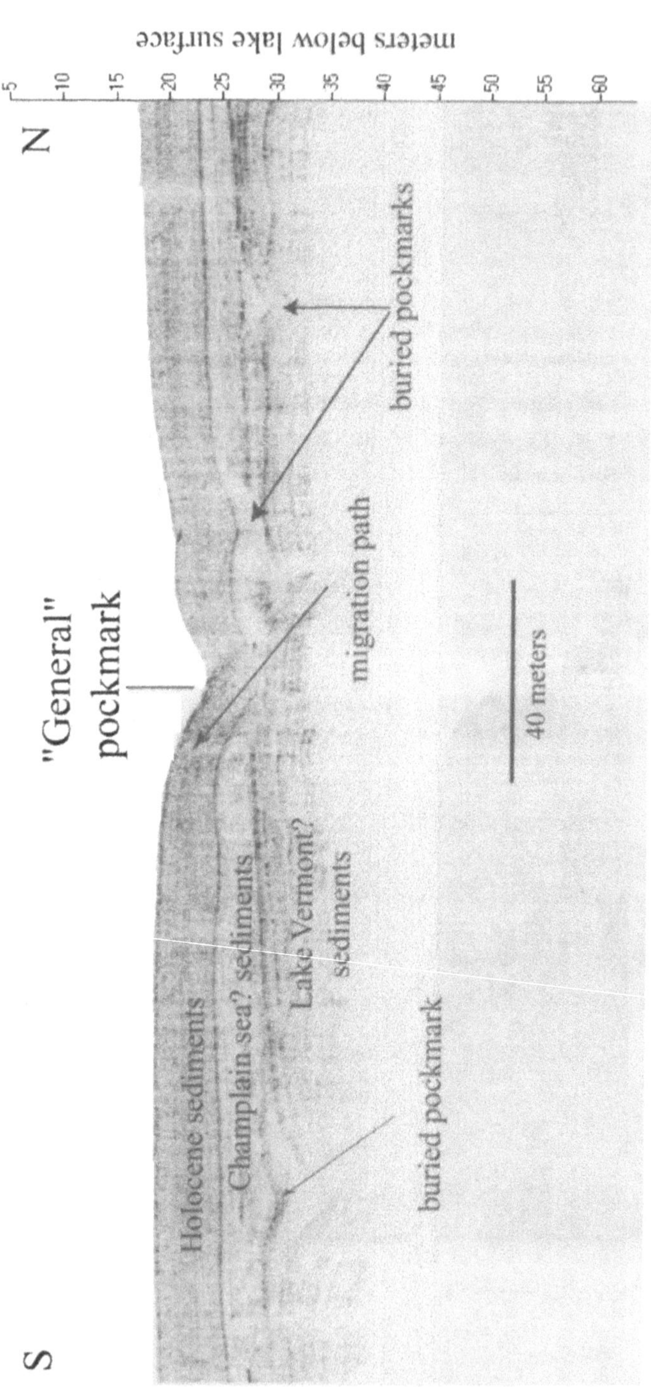

Figure 13. EdgeTech FM-2-16 kHz swept frequency (Chirp) seismic profile south to north through the large pockmark. Notice the field of buried pockmarks. The "General" appears to have existed for a long period and showed a southward migration.

The evidence suggests that the most likely cause for the pockmark formation is groundwater release. The mountains that border both sides of the lake provide sufficient hydraulic head to drive groundwater flow in the region. There are also significant faults throughout the region to channel the flow of groundwater. It is common to see water spurts (6 to 8 inches) along the mullions at the base of the Champlain Thrust exposed at Lone Rock Point north of Burlington Bay (Coish, R., Pers. Comm.; Figure 2).

For large-scale circulation patterns of the lake, the temperature difference between proximal sensors (discussed above) should be minimal. These conditions existed throughout the 37-day record (Figure 11), however there were anomalous thermal differences up to 6 °C. While event c (Figure 11) might reflect gradual warming of cooler water in the central core of the pockmark, the lag time between the onset of the temperature increase at the mooring and the final merging of the rim and mooring temperature traces appears to be excessive (2.7 days). In other instances (Figure 11 labeled a and e), rim temperatures remained virtually constant for periods up to ~ 6 days. In two other episodes, rim temperatures actually decreased over time (Figure 11 labeled d and b). The combination of these large thermal differences at nearly the same depth, a small separation span of ~20 m, and the lower than expected temperatures of the rim over long periods of time, suggest the input of water from an outside source (i.e., groundwater). Between mid-October and mid-May, thermal difference events rapidly become smaller in amplitude and less frequent. Only a single case in the middle of wintertime conditions showed a warmer rim temperature of ~0.15 °C for a period of ~3 days.

Why the major thermal-difference events are associated with internal seiche activity and why there was paucity of thermal events from mid-fall to mid-spring can only be speculated. Internal seiche activity may provide the best opportunity for observation of upwelling groundwater simply due to the large thermal excursions that are created by the oscillating metalimnion. In other words, the warmer the temperature above the sediment water interface, the more readily apparent is the groundwater signal. The lack of significant events after the onset of non-stratified conditions may be attributed to 1) our lack of knowledge pertaining to temporal variation of groundwater temperature at depth (i.e., warmer or colder than the overlying lake water), 2) the change of groundwater thermal characteristics as it moves from the confining aquifer to the sediment-water interface, 3) a lack of a significant hydraulic head for a majority of winter months, and 4) varying upwelling speeds.

Rise et al. (1999) demonstrated that seabed pockmarks form as fluid flows up dip along permeable layers where the Quaternary sediment is the thinnest, and therefore the internal strength of the sediment is most readily overcome. Based on the seismic profiles obtained in the region, it appears that the "General" pockmark has, over time, migrated along an inclined path, with an up-dip direction towards the south. The presence of other similarly shaped features (depressions) found in the deeper layers within the vicinity of the "General" (Figure 13) further suggests that pockmarks were active within this region over the past several thousand years.

5.2. Vertical Flow Event

The alignment of pockmarks with basement faults and apparent groundwater migration along these faults, it is assumed that groundwater upwelling through the pockmark would be captured by vertical current velocity data obtained from the ADCP, particularly if there were explosive or high discharge events. The event observed from September 16th through the 18th in the ADCP and the vertical temperature chain data was exactly the type of explosive event that was expected: a large cold thermal spike with associated vertical current velocities and active sediment resuspension (Figures 8 and 12).

During the 15th and 16th of September, winds were building from the north with average daily speeds of 0.7 to 3.9 m/s, respectively. During this time, stable stratification and bi-directional flow were occurring within the water column and in agreement with internal standing wave (seiche) dynamics. During the 17th, average daily wind speed rose to 10 m/s from the north (340°) with gusts as high as 21.6 m/s. These exceptionally strong winds rapidly forced warm epilimnic water to the south while deep, cold hypolimnic water from the central lake was brought in to replace the loss of surface water. The movement of water was so rapid that the metalimnion was displaced from ~24m to less than 3m within 4-5 hours. Judging from the rapid ascent of the metalimnion as well as the duration of the wind event, hypolimnic water most likely outcropped at the surface.

On September 18th the wind was still out of the north (350°), but had abated to 2.6 m/s, thus, permitting the surface and internal lake systems to relax as currents rapidly switched to the northwest. As seen in Figure 14, the metalimnion was depressed well past its pre-event depth to be missing from the "General" pockmark thermal records. Most likely, the metalimnion intersected the sediment-water interface west of the site in deeper water. Additionally, the cold thermal spike (previously discussed, Figure 8) was associated with negative (downward) vertical velocities while upward velocities (September 18 – Figure 8) were associated with a return of warm water into the pockmark. Located to the south, data from a mooring in the Shelburne Trough shows a similar thermal signature (Figure 14) when the hypolimnion approached the surface but did not occupy the entire water column as was observed over the general. Based on the occurrence of the cold water pulse at both sites, as well as the marked correlation of longer period variations of the metalimnion between the two sites, it was apparent that the September 17th cold water pulse event was not due to an (Figure 12) resuspension event but rather hypolimnic water infiltrating the entire bay from a high amplitude, wind-forced seiche event. The turbidity captured by the bottom photographs can be attributed to a high velocity component of the internal seiche flooding the bay carrying in resuspended sediments.

The reversal of vertical currents associated with the event on September 17 can be understood by the placement of the mooring in the pockmark. The mooring was located in the northwestern sector of the pockmark at its deepest point. As the current was flowing toward the southeast it flowed along the lake bottom (~21 m) until the pockmark was encountered (~28 m). Resulting flow down into the pockmark was observed by the ADCP as a negative (downslope) vertical current (Figure 8). Conversely, when the bottom current switched flow direction toward the northwest during the relaxation phase, flow was directed upward and out of the pockmark at the mooring, thus, creating a positive (upslope) vertical current. The opposite would have been observed if the mooring had been placed near the southern rim of the pockmark.

Although the pockmark is not yet positively connected to groundwater flowing out of the lakebed, the physical property data from the cores, ^{210}Pb and ^{137}Cs radionuclide analyses, thermal observations of the water column directly above the pockmark and video surveys suggest that this pockmark is active. Comparison of magnetic susceptibility and physical properties in the top 50 cm of piston cores (Figures 6 and 7), reveal that the cores from within the pockmark show higher magnetic susceptibility, lower water content and higher bulk density than the cores from outside the pockmark. Pedersen's (1992) grain size data show that the inside of the pockmark has coarser grain size as well as higher percentage of sand than the outside area (Figure 6). These data suggest that some process continues to winnow the fine-grained sediments from within the pockmark. The ^{210}Pb and ^{137}Cs data indicate that no modern sediment has accumulated on the rim and only a slight accumulation of sediment occurred at the deep site the pockmark (Table 1). The horizontal and vertical temperature chain data suggests that sporadic pulses of anomalous temperature water are systematically located from the outer perimeter of the pockmark to the midpoint of the rim; the rim being more dominant (Figure 10).

5.3. Conceptual Models of Pockmark Dynamics

In order to account for the observations within this pockmark, four conceptual models are proposed: groundwater seepage, contour-following currents, flow separation, and Taylor column spin up.

The first model requires low volume and sporadic seepage of groundwater from the rim and to a lesser extent, the base of the pockmark. This model can explain the temperature anomalies observed throughout the year. If the seepage of groundwater is strong enough it will suspend the fine-grained sediments leaving coarser-grained material inside the pockmark. This loss of fine grained, recent sediments by groundwater upwelling therefore explains the absence of modern day accumulation from the rim cores as documented by ^{210}Pb and ^{127}Cs. Watzin et al. (this volume) noticed distinct water chemistry variations within and outside the pockmark. Most notably was the increase of Ca^{++} within ~0.5 m of the sediment/water interface along the rim of the pockmark. These statistically significant values of Ca^{++} and Mg^{++} indicate that both were being released near the sediment/water interface. In previous cores from this area, Chase and Hunt (1972) showed that elevated Ca and Mg ions in interstitial water from the Champlain Sea sediment were due to the original saline environment of the sediments. As a result, groundwater migrating up through these sediments would most likely become enriched with Ca and Mg ions. In addition, the Paleozoic carbonate basement rocks should not be neglected as a source for these ions.

During low velocity flow, currents following bathymetric variations along the lake bottom (2nd model) could account for the observed positive and negative values of vertical currents located at the rims of these features. As mentioned earlier, the positioning of the ADCP would have significant bearing on whether upwards or downwards velocity was observed during any specific direction of current flow. In the case of this setting, the instrument would observe downslope movement with southerly flow and upward velocity with northerly flow. This type of flow regime, however, would not have the capacity to preferentially restrict (deposit/erode) sedimentation within or outside the pockmark. Thus it cannot explain the absence of modern accumulation of sediments at the rim along with minimal accumulation over its deep central floor. Flow down the rim, along the bottom and up the other side of the pockmark would tend to disperse the loose debris instead of concentrating it at the center of the pockmark.

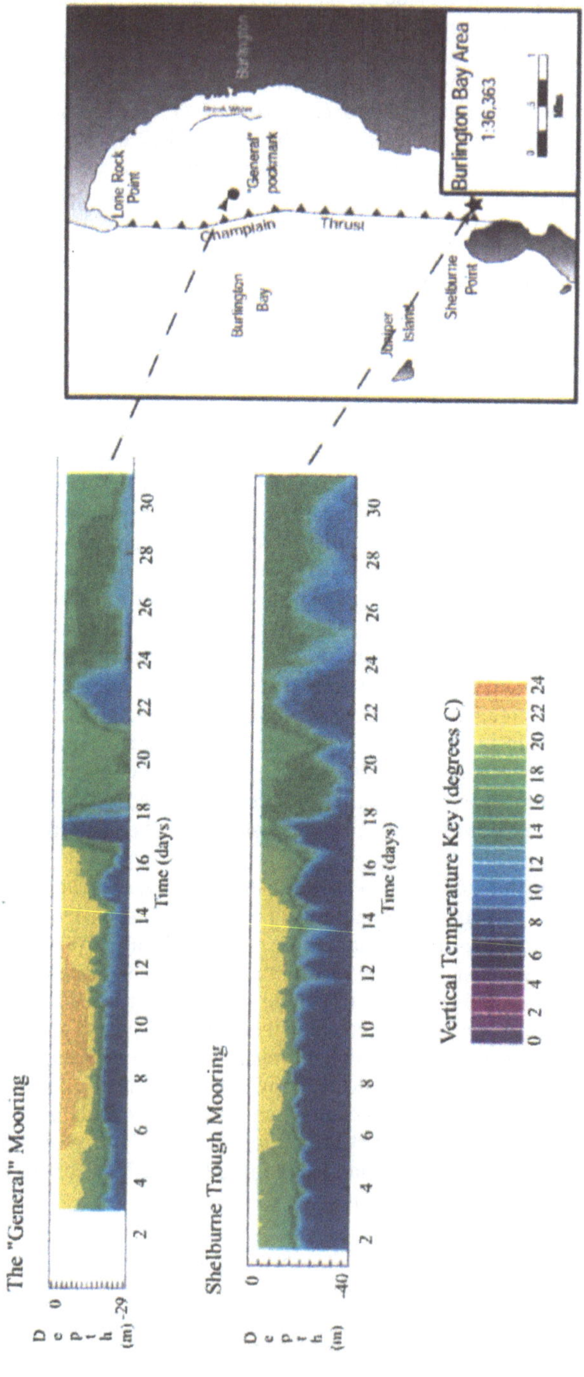

Figure 14. September 1999 vertical temperature data for the pockmark mooring and Shelburne Trough mooring. Each data set showed similar temperature trends for the September 17th thermal event.

Flow separation (3rd model) can create back eddies within the pockmark if the current across its top rim is sufficiently fast enough to no longer follow bottom topography. Depending on the direction of the current, this model could explain both positive and negative vertical currents recorded by the ADCP, although they would be more complex than the bottom following (lower flow) system described above. The lack of recent accumulation of sediments within the rim of the pockmark can be accounted for by virtually any type of flow that will take suspended fines and move them to shallower depths, there to be dispersed by the mean flow away from the pockmark (e.g. Josenhans et al., 1978). As with the previous model, this case cannot address the concentration of loose debris at the center of the pockmark, or the slight accumulation of sediments on the central deep floor of the feature. It should also be mentioned that high velocity flows, in excess of 20-30 cm/s, while they do occur in this region, are not the norm. As such, the use of flow separation in the modification/maintenance of pockmarks may be very limited.

A fourth model, Taylor column spin up based on the concept of conservation of potential vorticity, is more speculative but has the ability to explain the coalescing of loose debris at the center of the pockmark floor. Conservation of potential vorticity states that the sum of relative and planetary vorticity divided by the depth of the fluid is equal to a constant (Pond and Pickard, 1983). In this specific case, the Earth's rotational component or planetary vorticity can be considered to be constant for small scale, short duration flow features, thereby leaving relative vorticity and thickness of the water column as the only variables. As the thickness of the water column increases, relative vorticity must also increase to maintain the balance between terms (i.e., the constant). As a result, a water column moving over a depression must stretch (increase) and concurrently increase its vorticity (spin about an axis in a counterclockwise fashion) to maintain the balance between the two terms. The creation of a rotating body of fluid over the pockmark represents a cyclostrophic setting. Here, the temporal aspects of the flow are significantly shorter than the inertial period ($1/f$) where f, the Coriolis parameter, is a function of the Earth's angular velocity and latitude. The resulting balance of forces within these small-scale short-lived features (e.g., water spouts and dust devils) would be between the internal pressure gradient and centripetal components. In essence, the resulting effect for this specific application of a pockmark would be to set up a counterclockwise rotating column of fluid above the feature that would be in cyclostrophic balance. Although the relational aspects of the conservation of potential vorticity indicate that the rotation of the flow field will be counterclockwise, cyclostrophic flow can be in either direction (i.e., clockwise or counterclockwise; Neumann and Pierson, 1966). The internal pressure gradient provides the driving force for the feature. In the case of the pockmark, the lateral differences in temperature between the mooring and the rim would be sufficient to provide a pressure gradient within the depression of the pockmark.

A unique byproduct of cyclostrophic flow with bottom friction is the creation of an inward directed flow of water, within the bottom boundary layer, to the center of the feature. Induced vertical circulation would then be created at the center of the feature while downward flow would be located on the outside perimeter. Consequently, any loose or light debris could then be forced to the center of the pockmark and away from the feature's lateral boundaries, as do tea leaves at the bottom of a stirred cup of tea. This type of central accumulation of material along the bottom will occur if the feature spins clockwise or counterclockwise. In addition, cyclostrophic flow can remove fine-grained sediments where they are resuspended on the rim of the pockmark, and cause an accumulation of sediment in the deep central floor of the feature due to the net accumulation of debris at the center. In such a scenario, it is possible that a low percentage of fine-grained sediments are directed towards the center of the pockmark.

The present day hydrodynamics and physical processes occurring at the "General" are best modeled using a combination of the various models since no single case can account for all the observed characteristics. The anomalous thermal signatures associated with the base and rim of the pockmark suggests sporadic groundwater seepage (model 1). At this time, no positive correlation exists between precipitation at the Burlington International Airport and the sporadic groundwater seepage (Gutierrez, 2001). This seepage accompanied by current flow along the bottom could account for the positive and negative values of vertical current flow as recorded by the ADCP. Furthermore, if the groundwater seepage were strong enough to suspend fine-grained sediment above the floor, any current over the pockmark could remove the fine-grained sediments, leaving behind coarser grained sediments and cause no modern day accumulation (models 2, 3 and 4). Finally, the only method that could account for the consistently observed concentration of loose debris in the center of the pockmark would be that of cyclostrophic flow modified by bottom boundary layer frictional dynamics.

5.4. Water Budget

Shanley and Denner (1999) state that the high watershed to lake ratio indicates that surface water input would dominate over any groundwater input that is supplied from the fractured bedrock. In an attempt to better quantify this statement, pockmarks (in general) were provided temporal and spatial characteristics. Unfortunately, very little is known about upwelling rates of groundwater-generated pockmarks. As such, these terms are purely speculative and can only begin to quantitatively assess the importance of these features. First, a pockmark was defined to have a dominant area of upwelling (i.e., the mid to lower potion of the rim and a portion of its perimeter floor) as well as an associated efficiency (i.e., the percentage of the rim that has active upwelling). Temporal efficiency or the percentage of time that the feature is active over a given timeframe must also be considered as well as the vertical velocity of water moving through the sediment-water interface. For lack of any definitive information on groundwater flux into the pockmark, a conservative upwelling speed of 0.1 cm/s with temporal and spatial efficiency of 10% and 50%, respectively, provide an average outflow of 0.05 m^3/s for the "General" pockmark. Using this method and applying it to the known pockmarks in Lake Champlain (adjusting for size and assuming all are active), a total of 7.1 m^3/s groundwater influx for Lake Champlain can be estimated. This amount represents ~2.5% of mean flow of the lake discharge through the Richelieu River. If these crude estimates are correct, then indeed, the contribution of pockmarks maybe small to the overall inflow/outflow to the lake. On the other hand, if the groundwater upwelling velocity or efficiencies of these features are underestimated, their significance to the water budget may be of importance.

6. SUMMARY

The escape of groundwater is the most plausible cause for the formation of the large pockmark found in Burlington Bay. The pockmark is apparently still active and has migrated southward through time. Older buried pockmarks show that groundwater influx has had a long-term presence in Burlington Bay. Several models were presented to explain observations and characteristics of this feature. No single model accounts for all the observations. Sporadic groundwater seepage must be associated with some component of horizontal flow to remove fine resuspended sediment from the interior of the pockmark to the surrounding region. Either slow contour flowing currents or fast flowing currents that cause flow separation can remove the fine resuspended sediment. To explain the concentration of for loose debris located in the center of the deep floor, frictionally balanced cyclostrophic circulation created from Taylor column spin up over the pockmark is a suggested mechanism. Crude estimates of the volume flux of groundwater into Lake Champlain from all known pockmarks show it is a small percentage of the Richelieu River's mean outflow from the lake, however changes in groundwater upwelling rates and efficiency could significantly change the initial estimates.

Further investigations must be completed to better ascertain the influx of groundwater seepage as well as its chemical composition. A more in depth look at the relationship of precipitation and the sporadic nature of the groundwater upwelling needs to occur. It may prove beneficial to extend ground water models of the Burlington area to Burlington Bay of Lake Champlain. A closer spaced 2-D or 3-D seismic survey would aid in ascertaining the geometry of the "General" as well as identifying other buried features and orientations. Finally, a more sophisticated array of temperature and current meter sensors would provided needed information on the circulation dynamics as well as a better understanding of upwelling regions within the feature.

7. ACKNOWLEDGMENTS

This study was supported through a grant from the Lintilhac Foundation. We wish to acknowledge Captain Fred Fayette of the research vessel *Neptune* for his introduction and assistance for the initial investigations of these large pockmarks. We also thank Captain Dick Furbush of the UVM Research Vessel *Melosira* for assistance with the mooring deployment and piston cores. Art Cohn, Jonathan Eddy and Dave Andrews provided SCUBA assistance for the deployment of the horizontal t-chain and the underwater video surveys. Dr. Richard Bopp of Rensselaer Polytechnic Institute provided the ^{210}Pb and ^{137}Cs analyses. 2-D and 3-D modeling was accomplished using earthVisions, of Dynamic Graphics, Inc., Alameda, CA. Finally we wish to acknowledge the undergraduate students whose initial work on pockmarks paved the way – Rob Pedersen, Nichole North and Shelley Sayward.

REFERENCES

Acosta, J., 1984, Occurrence of acoustic masking in sediments in two areas of the continental shelf of Spain: Ria de Muros (NW) and Gulf of Cádiz (SW), *Mar. Geol.* **58**:427-434.

Barnhardt, W. and Kelley, J., 1995, Carbonate accumulation on the inner continental shelf of Maine: a modern consequence of late Quaternary glaciation and sea-level change, *J. of Sed. Res.*A**65**:195-207.

Barnhardt, W., Belknap, D. and Kelley, J., 1997, Stratigraphic evolution of the inner continental shelf in response to late Quaternary relative sea-level change, northwestern Gulf of Maine, *GSA Bull.* **109**:612-630.

Berkson, J. M., 1973, Possible syneresis origin of valleys on the floor of Lake Superior. *Nature* **254**:89-91.

Blasco, S. and Keyes, D., 2000, Lakebed disturbance features, Long Point area, eastern Lake Erie, Great Lakes, great science, great cities, *IAGLR* **42**:A9.

Blasco, S., Harrison, P., Keyes, D., Harmes, R., and Folkes, P., 1998, Distribution, age and origins of lakebed instability features, Georgian Bay, southern Ontario, *Abstracts with programs – GSA*, 30:296.

Bouriak, S., Vanneste, M., and Saoutkine, A., 2000, Inferred gas hydrates and clay diapirs near the Storegga Slide on the southern edge of the Voring Plateau, offshore Norway, *Mar. Geol.* **163**:125-148.

Chapman, C. H.,1942, Late-glacial and post-glacial history of the Champlain valley, Report of the State Geologist of Vermont.

Chase, J. and Hunt, A., 1972, Sub-bottom profiling in central Lake Champlain- A reconnaissance study, *Proc. 15th Conference of Great Lakes Res.*:317-329.

Cole, D., Stewart, S. A. and Cartwright, J. A., 2000, Giant irregular pockmark craters in the Palaeogene of the outer Moray Firth basin, UK North Sea, *Mar. and Petrol. Geol.* **17**:563-577.

Connally, G.G. and Sirkin, L. A., 1973, Wisconsinan history of the Hudson-Champlain Lobe: The Wisconsinan Stage, *GSA Memoir* **136**:47-69.

Corliss, B. H., Hunt, A., and Keigwin Jr., L. D., 1982, Benthonic foraminiferal faunal and isotopic data for the postglacial evolution of the Champlain Sea, *Quat. Res.* **17**:325-338.

Demyttenaere, R., Ibrahim, A., Tromp, J-P., and Zulkifli, B. A., 2000, Brunei deep water exploration; from sea-floor images to depositional models in a slope turbidite setting, *AAPG Bull.* **84**:1418.

Fader, G. B. J., 1991, Gas-related sedimentary features from the eastern Canadian Continental Shelf, *Cont. Shelf Res.*, **11**:1123-1153.

Fleischer, P., Orsi, T.H., Richardson, M.D. and Anderson, A. L., 2001, Distribution of free gas in marine sediments: A global overview, *Geo-Mar. Letters* **21**:103-122

Flood, R. D., 1981, Pockmarks in the deep sea, *EOS, AGU* **62**:304.

Flood, R.D. and Johnson, T.C., 1984, Side-scan targets in Lake Superior-evidence for bedforms and sediment transport. *Sedimentology* **31**:311-333.

Foland, S., Maher, N., and Yun, J., 1999, Pockmarks along the California continental margin; implications for fluid flow, *AAPG Bull.* **83**:687-688.

Garcia-Gil, S., Vilas, F., and Garcia-Garcia, A., 2002, Shallow gas features in incised-valley fills (Ria de Vigo, NW Spain) :a case study, *Cont. Shelf Res.* **22**:2303-2315.

Grant, A. C. and Morrison, M. L., 1998, Report on survey on pockmarks in southern Laurentian Channel, *Open File Report – Geological Survey of Canada*, Report 3596, 278pp.

Gutierrez, J., 2001, The General pockmark in Burlington Bay, Lake Champlain, Unpublished senior thesis, Middlebury College, Middlebury, VT., p.63.

Harrington, P., 1985, Formation of pockmarks by pore-water escape, *Geo-Mar. Letters* **5**:193-197.

Hillaire-Marcel, C., 1974, Le deglaciation au nordouest de Montreal: Donneees radiochronologiques et faits stratigraphiques, *Revue Geographique Montreal* **28**:407-417.

Hovland, M., and Judd, A. G., 1988, *Seabed Pockmarks and Seepages: Impact on Geology, Biology and the Marine Environment*: Graham and Trotman, London, 293 p.

Hunt, A. S., 1972, Bottom sediments of Lake Champlain 1965-1971, *GSA Bull.* **9**:630-639.

Hunt, A. S., 1977, Sediment thickness, Eastern Lake Champlain, Vermont Resources Center and U.S. Department of Interior.

Hunt, A. S., 1981, Detailed investigation of suspected Holocene fault movement in Central Lake Champlain, Vermont:1-47.

Hunt, A.S., Henson, E. B., and Bucke, D. P., 1972, Sedimentological and limnological studies of Lake Champlain, *Annual Meeting - New England Intercollegiate Geological Conference, Guidebook*, **64**:407-426.

Johnson, T.C., 1980, Lake-glacial and post glacial sedimentation in Lake Superior based on seismic reflection profiles, *Quat. Res.* **13**:380-391.

Johnson, T. C., Flood, R. D., Busch, W. H., and Halfman, J. D., 1984, Effects of bottom currents and fish on sedimentation in a deep-water, lacustrine environment, *GSA Bull.* **95**:1425-1436.

Josenhans, H. J., King, L.H., and Fader, G.B.J., 1978, A side-scan sonar mosaic of pockmarks on the Scotian Shelf: *Can. J. of Earth Sci.* **15**:831-840.

Judd, A.G. and Hovland, M., 1992, The evidence of shallow gas in marine sediments, *Cont. Shelf Res* **12**:1081-1095.

Kelley, J. T., Dickson, S. M., Beknap, D. F., Barnhardt, W. A., and Henderson, M., 1994, Giant sea-bed pockmarks: evidence for gas escape from Belfast Bay, Maine: *Geology* **22**:59-62.

King, L. H., and MacLean, B., 1970, Pockmarks on the Scotian Shelf: *GSA Bull.* **81**:3141-3148.

Manley, P.L., Manley, T. O., Saylor, J. H., and Hunkins, K. L., 1999, Sediment deposition and resuspension in Lake Champlain, *In*: Manley, T.O. and P.L. Manley (eds.), *Lake Champlain in Transition: From Research Toward Restoration*, Water Science and Application, AGU **1**:157-181.

Manley, T. O., Hunkins, K. L., Saylor, J. H., and P. Manley, 1999, Aspects of summertime and wintertime hydrodynamics of Lake Champlain, *In*: Manley, T.O. and P.L. Manley (eds.), *Lake Champlain in Transition: From Research Toward Restoration*, Water Science and Application, AGU 1:67-116.

Myer, G. E., and Gruendling, G. K., 1979, Limnology of Lake Champlain. Lake Champlain Basin Study, New England River Basin Commission. 417 pp.
Neumann, G. and W. J. Pierson, Jr., 1966, *Principles of Physical Oceanography*, Prentice-Hall, Inc., 545 pp.
North, N. S., 1997, A pockmark field in the Cumberland Head area, Lake Champlain: unpublished senior thesis, Middlebury College, 55 p.
Paull, E. K. and Dillion, W. P., 1981, Appearance and distribution of the gas hydrate refection in the Blake Ridge region, offshore southeastern United States, United States Geological Surveys.
Paull, C., Ussler III, W., Maher, N., Greene, H.G., Rehder, G., Loernson, T., and Lee, H., 2002, Pockmarks off Big Sur, California, *Mar. Geol.* **181**:323-335.
Pedersen, R. C., 1992, Pockmark formation in Burlington Bay: unpublished senior thesis, Middlebury College, 63 p.
Pickrill, R.,1993, Shallow seismic stratigraphy and pockmarks of a hydrothermally influenced lake, Lake Rotoiti, New Zealand, *Sedimentology* **40**:813-828.
Platt, J., 1977, Significance of pockmarks for engineers, *Offshore Engineer*, August, Letters and Books Section, pp. 45.
Pond, S and G. Pickard, 1983, *Introductory Dynamical Oceanography*, 2nd edition, Pergamon Press, 329 pp.
Rise, L., Saettem, J., Fanavoll, S., Thorsnes, T., Ottesen,D., and Boe, R., 1999, Sea-bed pockmarks related to fluid migration from Mesozoic bedrock strata in the Skagerrak offshore Norway, *Mar. and Petr. Geol.* **16**:619-631.
Sayward, S. E., 1996, The hydrodynamic system of Burlington Bay, Lake Champlain, Vermont: unpublished senior thesis, Middlebury College, 54 p.
Scalon, K. and Knebel, H., 1989, Pockmarks in the floor of Penobscot Bay, Maine, *Geo-Mar. Letters* **9**:53-58.
Shanley, J. B. and Denner, J. C., 1999, The hydrology of the Lake Champlain Basin, , *In*: Manley, T.O. and P.L. Manley (eds.), *Lake Champlain in Transition: From Research Toward Restoration*, Water Science and Application, AGU **1**:41-66.
Stanley, R.S., 1987, The Champlain thrust fault, Lone Rock Point, Burlington, Vermont, *in* Roy, D.C., ed., *Northeastern section of the Geological Society of America: Centennial field guide*: Boulder, Geological Society of America:225-228.
Solheim, A. and Elverhøi, A., 1985, A pockmark field in the Central Barents Sea; gas from a petrogenic source?, *Polar Res.* **3**:11-19.
Teller, J.T., 1987, Proglacial lakes and the southern margin of the Laurentide Ice Sheet, in Ruddiman, W. F. and H.E. Wright, J., eds., *North America and adjacent oceans during the last deglaciation*: Boulder, Geological Society of America:39-69.
Thibodeau, M., 1996, The origin and distribution of methane in Lake Champlain sediments: *Northeastern Geology and Environmental Sciences* **18**:130-136.
Thompson, S. D., 1994, Documenting the effects of the internal seiche in Lake Champlain on a Shallow Bay: Thompson's Point Bay, Lake Champlain, Vermont, : unpublished senior thesis, Middlebury College, 54 p.
Watzin M., P. Manley., T. O. Manley, and S. A. Kyriakeas, Lakebed pockmarks in Burlington Bay, Lake Champlain II. Habitat Characteristics and Biological Patterns, This volume.
Whiticar, M., 2002, Diagenetic relationships of mathanogenesis, nutrients, acoustic turbidity, pockmarks and freshwater seepages in Eckernförde Bay, *Mar. Geol.* **182**:29-53.
Whiticar, M. and Werner, F., 1981, Pockmarks; submarine vents of natural gas or freshwater seeps? *Geo-Mar. Letters* **1**:193-199.

LAKEBED POCKMARKS IN BURLINGTON BAY, LAKE CHAMPLAIN II. HABITAT CHARACTERISTICS AND BIOLOGICAL PATTERNS

Mary C. Watzin[1], Patricia L. Manley[2], T. O. Manley[2], Sofia A. Kyriakeas[1]

ABSTRACT

Pockmarks are common features on the bottom of Lake Champlain, especially in the central part of the lake, between Burlington, Vermont and Plattsburgh, New York. Pockmarks have the potential to provide a unique bottom environment for biota because of the different water quality and habitat conditions that may exist inside and outside the pockmark. In Burlington Bay, the "General" is a large pockmark, roughly 40 meters in diameter and 4 meters in depth, located in about 28 meters of water. We collected benthic invertebrate samples inside and outside the "General" pockmark, and collected sediment, pore and overlying water for analysis. We also caged zebra mussels inside and outside the pockmark to determine whether varying water quality or food availability in the pockmark would influence these mussels.

The concentrations of Al, Fe, Ca, K, Mg, Mn, S, Si, and Zn in the overlying water were higher inside the pockmark than outside, with generally highest concentrations near the outside rim of the pockmark rather than at its deepest point. The benthic invertebrate density also varied among sites, with generally more invertebrates inside than outside the pockmark. Clinkers, the remains of coal combustion, were common in the sediment both inside and outside the pockmark. Although clinker density was highly variable, in general, fewer invertebrates were found where the density of clinkers was high. The zebra mussels inside the lip of the pockmark grew less, but had higher shell weight to body weight ratios than those outside the pockmark. This suggests that food might be limited by the water flow pattern in the pockmark, but these mussels were able to take advantage of additional calcium in that water. Mean calcium concentration outside was 14.23mg/L, while inside it was 15.14 mg/L. It also suggests that zebra mussels in the broader lake may be calcium limited.

[1] Rubenstein Ecosystem Science Laboratory, School of Natural Resources, University of Vermont, Burlington, VT 05401
[2] Geology Department, Middlebury College, Middlebury, VT 05753

1. INTRODUCTION

In Lake Champlain, side-scan sonar surveys have shown the existence of large fields of pockmarks in Burlington Bay and Cumberland Bay areas (Manley et al., this volume). Pockmarks are crater-like depressions usually formed by upward migration of fluid through a soft sedimentary bottom (Hovland and Judd, 1988; King and MacLean, 1970). Side-scan sonar mosaics of Burlington Bay have shown pockmarks ranging in size from less than 1 m in diameter to over 30 m in diameter (Manley et al. this volume). They exist as single features and as linear chains, often associated with the Champlain thrust fault and other basement faults. The linear arrangement of the pockmarks and their alignment suggests that the large pockmarks might be created by groundwater migrating upward and through the lake sediment. The "General" (named by us) is the largest of these pockmarks. It has a diameter of about 40 m and a depth of about 4 m, and is nearly circular in shape. It has a flat bottom with the deepest section, at about 28 m, at the edge of the northern rim.

Project GULP (Groundwater Upwelling in Large Pockmarks) was initiated in 1999 to investigate the hydrology and ecology associated with the General. The purpose was to determine if this feature was active, and if so, what were the main constituents of the upward migrating gas or fluid and what were the effects on the local bottom communities. Manley et al. (this volume) summarizes the hydrodynamic and geophysical data collected. This paper focuses on the water quality and biological characteristics of the pockmark environment.

Because active upwelling would mix groundwater, with potentially different characteristics, into the overlying lake water, the first objective of the biological portion of the study was to document the water chemistry inside and outside the pockmark. Because different water chemistry, and, potentially, a different sediment structure inside and outside the pockmark would provide different habitat characteristics, the second objective of the study was to survey the benthic communities inside and outside the pockmark and compare them.

In the first year of the study, zebra mussels were encountered in debris removed from the pockmark. In Lake Champlain, calcium concentrations are low for zebra mussels; in fact mean summer calcium concentrations of 15.7 mg/L (Meyer and Gruendling, 1979; Eliopoulos and Stangel, 1999) are at the lowest end of the acceptable range for zebra mussels (Mellina and Rasmussen, 1994). In general, zebra mussels do not flourish unless the average calcium concentration exceeds 28 mg/L (Ramcharan et al., 1992). A model based on data from over 250 lakes in Europe predicts that zebra mussel densities in Lake Champlain should be constrained by its calcium concentration (Ramcharan et al., 1992). In fact, Lake Champlain has the lowest calcium environments in North America known to support adult zebra mussel populations (Eliopoulos and Stangel, 1999). In laboratory experiments, others have found that when calcium concentrations dip below 20 mg/L, adult zebra mussels actually lose calcium from their shells (Hincks and Mackie, 1997). In nearby south Burlington Bay, the majority of adult zebra mussels sampled now have thin shells, and many shells actually have holes in them (Watzin et al., 2003), suggesting significant calcium limitation.

If calcium concentrations were higher in upwelled groundwater, the pockmark offered the opportunity to explore the possibility that calcium was limiting to zebra mussels. Therefore, our third objective was to compare zebra mussel growth rates and shell thickness inside and outside the pockmark in order to determine whether zebra mussels in this location could extract the additional calcium from groundwater and accrue thicker shells.

2. METHODS

2.1. Study Site

The General is located in about 28 m of water, west of the Burlington breakwater (Figure 1). It is the largest pockmark documented to date, with a diameter of 40 meters. The sampling plan was designed to characterize the water quality, physical environment, and biota both inside and outside the pockmark.

2.2. Water Samples

To characterize water quality, water samples were collected by divers using a hand-held bow-like devise that simultaneously collected five 60-ml water samples at the sediment-water interface (0cm), 10 cm above the sediment-water interface, 20 cm above the sediment-water interface, 30 cm above the sediment-water interface, and 50 cm above the sediment-water interface in a vertical profile. Three such vertical sets of samples were collected outside the pockmark, three just inside the lip of the lip of the pockmark, and three in the deepest part of the pockmark. To minimize disturbance of the water column prior to sampling, samples were collected while swimming a line from outside the pockmark into the deepest part of the pockmark. All samples were collected on August 16, 2000.

All water samples were brought to the surface, transferred to 125 ml sample bottles, preserved with 1 ml concentrated nitric acid, and kept at 4°C until analysis. Trace elements concentrations were determined using a Perkin Elmer 3000DV Inductively Coupled Plasma Atomic Emission Spectrometer (ICP) (American Public Health Association, 1995). Estimated detection limits vary by element, but were generally between 2 and 10 μg/L for the elements of concern (Al, Ca, Cu, Fe, K, Mg, Mn, Na, P, S, Si, Zn).

After log transformation to homogenize variances where necessary, data were analyzed using two-way analysis of variance with sample location and depth above the sediment water interface as the main effects. When a significant difference was found, Student-Newman-Keuls (SNK) comparisons were used to find significant differences. For some analytes, homogeneity of variance could not be achieved, even with a log transformation; therefore, analysis of variance was performed on ranks, followed by a Wilcoxon test for significantly different variables within locations. All statistical analyses were run on SAS software (Statistical Analysis Systems Institute, 1988) using an experiment-wide error rate of 0.05.

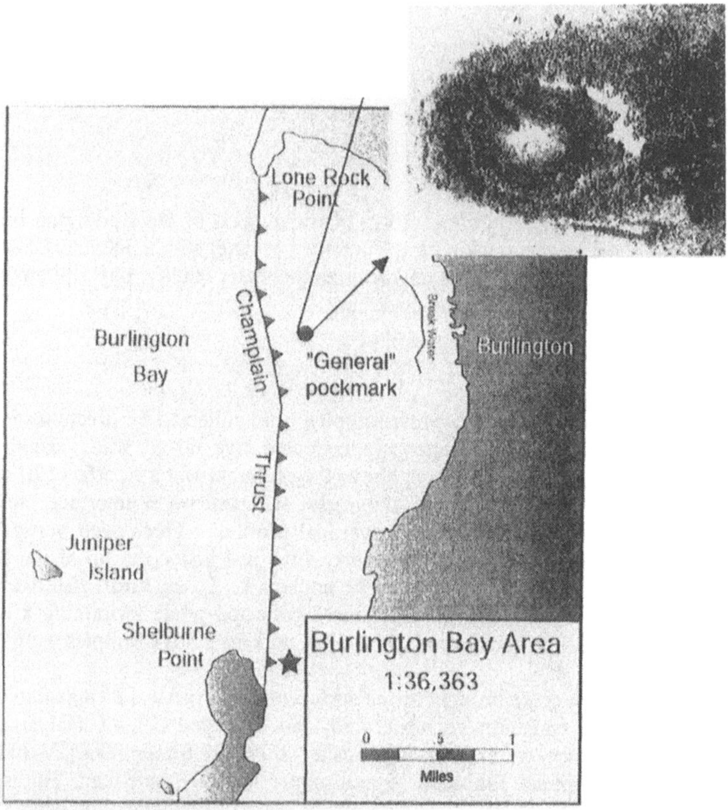

Figure 1. The location of the General Pockmark in Burlington Bay, Lake Champlain, and bottom seismic profile illustrating the shape. The central light core is the deepest portion of the feature.

2.3. Sediment Samples for Porewater Analysis

Sediment samples were collected for porewater analysis using a large Ponar grab sampler. Duplicate samples were collected outside the pockmark, inside the lip of the pockmark, and in the deepest hole (Table 1). After draining the water from the top of the grab sample, the sample was dumped into an acid washed HDPE bucket, thoroughly homogenized by stirring, and then scooped into an acid washed HDPE bottle. Bottles were packed leaving no head space, capped, sealed with parafilm, and kept at 4 °C until analysis.

Pore water was extracted by high-speed refrigerated centrifuge (30,000 rpm). Pore water samples were then acidified using concentrated nitric acid and analyzed by ICP as described above.

One-way analysis of variance was used to look for significant differences between concentrations of trace elements in the pore water at the three sample locations. When a significant difference was found, SNK comparisons were used to find significant differences. Because homogeneity of variance could not be achieved for all analytes even with log transformation, a test on ranks (Wilcoxon test) was used for some elements (Mg, Mn, S, Si).

Table 1. Collection locations, dates, and depths for benthic invertebrate and porewater samples.

Sample ID	Date	Depth (m)
	Benthic Invertebrate Samples	
Outside Pockmark		
1	8/16/2000	21.64
2	8/16/2000	21.64
3	8/16/2000	21.64
4	8/16/2000	21.95
Inside Lip		
1	8/30/2000	24.99
2	8/30/2000	24.99
3	8/30/2000	24.38
4	8/30/2000	24.99
Deep Hole		
1	8/16/2000	26.21
2	8/16/2000	26.21
3	8/16/2000	25.91
4	8/16/2000	28.04
5	8/16/2000	26.21
6	8/16/2000	27.13
	Porewater Samples	
Outside Pockmark		
1	8/16/2000	21.64
2	8/16/2000	21.03
Inside Lip		
1	8/30/2000	24.08
2	8/30/2000	24.38
Deep Hole		
1	8/16/2000	27.43
2	8/16/2000	25.91

2.4. Benthic Invertebrates

To address objective two, bottom samples were collected for benthic invertebrates using a Petite Ponar grab sampler, which collects a sample of 240 cm^2. Four replicate samples were collected outside the pockmark, four samples inside the lip of the pockmark, and six samples in the deepest hole (Table 1). Petit Ponar samples were emptied onto a 500 μm sieve, washed clean of fine sediment, and preserved in 10% buffered formalin with Rose Bengal. In the laboratory, invertebrates were picked from the residue and identified to major taxonomic group using a dissecting microscope. Because so many pieces of coal or large clinkers (the hard residue left after combustion) were noticed in the sample residues, this material was saved, dried at 105 °C for one hour, and weighed.

After log transformation to correct for heterogeneity of variance in the abundance data, analysis of covariance was used to test for significant differences between locations, using clinker weight as a covariate. Separate analyses were performed on each major taxonomic group.

2.5. Zebra Mussel Experiment

Because unpublished data suggested that groundwater percolating into Lake Champlain is higher in calcium than the ambient lake water, and the zebra mussels in Lake Champlain might be calcium limited, in addition to measuring calcium, we used zebra mussels as a biological indicator of the long-term, average calcium concentration in the pockmark.

We constructed eight approximately 12 cm by 30 cm cages of VEXAR plastic mesh. On April 14, 2000, zebra mussels were collected from nearby Blodgett's Reef, at about 10 m water depth, living on native mussel (*Elliptio complanata*) shells. Mussels were gently removed from the native mussels shell, being careful not to tear out the byssal gland. Similarly sized individuals were chosen to avoid potential age-related differences in calcification rates (Rhoads and Lutz 1980). One hundred ten similarly sized mussels were measured along the longest axis and placed into each cage. Analysis of variance was used to look for significant differences in starting shell lengths between cages; no significant differences were found.

A subsample of 50 mussels from our source population was also collected to calculate initial shell weight to body weight ratios. Each mussel from the subsample was carefully dissected, separating the soft tissue from the shell. Both soft tissue and shells were dried at 105 °C for 24 hours, and weighed. These dry weights were used to calculate the ratio of shell weight to soft body weight.

After zebra mussel additions, cages were kept in laboratory aquaria in Lake Champlain water for 36 hrs to make sure that all mussels would live. On the morning of August 16, 2000, divers placed the cages of zebra mussels on the lake bottom. Four cages were placed outside the pockmark and four cages were placed inside the lip of the pockmark. Cages were not placed in the deepest hole of the pockmark because we were concerned that oxygen concentrations there might be very low when groundwater was actively discharging.

Cages were retrieved on June 25, 2001. Each cage was opened, live mussels were separated from dead mussels, and all live mussels were measured along the longest axis. Then each mussel was carefully dissected, separating the soft tissue from the shell and dry weights calculated as described above.

Because zebra mussels were not tracked individually, a mean mussel length was calculated for each cage, and then zebra mussel growth was calculated as the change in length in each cage over the entire experimental period. Likewise, an average shell weight:soft body weight ratio was calculated for each cage by averaging the individual mussel values in each cage. These numbers were then compared using analysis of variance procedures with location as the main effect.

3. RESULTS

3.1. Water

The results of the two-way analysis of variance are presented in Table 2. Both location and distance above the sediment surface (depth) were significant for some elements (Al, Fe, Mg, Mn, S, Si), but there was no uniform response. The concentration of most elements tended to decrease with distance above the surface of the mud, but these differences were only significant for Al, Fe, Mn, Mg, P, S, and Si.

The differences among locations were harder to characterize. No element showed highest concentration outside the pockmark. Two elements, Zn and Mn, had highest concentrations in the deepest part of the pockmark, which then decreased along the transect inside the pockmark at the lip and again outside the pockmark. Many elements showed their highest concentration at the site inside the lip of the pockmark, with mid-level concentrations at the deep site, and lowest concentrations outside the pockmark. These included Al, Fe, Ca, K, Mg, S, and Si. For some of these elements, the difference between the deepest part of the pockmark and the outside location were minor (K, Mg, S), while for others, the difference was also significant (Fe, Si) (Figure 2).

For calcium, of particular importance for zebra mussels, concentrations were highest inside the lip of the pockmark, and least outside the pockmark (Figure 3). Cu, Na, and K showed no significant difference among locations.

The differences among concentrations in the pore water samples, which were not all collected on the same day as the water samples, did not follow the same patterns (Table 3). In these samples, Na, Mg, and S showed highest concentrations in the deepest part of pockmark, decreasing outward (Figure 4). Only Mn was highest at the inside location, and Fe, K, P, Si, and Zn showed highest concentrations outside the pockmark (Figure 4). Calcium showed a marginally significant difference between the inside and the deep and outside locations, with highest concentrations outside and in the deep hole.

3.2. Benthic Invertebrates

The abundance of all the invertebrate groups in the samples was generally low (36-373/sample) but all groups showed higher densities inside the pockmark compared to outside the pockmark (Table 4). For all groups but the nematodes, highest abundances were found at the deepest location in the pockmark. These differences were significant, however, only for total abundance and oligochaetes (Figure 5). Molluscs in the samples were primarily zebra mussels attached to clinkers or other bottom debris.

Clinkers, and sometimes uncombusted coal fragments as well, were found in all the samples, but the mass of this material was highly variable. In general, there were more clinkers and coal debris outside the pockmark, where invertebrate abundances tended to be lower, than inside the pockmark, (Table 4), but clinkers were never significant as a covariate in the statistical analyses, probably because of the extreme variability.

Table 2. Results of two-way ANOVA on mean trace element concentration (µg/L) in the overlying water samples. Var = variable, Locations: "outside" = outside the pockmark, "inside" = inside the rim, "deep" = the deepest part of the pockmark Depths: "0" = sediment-water interface, "10-50" = distances in cm above the sediment

Var	Effect	P-value	Mean Outside	Mean Inside	Mean Deep	Mean 0	Mean 10	Mean 20	Mean 30	Mean 50
Two-way ANOVA with location and depth as main effects										
Al	location	0.0075	0.130	1.412	0.631	2.107	0.511	0.278	0.135	0.396
	depth	0.0509								
	loc*dep	0.9214								
Fe	location	0.0093	0.335	3.702	1.796	5.881	1.333	0.717	0.344	0.964
	depth	0.0325								
	loc*dep	0.9717								
Zn	location	0.0446	0.047	0.058	0.102	0.106	0.073	0.054	0.055	0.071
	depth	0.3067								
	loc*dep	0.1848								
Mn	location	0.0044	0.055	0.424	0.432	1.118	0.185	0.098	0.043	0.069
	depth	0.0085								
	loc*dep	0.9130								
Two-way ANOVA on ranked data with location and depth as main effects										
Ca	location	0.0249	14.231	15.142	14.437	15.478	14.544	14.349	14.226	14.232
	depth	0.2865								
	loc*dep	0.8327								
Cu	location	0.0948	0.015	0.019	0.011	0.012	0.014	0.021	0.007	0.019
	depth	0.8720								
	loc*dep	0.7371								
K	location	0.0325	0.887	1.272	0.930	1.310	0.997	0.961	0.960	0.834
	depth	0.2114								
	loc*dep	0.8746								
Mg	location	0.0068	3.544	4.046	3.728	4.293	3.708	3.597	3.543	3.643
	depth	0.0373								
	loc*dep	0.7531								
Na	location	0.7075	6.954	7.033	7.015	7.037	7.021	6.942	6.917	7.091
	depth	0.2085								
	loc*dep	0.6087								
P	location	0.5618	0.015	0.188	0.080	0.325	0.045	0.022	0.000	0.048
	depth	0.0104								
	loc*dep	0.1264								
S	location	0.0048	3.583	4.095	3.769	4.342	3.759	3.646	3.580	3.672
	depth	0.0424								
	loc*dep	0.7455								
Si	location	0.0035	1.191	3.789	2.201	5.032	2.010	1.533	1.229	1.784
	depth	0.0263								
	loc*dep	0.9258								

Table 3. Results of ANOVA on pore water collected inside and outside the pockmark. "outside" = outside the pockmark, "inside" = inside the rim, "deep" = the deepest part of the pockmark"

Variable	Effect	P-value	Mean Outside	Mean Inside	Mean Deep
One-way ANOVA with location as main effect					
Al	location	0.053	0.997	0.792	0.994
Ca	location	0.083	25.845	25.171	25.838
Cu	location	0.006	0.019	0.013	0.020
Fe	location	<0.0001	18.415	16.481	11.649
K	location	0.004	2.148	2.303	3.043
Na	location	<0.0001	8.723	11.799	18.747
P	location	0.046	0.862	0.852	0.616
Zn	location	0.014	0.014	0.005	0.002
One-way ANOVA on ranks with location as main effect					
Mg	location	<0.0001	6.122	6.301	7.527
Mn	location	<0.0001	5.766	6.294	3.779
S	location	0.0004	6.317	6.510	7.789
Si	location	0.0022	18.404	18.022	13.872

Table 4. Results of ANOVA on invertebrate abundances inside and outside the pockmark. "SE" = standard error, "outside" = outside the pockmark, "inside" = inside the rim, "deep" = the deepest part of the pockmark

	Outside		Inside		Deep		
	Mean	SE	Mean	SE	Mean	SE	P-value
Total	136.5	30.5	68.3	7.7	38.5	9.1	0.0245
Chironomids	49.0	12.8	19.3	3.5	15.0	3.9	0.1345
Oligochaetes	45.5	12.4	15.5	2.7	10.5	1.9	0.0507
Nematodes	18.3	5.8	26.3	8.1	8.5	2.6	0.2795
Molluscs	17.7	5.2	7.0	2.5	4.5	0.9	0.0964
Amphipods	6.0	4.0	0.3	0.3	0.0	0.0	0.5135
Clinkers	11.9	2.5	32.6	32.6	33.4	8.0	0.1362

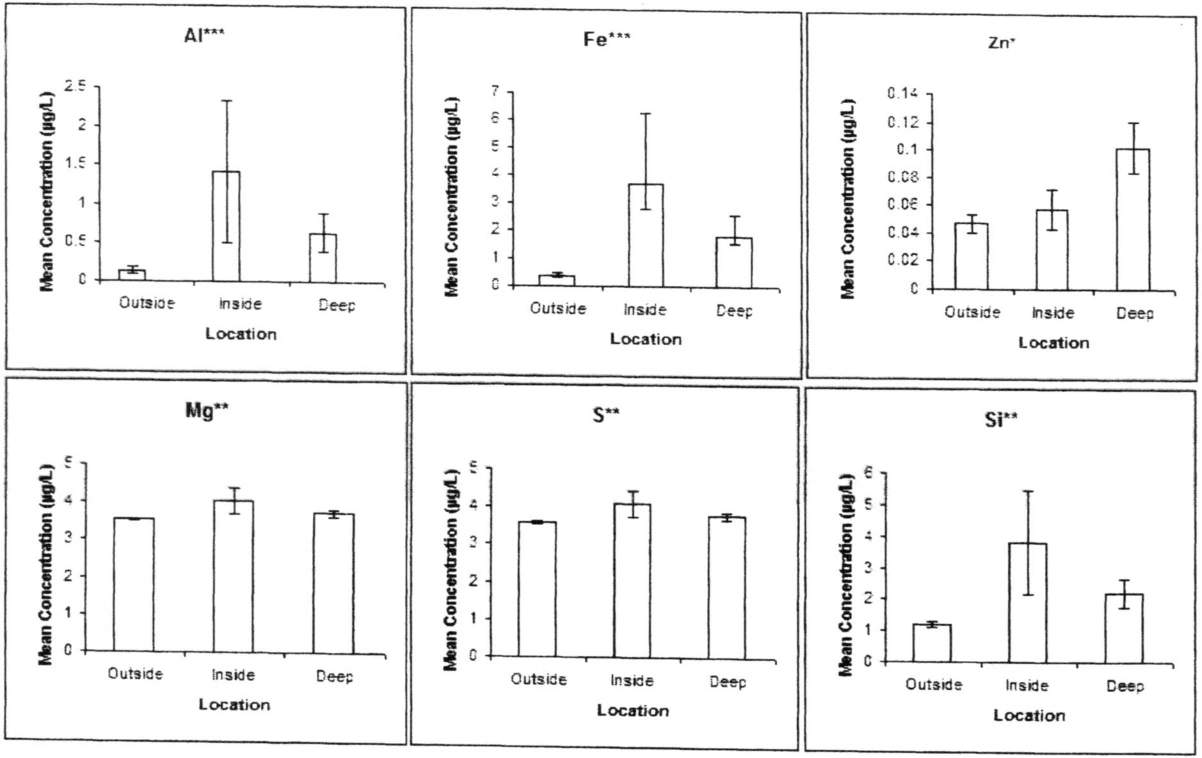

Figure 2. Mean concentrations of elements in overlying water that differed significantly inside and outside the pockmark. "Outside" = concentrations (µg/L) outside the pockmark, "Inside" = concentrations (µg/L) inside the rim of the pockmark, "Deep" = concentrations (µg/L) in the deepest part of the pockmark. *** $p \leq 0.001$, ** $p \leq 0.01$. Bars represent standard error.

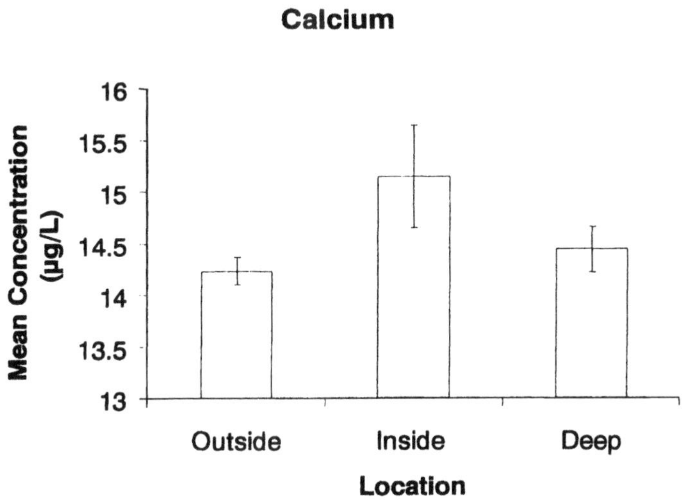

Figure 3. Mean concentrations of calcium in the overlying water inside and outside the pockmark. "Outside" = concentrations (µg/L) outside the pockmark, "Inside" = concentrations (µg/L) inside the rim of the pockmark, "Deep" = concentrations (µg/L) in the deepest part of the pockmark. Bars represent standard error.

3.3. Zebra Mussel Experiment

Survival in all cages was very high; however, there was a small, but statistically significant difference ($p<0.01$) between the number of dead zebra mussels in cages inside (mean = 1, standard error =0.070) and outside (mean = 6.25, standard error =3.63) the pockmarks.

At the beginning of the experiment, the average shell weight to body ratio of the zebra mussels was 18.0 ($\pm$0.206). At the end of the experiment, all mussels had lost shell mass. Much more significantly, however, there was a highly significant effect of location on both the growth and shell weight: body weight ratios at the end of the experiment (Table 5). The mussels inside the pockmark showed significantly less growth, but a significantly higher shell weight: body weight ratio (Figure 6) than those living outside the pockmark. In other words, the mussels inside the pockmark grew less, but maintained significantly thicker shells than the mussels outside the pockmark.

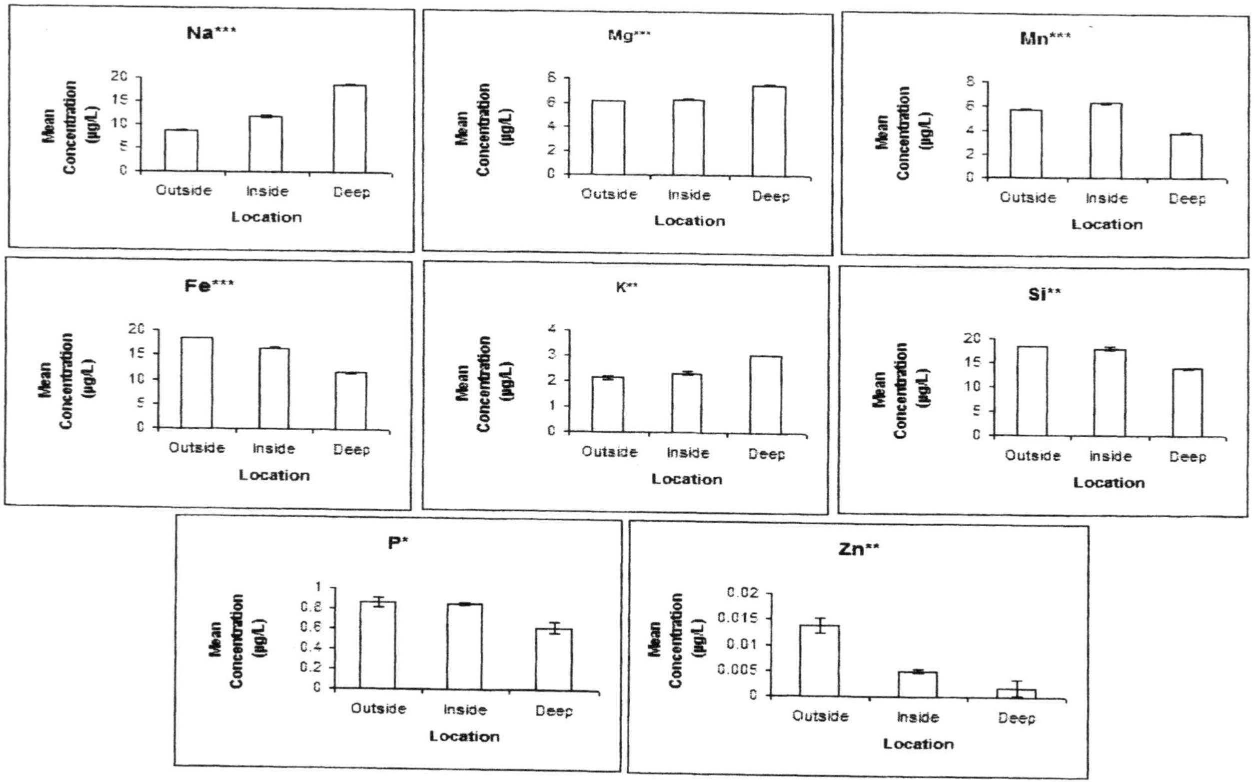

Figure 4. Mean concentrations of elements in pore water that differed significantly inside and outside the pockmark. "Outside" = concentrations (µg/L) outside the pockmark, "Inside" = concentrations (µg/L) inside the rim of the pockmark, "Deep" = concentrations (µg/L) in the deepest part of the pockmark. *** $p \leq 0.001$, ** $p \leq 0.01$, * $p \leq 0.05$. Bars represent standard error.

Table 5. One-way ANOVA on zebra mussel growth and shell weight:body weight ratio inside and outside the pockmark with location as the main effect. SE = Standard error

	Inside		Outside		
Variable	Mean	SE	Mean	SE	P-value
growth (mm)	1.85	0.362	3.09	0.114	0.0001
shell weight:body weight	10.9	0.699	6.30	0.102	0.0001

4. DISCUSSION

The overlying water chemistry data clearly show a different water quality inside the pockmark compared to outside, with many elements showing higher concentrations inside the pockmark. These elements, including Ca and Mg, would likely increase in concentration as groundwater moved through subsurface limestone and dolomite, which is common in the Lake Champlain Basin (Meyer and Gruendling, 1979). Many elements that were enriched inside the pockmark showed their highest concentration not in the deepest portion of the pockmark, but just inside the surrounding rim. This pattern is consistent with the thermal data recorded in the pockmark, which seems to indicate intermittent discharge of groundwater near the rim of the pockmark (Manley et al., this volume). The intermittent flow pattern suggested by the thermal data may also explain the inconsistency between the overlying water and porewater results. These two sets of samples were collected on different days, separated by about a week. It is possible that the pockmark was discharging groundwater on the first sampling trip, when the overlying water samples were collected, but not discharging on the later date, when the porewater samples inside the rim were collected, but this cannot be known.

The differing water chemistry and the variations in bottom sediment and accumulated organic matter in the pockmark likely do provide different habitat conditions for benthic invertebrates. We found highest densities of most invertebrate groups in the deepest part of the pockmark. Bottom photography documented the accumulation of mats of decaying aquatic vegetation (primarily *Potomogeton* spp. from the nearshore environment) in the deep hole and the ^{210}Pb data suggest a slight accumulation of fine sediments in the deepest portion of the pockmark (Manley et al., this volume). No sediment accumulation occurred in the perimeter of the pockmark, near the rim. The accumulations of decaying organic matter and fine organic matter in the sediments would provide a food resource to benthic organisms. Differences were most striking for the oligochaetes and chironomid larvae, both mostly deposit feeding taxa known to respond to organic enrichment (Hilsenhoff, 1988; Barton, 1988; Rosenberg and Resh, 1993; Reynoldson et al., 1995).

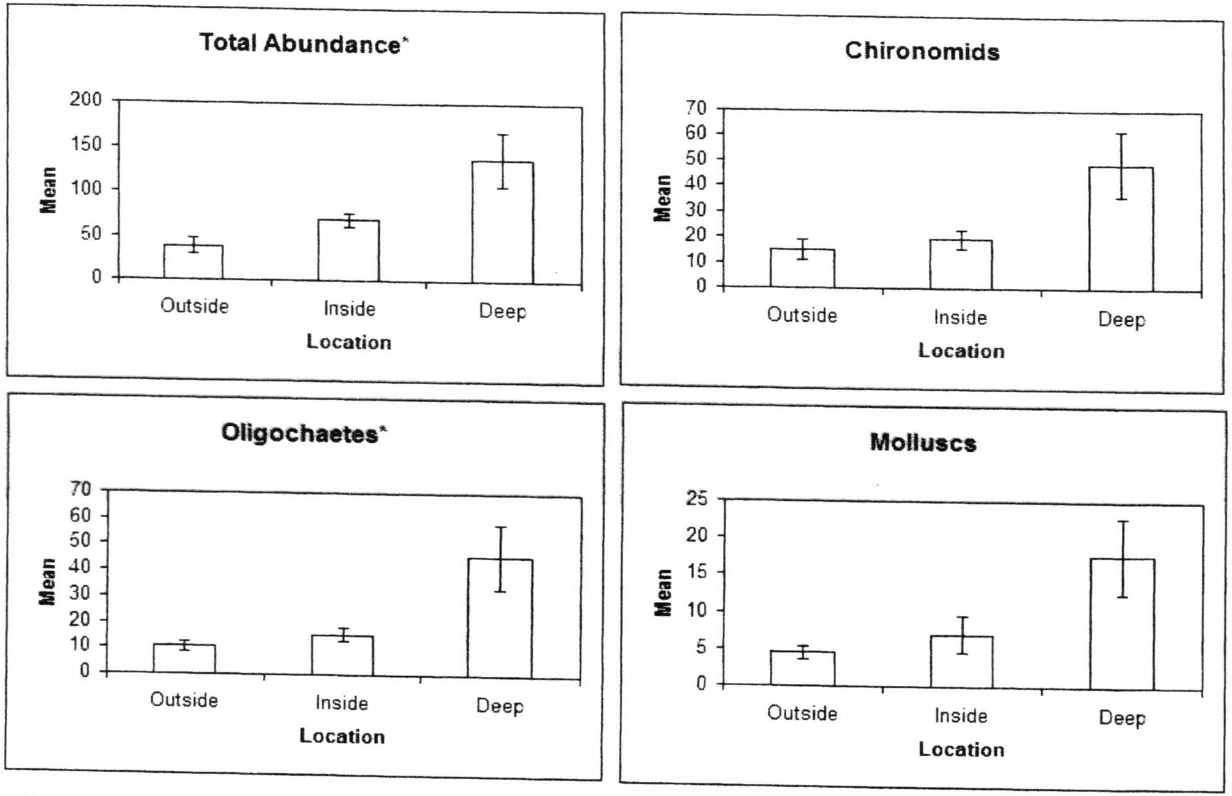

Figure 5. Mean abundance of benthic invertebrates in samples collected inside and outside the pockmark. "Outside" = outside the pockmark, "Inside" = inside the rim of the pockmark, "Deep" = the deepest part of the pockmark. Bars represent standard error.

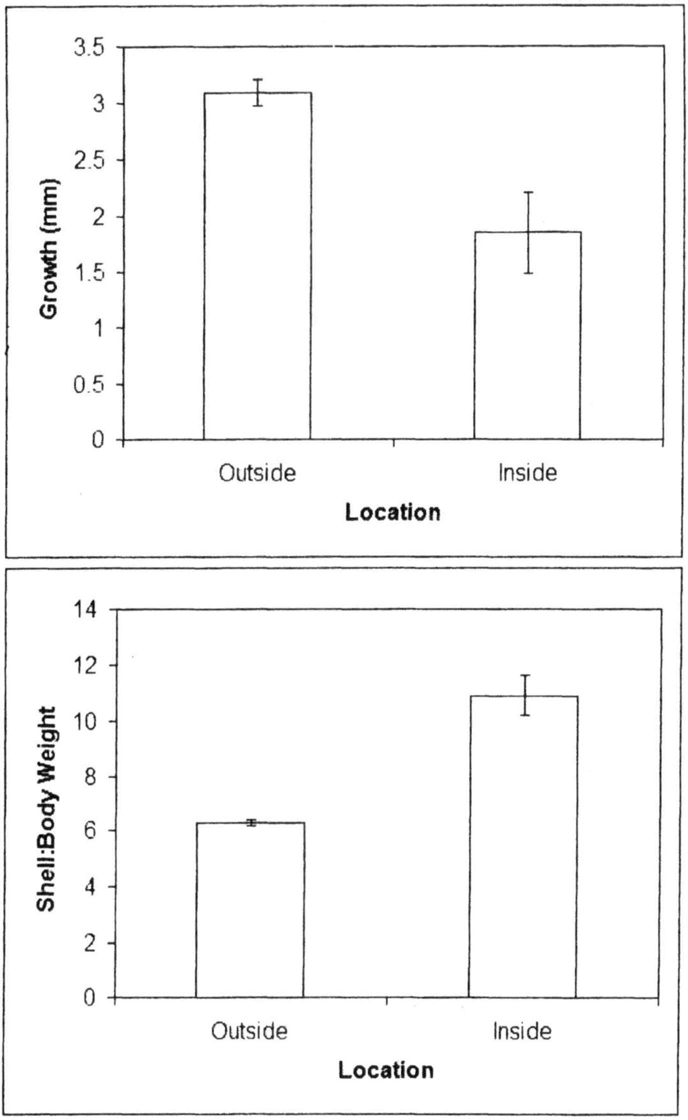

Figure 6. Mean growth over 10 months and final shell weight: soft body weight ratio of zebra mussels caged inside and outside the pockmark. "Outside" = outside the pockmark, "Inside" = inside the rim of the pockmark. Bars represent standard error.

The large mass of clinkers found at our site was unexpected. Clinkers were found both at the sediment water interface and buried underneath the sediment surface. They are the legacy of the steamboat era on Lake Champlain (1809-1930) when literally tons of coal were burned every day on each steamboat on the lake (Ross, 1997). Although clinkers (and coal) were not significant as a covariate in our statistical analyses, there was generally a negative association between clinker density and the density of most invertebrate groups. The exception to this pattern was zebra mussels. Zebra mussels were found at our study site either attached to clinkers or to other debris on the surface of the sediments. In this case, clinkers made a habitat that is not generally suitable for zebra mussels habitable by providing a firm attachment surface that would keep the mussel's filter feeding siphons out of the mud.

Although unique communities have been documented around pockmarks and groundwater seeps in the sea for several decades, including the unique hydrothermal vent communities (Corliss et al., 1979; Davis and Spies, 1980; Horikoski and Ishii, 1985; Hovland and Thomsen, 1989), only anecdotal attention has been paid to the ecology of these features in lakebeds. In several lakes, groundwater seeps have been shown to be a significant source of nutrients that may stimulate algal growth (Brock et al., 1982; LaBaugh et al., 1995; Hagerthey and Kerfoot, 1998), but the benthos around these seeps have not been investigated, perhaps because large pockmarks are not generally associated with the seepage areas. Johnson et al. (1984) postulated that fish may help form these features through their nesting activity, although this is not likely at the depths of the General pockmark. The extent of the pockmark fields in Lake Champlain suggests that they may be a significant source of habitat heterogeneity in the central portion of the lake.

Another comparison might be drawn between the pockmark communities and the sometimes unique communities found in springs and spring-fed streams (Resh, 1983; Williams and Danks, 1991; van der Kamp, 1995; Ferrington et al., 1995; Williams and Williams, 1998; McCabe 1998). Although it is difficult to generalize about the wide range of patterns that have been found at these sites, both species richness and relative abundance are influenced by the presence of a spring. Some species are found only near the source of the spring, while others are found only downstream of the spring. Increased habitat heterogeneity, especially changes in water temperature and water quality may be driving at least some of these responses (Glazier, 1991; van Everdingen, 1991; Webb et al., 1998).

The results of the zebra mussel caging experiment provide evidence of active discharge of groundwater containing elevated concentrations of calcium. On our one sampling date, we found a small, but statistically significant difference in the calcium concentration in the overlying water inside and outside the pockmark. The thicker shells that the mussels inside the pockmark had at the end of the experiment compared to those outside suggest that these individuals were able to take advantage of the enriched calcium they experienced when groundwater was actively discharging in the pockmark.

Although zebra mussels inside the pockmark ended the experiment with thicker shells, zebra mussels both inside and outside the pockmark lost shell weight over the course of the experiment. The concentrations of calcium measured in the overlying water column in both locations are extremely low for shell calcification (Vinogradov et al., 1993). This, combined with the depth of the site and the general lack of suitable hard substrate (Mellina and Rasmussen, 1994) may explain their low abundance in the benthic samples.

The slower growth rate of the mussels inside the pockmark suggests that habitat conditions overall were more stressful in this location. The hydrodynamic investigations

(Manley et al., this volume) suggest that the highest upward flow rates are occurring inside the rim of the pockmark, exactly the location where we placed the mussel cages. This water flow pattern would move particles up and out of the pockmark, including biological particles, the plankton or detritus in the water column that would serve as food for these filter feeders. Likewise, when groundwater was actively discharging inside the pockmark, access to planktonic food would be more limited. Zebra mussels rely on regular water flow past their siphons to provide access to food.

5. CONCLUSIONS

The results of the water quality and biological investigations appear to be fully consistent with the hydrodynamic patterns postulated for the site (Manley et al., this volume). Measured differences in the concentrations of Al, Fe, Ca, K, Mg, Mn, S, Si, and Zn suggest that groundwater is actively discharging in the General pockmark. The differing water chemistry and the variations in sediment organic matter inside and outside the pockmark provided different habitat conditions, and, therefore, each site supported a slightly different benthic community. Zebra mussels responded to the increased calcium, but likely reduced food inside the pockmark with a slower overall growth rate, but thicker shells. The significance of this feature and others like it within Lake Champlain warrants additional attention.

5.1 Acknowledgments

This study was supported by a grant from the Lintilhac Foundation. We thank Art Cohn, Pierre LaRoque, and Dave Andrews for SCUBA assistance collecting the overlying water samples and placing and retrieving the zebra mussel cages. We also thank Dick Furbush, Captain of the R/V *Melosira*, Matthew Weitzman and Bob Acabbo for assistance in the field, and Matthew Weitzman, Jamie Rowder, and Beth Lancaster for assistance in the laboratory.

REFERENCES

American Public Health Association. 1995. Standard Methods for the Examination of Water and Wastewater. 19th Edition. American Public Health Association. Washington DC.

Barton, D.R. 1988. Distribution of some common benthic invertebrates in nearshore Lake Erie, with emphasis on depth and type of substratum. J. Great Lakes Res. 14:34-43.

Brock, T.D., D.R. Lee, D. Janes, and D. Winek. 1982. Groundwater seepage as a nutrient source to a drainage lake: Lake Mendota, Wisconsin. Water Res. 16:1255-1263.

Corliss, J.B., J. Dymond, L.I. Gordon, J.M. Edmond, R.P. von Hersen, R.D. Ballard, K. Green, D. Williams, A. Bainbridge, K. Crane, and T.H. van Andel. 1979. Submarine thermal springs on the Galapagos Rift. Science 203:1073-1083.

Davis, P.H. and R.B. Spies. 1980. Infaunal benthos of a natural petroleum seep. Study of community structure. Mar. Biol. 59:31-41.

Eliopoulos, C. and P. Stangel. 1999. Lake Champlain 1998 zebra mussel monitoring program. Final Report, June 1999. Vermont Department of Environmental Conservation, Waterbury, VT. 32 pp. + appendices.

Everdingen, R.O. van. 1991. Physical, chemical and distributional aspects of Canadian springs. Memoirs Entomolog. Soc. Canada 155:7-28.

Ferrington, L.C. Jr., R.G. Kavanaugh, F.J. Schmidt, and J.L. Kavanaugh. 1995. Habitat separation among Chironomidae (Diptera) in Big Springs. . Pp 152-165 in: L.C. Ferrington, Jr. ed, Biodiversity of aquatic insects and other invertebrates in springs. J. Kansas Entomolog. Soc. Spec. Publ. 1.

Glazier, D.S. 1991. The fauna of North American temperate cold springs: patterns and hypotheses. Freshw. Biol. 26: 527-542.

Hagerthey, S.E. and W.C. Kerfoot. 1998. Groundwater follow influences the biomass and nutrient ratios of epibenthic algae in a north temperate seepage lake. Limnol. Oceanogr. 43:1227-1242.

Hilsenhoff, W. L. 1988. Rapid field assessment of organic pollution with family level biotic index. J.North Amer. Benthol. Soc. 7:65-86.

Hincks, S.S. and G.L. Mackie. 1997. Effects of pH, calcium, alkalinity, hardness, and chlorophyll on the survival, growth, and reproductive success of zebra mussel (*Dreissena polymorpha*) in Ontario lakes. Can. J. Fish.Aquat. Sci. 54:2049-2057.

Horikoshi, M. and T. Ishii. 1985. Cold seep communities at 3800 and 5850 m off Japan. Deep Sea Notes: 16-17.

Hovland, M and A.G. Judd. 1988. Seabed pockmarks and seepages: impact on geology, biology and marine environment. Graham and Trotman, London. 293 pp.

Hovland, M. and E. Thomsen. 1989. Hydrocarbon-based communities in the North Sea? Sarsia: 74:29-42.

Johnson, T.C., R.D. Flood, W.H. Busch, and J.D. Halfman. 1984. Effects of bottom currents and fish on sedimentation in a deep-water, lacstrine environment. Geol. Soc. Amer. Bull. 95:1425-1436.

Kamp, G. van der. 1995. The hydrogeology of springs in relation to the biodiversity of spring fauna: a review. Pp 4-17 in: L.C. Ferrington, Jr. ed, Biodiversity of aquatic insects and other invertebrates in springs. J. Kansas Entomol. Soc. Spec. Publ. 1.

King, L.H. and B. MacLean. 1970. Pockmarks on the Scotian Shelf. Geol. Soc. Amer. Bull. 81:3141-3148.

LaBaugh, J.W., D.O. Rosenberry, and T.C. Winter. 1995. Groundwater contribution to the water and chemical budgets of Williams Lake, Minnesota, 1980-1991. Can. J. Fish. Aquat. Sci. 52: 754-767.

Manley, P.L., T.O. Manley, M.C. Watzin, and J. Gutierrez. 2002. Lakebed pockmarks in Burlington Bay, Lake Champlain. I. Hydrodynamics and implications of origin. This volume.

McCabe, D.J. 1998. Biological communities in springbrooks. Pp 221-228 in: L. Botosaneanu, ed., Studies in Crenobiology: the biology of springs and springbrooks. Backhuys Publishers, Leiden.

Mellina, E. and J.B. Rasmussen. 1994. Patterns in the distribution and abundance of zebra mussel (*Dreissena polymorpha*) in rivers and lakes in relation to substrate and other physicochemical factors. Can. J. Fish. Aquat. Sci. 51: 1024-1036.

Meyer, G.E. G.K. Gruendling. 1979. Limnology of Lake Champlain. Lake Champlain Basin Study, New England River Basin Commission. 417 pp.

Ramacharan, C.W., D.K. Padilla, and S.I Dodson. 1992. Models to predict potential occurrence and density of the zebra mussel, *Dreissena polymorpha*. Can. J. Fish. Aquat. Sci. 49:2611-2620.

Resh, V.H. 1983. Spatial differences in the distribution of benthic macroinvertebrates along a springbrook. Aquat. Insects 5: 193-200.

Reynoldson, T.R., R.C. Bailey, K.E. Day, and R.H. Norris. 1995. Biological guidelines for freshwater sediment based on Benthic Assessment of Sediment Toxicity (the BEAST) using a multivariate approach for predicting biological state. Aust. J. Ecol. 20: 198-219.

Rhoads, D.C and R.A. Lutz. 1980. Skeletal growth of aquatic organisms. Plenum Press, New York. 594 pp.

Rosenberg, D. M., and V. H. Resh, eds. 1993. Freshwater Biomonitoring and Benthic Macroinvertebrates. New York: Chapman and Hall, 488 pp.

Ross, O. 1997. The steamboats of Lake Champlain 1809-1930. Vermont Heritage Press. 184 pp.

SAS Institute Inc. 1988. SAS/STAT™ User's Guide, Release 6.03. Cary, North Carolina, SAS Institute Inc.

Vinogradov, G.A., N.F. Smirnova, V.A. Sokolov, and A.A. Fruznitsky. 1993. Influence of chemical composition of the water on the mollusk *Dreissena polymorpha*. Pp 283-293 in: T.F. Nalepa and D.W. Schlosser, eds., Zebra mussels: biology, impacts and control. Lewis Publishers. Boca Raton, FL.

Watzin, M.C., J. Rowder, B. Lancaster, R Acabbo, and L Bronson. 2003. Zebra mussels, shipwrecks, and the environment – will fish become significant predators on zebra mussels in Lake Champlain? Final Report to the Argosy Foundation. Boston, MA. 27pp.

Webb, D.W., M.J. Wetzel, P.C. Reed, L.R. Phillippe, and T.C. Young. 1998. The macroinvertebrate biodiversity, water quality, and hydrogeology of ten karst springs in the Salem Plateau section of Illinois, USA. Pp 39-48 in: L. Botosaneanu, ed., Studies in Crenobiology: the biology of springs and springbrooks. Backhuys Publishers, Leiden.

Williams, D.D. and H.V. Danks. 1991. Arthropods of springs: introduction. Mem. Entomol. Soc. Canada 155:3-5.

Williams, D.D. and N.E. Williams. 1998. Invertebrate communities from freshwater springs: what can they contribute to pure and applied ecology? Pp 251-261 in: L. Botosaneanu, ed., Studies in Crenobiology: the biology of springs and springbrooks. Backhuys Publishers, Leiden.

HYDRODYNAMICS OF THE SOUTH MAIN LAKE AND SOUTH LAKE, LAKE CHAMPLAIN

T. O. Manley*

ABSTRACT

Thermal data were collected in the South Lake of Lake Champlain during a period of three years, 1997-1999, from early June to late October. Water level data available for analysis were collected in 1997 and 1999 at Port Henry, NY and Whitehall, NY. Thermal observations varied from 4 to 8 sites evenly distributed along the South Lake. Both deep and surface temperatures were monitored at a nominal sampling rate of 30 minutes. Water level data are consistent at the two sites and were strongly correlated with north-south wind forcing. Surface temperatures had temporal variations ranging from diurnal to interannual. North of Five Mile Point, internal seiche dynamics of the Main Lake were observed as warm pulses of water immediately followed by a hypolimnic internal gravity current. These events are referred to as "thermal hooks" due to their unique signature when plotted. These thermal hook events, which number from 4-6 per observational period per year, were estimated to propagate rapidly past the Crown Point Bridge at maximum speeds of 42 cm/s. When entering shallower water, propagation speed drops rapidly to approximately 8 cm/s. These numbers agree well with observed propagation times from the Crown Point Bridge to Five Mile Point. Past Five Mile Point, these thermal hook pulses can no longer be accurately defined although there might be indications that they did propagate past this site. In 1997, an ADCP was placed at a depth of 6.4 m at the Crown Point Bridge for a period of 2 weeks. Results showed a very dynamic bi-directional velocity of water into and out of the South Lake along with a strong diurnal signal associated with echo intensity, which was believed to be associated with vertical migration of zooplankton or slightly larger species. During this 2-week time period, the velocity structure of a thermal hook event was documented and agreed well with expected current velocities. Spectral and cross-spectral analyses show that the South Lake does have preferred oscillations and that these are of long period (1-7 days) and driven by the atmosphere. Data from the met station at the International Paper Company

* Middlebury College, Dept. of Geology, Bicentennial Hall, Middlebury, VT 05753

(1999) showed the inability of the Main Lake met stations (Colchester Reef and Burlington International Airport) to adequately define proper winds conditions existing over the South Lake.

1. INTRODUCTION

With an average depth of 2.7 m and width 1.08 km, the South Lake possesses completely different characteristics over its 53 km length than that of the Main Lake which has an average depth and width of 31 m and nearly 7 km, respectively (Meyer and Gruendling, 1979). While the volume of the South Lake comprises only 0.6 percent of the total volume of Lake Champlain (i.e., 0.156 cubic kilometers), its drainage basin of approximately 3000 square kilometers comprises approximately 15 percent of Lake Champlain's total drainage basin (Henson, 1972; Meyer and Gruendling, 1979). From these statistics as well as the use of Figure 1, it can be observed that the South Lake represents more of a river-like environment when compared to the broader reaches of Lake Champlain.

Without exception, the internal seiche of the Main Lake, which propagates along the interface between the upper mixed layer (epilimnion) and the deeper cold water (hypolimnion), is the most dynamic feature existing within the lake (Manley et al., 1999). Hidden from view, it rarely has any surface expression, yet it possesses the capability of moving several cubic kilometers of water within the lake every 2-3 days (Manley et al., 1999). The Main Lake, north of Thompsons Point, has been monitored for over five years, yet evidence suggests that the region between Thompson's Point and the Crown Point Bridge may control a significant amount of the internal mixing found in the lake (Hunkins et al., 1998; Manley et al., 1999).

In 1997, NOAA funded a research program within the southern portion of the Main Lake (hereto referred to as the South Main Lake) with specific goals of looking for evidence of breaking internal seiches, hypolimnic outcropping at the surface, shear instabilities, and sediment resuspension. During this program, eight subsurface moorings were deployed near the end of May (Figure 2). The five 'deepest' moorings were equipped with Acoustic Doppler Current Profilers (ADCP) capable of profiling a majority of the water column, 40 m thermistor chains that spanned the metalimnion, sediment traps, and one bottom photographic system. The three shallowest moorings (closest to the Champlain bridge) were only equipped with temperature probes and sediment traps.

Results from the 1997 South Main Lake program defined the presence of both linear and non-linear internal seiche regimes (Klein, 1998). During strong non-linear events, the propagation of the deep hypolimnic water formed a steep-walled internal gravity current as it progressed southwards from Potash Point to the Crown Point Bridge. Figure 3 shows the thermal variability of the water column through time at the Mullen Bay and Crown Point bridge sites. For most of the deep hypolimnic cooling events occurring at Mullen Bay (i.e., those green to blue colored pulses in the deep water seen in Figure 3), there is an accompanying event observed at Crown Point after a brief phase lag. What is noteworthy is the change in physical structure of the leading edge of the wave over this distance. The wave front has the appearance of becoming much steeper. As a result of these large southerly excursions of deep hypolimnic water into the shallow water regimes directly north of the Crown Point bridge, near surface temperatures were also observed to

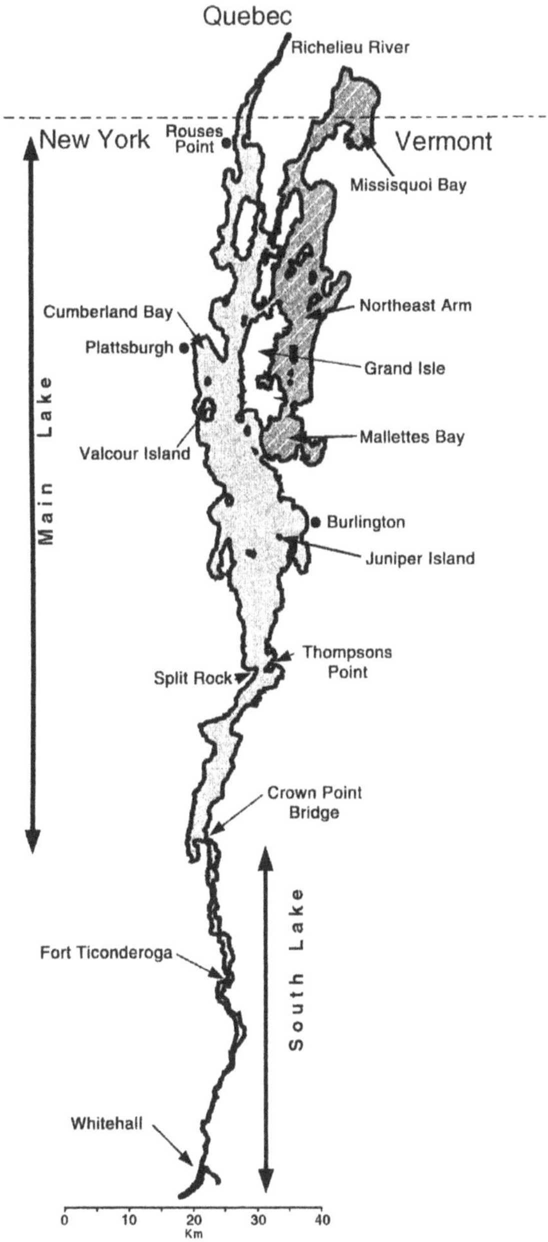

Figure 1. Lake Champlain and its major divisional boundaries (adapted from Manley et al., 1999).

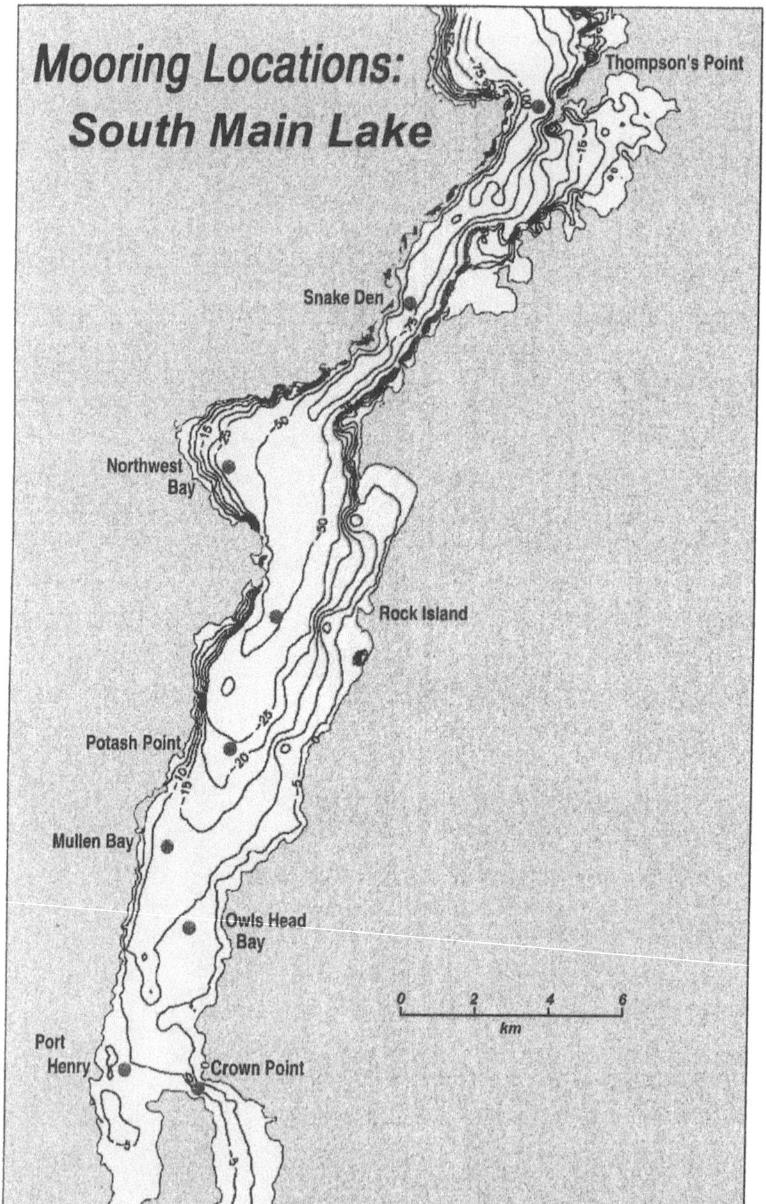

Figure 2. Bottom bathymetric map of the South Main Lake showing the locations of the eight subsurface moorings during the 1997 survey. Depths are in meters (adapted from Klein, 1998).

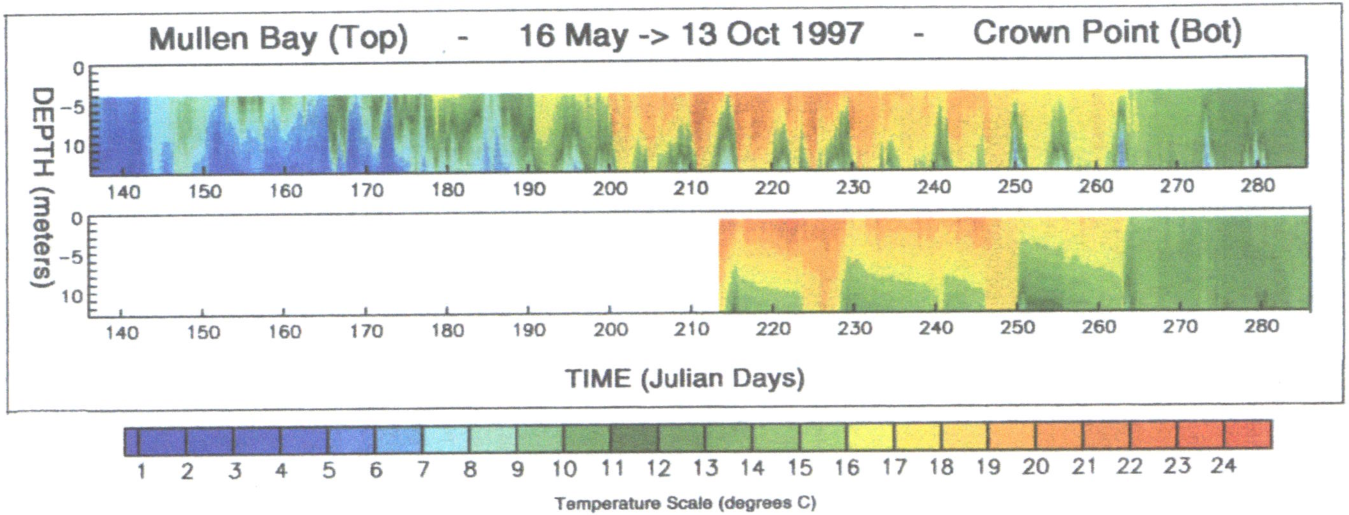

Figure 3. Gridded thermal data from the Mullen Bay and Crown Point bridge sites showing the passage of several internal gravity currents as they propagated southwards. While not all hypolimnic intrusions observed at Mullen Bay propagated southwards to the Crown Point bridge, several large events did. Note that transitional difference in structure of the internal gravity current's leading edge between the two different sites.

decrease, because of the removal of warm surface water during these events. Directly north of the Crown Point Bridge, a remnant cold pool of water was found to exist for some time following each one of these southern propagating hypolimnic events.

Analysis of currents within the 1997 South Main Lake data set confirmed those findings of Klein (1998), but additionally presented evidence of a net counter-clockwise circulation existing within the South Main Lake (Osterberg, 1999). Osterberg (1999) further showed that the presence of a second internal vertical mode was most likely the result of northward propagating internal surges created by strong northerly winds that rapidly depressed the metalimnion in the South Main Lake.

South of the Crown Point Bridge (i.e., the South Lake), few hydrodynamic investigations have taken place and of these, all have been short-term and/or spatially restricted. As a result of this general lack of long-term observations within the South Lake, a 3-year program (1997-1999) was initiated to critically investigate the thermal variability and associated forcing functions existing within the boundaries of Lake Champlain known as the South Lake. Specifically, this program was designed to determine the extent of the Main Lake's internal seiche dynamics within the shallow South Lake, lake level variability and potential correlation with atmospheric forcing. Currents within the South Lake could not be monitored consistently due to lack of funding. As such, the thermal observations when combined with lake level variations and atmospheric forcing could only be used to suggest general motion within the water column.

During the 1997 South Main Lake pilot program, four evenly spaced sites were instrumented with temperature sensors based on the location of navigational aids as well as their proximity to the deep center channel. These sites were located at the Crown Point Bridge, Five Mile Point, Benson landing and Whitehall. In subsequent years (1998, 1999), four additional intermediate sites were added. These eight sites, from north to south, were the Crown Point Bridge, Putnam Creek, Five Mile Point, Catfish Bay, Gourlie Point, Benson Landing, Dresden Narrows, and Whitehall. Figure 4 shows the location of these various sites within the South Lake.

In 1998 and 1999, all sensors were placed as close to their 1997 locations as possible in order to maintain a consistent spatial data set over time. Realistically, variations in spatial positioning could be primarily attributed to the location of the navigational aid used. Frequently, the U.S. Coast Guard services these navigational aids and hence, their repositioning could be slightly off from year to year. In general, it is believed that the location of these temperature sensors during the several years of observations has been within a maximum range circle of 25 m. Satellite positioning was not required in this particular program because of the consistency by which the U.S. Coast Guard places their navigational aids.

2. INSTRUMENTATION

Since it was important to monitor both the deep and surface water temperatures, a system was devised for both easy deployment and retrieval in near-zero visibility under water conditions. This system consisted of two separate components; the navigational aid itself and a separate anchor pad with standoff. For each navigational aid used, a temperature sensor was secured to the uppermost chain-link below the surface buoy. Depending on the type of buoy present, the depth of the temperature sensor could range from 0.9 to 1.8 m. The depth gauge of the SCUBA diver who installed them often documented the depth at which the sensors were placed.

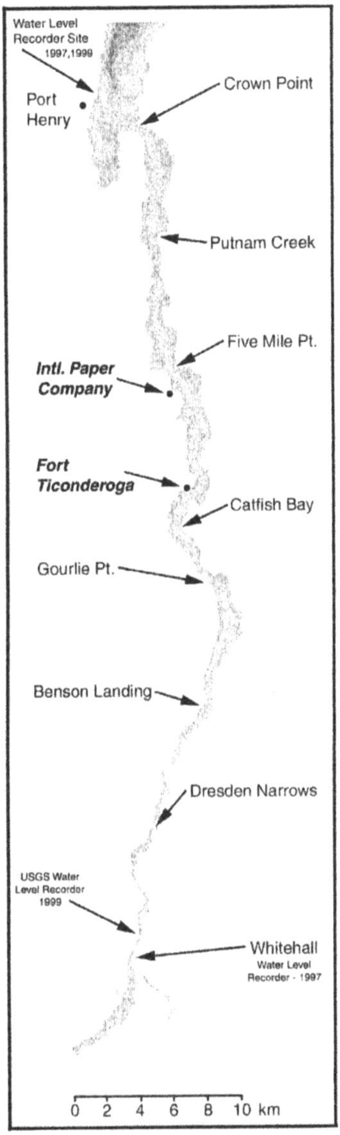

Figure 4. Location of instrumented sites during the summers of 1997, 1998, and 1999 within the South Lake.

In order to monitor the deep temperatures of the main channel above the sediment water interface, a standoff anchor pad system was fabricated. The system was composed of a 12 in. high stainless steel frame (to which the temperature sensor was attached) and a tractor wheel weight (forming its base). The total weight of the system (approximately 85 pounds in air; 65 pounds in water) prohibited easy movement of the temperature sensor. In specific cases, two tractor wheel weights were utilized to provide greater immobility where potential anchor-drag situations were a higher probability.

Once the surface temperature sensor was installed, a small diameter stainless steel wire rope was looped over the navigational aid and permitted to sink to the bottom thereby wrapping itself around the anchor pad. The other end of the wire rope was securely attached to the standoff. The boat was utilized to stretch the ground line into the deepest part of the center channel and deploy the standoff system through the use of a rope. The depth of the sensor was recorded as the measured length of the drop-line or via a portable echo sounder.

In low-visibility underwater conditions, a SCUBA diver could locate the ground line by following the navigational aid's chain down to the bottom. The diver would then cut the ground line and then bring it to the surface. Once the diver was back on board, the standoff system would be recovered.

The temperature sensors used for this program were internally recording Seamon Minis manufactured by Hugrun. Sampling accuracy is 0.1 °C with measurement precision of 0.01 °C. Longterm stability of these sensors suggests calibration every 5-8 years or during battery replacement at the factory (which ever is shorter). The sensors are totally programmable as to sampling rate from a few seconds to that of weeks. Data collected in 1997 and 1998 program were taken every 30 minutes; however, the sampling rate was increased to every 10 minutes in 1999 in order to capture higher frequency components.

During 1997, two water level recorders were installed by NYDEC in an effort to gain an initial understanding of lake level fluctuations at the northern and southern ends of the South Lake. The first water level recorder was installed at a boat landing facility at Stevens Point (Port Henry). The second water level recorder was installed on a navigational pier at the Whitehall site (see Figure 4). On September 14, 1998, the USGS completed the installation of satellite-linked water level recorders near Whitehall and Port Henry. Both sites remained operational until June 2000.

In 1997, a meteorological station was installed in Port Henry in an effort to gain more detailed information on localized wind forcing for the South Lake. Unfortunately, the meteorological station did not function properly and, as result, no data were collected during the year. During the 1998 field season, the International Paper Co. (IPC) permitted NYDEC and USGS to use one of their towers for a meteorological station (wind speed and direction only). It was not until January 9th, 1999 that the IPC met station became operational.

Due to the lack of meteorological observations in the South Lake prior to 1999, data from the Burlington International Airport (BIA) and Cholchester Reef (CR) were utilized in the analysis of thermal and water level data for 1997 and 1998. Both BIA and CR are both located near the central portion of the Main Lake on its eastern perimeter Figure 1). BIA is located several miles inland of the city of Burlington while CR is located above the water surface at the USCG Cholchester Reef site. Of these two sites, BIA produced the most consistent data and is routinely used in all three years. During 1998, CR had 14 days of non-recorded information during two data gaps. This increased to roughly 30 days during the 1999 field season. IPC data possessed only 2 data gaps in 1999 that comprised a total of 10 days. As a consequence, CR was not utilized during 1999. This paper provides as detailed an analysis as possible for each of the years 1997 to 1999 given the lack of many critical observations within the South Lake during 1997 and 1998.

While water velocity observations could not be obtained for a vast majority of the time, the Crown Point Bridge did become a testing site in 1997 for a new side-looking ADCP made by RD Instruments. The unit was attached to the corrugated-steel piling structure of the western central peer at a depth of 21 feet and faced directly across channel towards the eastern central pier of the bridge. The period of observation extended from 11-24 Sept. The purpose of this instrument was to monitor velocity and echo intensity across the channel at this one specific depth. Correlations with other measurements taken at the Crown Point Bridge will be provided in more detail later on.

3. OBSERVATIONS

3.1. General Information

During each of the field seasons, the data collection window extended from mid-June to mid-October. Due to funding and scheduling issues, each year comprised a slightly different window of time, yet still overlapping. In 1997, data were collected from August 1 (day 213) to October 28 (day 302). In 1998, data were collected from July 9 (day 190) to October 15 (day 288). In 1999, they were collected from June 10 (day 161) to October 2 (day 275). For analysis purposes, the Gregorian calendar is often converted into a sequential Julian day system (relative to the beginning of that year). Since 1997, 1998, or 1999 are not leap years, Table 1 can provide a simple cross-reference between Gregorian and Julian dates.

Table 1. Gregorian and Julian day cross-reference list for the years 1997, 1998, and 1999.

Month	*Day*	*Julian Day*
January	1	1
February	1	32
March	1	60
April	1	91
May	1	121
June	1	152
July	1	182
August	1	213
September	1	244
October	1	274
November	1	305
December	1	335

3.2. Meteorological Data

Table 2 provides general statistics for all three years at the three meteorological sites. Comparisons of CR and BIA during 1997 and 1998 show that CR is consistently higher in velocity by a factor of two to three. Wind direction, however, appears to be similar when looking at the average, but slightly different (~10 degrees) when comparing medians.

It has often been noticed that BIA/CR wind data, while defining characteristics of the central Main Lake, do not adequately represent observations at the far reaches of Lake Champlain. The implementation of the IPC meteorological station represents the first opportunity to directly compare between winds observed in the South Lake and those at BIA. Using the median and average values found in Table 2, BIA shows a significant variation in wind direction (30 and 20 degrees, respectively) greater than that reported by IPC. This is also confirmed in Figure 5 (a & b) through the use of polar histograms for wind direction as well as standard histograms for speed (c & d). Additionally, there were many instances where component wind speeds reported at IPC were opposite to that observed that BIA. Although the speed histograms appear to be similar between BIA and IPC, it is important to note that these differences, particularly at lower velocities, may become significant with respect to future numerical simulations of the South Lake. In recent numerical simulations of Lake Champlain (Hunkins et al., 1999), it was evident that the lack of credible low-wind speeds and directions failed to produce observed circulation patterns (Dr. Dan Mendelson, personal communication).

Spectral analysis of the various components of wind speed as well as wind speed magnitude also showed consistent variations between BIA and IPC. Typically, BIA preceded most of the wind events at IPC by approximately 3 hours in the north-south component and as much as 8 hours in the east-west component. With respect to the magnitude of wind speed, spectral analysis indicated that for the major spectral components, IPC lagged behind BIA from 0.5 hours to 6.2 hours. Additionally, IPC data consistently provided better spectral comparisons with South Lake data than that of BIA (e.g., cross-spectral analysis utilizing BIA wind forcing often showed weak or no correlation with several measured parameters, however, IPC provided significant correlations). While the observed directional shift at IPC is most likely due to stronger topographic steering in the vicinity of the South Lake, it is important to restate that these variations in wind speed and direction should be considered significant when future numerical simulations of this region are attempted.

3.3. Water Level Data

As previously mentioned, 2 sites were maintained during most of this time. The first was located north of the Crown Point Bridge at the Stevens Point docking facility at Port Henry, while the second was located at the Whitehall site. This Whitehall site, properly referred to as U.S. Coast Guard navigational aid No. 32, is a solid pier structure. In 1997, the water level gauge itself was attached to a double-weight standoff pad along with a shallow temperature sensor. At Stevens Point, the water level recorder was attached to one of the metal bulkheads. Both sites measured water level relative to their own datum plane and therefore, must be considered relative observations. In 1998, the USGS took over responsibility for the water level stations at Stevens Point and the Whitehall site. In

Table 2. Statistics from meteorological stations within the Champlain Valley during 1997, 1998, and 1999 for BIA, CR, and IPC. Fields are number of observations, median, average, standard deviation, variance, minimum, and maximum.

Year	*Site*	*Parameter*	*Num.*	*Med.*	*Avg.*	*Std. Dev*	*Var.*	*Min.*	*Max.*
1997	BIA	Air Temp.	2065	15.5	14.4	6.9	46.9	-4.4	32.2
		Speed	1776	3.1	3.6	1.8	3.3	1.3	10.7
		Direction	1776	60.7	36.7	-	-	4.6	355.8
		Speed (E-W)	1776	0.3	0.3	1.9	3.9	-7.2	6.4
		Speed (N-S)	1776	0.2	0.5	3.5	12.0	-9.1	10.3
1997	CR	Air Temp.	1879	16.9	15.5	5.7	31.9	0.2	29.7
		Speed	1879	5.1	5.5	2.8	8.0	0.2	13.6
		Direction	1879	47.5	45.3			0.0	360.0
		Speed (E-W)	1879	1.2	1.1	2.9	8.6	-13.0	13.6
		Speed (N-S)	1879	1.1	1.1	5.2	26.6	-12.1	12.2
1998	BIA	Air Temp.	2329	18.2	17.9	5.6	31.0	-0.74	31.2
		Speed	2329	2.7	3.0	2.0	4.1	0.01	10.4
		Direction	2329	69.4	45.3			0.1	360.0
		Speed (E-W)	2329	0.5	0.6	1.5	2.2	-4.3	9.8
		Speed (N-S)	2329	0.2	0.6	3.2	10.1	-6.6	9.7
1998	CR	Air Temp.	2023	19.2	18.4	4.4	19.7	4.8	27.9
		Speed	2023	5.2	5.4	3.0	8.9	0.2	16.5
		Direction	2023	41.2	42.7			0.1	359.7
		Speed (E-W)	2023	1.8	1.8	2.5	6.0	-5.9	13.6
		Speed (N-S)	2023	2.1	1.9	5.0	25.3	-10.0	16.3
1999	BIA	Air Temp.	3145	19.6	19.0	7.0	48.5	-3.3	35.2
		Speed	3145	2.7	3.1	2.2	5.0	0.0	13.6
		Direction	3145	37.8	37.4			0.0	360.0
		Speed (E-W)	3145	0.3	0.6	1.6	2.4	-4.2	6.7
		Speed (N-S)	3145	0.3	0.7	3.4	11.4	-13.1	10.1
1999	IPC	Air Temp.	2905	N/A	N/A	N/A	N/A	N/A	N/A
		Speed	2905	2.6	2.9	1.9	3.8	0.0	15.2
		Direction	2905	5.2	17.9			0.0	360.0
		Speed (E-W)	2905	0.1	0.3	1.4	2.1	-3.2	7.3
		Speed (N-S)	2905	1.2	0.8	3.1	9.6	-15.2	8.5

contrast to the 1997 observations, these new sites were referenced to the standard lake level of 93 feet defined on all Lake Champlain NOAA charts.

Visual comparison between the east-west and north-south wind velocities obtained at the CR meteorological station show a very strong correlation between north-south wind forcing and that of water level variations observed at both sites in 1997 (Figure 6a). East-west wind forcing, as seen in Figure 6b, does not show such a strong correlation nor, would one expect such a correlation over the narrower axis of this river-like environment.

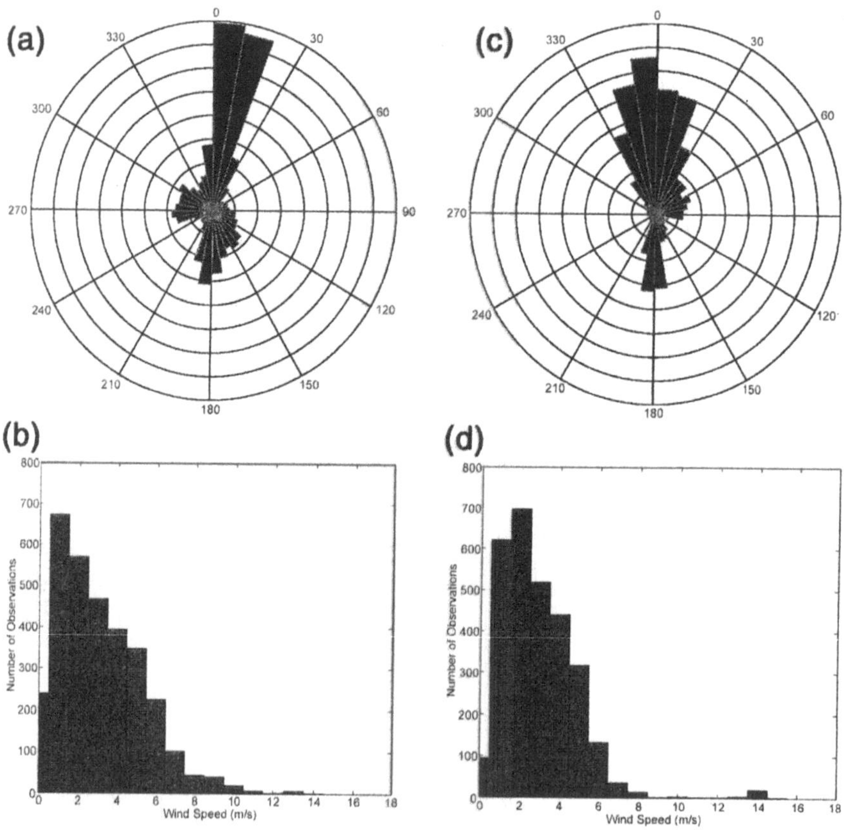

Figure 5. Visual comparison between the polar histograms of directional frequency as well as standard speed histograms for Burlington International Airport (a & b) and International Paper Company (c & d).

As can be seen in the 1997 and 1999 water level data (when both sites were operational; Figures 6 and 7), lake-level oscillations were almost consistently higher at Whitehall when compared to those at Port Henry. While many of the higher frequency events observed at Whitehall can be seen to exist at Port Henry, those observed at Port

Henry were 2-5 times smaller. Two individual events at days 220 (lowering) and 226 (rising; Figure 7) show a consistent lag of approximately 3 hrs from Port Henry to Whitehall. A simple calculation for the natural resonant period of the South Lake shows 5.7 hours using an average depth of 2.7 m and length of 53 km. Half of this period (~2.9 hours) would represent the one-way travel time from Port Henry to Whitehall. This agrees well with observations.

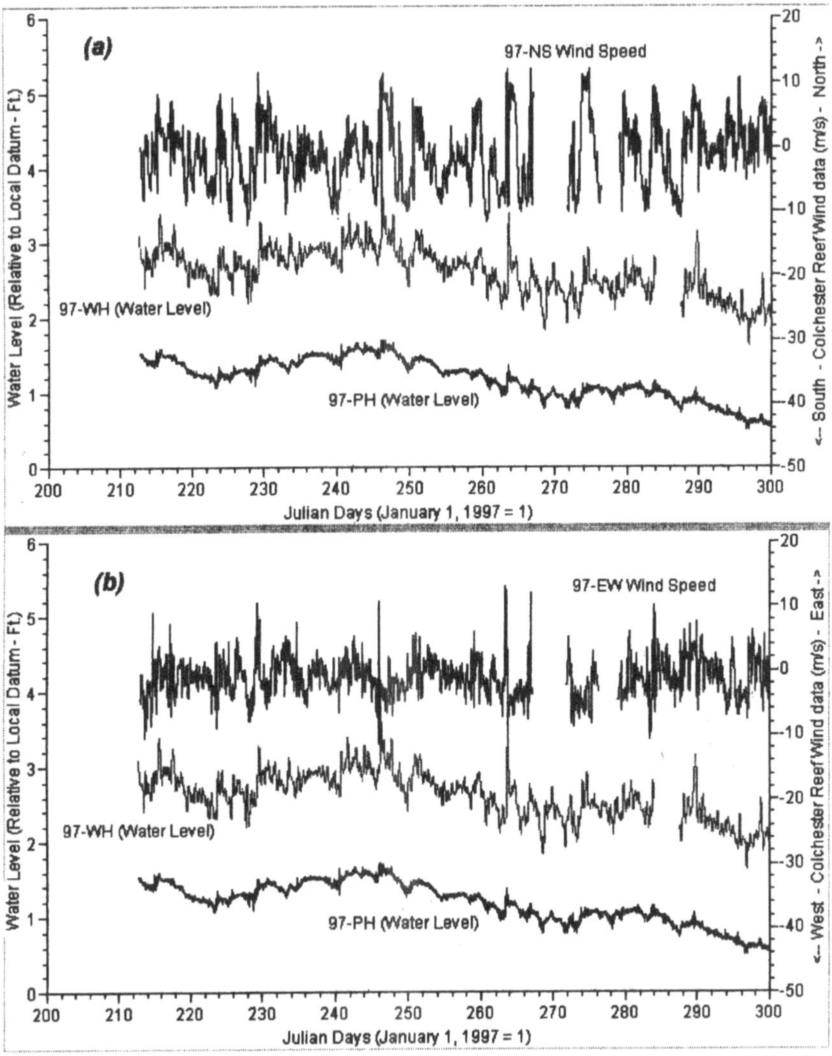

Figure 6. Water level data collected during 1997 at Port Henry and Whitehall shown with (a) N/S and (b) E/W wind velocity data from Colchester Reef.

The elevated water levels at Whitehall with respect to Port Henry also support previous observations by Myer and Gruendling (1979) of higher lake setups to be a function of narrowing and shoaling of the northern sector of the lake above Valcour Island. The South Lake represents a similar environment except the narrowing and shoaling are more extreme when compared to the north end of lake. Hence, the South Lake appears to amplify water level variations as they propagate southward from Port Henry.

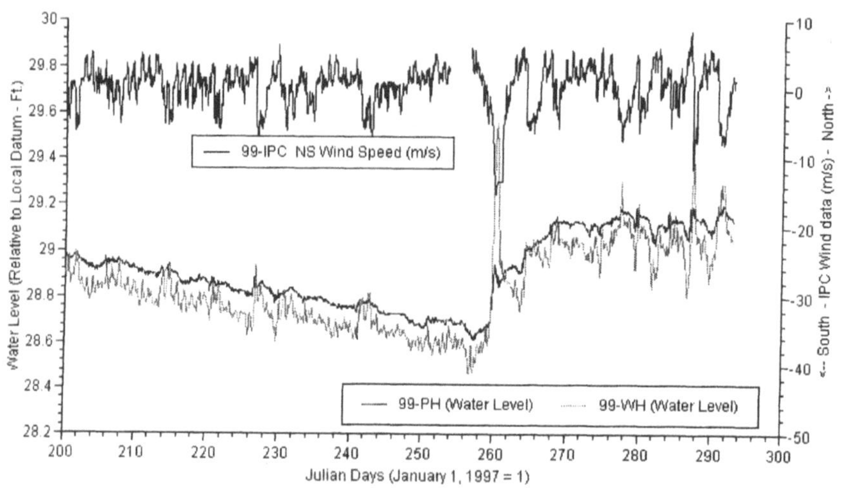

Figure 7. Water level data collected during 1999 at Port Henry and Whitehall shown with N/S wind velocity data from IPC.

During extreme wind forcing events, this effect can be quite obvious. Cases in point would be at days 226, 242, 260 and 287 (Figure 7). The largest of these was caused by hurricane Floyd as it moved through the Champlain Valley over the 16th and 17th of September, 1999. With maximum hourly averaged wind speeds of 13.6 and 15.2 meters per second at BIA and IPC, respectively, South Lake levels at Whitehall rose rapidly to heights of 0.97 meters above the preceding normal levels in a matter of 1.04 days. Due to the heavy rains associated with hurricane Floyd, water levels did not return back to their previous norm but rather 0.34 meters higher than the preceding normal levels. Over the course of the next several weeks, water levels consistently rose throughout the South Lake system. It should also be noted for completeness that the USGS water level recorders do not appear to have the same datum level in that Whitehall consistently shows a lower lake level than Port Henry. If such a case were true, the long-term hydraulic head would force a southward flowing current throughout the entire South Lake (i.e., opposite from what it is). It is estimated that this error is approximately ~0.2 ft.

Spectral analysis on both water level recorders show a dominance of energy with periods greater than 1 day and not that of the South Lake's natural resonant period of approximately six hours. While spectral peaks will vary from season to season and year to year depending upon atmospheric forcing (Manley et al, 1999), several dominant periods (the top three spectral peaks) were observed in both the 1997 and 1999 data sets. These were 2.37-2.67, 1.78, and 1.52 days. 1.07 days was a strong spectral peak in 1997 but was eighth in 1999. Other spectral peaks outside of the most common ones previously defined were 5.33-7.11, 2.13, and 1.33 days. Since the natural resonant period of the South Lake is on the order of six hours, these previously defined dominant spectral peaks characterize external or atmospherically forced oscillations (Manley et al., 1999). Many of these spectral peaks are observed in the meteorological information, particularly those of 2.67-5.33, 1.78,1.52,and 1.02-1.07 days. Therefore, it appears that water levels within the South Lake were forced by longer period atmospheric events greater than 1 day.

Although it is possible that the South Lake does possess an internal natural resonant period, it has not appeared as an obvious signal in any of the thermal records obtained within this region. Given that the average depth of the thermocline is 10 m, the thermocline shore would have its maximum extent some 9 km to the south of the Crown Point Bridge (i.e., Putnam Creek). Extension of the thermal properties of Lake Champlain (Manley et al., 1999) to this region would indicate that the fundamental period for such an internal standing wave to be approximately 1.7 days. While this appears to fit within the 1.52-1.78 day spectral peak band, it is unlikely that this internal seiche period is the cause of this widely observed spectral band. If it were the case, then only observations taken from Putnam Creek northwards would possess this period and not those stations to the south. In reality, this spectral band is widely observed throughout the South Lake as far down as Whitehall. Atmospheric winds would therefore represent the primary driving force for the dominant modes of oscillation found in the South Lake.

3.4. Thermal Information

Two-letter nemonic codes have been given to each of the site locations: Port Henry (PH), Crown Point (CP), Putnam Creek (PC), Five Mile Point (FM), Catfish Bay (CB), Gourlie Point (GP), Benson Landing (BL), Dresden Narrows (DN), and Whitehall (WH). The legend of each plot defines the colors used, the specific year, site location, and depth. Only one temperature sensor failed during the three years of observations. This was due to a flooding of the deep temperature sensor at Putnam Creek in 1999.

3.4.1. Surface Temperatures

Virtually all surface temperatures recorded in the South Lake from years 1997 to 1999 have strongly coherent temperatures and variability. As might be expected, air temperature variations are directly correlated with the surface observations, albeit with some associated phase lag. Figures 8a, 8b, and 9 provide an overview of all shallow surface temperatures taken within the South Lake over the years 1997, 1998, and 1999, respectively.

Cross-spectral analysis between BIA air temperature and the top temperature sensors at Crown Point and Whitehall clearly show the dominance of diurnal forcing. Other strong cross-spectral peaks appear at 5.33, 1.33, and 0.85-0.89 days. Cross correlation with these top temperatures and north-south directed winds at BIA showed very weak correlations. Using IPC wind data however, the dominant significant peak appeared at 1.12 days. This can be explained by the fact that wind often possesses a diurnal component that is driven by the relative difference in heating of the land and water surfaces. As one moves to the deeper temperature sensors, cross spectral analysis shows weaker and weaker correlations with increasing depth. A case in point would be that of the Crown Point bottom temperature sensor, which shows no cross-spectral correlations with air temperature.

When compiling the auto spectral analyses over three years at the eight various sites along the South Lake, a compilation of the five highest peaks for each analysis created a list of 25 dominant periods. These ranged from 0.59 to 5.33 days. In order to assimilate these into a meaningful set of observations that reflect the larger South Lake domain, a summary table was created to reflect the total number of times any given particular period was observed divided by the number of spectral analyses. In this particular case, a total of 20 spectral analyses were completed for the top temperature sensors over three years of observations. Table 3 defines these results in terms of percent occurrence for given period. As expected from previous observations, the diurnal spectral peak was found within the first five dominant spectral peaks of each individual analysis. Clearly, it represents the most dominant mode of oscillation within the South Lake. Following the diurnal period with occurrence percentages of 40 and 35 percent are the groups of 3.52, 1.52 and 5.33, 1.33, and 0.50 days, respectively. These are less robust signals but are nonetheless significant modes that are atmospherically driven.

All years show similar diurnal to seasonal trends. Although they are recorded during slightly different time periods within the year, the general trend of decreasing temperatures through time is apparent. On most occasions, surface temperatures at all sites become very similar and highly coherent with respect to variability past Julian day 220. Prior to this time, there tends to be more variability as well as a wider range of temperatures being observed within the South Lake. Since surface temperatures were all taken within a relatively narrow vertical range of 0.6 m (i.e., from 1.2 m to 1.8 m; Whitehall being the exception at 0.9 meters), there is some indication that these variations in temperature may reflect an environmental segmentation of the South Lake. This segmentation could be produced in a wide variety of ways, some of which would be due to natural restrictions, varying bottom topography, and/or the presence of river input. The proximity to the South Main Lake also plays an important role in the variability of surface temperatures as can be seen in Crown Point and Putnam Creek variability. Temporal scales of synoptic atmospheric motion can also be observed in these records by longer-term oscillations (5-10 days) of lake surface temperature.

It should also be noted that in 1998, the surface temperature sensor at Benson Landing (serial number 213) was inconsistent with the deep sensor at the site (i.e., 3.7 m) as well as immediately to the north (Gourlie Point) and south (Dresden Narrows). When compared to the surface temperatures at Gourlie Point and Benson Landing, it appeared as though there was a consistent offset of 7 °C between days 202 – 276. The preceding record from day 161.75 – 201.95 was anomalously high even after the 7 °C offset had been applied. While high and low frequency events within the corrected record match the nearby surface sensors at Gourlie Point and Dresden Narrows, the Benson Landing surface sensor was included in the data for event timing purposes only. One can see the close mapping of this sensor with the other surface sensors in Figure 10a.

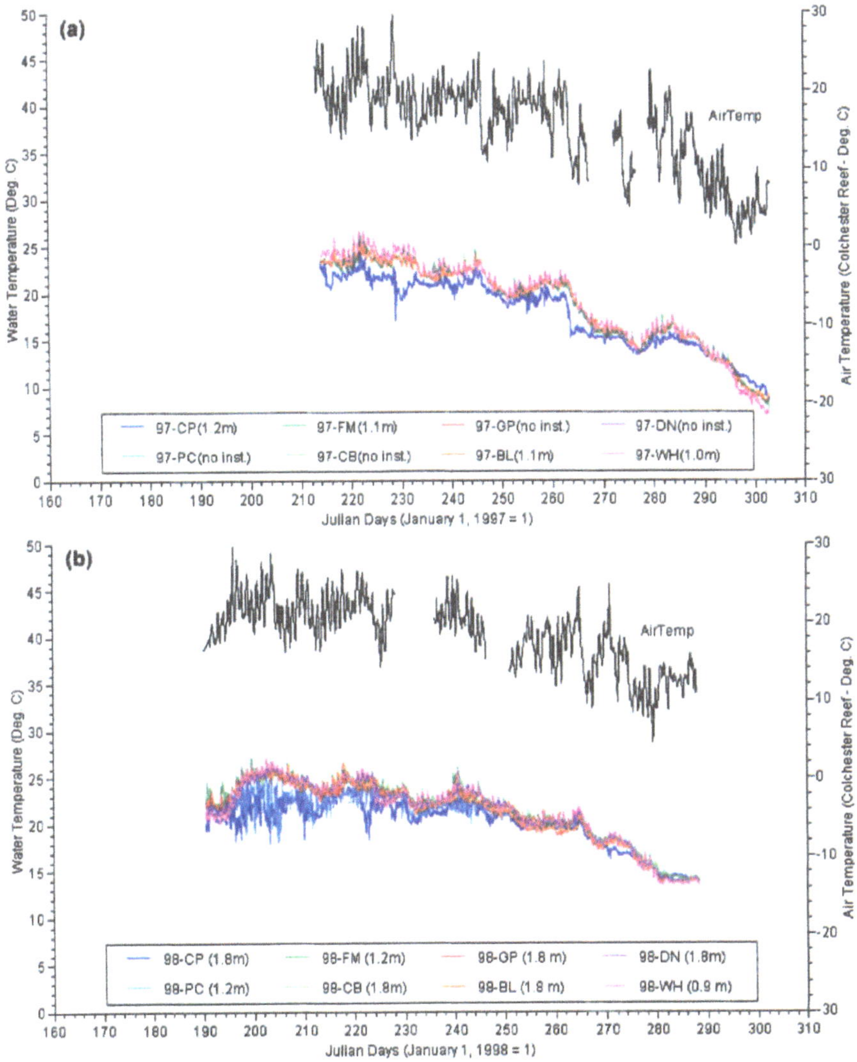

Figure 8. (a) Surface temperature observations at all sites during 1997. Air temperature from the Colchester Reef meteorological station is also included. Note that in this year, temperature observations were not made at Putnam Creek, Catfish Bay, Gourlie Pt., or Dresden Narrows. (b) Surface temperature observations at all sites during 1998. Air Temperature was from CR.

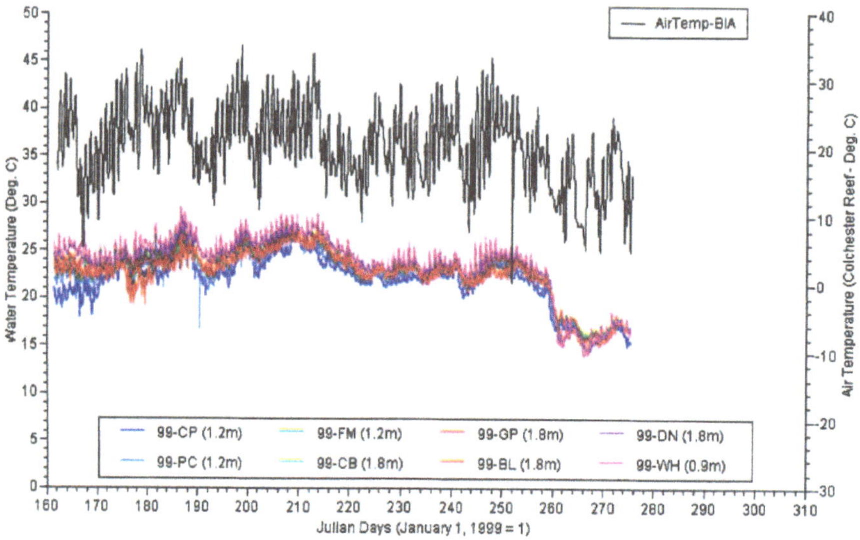

Figure 9. Surface temperature observations at all sites during 1999. Air Temperature was from CR.

Table 3. Dominance of auto-spectral peaks over the entire South Lake for near surface temperature sensors. All eight observational sites were included in the analysis.

period (days)	number of occurrences	maximum number of potential occurrences	occurrence percentage
1.02	20	20	100
3.52 1.52	8	20	40
5.33 1.33 0.50	7	20	35

Diurnal variations of the surface water are apparent in Figure 10a, which is an expanded view of the 1998 data taken at the 8 sites along the South Lake. Atmospheric temperature variations are also shown on the plot and are coherent with those observed at virtually all other sites. Outside of the proximal sites to the Main Lake (Five Mile Point, Putnam Creek, and Crown Point), the strongest response to atmospheric forcing can be observed at Whitehall. This can be attributed to be very close proximity of the temperature sensor to the surface (0.9 m). All other surface temperature sensors, when compared to Whitehall, displayed smaller amplitudes, which were directly related to their increased depth away from the surface.

3.4.2. Deep Temperatures

Deep temperatures observed in the South Lake for all years have a much wider range of variability, both temporally and thermally. Figures 10b, 11a, and 11b provide information at all sites during the years 1998, 1997, and 1999, respectively. In 1997, the first indication of strong thermal differences along the lake bottom became apparent. This was evidenced by the large variation seen between Crown Point and Benson Landing. A more detailed comparison of the 1997 Benson Landing and Crown Point thermal data can be seen in Figures 12a and 12b, respectively. At Benson Landing, both the deep and surface temperatures mimic each other to a high degree. Diurnal variability is much more evident in the surface temperature sensor; however, it does also appear in the deeper data. Crown Point thermal data possesses a higher degree of thermal variability, both in the surface and deep layers prior to Julian day 265. After that, seasonal cooling reduces the temperature difference between the two sensors and enhanced similarity becomes apparent. The very strong thermal differences observed between the upper and lower sensors defines two very distinct thermal regimes, a warm epilimnion and a colder metalimnion / hypolimnion. It is more probable that the deep sensor, on average, is within the metalimnion.

Dominant modes of preferred oscillation were also calculated for the bottom temperature sensors within the South Lake. Table 4 provides these results. Since the South Lake is predominantly shallow, the dominance of the 1.02 day spectral peak is understandable. Periods of 1.78, 2.67, 1.33, 1.19 and 0.5 days have similarly been observed in the auto-spectra of Colchester Reef, BIA, or IPC. The 4.27 day signal is strongest north of Catfish Bay and may reflect the oscillatory motion of the uninodal internal seiche within the Main Lake. Additionally, the 2.67 and 1.94 day modes may also reflect other dominant modes of the Main Lake's internal seiche at periods of 2.7 and 1.8 days (Manley et al., 1999).

The last remaining spectral peak of 1.42 days requires some additional elaboration. Unlike the longer period signal of 4.27 days, which appears only in the northern portion of the South Lake, the 1.42 day peak appears throughout the entire South Lake. While this 1.42 day period has been observed within the Main Lake by Manley et al. (1999), it is strongest during wintertime and most likely associated with an atmospheric spectral peak at 1.5 days. Closer inspection of the spectral records shows that 80% of 1.42 day peaks occurred during 1998, and it was only during that year that the 1.42 day spectral peak occurred at any of the meteorological stations, specifically CR. The remaining 20% of observations with 1.42 days may be forced with the more commonly observed 1.52 day spectral peak frequently observed at the meteorological stations. In brief, the South Lake appears to have a preferred response at 1.42 days, but may not be frequently observed due to the fact that atmospheric forcing may not possess that specific component on an annual basis.

What is most striking are the large thermal oscillations observed in the deep center channel at the Crown Point bridge. These thermal oscillations represent bi-modal variability (i.e., increases as well as decreases in temperature). Invariably, rapid decreasing temperatures immediately followed a warming event. These rapid drops in

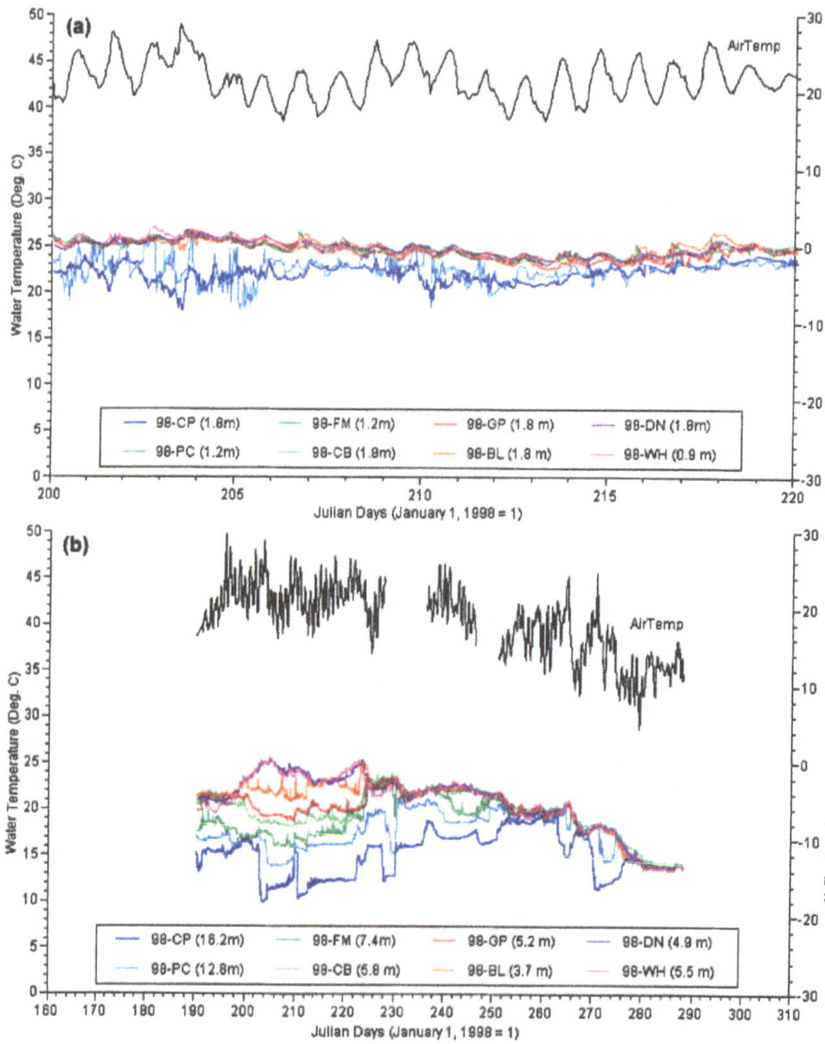

Figure 10. (a) Surface temperature observations at all sites during 1998, however, expanded over a 20-day period ranging from mid- to late-August. (b) Deep temperature observations at all sites during 1998. Air temperature was from the Colchester Reef meteorological station.

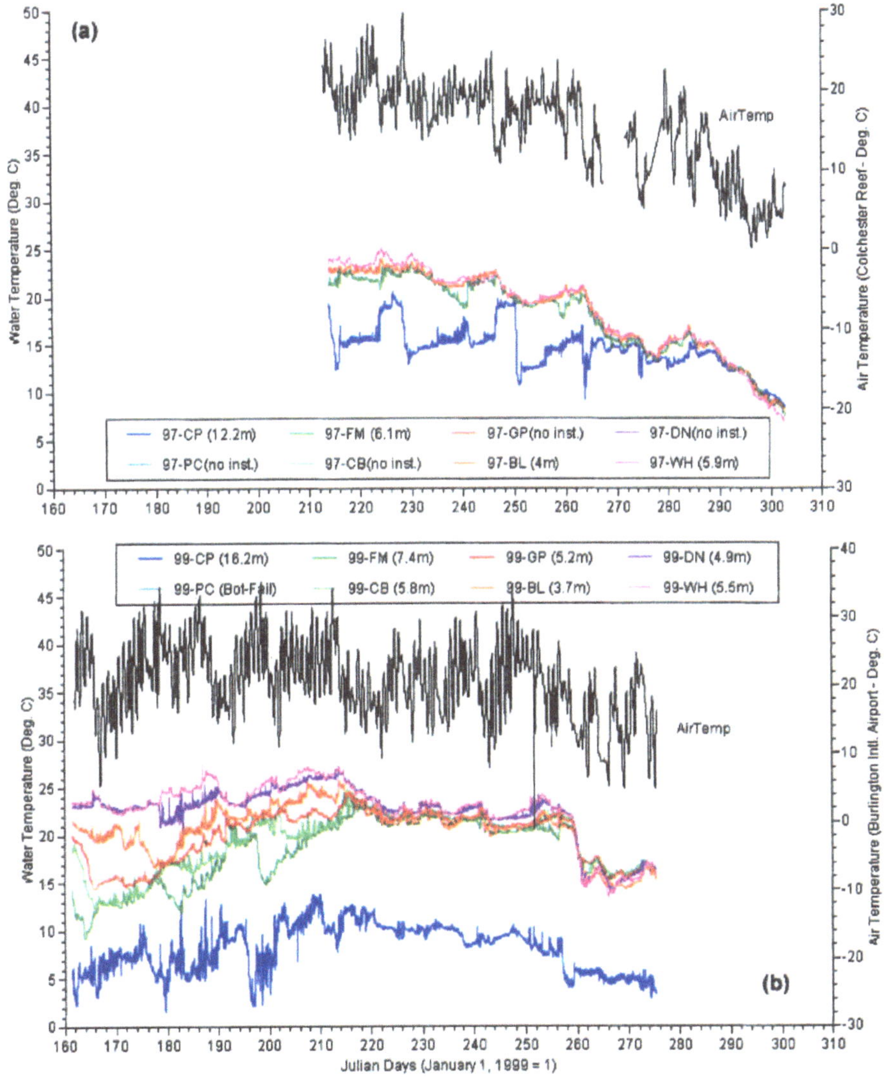

Figure 11. (a) 1997 deep temperatures at all sites. Air temperature from Colchester Reef is provided. Note that in this year, temperature observations were not made at Putnam Creek, Catfish Bay, Gourlie Point, and Dresden Narrows. (b) Deep temperature observations at all sites during 1999. Air temperature was from the Burlington International Airport meteorological station.

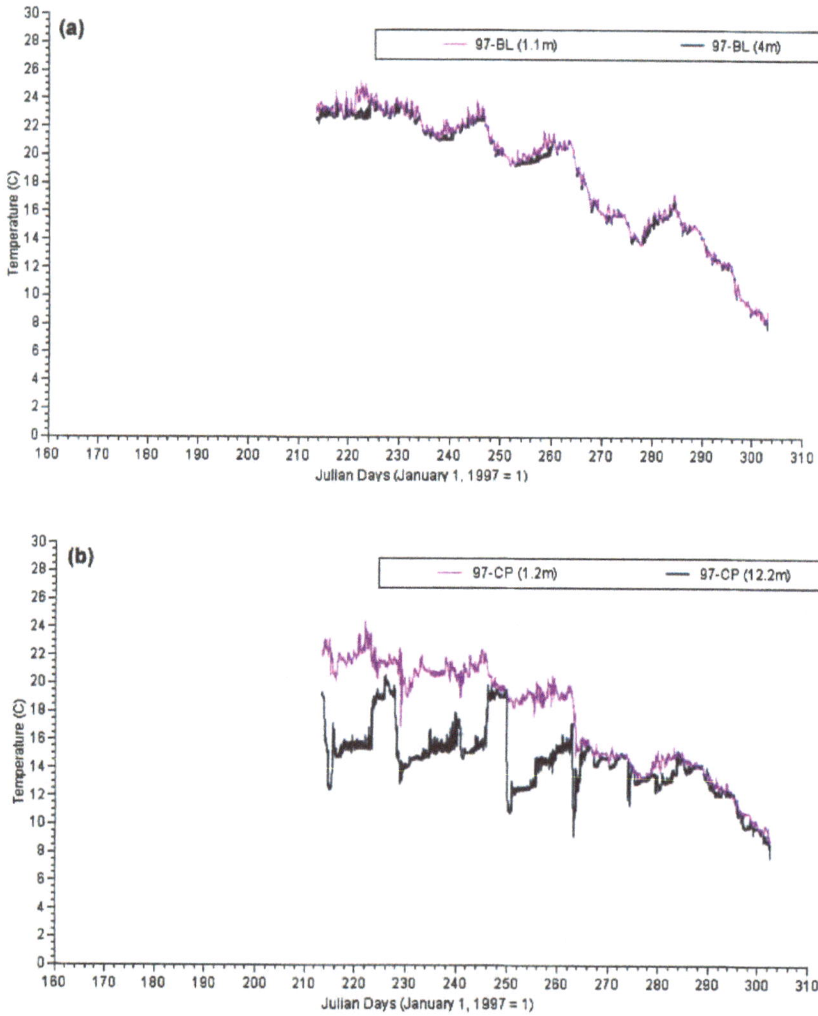

Figure 12. (a) Benson Landing thermal data for 1997. Surface and deep temperatures were taken at 1.1 m and 4 m, respectively. (b) Crown Point thermal data for 1997. Surface data was taken at 1.2 m while deep data was observed at 12.2 m.

Table 4. Dominance of autospectral peaks over the entire South Lake for Bottom temperature sensors. All eight observational sites are included.

period (days)	number of occurrences	maximum number of potential occurrences	occurrence percentage
1.02	15	19	79
1.42	10	19	53
1.94 1.78	7	19	37
2.67 1.19	6	19	32
4.27	5	19	26
1.33 0.50	4	19	21

temperature, when viewed with expanded time scales (see Figure 13a), has unique thermal signatures in the shape of a fishhook. An example of one such feature can be found at day 203 in Figure 10b. Due to their unique signatures, these particular features, which include the preceding thermal rise in temperature, will be referred to as "thermal hooks". The rapid thermal drop in temperatures within these thermal hooks can approach 10 ^{0}C in just a few hours (e.g., Figure 13a at day 250). For the Crown Point event, a large thermal increase of 4 °C was observed at 12.2 m (days 246-250) prior to the 9 °C drop in temperature (day 250-250.8) associated with the leading edge of the hook. The trailing edge of the event (day 250.8-251.2) warms by approximately 2 °C and subsequently maintains that temperature for a lengthy period of time afterwards (days 251.2-256). Three additional thermal hook events are provided in Figures 13a, 14a, and 14b. There appears to be, on average, 6 thermal hooks per observational period. Specifically, 6 moderate to strong events were observed in 1997, 8 events were observed in 1998, and 4 events were observed and 1999.

Inter-annually, thermal variations can also be observed over the entire South Lake. For example, in 1997, observations from Benson Landing to Whitehall were thermally consistent to within ~ 2 ^{0}C. On the other hand, in 1998, a much wider range in temperature (maximum variation of ~11 ^{0}C) was evident at virtually all sites until atmospheric cooling dominated the thermal signatures in the South Lake. The same can also be stated for 1999, except that after day 220, the thermal signatures at all the sites were very consistent.

3.5. 1997 ADCP Water Velocity Measurements

As previously mentioned, the RDI doghouse was situated on the western central pier of the Crown Point Bridge looking directly across channel to the eastern central pier. Measurements of along-channel and cross-channel water velocities as well as echo intensities were measured every 10 minutes for approximately 14 days. The bin-cell size was set at 1 meter with an initial offset of 2 meters. While there was a myriad of ways to display the data, the most useful for this report was smoothed, beam-averaged center-channel observations. The term beam-averaged reflects a 10-minute average of the four independent beams into a final output of velocities (along- and cross-channel) in 1 m bins ranging from the sensor head. Additionally, only those observations taken from the center channel (i.e., approximately 5 to 50 meters from the ADCP) were used. Echo intensities suggested that at a depth of 21 feet and at 60 meters out, bottom topography or the eastern pier was being detected. Hence, the use of only those observations recorded within the central channel. Since observations were obtained every 10 minutes, an 8-hour boxcar filter was utilized on the data and subsequently sub-sampled every hour. The entire data set extended from midmorning on day 253 (11 September) to midday on day 267 (24 September).

Figure 15 shows the results of the parameters monitored during this time period. Cross-channel velocity was the weakest of the two velocity components. The along-channel velocity possessed both higher amplitudes as well as bi-directional characteristics. Basic statistics for along-channel, cross-channel, and echo intensity are provided in Table 5. From this table, it can be seen that both velocity components had a median and average that were positive; reflecting an eastward directed cross-channel component as well as a northerly along-channel velocity. It should also be noted that the cross-channel velocity was routinely positive regardless of whether velocity was into or out of the South Lake (i.e., in the along-channel direction). While there is no historical information on velocity at the Crown Point Bridge, there is nevertheless a well-defined net northward velocity in the South Lake (Myer and Gruendling, 1979). This longer-term northward velocity in the South Lake agrees well with the along-channel average shown in Table 5.

Echo intensity, which is an indirect measure of the quantity of backscatterers in the water column, possessed a very strong diurnal signal. Although it is impossible to define whether these backscatterers are biological and/or sedimentological, the diurnal signal suggests that there may be a strong biological component. Diurnal migrations have been documented within Lake Champlain and specifically associated with upward looking ADCP measurements in the South Main Lake (Osterberg, 1999), and in Shelburne Bay (Sardilli, 1999).

Auto-spectra of the various parameters also show the dominance of long period oscillations (Table 6). The strongest component, in both the cross-channel and along-channel directions, reflects the more synoptic atmospheric pattern of 7.11 days. Most of the remaining spectral periods have been previously observed in auto- and cross-spectra with temperature loggers and atmospheric wind forcing with the exception of the 0.31 day (7.44 hours) peak found in the cross-channel component. Presently, this specific component is not well documented in any of the literature relating to Lake Champlain. It exceeds the longest surface seiche period of 4.2 hours by a factor of almost 2 and can therefore not be considered a component associated with surface oscillations of lake. Nor does it appear likely that this represents an east-west internal seiche oscillation at a latitude of Port Henry since the bottom bathymetry appears to be too shallow for this to exist. Investigation into the auto spectral records for BIA does, however, show the 0.30-0.32 day peak as the 13th strongest for both the east-west and north-south wind-speed components. Considering the removal of virtually all other forcing components, it becomes more plausible that this 0.31 spectral peak is atmospherically forced.

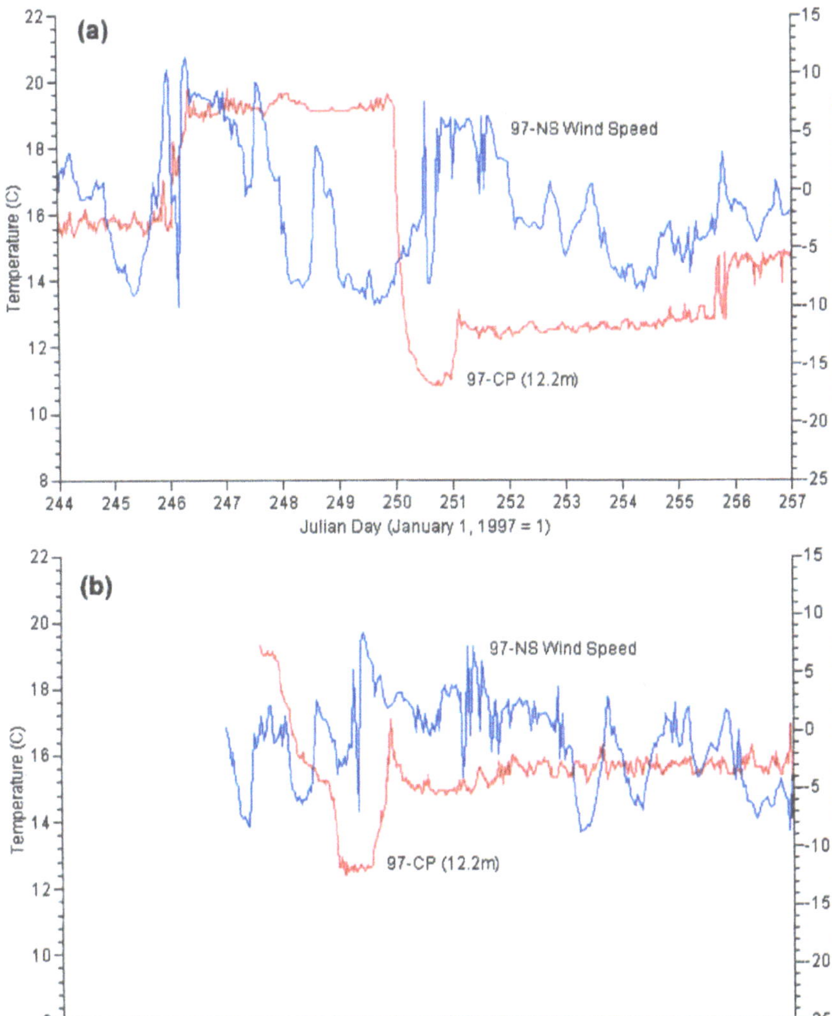

Figure 13. Expanded view of (a) thermal hook event number 3 for 1997 and (b) thermal hook event number 1 for 1997. N/S wind speed from Colchester Reef is provided for comparison. See text for further details.

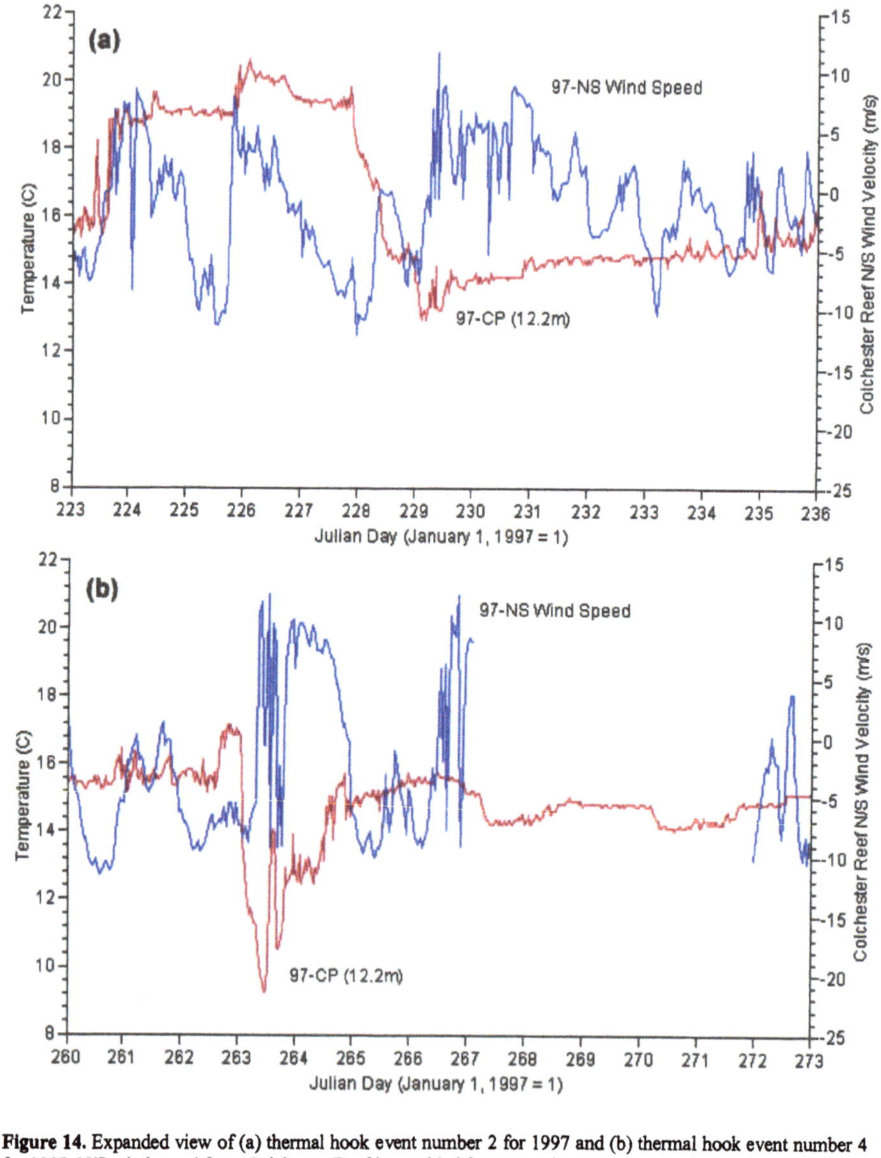

Figure 14. Expanded view of (a) thermal hook event number 2 for 1997 and (b) thermal hook event number 4 for 1997. N/S wind speed from Colchester Reef is provided for comparison. See text for further details.

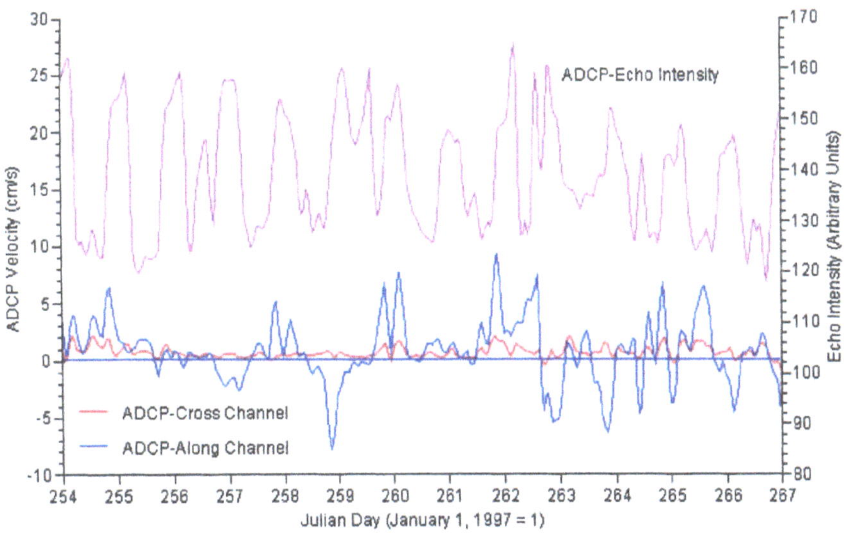

Figure 15. Doghouse records of along-channel, cross-channel velocity as well as echo intensity for the duration of its deployment.

Table 5. Statistical information obtained from the ADCP doghouse located on the western pier the Crown Point bridge at a depth of 21 feet. Results were from data collected within the center channel (5-50 meters). Velocities are in cm per second. Echo intensities are in arbitrary units.

Year	*Parameter*	*Num.*	*Med.*	*Avg.*	*Std. Dev*	*Var.*	*Min.*	*Max.*
1997	Speed (cross-channel)	313	0.5	0.6	0.5	2.5	-1.2	2.1
	Speed (along-channel)	313	0.6	0.6	2.8	7.7	-8.0	9.1
	Echo intensity	313	138	139	12	132.	117	165

The along-channel averaged-velocity data show no strong visual correlation with either the top (Figure 16a) or bottom (Figure 16b) temperature loggers at the Crown Point Bridge. Nor is there any apparent correlation with north-south winds recorded at Cholchester Reef (Figure 17). While this initially appears to be an unlikely situation, the answer appears to be associated with the depth of observation. As noted earlier, the depth of the ADCP was at 21 feet (6.4 meters) and therefore, most likely residing within the boundary of the metalimnion. As a result of this positioning, both upper and lower velocities (which may be opposed depending on the circumstances) would be observed by this instrument depending upon the position of the thermocline.

Table 6. First five significant periods (in descending order of power spectra magnitude) obtained from auto-spectral analysis of doghouse observations. Periods are given in days.

Year	*Parameter*	*First*	*Second*	*Third*	*Fourth*	*Fifth*
1997	Speed (cross channel)	7.11	1.78	2.67	0.31	1.42
	Speed (along channel)	7.11	1.94	2.37	3.56	0.85
	Echo intensity	0.97	2.67	0.85	0.50	1.33

By chance, one of the thermal hook events occurred during the time period that the ADCP was located at the Crown Point bridge site. This particular event occurred near day 263 (Figure 16b & 17). The same deep temperature at Crown Point (12.2 meters) was also plotted against along-channel velocity in Figure 16b. As with all thermal hook events, there is typically an increased temperature jump prior to the rapid drop in temperature. This warming event, labeled A in Figure 16b, is in response to southerly-directed winds over the Main Lake. This forces warmer epilimnic water to the southern portion of the lake as well as depressing the thermocline to deeper levels. In the case of the Crown Point region, a significant portion of the metalimnic and hypolimnic waters must have been removed to account for this increased temperature in the deep water. Associated with this warmer epilimnic advance is a net southerly velocity. This southerly velocity is well correlated with the warm intrusion of epilimnic waters into the deep layers as can be seen in Figure 16b (labeled B). After a rapid change in wind direction from south-directed to north-directed (Figure 17), temperature dropped rapidly. During this transition period, water velocity shifted from south to north, in keeping with the onset of a southward propagating deep gravity current (not observed by the ADCP) and the compensating northward velocity (observed) of epilimnic water out of the South Lake. These observations also agree with Klein's (1999) 1997 observations of a deep hypolimnic pulse propagating past the shelf break at day 261.625, Mullen Bay at day 262.625, and then past the Crown Point bridge at day 263.333.

4. ANALYSIS

Thermal variations at or near the lake surface are strongly linked to atmospheric forcing, specifically heating and cooling. In the deeper layers of the northern portion of the South Lake, heating and cooling cannot account for the large thermal variations associated with these thermal hooks. From the combined thermal information over the past three years, it is clear that these thermal hooks represent events forced by Main Lake dynamics. When looking more specifically at these events with expanded temporal scales (Figures 11b, 12a, and 12b), a progression of each event to the south can be observed.

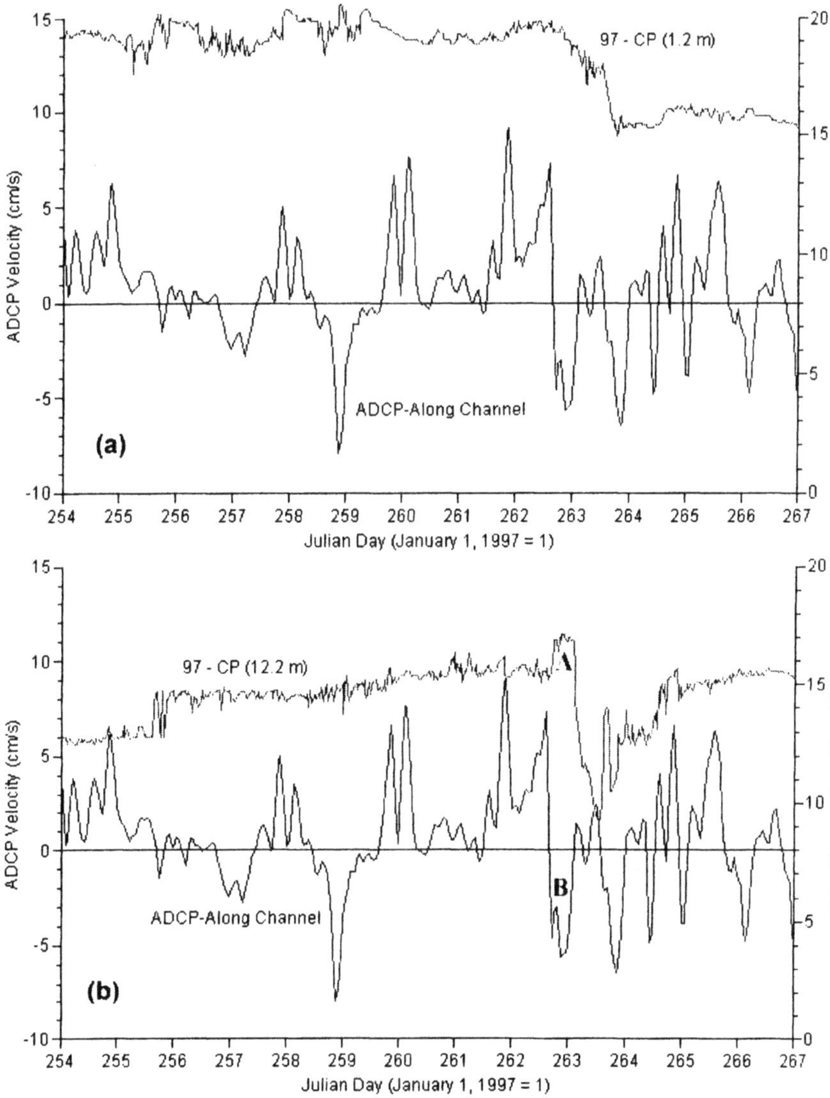

Figure 16. **1997 ADCP derived along-channel velocity compared to (a) the top and (b) bottom Crown Point temperature logger.**

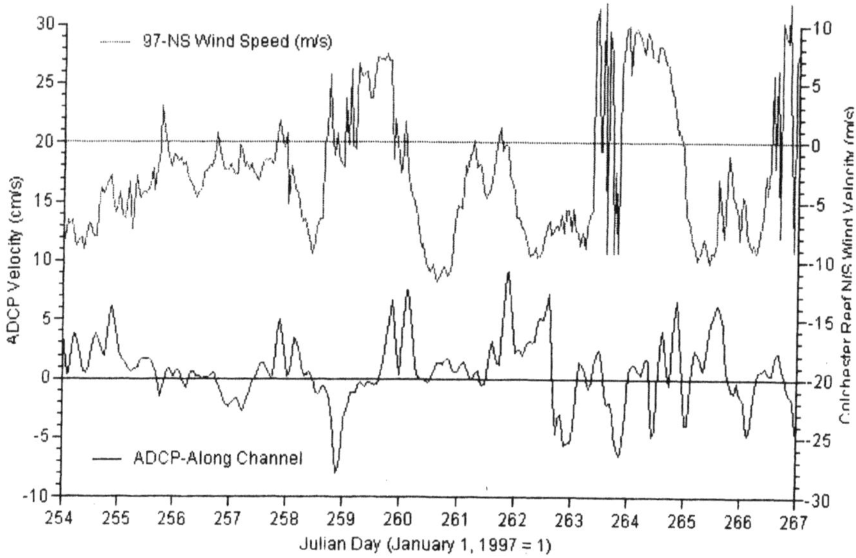

Figure 17. 1997 ADCP derived along-channel velocity compared to that of the north-south wind speed component at Colchester Reef.

From day 227 to day 230, a thermal pulse starting at the Crown Point Bridge can be seen to propagate southwards through the Putnam Creek site all the way to Five Mile Point (Figure 18; A, B, & C, respectively). The change in thermal signature over this distance of approximately 15 km was primarily due to turbulent mixing and diffusion along its path. Very few thermal pulses have been observed to extend past Five Mile Point. Even then, it is not apparent that these anomalies past Five Mile Point represent those passing through the Crown Point Bridge.

Figure 19a shows two thermal hook pulses at Crown Point during days 203-205, and 210-211. The first event was observed to propagate all the way into five Mile Point (A, B, & C), however, the second event (D) had a significantly reduced thermal signature at Putnam Creek and none at Five Mile Point. Using the coldest value of the initial drop in temperature as a unique marker of the first event, propagation time from Crown Point to Putnam Creek was 33 hours (1.375 days). From Putnam Creek to Five Mile Point was an additional 31.92 hours (1.333 days).

Figure 19b also shows two more thermal hook pulses propagating southwards towards Five Mile Point. These particular pulses are unique in their thermal signatures and cannot be mistaken for other dynamic processes. It is worthwhile noting that both pulses propagated past Putnam Creek, but never reached Five Mile Point.

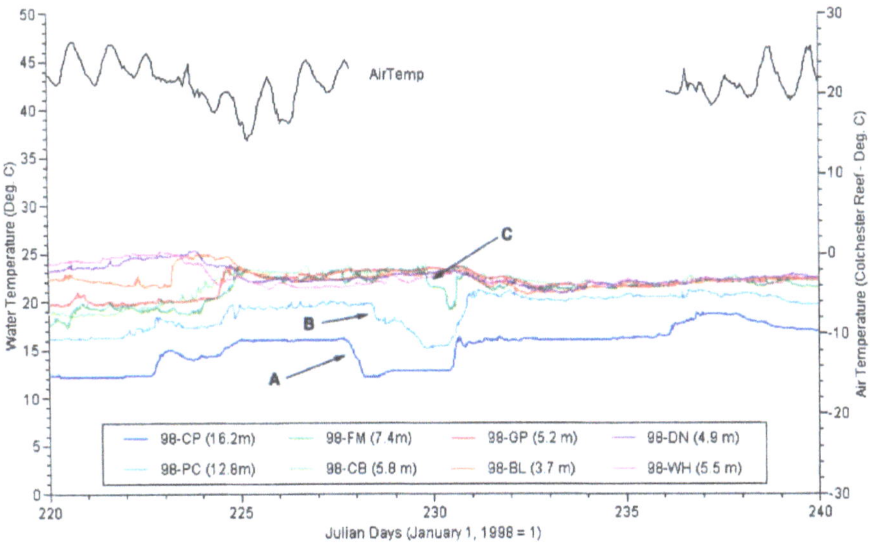

Figure 18. Bottom thermal data observed from day 220-240 during 1998. Note the large thermal hook pulse as it propagates from Crown Point (A) through Putnam Creek (B) and finally, at Five Mile Point (C).

5. DISCUSSION

From these observations, it is apparent that Main Lake forcing does have a significant impact on the northern portion of the South Lake. When combined with meteorological wind forcing (obtained from CR, BIA, and IPC), there is a consistent picture of strong north winds followed by relaxed or southerly winds. Northerly winds force warm epilimnic surface water to the south and hence, drop the thermocline in the northern portion of the South Lake dramatically. This in turn causes a large thermal increase in the deep-water observations. After the wind relaxes or reverses direction, warm surface water will be removed from the region as part of the relaxation phase of the Main Lake internal seiche and/or forced strongly to the north by southerly winds. It should be stressed that not all reversing wind events force hypolimnic water into the South Lake region. It takes not only proper wind forcing but just as importantly, the proper phase relationship with the internal seiche, internal stratification, and potentially other internal dynamics within the Main Lake such as internal surges (Dr. K. Hunkins, recent unpublished modeling efforts in Lake Champlain). If this were not the case, then one would expect an internal gravity current propagating through the Crown Point Bridge approximately every 4.25 days in association with the uninodal internal seiche. When the proper conditions are met, however, metalimnic and hypolimnic water will then invade the deeper portions of the northern sector of the South Lake as internal gravity currents (Klein, 1998). Internal gravity currents have also been observed in the northern part of Lake Champlain (Saylor et al., 1999) but were much larger (lateral and vertical extent) than the thermal hook pulses observed in the South Lake.

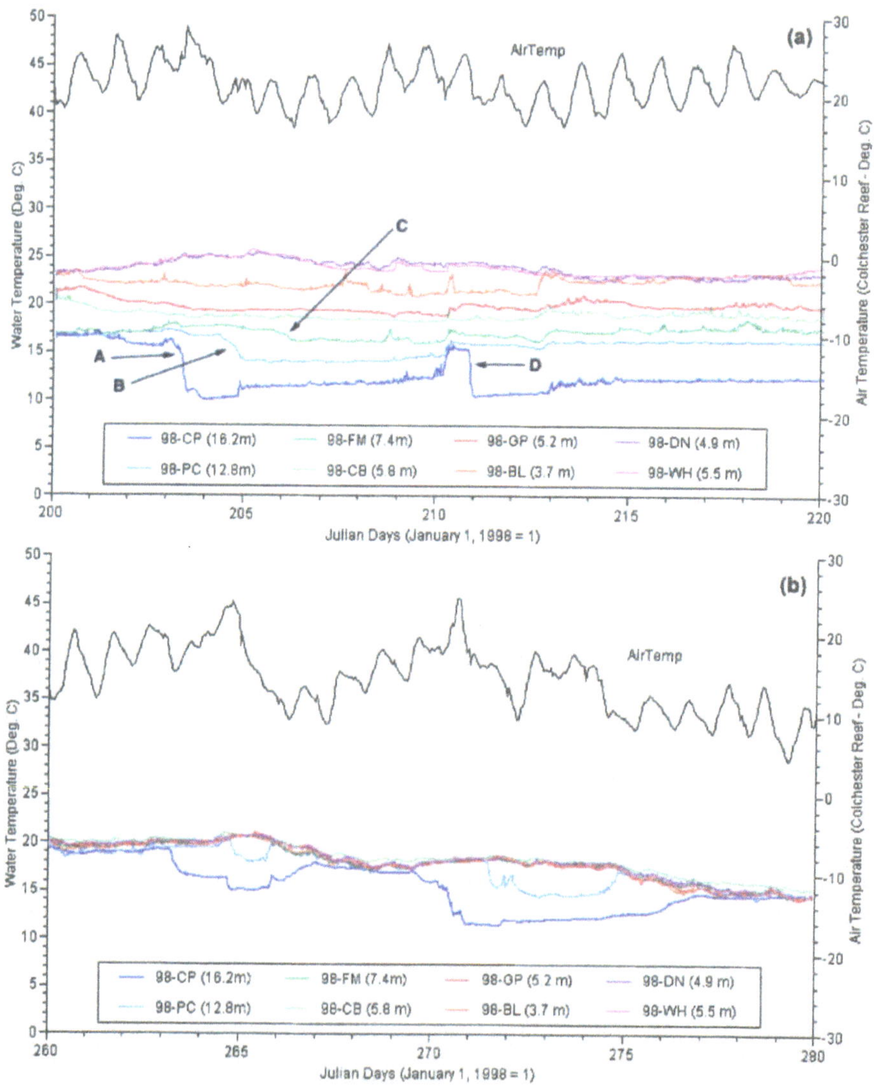

Figure 19. Bottom thermal data observed from day 200-220 during 1998. Two thermal hook pulses can be observed within this time. The first pulse propagates southward to Five Mile Point, however, the second one did not appear at any sites upstream. (b) Bottom thermal data observed from day 260-280 during 1998. Two broad thermal hook pulses can be observed within this time frame.

Klein (1998) indicated that these non-linear internal gravity currents change shape as they approach shallower water, almost appearing as a near-vertical walls of cold water (see Figure 3) propagating along the bottom sediment-water interface at maximum observed speeds of 42 cm/s. The thermal hook pulses that were recorded at the Crown Point bridge site were primarily what one would expect to find with such an internal gravity current propagating past the site. The rapid drop in temperature would represent the transitional boundary between the warm epilimnic water being displaced by the gravity current as it travels southwards along the bottom of the lake. The coldest water would represent the least altered core of the hypolimnic water that was lifted onto the shallow shelf of the South Main Lake. The slightly warmer trailing water would likely represent a combination of altered core water due to mixing as well as slightly warmer hypolimnic water derived from a shallower depth after the initial creation of the gravity current.

Yih (1980) developed a simplified equation of the propagation velocity (U) for internal gravity currents within a rectangular shaped basin as

$$U = .707\sqrt{(gd(\rho_2 - \rho_1)/(\rho_2 + \rho_1))}$$

where g is gravity, d is the average depth, and ρ_1 and ρ_2 are densities of the upper and lower layers, respectively. Using estimated average depths of the channel from Crown Point to File Mile Point as well as representative temperatures (densities) observed during the passage of the first thermal hook pulse in Figure 18, the average propagation speed was determined to be 10.7 cm/s. This produced a modeled travel time from the Crown Point Bridge to Five Mile Point of 39 hours. While the calculated time is less than the observed propagation time for the event by some 9 hours, it appears to be well within the variability of accepted parameters of depth, bottom topography, channel restrictions, density variations along its path, and the simplified nature of this equation.

6. CONCLUSIONS

The South Lake, defined from a northern boundary of Crown Point Bridge to that of Whitehall at its southern boundary, represents several different hydrodynamic regimes. North of Five Mile Point, Main Lake internal seiche dynamics as well as local atmospheric forcing control flow dynamics. These combined forcing functions produce rapid influx of warm epilimnic water (thermocline depression) immediately followed by rapid intrusion of metalimnic/hypolimnic water. These intrusions represent internal gravity currents with steep leading faces as they passed the Crown Point Bridge. The characteristic thermal signature of these events passed through the Crown Point Bridge is that of a "thermal hook". Observations showed that the hypolimnic portion of the thermal hook pulse propagated at an average speed of approximately 8 cm/s over the distance from the Crown Point Bridge to Five Mile Point. Higher speeds, possibly approaching and upper limit of 42 cm/s (~ 1 knot) (Klein, 1998) might be expected at the Crown Point site. The average observed propagation velocity between Crown Point and Five Mile Point agrees well with the simplified model of internal gravity currents existing within a rectangular basin (Yih, 1980). ADCP data collected at the Crown Point Bridge indicated a maximum channel-averaged southward velocity of 5.7 cm/s during the southerly flow of epilimnic water past the Crown Point Bridge. It did not capture the deep southward velocity of the gravity current due to its shallow depth (6.4 m), but did define the time when flow reversal occurred.

Although it was expected that surface temperatures over the South Lake might be more uniform, this was not always the case. In 1997, Surface temperatures (excluding Crown Point and Putnam Creek) were very similar over the entire South Lake, however, in 1998 and 1999, surface temperatures varied considerably. These variations observed in the surface waters may indicate that the South Lake could be subdivided into separate environmental compartments.

Water level variations observed in 1997 and 1999 at Port Henry and at Whitehall showed very consistent results with each other. Visual correlation with north-south winds obtained at the CR/BIA and IPC meteorological station's was very strong. East-west wind forcing possessed much less correlation with water level data at both sites. The more restricted and southerly site, Whitehall, had much larger variations in water level than that of Port Henry thus giving the data a more noisy appearance. As previously noted, these amplified signals of water level at Whitehall were most likely due to the shoaling and restricting affects of the narrow river environment and not due to a change in gauges.

The side looking ADCP mounted at the Crown Point bridge site at a depth of 6.4 meters provided a unique view of the integrated current structure across the center channel. The most dominant signature was that of along-channel velocity which possessed significant variability in flow direction. Cross-channel velocity, as expected, was small and not seriously investigated in this report. Echo-intensities, which define the amount of particulates within the water column (backscatterers), showed an exceedingly strong diurnal component, which implied some type of biological activity.

Auto-and cross-spectral analyses were completed on all observations over all years. The most significant outcome was that the South Lake has preferred modes of oscillation that are directly linked to atmospheric forcing. Other than semi-diurnal and diurnal forcing events, oscillations within the South Lake range from 1.42 - 7.11 days. The 7.11 day spectral peak was the largest observed in both the along-channel and cross-channel velocities observed by the side-looking ADCP.

With regards to lake water quality, only a few speculations can be made at this time. First, it is possible that the bottom velocities obtained during these thermal hook events may have the capability to resuspend sediments and subsequently redistribute them both to the south and possibly back to the north (relaxation phase). In other circumstances where bottom velocities would not reach an erosional threshold, it is possible that clearer (less turbid), deep, hypolimnic water may be evident along the entire path length of the event. It should also be made clear that these intrusions are deep and often times well below the level of surface Secchi disk readings, and therefore, would not have a significant impact on the public's perception of water clarity. To document whether or not these cases exist, longterm monitoring through the use of self-contained ADCPs, temperature sensors, and transmissometers would have to be made.

Recommendations for future Work relating to the South Lake would be that of a more rigorous set of water velocity observations in conjunction with thermal and water level variations. Meteorological observations for the South Lake should not rely upon CR or BIA due to their inability to properly define wind speed and direction within this reach of Lake. IPC or a similar site must be considered essential. If at all possible, water level observations should be completed along the South Lake rather than at the two endpoints. This would better afford any future researchers the ability to properly define propagation speeds of rapid changes in water level through the South Lake. The final culmination to such a research program within the South Lake would be that of a modeling effort that could eventually be linked to Main Lake dynamics.

7. ACKNOWLEDGMENTS

Special thanks are given to Dr. Scott Quinn for coordinating all aspects of this program (boats and personnel) with NYDEC as well as my diving partner, Mr. Rob Burnham ("The Groper"), who could find anything in zero visibility conditions. My complements to both Scott and Rob for maintaining a continual level of enthusiasm and joviality while out in the field, even under somewhat adverse conditions. Special thanks also go to Mr. Leigh Walrath, and Ms. Joanna Clifford for providing additional support on many of our South Lake cruises. 2-D and 3-D modeling was accomplished using earthVisions, of Dynamic Graphics, Inc., Alameda, CA. It should be noted that Mr. Lee Gordon of RDI processed the ADCP data. These data were later provided in MATLAB format.

REFERENCES

Henson, E. B., 1972, Summary of limnological conditions of southern Lake Champlain (19 66-1972), University of Vermont, Burlington, VT, unpublished report.

Hunkins, K., T. O. Manley, P. Manley and J. Saylor, 1998, Numerical studies of the four-day oscillation in Lake Champlain, J. Geophys. Res., 103, 18,425-18,436.

Hunkins, K., D. Mendelsohn, and T. Isaji, 1999, Numerical hydrodynamic model of Lake Champlain, *Lake Champlain in Transition; from Research Towards Restoration*, AGU Water Science and Application Monograph No. 1, T. O. Manley and P. L. Manley, editors, 67-117.

Klein, L., 1998, Effects of the internal seiche in the South Main Lake of Lake Champlain, undergraduate thesis, Middlebury College, Dept. of Geology, 44 pp.

Manley, T. O., K. L. Hunkins, J. H. Saylor, G. S. Miller and P. L. Manley, 1999, Aspects of summertime and wintertime hydrodynamics of Lake Champlain, *Lake Champlain in Transition; from Research Towards Restoration*, AGU Water Science and Application Monograph No. 1, T. O. Manley and P. L. Manley, editors, 67-117.

Meyer, G., and G. K. Gruendling, 1979, Limnology of Lake Champlain, Lake Champlain Basin study No. 30, New England river basins commission, Boston, Massachusetts 02109.

Osterberg, E., 1999, Hydrodynamics of the South Main Lake of Lake Champlain, undergraduate thesis, Middlebury College, Dept. of Geology, 106 pp.

Sardilli, D., 1999, The Internal Dynamics of Shelburne Bay, Vt., undergraduate thesis, Middlebury College, Dept. of Geology, 62 pp.

Saylor, J. H., G. Miller, K. Hunkins, T. O. Manley and P. Manley, 1999, Gravity currents and internal bores in Lake Champlain, *Lake Champlain in Transition; from Research Towards Restoration*, AGU Water Science and Application Monograph No. 1, T. O. Manley and P. L. Manley, editors, 135-156.

Yih, C. S., Stratified Flows, 1980, Academic Press, New York.

A REVIEW OF VORTICITY WAVES IN LAKES

Kenneth Hunkins*

ABSTRACT

Long-period wave motion in lakes may be of two types, either internal gravity waves or vorticity waves. Internal gravity waves have long been identified as an important component of long-period motion in large stratified lakes. Less attention has been given to the identification of vorticity waves which may theoretically exist in unstratified as well as stratified lakes. So far vorticity waves have only been convincingly identified only in Lake Michigan. Here we examine some elementary models of vorticity waves in channels and rectangular basins. The characteristics of vorticity waves in these idealized models are discussed to provide guidance for a search in Lake Champlain for these waves.

1. INTRODUCTION

Currents with time scales longer than one day are prominent in continuous records of current velocities in many large lakes, including Lake Champlain. Such motions are strongly influenced by winds, bottom topography and the Earth's rotation. At the longest time scales these currents approach steady state and transport water with dissolved and suspended matter over long distances. There is a practical need for an understanding of these long-period motions. A water parcel containing plankton, natural substances or manmade pollutants may be carried long distances. For example, in two days time a water parcel moving at a maximum wave speed of 25 cm/s will travel nearly 30 km. Long-period waves deserve further study in order to better predict their motions for practical applications as well as for their intrinsic interest. One approach, and the one used here, is the examination of resonant oscillations in an idealized basin with some resemblance to an actual lake basin. The wave theory used is based on the shallow-water

* Lamont-Doherty Earth Observatory of Columbia University
Palisades, NY 10964

equations, also known as the long-period or Laplace equations. Long-period waves in large lakes can be either of two distinct physical types, gravity waves or vorticity waves, depending on the restoring force. Currents with a period of four days which are observed in Lake Champlain have been identified as wind-forced gravity waves. Gravity wave theory has been successful in explaining wind-driven motion with long periods in many lakes during summer stratified conditions (e.g. Hunkins et al, 1998; Mortimer, 1979). Less attention has been given to the possible presence of vorticity waves in long, narrow lakes such as Lake Champlain. Before attempting to detect the presence of vorticity waves in Lake Champlain with field observations it seemed desirable to review the characteristics of vorticity waves in idealized lakes.

2. BACKGROUND

For wave motion to occur there must be a force which tends to restore equilibrium in the medium. For vorticity waves the restoring force is the tendency for vorticity, the rotary motion of the flow, to be conserved, much as angular momentum is conserved for solid bodies. Imagine vertical water columns changing their rate of rotation as they move over changing water depths. As a column moves from deep to shallow water, for example, its radius is increased as it is squashed and its rate of rotation is slowed. Varying topography is required for the existence of these waves. Athough gravity waves may exist in basins with flat bottoms, vorticity waves exist only in basins with sloping bottoms. The Earth's rotation is also essential to their existence. Their periods are always greater than the Coriolis period (17 hours at the latitude of Lake Champlain). Gravity waves, in contrast, may exist at all frequencies and are only modified by rotation effects at low frequencies.

The physical mechanism of topographic vorticity waves is illustrated in Fig. 1.

A string of vertical water columns originally lies in a straight line parallel to the depth contours. The slope is bounded on one side by deep water and by shallow water on the other. Basic rotation is counter-clockwise (northern hemisphere). Initially the line of columns is displaced to form a sine wave. Those columns displaced into deep water are lengthened and their vorticity increases. They then have positive vorticity (counter-clockwise spin) relative to the basic rotation. Columns displaced into shallow water are shortened and develop a negative vorticity (clockwise spin). The induced motion is a train of gyres with alternating sign. Phase propagation is parallel to the depth contours with shallow water on the right-hand side when facing in the direction of propagation.

There is another type of vorticity wave with long periods which is found in the ocean and in the atmosphere but is not expected to be found in lakes. These are planetary vorticity waves, known also as Rossby waves, which require the change in planetary vorticity which occurs with latitude to provide a restoring force. These waves are not considered to be important in lakes since, unlike the ocean and atmosphere, lakes are quite limited in meridional extent. Thus there is little variation of planetary vorticity over lakes and planetary vorticity may generally be considered constant. The basic physical mechanisms and mathematical developments for the two types of vorticity waves are quite similar however (Veronis, 1966).

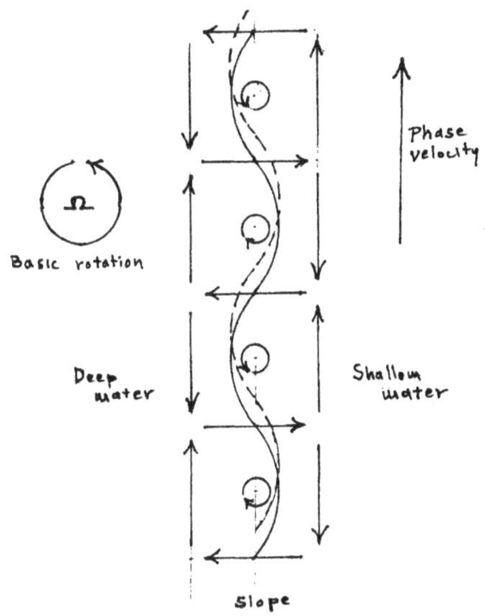

Figure 1. Schematic plan view of a topographic vorticity wave propagating along a sloping bottom. Water columns initially in a straight line are displaced into a sine wave. The columns are displaced horizontally up or down the slope inducing vorticity changes in the basic vorticity. Straight arrows indicate the induced motion. Transverse velocities displace the sine wave into the position shown by the broken line. The result is that phases travel in the direction indicated. (Adapted from Longuet-Higgins, 1972)

Although there is abundant evidence in the ocean for topographic vorticity waves in the form of trapped waves over continental slopes (Allen, 1980; Mysak, 1980; Brink, 1991), called continental shelf waves in this case, or around submarine plateaus (Hunkins, 1986), evidence for their existence in lakes is sparse. One of the most convincing demonstrations and analyses of vorticity waves in a lake was presented by Saylor et al (1980). They interpreted current oscillations of 100-hour period in the southern basin of Lake Michigan as first mode vorticity waves by comparing observations to results from a circular basin model with a power-law bottom profile. Rotary current spectra, distinguishing between clockwise and counter-clockwise motion, agreed with theory and provided convincing evidence of vorticity waves. Subsequent field studies in Lake Michigan have shown that vorticity waves are a common feature in that lake (Saylor, personal communication).

Investigations in smaller lakes have also produced evidence of vorticity waves but the evidence is not as convincing as in the case of Lake Michigan. Investigations on Lake Lugano in Switzerland revealed motions with a 74-hour period. This period was interpreted as a higher mode of small dimensions and long period (Mysak, 1985). The existence of such a complex and small-scale current regime over long time intervals seems problematical. Moreover, the ratio of breadth to length for the model basin was 2:3 rather than the actual 1:10. Stocker (1987) has presented a detailed theoretical study of topographic waves in channels in an attempt to understand vortex modes in long narrow lakes such as the subalpine lakes of Switzerland. These modes were shown to have a rich and complex structure in channels and closed basins with regular shorelines (Stocker and Hutter, 1987). Numerical studies show resonant modes trapped in bays in addition to the modes found analytically in straight channels. However observational evidence for such structure in lakes has been lacking so far.

The most distinctive feature of the current spectrum in Lake Champlain is a broad peak centered on a period of 100 hours. This baroclinic motion was interpreted as a fundamental wind-driven internal seiche and gravity wave theory appeared to provide adequate interpretation (Hunkins et al, 1998). Vorticity waves were originally dismissed as an explanation because there was an apparent mismatch in frequency and wavennumber based on the theoretical studies of others. The identification of vorticity waves requires extensive instrument arrays and detailed analysis of current records. Such records and analysis are not yet available for Lake Champlain so the presence or absence of vorticity waves in that lake has yet to be determined. As a basis for continued analysis of available data and for planning future field investigations, some elementary theoretical models of vorticity waves are explored here. First, plane waves over an exponentially-sloping bottom of infinite extent are examined. Then vorticity waves in an idealized channel and in a rectangular basin are analyzed.

3. BASIC EQUATIONS

The linear forced shallow-water equations of motion in a rectangular coordinate system are

$$\frac{\partial u}{\partial t} - fv + g\frac{\partial \zeta}{\partial x} = \frac{1}{\rho h}\tau_x \tag{3.1}$$

$$\frac{\partial v}{\partial t} + fu + g\frac{\partial \zeta}{\partial y} = \frac{1}{\rho h}\tau_y \tag{3.2}$$

and the equation of continuity is

$$\frac{\partial (uh)}{\partial x} + \frac{\partial (vh)}{\partial y} + \frac{\partial \zeta}{\partial t} = 0 \tag{3.3}$$

where u and v are depth-averaged velocities in the x (east) and y (north) directions respectively. The Coriolis parameter is, $f = 2\Omega \sin \phi$, where Ω is the angular frequency of the Earth and ϕ is latitude (positive in the northern hemisphere), h is mean depth, ζ is the disturbance height of the surface, g is the acceleration due to gravity and ρ is water density. Components of wind stress at the surface and friction at the bottom are represented by τ_x and τ_y.

Topographic vorticity waves are manifested primarily in the current field. Changes in surface height are small and it is possible in many cases to consider a surface fixed in elevation by a rigid lid. The rigid-lid approximation assumes that the divergence parameter, $\varepsilon = \dfrac{f^2 L^2}{gh}$, is small (Buchwald and Adams, 1968). The divergence parameter is the squared ratio of a characteristic length, L, to the radius of deformation, $\dfrac{\sqrt{gh}}{f}$. For barotropic motion in a long, narrow lake at mid-latitude such as Lake Champlain, we find $\varepsilon \ll 1$. The motion is then horizontally non-divergent and (3.3) simplifies to

$$\frac{\partial(uh)}{\partial x} + \frac{\partial(vh)}{\partial y} = 0. \tag{3.4}$$

Equation (3.4) allows velocities to be conveniently expressed in terms of a single dependent variable, the stream function for mass transport, ψ, where

$$\rho h u = \frac{\partial \psi}{\partial y} \quad \text{and} \quad \rho h v = -\frac{\partial \psi}{\partial x}. \tag{3.5}$$

Cross-differentiation of (3.1) and (3.2) and substitution from (3.5) yields a transport vorticity equation,

$$\frac{\partial}{\partial t}\left[\nabla^2 \psi - \frac{1}{h}\frac{\partial h}{\partial x}\frac{\partial \psi}{\partial x} - \frac{1}{h}\frac{\partial h}{\partial y}\frac{\partial \psi}{\partial y}\right] + f\left[\frac{1}{h}\frac{\partial h}{\partial x}\frac{\partial \psi}{\partial y} - \frac{1}{h}\frac{\partial h}{\partial y}\frac{\partial \psi}{\partial x}\right]$$

$$= -\left[\frac{\partial \tau_y}{\partial x} - \frac{\partial \tau_x}{\partial y}\right] + \frac{1}{h}\frac{\partial h}{\partial x}\tau_y - \frac{1}{h}\frac{\partial h}{\partial y}\tau_x \tag{3.6}$$

where $\nabla^2 = \dfrac{\partial^2}{\partial x^2} + \dfrac{\partial^2}{\partial y^2}$ denotes the horizontal Laplacian operator.

The water body extends infinitely in all horizontal directions. This equation represents the time rate of change in vertical vorticity of a water column as it moves over varying bottom topography.

When depth varies only in the x direction and wind stress acts only in the y direction, (3.6) reduces to

$$\frac{\partial}{\partial t}\left[\nabla^2 - \frac{h'}{h}\frac{\partial}{\partial x}\right]\psi + f\,\frac{h'}{h}\frac{\partial \psi}{\partial y} = \frac{h'}{h}\tau_y \tag{3.7}$$

where primes indicate differentiation with respect to x. Wind stress and bottom friction will be neglected in the following discussion of free waves.

4. TWO-DIMENSIONAL VORTICITY WAVES

Assume an exponential bottom profile sloping upward in the x (east) direction,

$$h = h_0 e^{-2\alpha x} \text{ and } \qquad \frac{h'}{h} = -2\alpha , \tag{4.1}$$

This particular profile reduces (3.7) to a partial differential equation with constant coefficients,

$$\frac{\partial}{\partial t}\left[\nabla^2 + 2\alpha\frac{\partial}{\partial x}\right]\psi - 2\alpha f\frac{\partial \psi}{\partial y} = 0. \tag{4.2}$$

To find the relationship between frequency and wavenumber, introduce a plane-wave solution,

$$\psi(x, y, t) = \psi_0 e^{i(kx+ly-\omega t)}, \tag{4.3}$$

into (4.2). Wavenumbers in the x and y directions are represented by k and l respectively and frequency by ω. The result is a complex dispersion equation,

$$-k^2 - l^2 + 2i\alpha k + 2\alpha l\frac{f}{\omega} = 0, \tag{4.4}$$

relating frequency to wavenumber. The imaginary term represents growth or decay resulting from motion over a sloping bottom. To obtain a dispersion equation in terms of real variables, introduce a complex wavenumber, $k = k_r + i\alpha$. In terms of real variables the relation between frequency and wavenumber then becomes

$$\omega = \frac{2\alpha fl}{k_r^{\,2} + l^2 + \alpha^2}. \tag{4.5}$$

One sees that frequency approaches zero at both very high and very low wavenumbers. At intermediate wavenumbers there is a frequency maximum. The wavenumber of the maximum is found by first differentiating (4.5) to give the group velocity,

$$U = \frac{\partial \omega}{\partial l} = \frac{2\alpha f \left(k_r^2 - l^2 + \alpha^2\right)}{\left(k_r^2 + l^2 + \alpha^2\right)^2}, \tag{4.6}$$

and setting the result equal to zero,

$$l = \sqrt{k_r^2 + \alpha^2}. \tag{4.7}$$

Substitute (4.7) into (4.5) to find the value of the maximum frequency (minimum period),

$$\omega_{\max} = \frac{\alpha f}{\sqrt{k_r^2 + \alpha^2}}. \tag{4.8}$$

Waves are thus limited to low frequencies, $\omega \leq f$.

Phase velocity is in the y direction, parallel to the contours,

$$c = \frac{\omega}{l} = \frac{2\alpha f}{k_r^2 + l^2 + \alpha^2}, \tag{4.9}$$

and is seen to be always positive, so waves travel in the y direction (north) with shallow water on their right-hand side as expected. Group velocity, given by (4.6), is zero at the frequency maximum. This means that for wave trains composed of a band of frequencies, energy does not propagate at the frequency maximum. Energy will remain at its original location without dispersing. This may have application to lakes since it predicts enhanced energy at the frequency maximum.

5. VORTICITY WAVES IN AN EXPONENTIAL HALF-CHANNEL

A channel with parallel sides and breadth, B, provides the simplest idealization for the topography of a long, narrow lake. To find the form of the stream function in the x direction for a channel oriented in the y direction, introduce a solution of the form,

$$\psi(x, y, t) = F(x)e^{i(ly - \omega t)}, \tag{5.1}$$

into (4.2) to reduce it to an ordinary differential equation,

$$\frac{d^2F}{dx^2} - \frac{h'}{h}\frac{dF}{dx} - \left[l^2 + l\frac{f}{\omega}\frac{h'}{h}\right]F = 0. \tag{5.2}$$

There can be no transport across the shoreline boundaries at x = (0, B). To satisfy these boundary conditions the stream function, ψ, must be constant on the boundaries. For simplicity, the constant is taken here to be zero. The boundary conditions to be satisfied are then

$$F(0)=F(B)=0\,.$$

To proceed further we need to specify a cross-sectional shape for the channel. The same exponential slope used earlier is employed again to define the bottom profile. The half-channel profile is represented by an exponential slope,

$$h=h_0 e^{-2\alpha x}\,.$$

Equation (5.2) becomes an ordinary differential equation with constant coefficients,

$$F''+2\alpha F'-\left[l^2-2\alpha l\frac{f}{\omega}\right]F=0 \tag{5.3}$$

This asymmetric profile with depth decreasing in the x (east) direction may be considered one-half of a symmetrical channel. It is an approximation to the eastern side of the Lake Champlain basin. For simplicity we analyze only this half channel and find that vorticity waves consist of gyres propagating in the positive y (north) direction. If we were to analyze another channel, the mirror image of this profile, forming the other side of a symmetric channel we would find gyres propagating in the negative y (south) direction. So in a symmetrical channel there will be vorticity waves traveling northward on one side and southward on the other side with shallow water on the right in both cases.

Substitute a trial solution, $F=e^{mx}$,

into (5.2) to give a quadratic characteristic equation,

$$m^2+2\alpha m-\left[l^2-2\alpha l\frac{f}{\omega}\right]=0 \tag{5.4}$$

with the solution, $$m=-\alpha\pm\sqrt{\alpha^2+l^2-2\alpha l\frac{f}{\omega}}\,. \tag{5.5}$$

In order to satisfy the boundary conditions we must have oscillatory solutions and the quantity under the radical must be negative. So we have $2\alpha l\frac{f}{\omega}>(\alpha^2+l^2)$

and m will be a complex conjugate. Therefore, it would be easier to understand if m were written as

$$m=-\alpha\pm i\sqrt{2\alpha l\frac{f}{\omega}-\alpha^2-l^2}\,. \tag{5.6}$$

The solution is then

$$F = e^{-\alpha x} e^{i\beta x} \tag{5.7}$$

where

$$\beta = \pm\sqrt{2\alpha l \frac{f}{\omega} - \alpha^2 - l^2} \tag{5.8}$$

This solution can be rewritten as

$$F = e^{-\alpha x}(A_1 \sin \beta x + A_2 \cos \beta x) \tag{5.9}$$

where A_1 and A_2 are constants.

To satisfy the boundary condition at x = 0, set $A_2 = 0$ and to satisfy the boundary condition at x = B, set

$$\beta = \frac{n\pi}{B} \tag{5.10}$$

where n = 1, 2, 3, ... is the mode number. Introduce (5.10) into (5.8) to form a dispersion equation relating frequency to along-channel wavenumber

$$\omega = \frac{2\alpha l f}{\alpha^2 + l^2 + \left(\frac{n\pi}{B}\right)^2} \tag{5.11}$$

Rewriting in terms of dimensionless variables, $\hat{\omega} = \frac{\omega}{f}$ and $\hat{l} = \frac{l}{\alpha}$,

the dispersion equation becomes

$$\hat{\omega} = \frac{2\hat{l}}{1 + \hat{l}^2 + \left(\frac{n\pi}{\alpha B}\right)^2} \tag{5.12}$$

With Lake Champlain in mind, choose the following values for the profile parameters:

$$h_0 = 100m\,,\ \alpha = 5x10^{-4}m^{-1}, \text{ and } B = 3{,}000m$$

The dispersion curve using these constants is depicted in Figure 2. The dispersion curve has two branches separated by a frequency maximum which is always less than the Coriolis frequency, f. The frequency maximum is found by differentiating (5.12) and setting the result equal to zero. Since the frequency maximum is predicted to be of

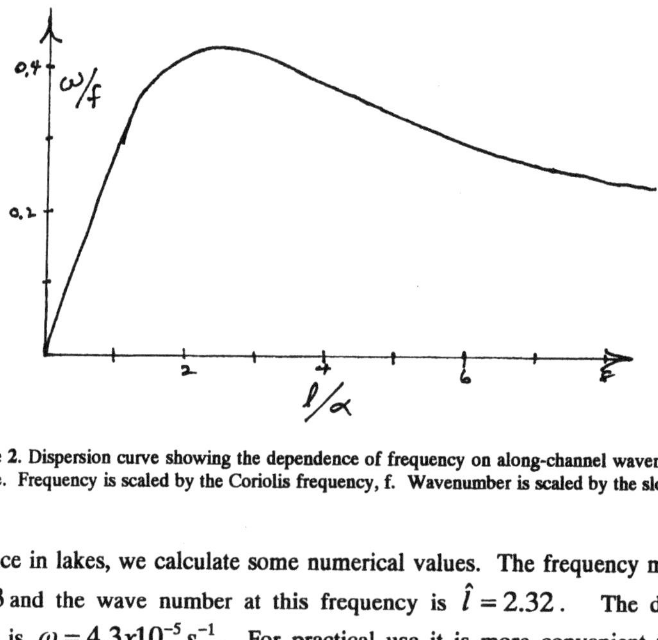

Figure 2. Dispersion curve showing the dependence of frequency on along-channel wavenumber for the lowest mode. Frequency is scaled by the Coriolis frequency, f. Wavenumber is scaled by the slope parameter, α.

importance in lakes, we calculate some numerical values. The frequency maximum is $\hat{\omega} = 0.43$ and the wave number at this frequency is $\hat{l} = 2.32$. The dimensional frequency is $\omega = 4.3x10^{-5}s^{-1}$. For practical use it is more convenient to consider period and wavelength. In terms of period, there is a minimum at

$$T_{min} = \frac{2\pi}{\omega} = 40.5 \text{ hours.}$$

The wavelength at the frequency maximum is $\lambda = \dfrac{2\pi}{l} = 5.41$ km.

The frequency maximum divides the dispersion curve into two branches. The high-wavenumber branch is dispersive with the unusual property that frequency decreases with increasing wavenumber. This means that the wavelength or gyre size decreases as the period grows longer. We have the paradox that at very long periods, approaching steady state, the wavelengths become very small. Group velocity is directed opposite to phase velocity. The low-wavenumber branch is non-dispersive with constant phase velocity.

For a wavelength of 100 km, approximating the length of the Main Lake, the period is 15.6 days with phase velocity, c = 7.1 cm/s. For the low wavenumber branch, group velocity, $U = \frac{d\omega}{dl}$, has the same value as phase velocity and is in the same direction.

Finally we examine the cross-channel shape of the stream function and the current velocity. The cross-channel shape of the stream function is given by

$$F = Ae^{-\alpha x} \sin \frac{n\pi x}{b} \tag{5.13}$$

and the stream function profile for the chosen parameters is depicted in Figure 3.

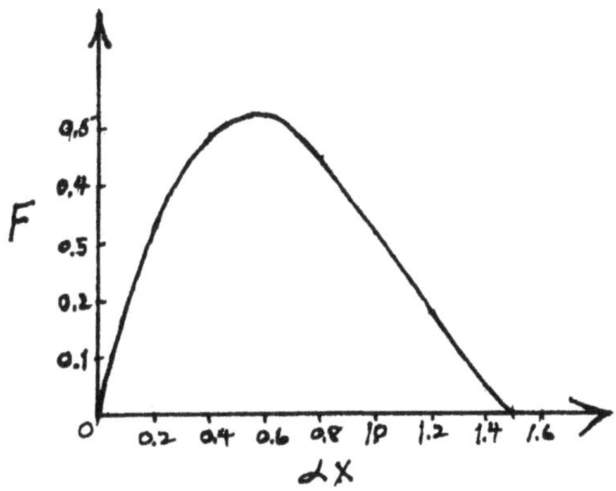

Figure 3 Profile of the stream function, F, as a function of scaled cross-channel distance for an exponential half-channel.

The stream function for mass transport can be calculated using (5.1) and (5.13). A contour plot of the stream function at an instant of time is shown in Fig. 4. Mass transport is along the contour lines. There are alternating clockwise and counterclockwise gyres. The gyres are nearly equidimensional at the cut-off frequency. They extend over the width of the half-channel, 3 km, and their y-dimension is one half wavelength or 2.70 km. Two wavelengths, consisting of four gyres, are depicted. The pattern is repeated endlessly along this infinite channel. This pattern is repeated in time with a period of 40.5 hours. Longer periods on the high wavenumber branch will have even shorter wavelengths. We have chosen the minimum period for this example. However the small spatial scale at these low frequencies may not exist in the presence of turbulent dissipation. It seems questionable whether such small features can persist in a lake over long periods of time. Only carefully planned observations over small scales and long time spans can settle the question of the existence of high-wavenumber topographic waves in lakes.

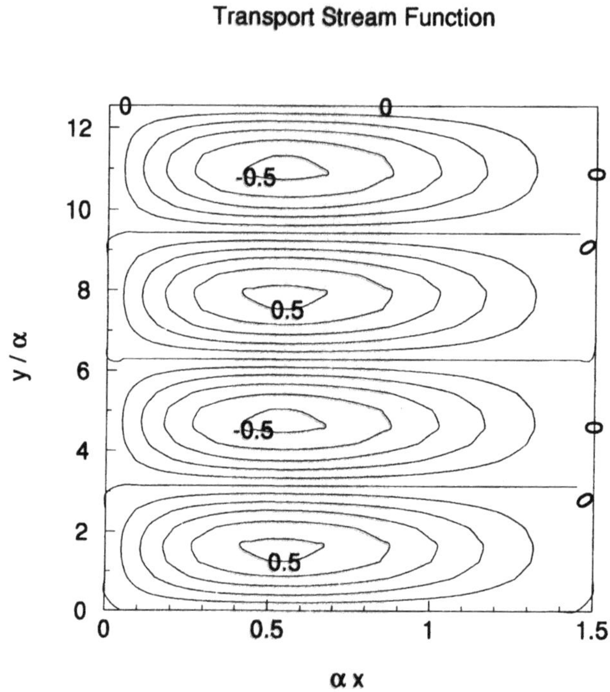

Figure 4 Contour plot at an instant of time for the stream function of mass transport in an exponential half-channel shoaling in the x direction. Flow is along the streamlines. Alternating clockwise and anti-clockwise gyres travel in the positive y direction parallel to the slope with shallow water on their right-hand side.

6. VORTICITY WAVES IN A CHANNEL WITH A GAUSSIAN PROFILE

A channel with a Gaussian depth cross-section is closer to the actual profiles of long narrow lakes than the exponential profiles used in the previous sections. Only free waves are considered and both surface and bottom stress are neglected. The bottom slopes only in the x (east-west) direction. The wave equation (3.7) becomes

$$\frac{\partial}{\partial t}\left[\nabla^2 - \frac{h'}{h}\right]\psi + \frac{h'}{h} f \frac{\partial \psi}{\partial y} = 0 \tag{6.1}$$

where $\rho hu = \frac{\partial \psi}{\partial y}$ and $\rho hv = -\frac{\partial \psi}{\partial x}$ as before. Primes indicate differentiation with respect to x.

For waves propagating in the y (north) direction,

$$\psi(x,y,t) = F(x)e^{i(ly-\omega t)},$$

(6.2)

and (6.1) is reduced to an ordinary differential equation to be solved for the transverse wave profile,

$$F'' - \frac{h'}{h}F' - \left[l^2 + l\frac{f}{\omega}\frac{h'}{h}\right]F = 0 \quad .$$

(6.3)

Select a Gaussian shape for the transverse bottom profile, $h = h_0 e^{-m^2x^2}$, so that $\frac{h'}{h} = -2m^2 x$. Substitution in (6.3) yields

$$F'' + 2m^2 xF' - \left[l^2 - 2m^2 l\frac{f}{\omega}x\right]F = 0$$

(6.4)

Introduce scales for the independent variable, $\hat{x} = mx$, and for the parameters, $\hat{l} = l/m$, and $\hat{\omega} = \omega/f$ so that (6.4) becomes the scaled equation,

$$\frac{d^2F}{d\hat{x}^2} + 2\hat{x}\frac{dF}{d\hat{x}} - \left[\hat{l}^2 - 2\frac{\hat{l}}{\hat{\omega}}\hat{x}\right]F = 0.$$

(6.5)

Henceforward drop the carets with the understanding that the symbols x, l and ω now represent the scaled quantities $\hat{x}$, $\hat{l}$ and $\hat{\omega}$. Rewrite (6.5) as

$$\frac{d^2F}{dx^2} + 2x\frac{dF}{dx} - \left[l^2 - 2\frac{l}{\omega}x\right]F = 0.$$

(6.6)

Represent the solution to (6.6) by a power series,

$$F(x) = a_0 + a_1x + a_2x^2 + a_3x^3 + \ldots\ldots$$

(6.7)

Substituting the power series into (6.6) and equating coefficients of like powers of x sets the following conditions on the coefficients.

$$a_2 = \frac{l^2}{2}a_0$$

$$a_3 = \frac{\left(l^2-2\right)a_1 - 2\frac{l}{\omega}a_0}{6}$$

$$a_4 = \frac{\left(l^2-4\right)a_2 - 2\frac{l}{\omega}a_1}{12}$$

Conditions on transport at the channel axis, x = 0, are no transport in the x direction and fixed transport in the y direction, u(0) = 0 and v(0) = a constant, for all values of y. This requires that $F\left(0\right)=0$ and $F'\left(0\right)=1$ for all y values. Thus reference to (6.7) shows that $a_0 = 0$ and $a_1 = 1$. The coefficients for the first five terms are

$$a_0 = 0$$

$$a_1 = 1$$

$$a_2 = 0$$

$$a_3 = \frac{l^2-2}{6}$$

$$a_4 = -\frac{l}{6\omega}$$

Substituting into (6.7)

$$F(x) = x + \frac{l^2-2}{6}x^3 - \frac{l}{6\omega}x^4 + \ldots.. \tag{6.8}$$

There is no transport across the shoreline at $x = \pm b$. Substitution of this boundary condition, $F\left(\pm b\right)=0$, into (6.8) yields a dispersion equation relating frequency and wavenumber,

$$\omega = \frac{b^3 l}{6 + b^2 l^2 - 2b^2}. \tag{6.9}$$

The dispersion curve (Figure 5) extends across all wavenumbers . Positive wavenumbers represent propagation in the positive y direction (north) on the positive x (east) side of the channel. Negative wavenumbers (not plotted) refer to propagation in the negative y direction (south) on the negative x (west) side of the channel. The dispersion curve for the negative wavenumbers is the mirror image, reflected in the x axis, of the illustrated dispersion curve.

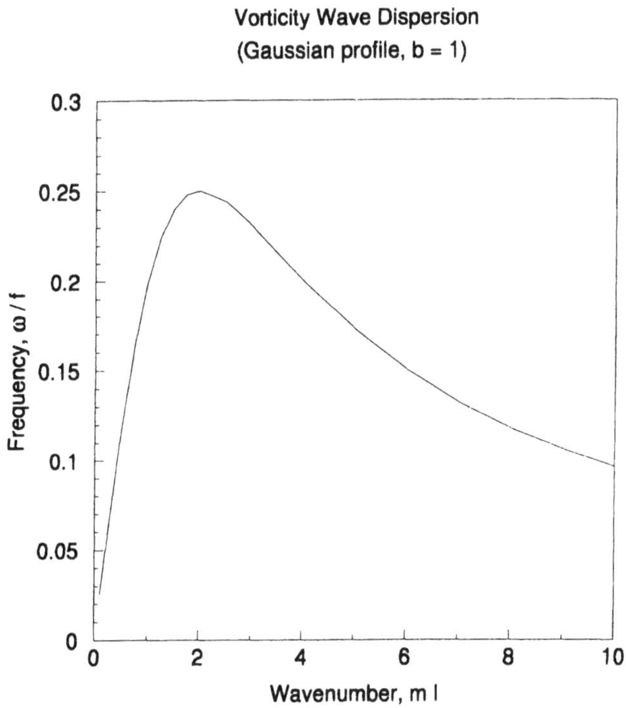

Figure 5 Dispersion curve plotting frequency versus wavenumber for a channel with Gaussian profile.

In calculating some numerical values for the Gaussian profile we again restrict ourelves to the frequency maximum. Take $m = 5x10^{-4} m^{-1}$ and b = 2 km as approximations to represent Lake Champlain as a channel. Proceeding as in section 5 we find that the frequency maximum corresponds to a wavelength of 6.28 km and a period of 69.8 hours.

$$\text{Phase velocity,} \quad c = \frac{\omega}{l} = \frac{b^3}{6 + b^2 l^2 - 2b^2}, \tag{6.10}$$

is positive (north) for $b > 0$ (eastern side) and negative (south) for $b < 0$ (western side). At the frequency maximum, c = 25 cm/s.

7. VORTICITY WAVES IN A RECTANGULAR BASIN

To investigate vorticity waves in closed basins, boundaries are added to the ends of the half-channel model. Consider a rectangular basin of length, L, and breadth, B, with the same exponential bottom slope used in the previous examples. Introduce a harmonic time dependence, $e^{-i\omega t}$, into the vorticity transport equation (4.2) to give

$$\left[\nabla^2 + 2\alpha \frac{\partial}{\partial x}\right]\psi - 2i\alpha \frac{f}{\omega}\frac{\partial \psi}{\partial y} = 0. \tag{7.1}$$

Separate the spatial variables by setting $\psi(x,y) = X(x)Y(y)$.
Differentiation and substitution into (7.1) gives

$$Y\frac{\partial^2 X}{\partial x^2} + X\frac{\partial^2 Y}{\partial y^2} + 2\alpha Y\frac{\partial X}{\partial x} - 2i\alpha\frac{f}{\omega}X\frac{\partial Y}{\partial y} = 0 \tag{7.2}$$

Divide by XY,

$$\frac{1}{X}\frac{\partial^2 X}{\partial x^2} + \frac{1}{Y}\frac{\partial^2 Y}{\partial y^2} + 2\alpha\frac{1}{X}\frac{\partial X}{\partial x} - 2i\alpha\frac{f}{\omega}\frac{1}{Y}\frac{\partial Y}{\partial y} = 0, \tag{7.3}$$

and separate (7.3) into a pair of ordinary differential equations, each equal to the separation constant, $-\lambda^2$,

$$\frac{\partial^2 X}{\partial x^2} + 2\alpha\frac{\partial X}{\partial x} + \lambda^2 X = 0 \tag{7.4}$$

$$\frac{\partial^2 Y}{\partial y^2} - 2i\alpha\frac{f}{\omega}\frac{\partial Y}{\partial y} - \lambda^2 Y = 0 \tag{7.5}$$

Substitution of a trial solution, $X = Ce^{px}$, into (7.4) gives

$$p^2 + 2\alpha p + \lambda^2 = 0 \tag{7.6}$$

and

$$p = -\alpha \pm \sqrt{\alpha^2 - \lambda^2}.$$

The boundary condition, $\psi = 0$ at x = 0 requires that $\lambda^2 \geq \alpha^2$. Therefore, a preferred form of the equation would be

$$p = -\alpha \pm i\sqrt{\lambda^2 - \alpha^2} \tag{7.7}$$

$$X = e^{-\alpha x}\left[C_+ e^{i\sqrt{\lambda^2-\alpha^2}x} + C_- e^{-i\sqrt{\lambda^2-\alpha^2}x}\right] \tag{7.8}$$

A trial solution for (7.5), $Y = De^{qy}$, yields

$$q^2 - 2i\alpha\frac{f}{\omega}q - \lambda^2 = 0 \tag{7.9}$$

$$q = i\alpha\frac{f}{\omega} \pm \sqrt{-\left(\frac{\alpha f}{\omega}\right)^2 + \lambda^2} \tag{7.10}$$

$$Y = e^{i\alpha\frac{f}{\omega}y}\left[D_+ e^{i\sqrt{\left(\alpha\frac{f}{\omega}\right)^2-\lambda^2}y} + D_- e^{-i\sqrt{\left(\alpha\frac{f}{\omega}\right)^2-\lambda^2}y}\right] \tag{7.11}$$

Take the real parts of the solution which satisfy the boundary condition, $\psi = 0$ at $x = 0$ and $y = 0$, to give

$$\psi = Ae^{-\alpha x}\sin\sqrt{\lambda^2-\alpha^2}x \cdot \sin\sqrt{\left(\alpha\frac{f}{\omega}\right)^2 - \lambda^2}y \cdot \cos\left(\alpha\frac{f}{\omega}y - \omega t\right) \tag{7.12}$$

The boundary conditions, $\psi = 0$ at $x = B$ and $x = L$, require that

$$\sqrt{\lambda^2 - \alpha^2} = \frac{n\pi}{B} \qquad \text{and} \qquad \sqrt{\left(\frac{\alpha f}{\omega}\right)^2 - \lambda^2} = \frac{m\pi}{L} \tag{7.13}$$

where the mode numbers, n and m, are integers.

Finally, we obtain the wave form characterized by these integers

$$\psi_{mn} = A_{mn} e^{-\alpha x} \sin\frac{n\pi x}{B} \cdot \sin\frac{m\pi y}{L} \cdot \cos\left(\alpha \frac{f}{\omega} y - \omega_{mn} t\right) \tag{7.14}$$

with a characteristic frequency

$$\left(\frac{\alpha f}{\omega_{mn}}\right)^2 = \left(\frac{n\pi}{B}\right)^2 + \left(\frac{m\pi}{L}\right)^2 + \alpha^2$$

or

$$\omega_{mn} = \frac{\alpha f}{\sqrt{\left(\frac{n\pi}{B}\right)^2 + \left(\frac{m\pi}{L}\right)^2 + \alpha^2}} \tag{7.15}$$

The dispersion relation is no longer a continuous curve as it was for the channel model. There are discrete frequencies dependent on the basin dimensions and the mode numbers.

The rectangular basin model assumes that there are vertical walls bounding the ends. In Lake Champlain, there are partially open boundaries at the ends of the main basin and this model may not be relevant for that particular lake. It may be that the channel model is preferable for Lake Champlain.

The straight boundaries in these simple models do not reproduce an actual shoreline with its bays and headlands. The numerical experiments of Stocker (1987) on vorticity waves in curved channels are relevant to studies of these waves in irregular channels. More detailed numerical experiments with irregular shorelines using finite-element models have shown that gyres may be trapped in bays.

8. IMPLICATIONS FOR FIELD OBSERVATIONS

Although application of the foregoing simple models to identification of vorticity waves in long narrow lakes can only be on a limited basis since the idealized model geometry merely suggests the shapes of real lakes, the models do demonstrate some of the essential charaacteristics of these waves. Several generalizations may be made on their basis which should assist a search for vorticity waves in Lake Champlain. These waves take the form of gyres traveling parallel to depth contours with shallow water on the right hand side when looking in the direction of travel. In a long narrow lake oriented in a north-south direction like Lake Champlain we expect to see gyres propagating northward on the eastern side and gyres propagating southward on the western side. Propagation speeds will be in the vicinity of tens of centimeters per second and periods will be several days. Vorticity waves will be most evident in current records and will only produce a weak signal in pressure records. Rotary spectra have been an important tool for identifying vorticity waves in Lake Michigan and should prove useful in analyzing current records from Lake Champlain.

The different models in sections 5, 6 and 7 show large differences between their calculated numerical values for wavelength and period at the frequency maximum. Evidently these values are quite sensitive to the exact shape and dimensions of the channel. A numerical model with a fine grid will be required for detailed comparisons between theory and lake observations in the future.

Identification on the basis of period alone during the summer season is not generally possible since the periods of vorticity waves in large, long and narrow lakes overlap the fundamental internal gravity wave period which ranges from tens of hours to several days in such lakes. This discussion of vorticity waves has only considered barotropic motions and applies to winter unstratified conditions as well as to summer stratified lake conditions. Internal gravity waves will not be present in the unstratified winter lake making it a likely season to search for vorticity waves unmasked by internal gravity waves. During the stratified season it should be possible to eliminate the baroclinic modes of internal gravity waves if a vertical string of two or more current meters are deployed and the outputs are averaged.

REFERENCES

Allen, J., Models of wind-driven currents on the continental shelf, *Annu. Rev.Fluid Mech., 12,* 389-433, 1980.

Brink, K., Coastal-trapped waves and wind-driven currents over the continentalshelf, *Annu. Rev. Fluid Mech., 23,* 389-412, 1991.

Buchwald, V. and J. Adams, The propagation of continental shelf waves, *Proc. Roy. Soc. A. 305,* 235-250, 1968.

Hunkins, K., Anomalous diurnal tides on the Yermak Plateau, *J. Mar. Res. 44,* 51-69, 1986.

Hunkins, K., T.O. Manley, P. Manley, and J. Saylor, Numerical studies of the 4-day oscillation in Lake Champlain, *J. Geophys. Res., 103 (C9),* 18,425-18,436, 1998.

Longuet- Higgins, M. S., Topographic Rossby waves, *Mem. Soc. Roy. des Sciences de Liege, 6 serie, tome II,* 11-16, 1972.

Longuet-Higgins, M. S., Some dynamical aspects of ocean currents, *QJRMS, 91,* 425-451, 1965.

Mortimer, C., Strategies for coupling data collection and analysis with dynamical modeling of lake motions, in *Hydrodynamics of lakes,* edited by W. Graf and C. Mortimer, Elsevier, Amsterdam, 183-223, 1979.

Mysak, L., Elliptical topographic waves, *Geophys. Astrophys. Fluid Dyn., 31,* 93-135, 1985.

Mysak, L., Topographically trapped waves, *Annu. Rev. Fluid Mech.,12,* 45-76, 1980.

Saylor, J., J. Huang, and R. O. Reid, Vortex modes in Southern Lake Michigan, *J. Phys. Oceanogr., 10,* 1814-1823, 1980.

Stocker, T., Topographic waves: Eigenmodes and reflections in lakes and semi-infinite channels, *Nr. 93 Mitteilungen der Versuchsanstakt fur Wasserbau, Hydrologie und Glaziologie an der E.T.H, Zurich,* 1987.

Stocker, T. and K. Hutter, Topographic waves in rectangular basins, *J. Fluid Mech., 185,* 107-120, 1987.

Veronis, G., Rossby waves with bottom topography, J. Mar. Res. 24, 338-49, 1966.

INDEX

MIX
Papier aus verantwortungsvollen Quellen
Paper from responsible sources
FSC® C105338
www.fsc.org

If you have any concerns about our products,
you can contact us on
ProductSafety@springernature.com

In case Publisher is established outside the EU,
the EU authorized representative is:
Springer Nature Customer Service Center GmbH
Europaplatz 3, 69115 Heidelberg, Germany

Printed by Libri Plureos GmbH
in Hamburg, Germany